The Blood–Brain Barrier

METHODS IN MOLECULAR MEDICINE™

John M. Walker, Series Editor

METHODS IN MOLECULAR MEDICINE™

The Blood–Brain Barrier

Biology and Research Protocols

Edited by

Sukriti Nag

*Division of Neuropathology, University of Toronto
and Toronto Western Research Institute and Department of Pathology,
University Health Network, Toronto, Canada*

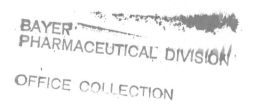

Humana Press ✳ **Totowa, New Jersey**

© 2003 Humana Press Inc.
999 Riverview Drive, Suite 208
Totowa, New Jersey 07512

www.humanapress.com

This publication is printed on acid-free paper. ∞
ANSI Z39.48-1984 (American Standards Institute) Permanence of Paper for Printed Library Materials.

Production Editor: Robin B. Weisberg.

Cover illustration: Merged confocal image depicting components of the blood–brain barrier. Factor VIII immunostaining shows the endothelium (green) of intracerebral vessels, which are surrounded by astrocytic processes (red). Image provided by Sukriti Nag.
Cover design by Patricia F. Cleary.

For additional copies, pricing for bulk purchases, and/or information about other Humana titles, contact Humana at the above address or at any of the following numbers: Tel.: 973-256-1699; Fax: 973-256-8341; E-mail: humana@humanapr.com; or visit our Website: www.humanapress.com

Printed in the United States of America. 10 9 8 7 6 5 4 3 2 1

Library of Congress Cataloging in Publication Data

The blood–brain barrier : biology and research protocols / edited by Sukriti Nag.
       p. ; cm. —(Methods in molecular medicine; 89)
    Includes bibliographical references and index.
    ISBN 1-58829-073-5 (alk. paper) 1-59259-419-0 (e-book)
    ISSN 1543-1894
      1.Blood–brain barrier—Laboratory manuals. I. Sukriti Nag. II. Series.
      [DNLM: 1. Blood–Brain Barrier—physiology. 2. Clinical Protocols. 3. Molecular Biology
      —methods. 4. Molecular Diagnostic Techniques—methods WL 200 B6547 2003]
    QP375.B528 2003
    612.8'24—dc21
                                                                2003041675

Preface

Blood–brain barrier (BBB) breakdown leading to cerebral edema occurs in many brain diseases—such as trauma, stroke, inflammation, infection, and tumors—and is an important factor in the mortality arising from these conditions. Despite the importance of the BBB in the pathogenesis of these diseases, the molecular mechanisms occurring at the BBB are not completely understood. In the last decade a number of molecules have been identified not only in endothelial cells, but also in astrocytes, pericytes, and the perivascular cells that interact with endothelium to maintain cerebral homeostasis. However, the precise cellular interactions at a molecular level in steady states and diseases have still to be determined. The introduction of new research techniques during the last decade or so provide an opportunity to study the molecular mechanisms occurring at the BBB in diseases.

The Blood–Brain Barrier: Biology and Research Protocols provides the reader with details of selected morphologic, permeability, transport, in vitro, and molecular techniques for BBB studies, all written by experts in the field. Each part is preceded by a review that emphasizes the advantages and pitfalls of particular techniques, as well as offering much relevant current information. The techniques provided will be helpful to both beginners in BBB research and those more experienced investigators who wish to add a specific technique to those already available in their laboratories. Although a number of in vitro techniques are included, it is suggested that they be complemented with data derived from in vivo studies to gain a truer picture of the biological process.

It is hoped that the methods described in *The Blood–Brain Barrier: Biology and Research Protocols* will aid researchers in the isolation of molecules not yet described, and increase our understanding of how they interact at the BBB to maintain cerebral homeostasis as well as of the mechanisms that result in BBB breakdown in diseases. Advances in technology will necessitate collaboration among researchers having expertise in many of these techniques to solve biological questions.

A greater understanding of the molecular mechanisms occurring at the BBB in diseases is also necessary in order to identify substances/molecules that can be targeted for pharmacological manipulation and/or gene therapy and to determine when therapeutic intervention can attenuate the disease process.

I would like to acknowledge Prof. John M. Walker for this opportunity and for his help and all the authors who have contributed their protocols.

This book is dedicated to Mohit Kumar and Labonya Nag.

Sukriti Nag

Contents

Contents

Contributors

N. JOAN ABBOTT • *Center for Neuroscience Research, King's College London, London, UK*

PETE ADAMSON • *Division of Cell Biology, Institute of Ophthalmology, University College London, UK*

DAVID D. ALLEN • *Department of Pharmaceutical Sciences, Texas Tech University Health Sciences Center, Amarillo, TX*

DAVID A. ANTONETTI • *Departments of Cellular and Molecular Physiology and Ophthalmology, The Penn State College of Medicine, Hershey, PA*

KAMLESH ASOTRA • *Cedars-Sinai Research Institute and Atheroscleosis Research Center, Cedars-Sinai Medical Center, Los Angeles, CA*

DAVID J. BEGLEY • *Center for Neuroscience Research, King's College London, London, UK*

KEITH L. BLACK • *Maxine Dunitz Neurosurgical Institute, Cedars-Sinai Medical Center, Los Angeles, CA*

INGOLF E. BLASIG • *Institute of Molecular Pharmacology, Berlin, Germany*

RUBEN J. BOADO • *Department of Medicine and Brain Research Institute, UCLA School of Medicine, Los Angeles, CA*

ROMÉO CECCHELLI • *Laboratoire de Physiopathologie de la Barrière hémato-encéphalique, Faculté des sciences Jean Perrin, Université d'Artois, Lens, France*

WAI CHAN • *Department of Pathology and Neurosurgery, New York University School of Medicine, New York, NY*

EAIN M. CORNFORD • *Department of Neurology and The Brain Research Institute, UCLA School of Medicine, Los Angeles, CA*

MARCIA E. CORNFORD • *Department of Pathology, Harbor-UCLA Medical Center, Torrance, CA*

PIERRE-OLIVIER COURAUD • *Institute Cochin, Department of Cellular Biology, Paris, France*

LUCA CUCULLO • *Division of Cerebrovascular Research, Department of Neurological Surgery, Cleveland Clinic Foundation, Cleveland, OH*

HAIQING DAI • *Department of Pharmaceutical Sciences, University of Nebraska Medical Center, Omaha, NE*

ANNETTE DAMERT • *Paul-Ehrlich Institute, Federal Agency for Sera and Vaccines, Langen, Germany*

SHAILESH Y. DESAI • *Division of Cerebrovascular Research, Department of Neurological Surgery, Cleveland Clinic Foundation, Cleveland, OH*

PAULA DORE-DUFFY • *Division of Neuroimmunology, Department of Neurology, Wayne State University School of Medicine, Detroit Medical Center, Detroit, MI*

KATERINA DOROVINI-ZIS • *Department of Pathology, Vancouver General Hospital, Vancouver, Canada*

LESTER R. DREWES • *Departments of Biochemistry and Molecular Biology, University of Minnesota, Duluth, MN*

WILLIAM F. ELMQUIST • *Department of Pharmaceutics, University of Minnesota, Minneapolis, MN*

BRITTA ENGELHARDT • *Max-Planck Institute for Physiological and Clinical Research, Bad Neuheim and Max Planck Institute for Vascular Biology, Münster, Germany*

FRANK M. FARACI • *Cardiovascular Center, University of Iowa College of Medicine, Iowa City, IA*

LAURENCE FENART • *Laboratoire de Physiopathologie de la Barrière hémato-encéphalique, Faculté des sciences Jean Perrin, Université d'Artois, Lens, France*

JOHN GREENWOOD • *Division of Cell Biology, Institute of Ophthalmology, University College London, UK*

REINER F. HASELHOFF • *Institute of Molecular Pharmacology, Berlin, Germany*

RICHARD A. HAWKINS • *Department of Physiology and Biophysics, Finch University of Health Sciences/The Chicago Medical School, North Chicago, IL*

MOHAMMED HOSSAIN • *Division of Cerebrovascular Research, Department of Neurological Surgery, Cleveland Clinic Foundation, Cleveland, OH*

HANH HUYNH • *Department of Pathology, Vancouver General Hospital, Vancouver, Canada*

SHIGEYO HYMAN • *Department of Neurology and The Brain Research Institute, UCLA School of Medicine, Los Angeles, CA*

DAMIR JANIGRO • *Division of Cerebrovascular Research, Departments of Neurological Surgery and Cell Biology, Cleveland Clinic Foundation, Cleveland, OH*

KELLY M. KIGHT • *Division of Cerebrovascular Research, Department of Neurological Surgery, Cleveland Clinic Foundation, Cleveland, OH*

EBERHARD KRAUSE • *Institute of Molecular Pharmacology, Berlin, Germany*

HEIKE KUSSEROW • *Institute for Pharmacology and Toxicology, Berlin, Germany*

HIROYUKI KUSUHARA • *Graduate School of Pharmaceutical Sciences, University of Tokyo, Japan*

JOSEPH C. LAMANNA • *Departments of Anatomy, Physiology and Neurology, Case Western Reserve University School of Medicine, Cleveland, OH*

MELANIE LASCHINGER • *Max-Planck Institute for Physiological and Clinical Research, Bad Neuheim and Max-Planck Institute for Vascular Biology, Münster, Germany*

STEFAN LIEBNER • *Vascular Biology, FIMO-FIRC Institute of Molecular Oncology, Milan, Italy*

ANDREA LIPPOLDT • *Nephrology Section, Department of Medicine, Hannover Medical School, Hannover, Germany*

ALBERT S. LOSSINSKY • *Neural Engineering Laboratory, Huntington Medical Research Institute, Pasadena, CA*

MATTEO MARRONI • *Division of Cerebrovascular Research, Department of Neurological Surgery, Cleveland Clinic Foundation, Cleveland, OH*

CHRISTIAN T. MATSON • *Departments of Biology, Biochemistry and Molecular Biology, University of Minnesota, Duluth, MN*

WILLIAM G. MAYHAN • *Department of Physiology and Biophysics, University of Nebraska Medical Center, Omaha, NE*

ARUMUGAM MURUGANANDAM • *Institute for Biological Sciences, National Research Council of Canada, Ottawa, Canada*

SUKRITI NAG • *Division of Neuropathology, University of Toronto and Toronto Western Research Institute, University Health Network, Toronto, Canada*

NAGENDRA NINGARAJ • *Maxine Dunitz Neurosurgical Institute, Cedars-Sinai Medical Center, Los Angeles, CA*

WILLIAM M. PARDRIDGE • *Department of Medicine and Brain Research Institute, UCLA School of Medicine, Los Angeles, CA*

DARRYL R. PETERSON • *Department of Physiology and Biophysics, Finch University of Health Sciences/The Chicago Medical School, North Chicago, IL*

RUKMINI PRAMEYA • *Department of Pathology, Vancouver General Hospital, Vancouver, Canada*

JANE ELIZABETH PRESTON • *Institute of Gerontology, King's College, London, UK*

MICHELLE A. PUCHOWICZ • *Departments of Anatomy and Physiology, Case Western Reserve University School of Medicine, Cleveland, OH*

ANDREAS REICHEL • *Research Pharmacokinetics, Schering AG, Berlin, Germany*

FRANÇOISE ROUX • *INSERM, Hôpital Fernand Widal, Paris, France*

RICHARD ROZMAHEL • *Department of Pharmacology, University of Toronto, Toronto, Canada*

RICHARD R. SHIVERS • *Department of Zoology, University of Western Ontario, London, Canada*

YUICHI SUGIYAMA • *Graduate School of Pharmaceutical Sciences, University of Tokyo, Japan*

JACQUELINE SHUKALIAK-QUANDT • *Department of Pathology, Vancouver General Hospital, Vancouver, Canada*

QUENTIN R. SMITH • *Department of Pharmaceutical Sciences, Texas Tech University Health Sciences Center, Amarillo, TX*

DANICA STANIMIROVIC • *Institute for Biological Sciences, National Research Council of Canada, Ottawa, Canada*

NATHALIE STRAZIELLE • *R&D in Neuropharmacology at INSERM U 433, Faculte de Medicine RTH Laennec, Lyon, France*

JAMSHID TANHA • *Institute for Biological Sciences, National Research Council of Canada, Ottawa, Canada*

TETSUYA TERASAKI • *Graduate School of Pharmaceutical Sciences, Tohoku University, Japan*

PETER VAJKOCZY • *Department of Neurosurgery, Mannheim Clinic, University of Heidelberg, Germany*

HARTWIG WOLBURG • *Institute of Pathology, University of Tübingen, Germany*

ELLEN B. WOLPERT • *Department of Ophthalmology, The Penn State College of Medicine, Hershey, PA*

DONALD WONG • *Department of Pathology, Vancouver General Hospital, Vancouver, Canada*

KUI XU • *Department of Anatomy, Case Western Reserve University School of Medicine, Cleveland, OH*

DAVID ZAGZAG • *Department of Pathology and Neurosurgery, New York University School of Medicine, New York, NY*

I

TISSUE TECHNIQUES

1

Morphology and Molecular Properties of Cellular Components of Normal Cerebral Vessels

Sukriti Nag

1. Introduction

The blood–brain barrier (BBB) includes anatomical, physicochemical, and biochemical mechanisms that control the exchange of materials between blood and brain and cerebrospinal fluid (CSF). Thus two distinct systems, the BBB and the blood–CSF barrier systems, control cerebral homeostasis. However, both systems are unique, the BBB having a 5000-fold greater surface area than the blood–CSF barrier (1,2). The concentrations of substances in brain interstitium, which is determined by transport through the BBB, can differ markedly from concentrations in CSF, the composition of which is determined by secretory processes in the choroid plexus epithelia (3). This review will focus on cellular components of cerebral vessels with emphasis on endothelium, basement membrane, and pericytes as well as the perivascular macrophage (**Figs. 1** and **2A**), which in light of new information is distinct from pericytes. This review deals less with pathogenesis and more with some of the molecules that have been discovered in these cell types in the past decade. Although astrocytes invest 99% of the brain surface of the capillary basement membrane and are important in induction and maintenance of the BBB, this topic will not be discussed and readers are referred to reviews in the literature (4–11).

2. Cerebral Endothelial Cells

Cerebral capillaries are continuous capillaries, their wall being composed of one or more endothelial cells. The endothelial surface of 1 g of cerebral tissue has been calculated to be approx 240 cm^2 (12). Cerebral endothelial cell surface properties such as charge and lectin binding are discussed in Chapter 9. Some of the markers specific for localization of cerebral endothelium and others that are ubiquitous, being present in endothelia of non-neural vessels, are given in **Table 1** (13–28).

2.1. Cerebral Endothelial Cells in Vasculogenesis and Angiogenesis

Vasculogenesis is the process whereby a primitive network is established during embryogenesis from multipotential mesenchymal progenitors. This occurs in the rat by embryonic d 10 after which the intraparenchymal network develops by sprouting from preexisting vessels, a process termed angiogenesis. Research in the past decade has led

From: *Methods in Molecular Medicine, vol. 89:*
The Blood–Brain Barrier: Biology and Research Protocols
Edited by: S. Nag © Humana Press Inc., Totowa, NJ

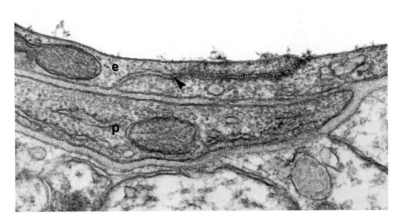

Fig. 1. Segment of normal cerebral cortical capillary wall consists of endothelium (e) and a
pericyte (p) separated by basement membrane. This rat was injected with ionic lanthanum, which
has penetrated the interendothelial space upto the tight junction (arrowhead). ×70,000.

to a greater understanding of cerebral angiogenesis during brain development *(28,29)*
and in pathological states, including neoplastic *(28)* and non-neoplastic conditions *(30)*.
During angiogenesis, endothelial cells participate in proteolytic degradation of the basement membrane and extracellular matrix and migrate with concomitant proliferation and
tube formation. Subsequent stages of angiogenesis involve increases in the length of individual sprouts, the formation of lumens, and the anastomosis of adjacent sprouts to form
vascular loops and networks. An integral component of angiogenesis is microvascular
hyperpermeability, which results in the deposition of plasma proteins in the extracellular space forming a matrix that supports the ingrowth of new vessels *(31,32)*.

Endothelial proliferation is tightly regulated during brain development *(33)*. In the
mouse brain, for example, endothelial turnover and sprouting are maximal at postnatal d 6–8 *(34)*. Proliferation then slows and the turnover is very low in the adult brain
(35). However, endothelial cells in the adult are not terminally differentiated and post-
mitotic cells and when stimulated such as occurs during wound healing or tumor growth,
they can rapidly resume cell proliferation giving rise to new capillaries.

Fig. 2. *(see facing page)* (**A**) A cryostat section shows perivascular macrophages using anti-
ED2 antibody. The inset shows these cells at higher magnification. Note that these cells are associated with vessels, which have the caliber of veins and not capillaries. (**C,D**) Merged confocal
images of normal rat brain dual labeled for Ang-1 and Ang-2 proteins. Normal vessels show
endothelial localization of Ang-1 (green) only in rat brain (**B**) and choroid plexuses (**C**) and there
is no detectable localization of Ang-2. Note the granular immunostaining in choroid plexus
epithelial cells indicating colocalization of Ang-1 and Ang-2 (yellow). (**D**) Cultured cells derived
from cerebral microvessels show adherance of antibody-coated ox red blood cells forming rosettes
indicating presence of Fc receptors. Note that many of these cells contain Factor VIII indicating
that they are endothelial cells (arrowheads). (**E**) Electron micrograph demonstrating that the
cells to which antibody-coated ox red cells have adhered also show cytoplasmic Factor VIII
immunostaining indicating its endothelial nature. Scale bar A–C = 50 µm; Inset = 25 µm;
D × 100; E × 8000.

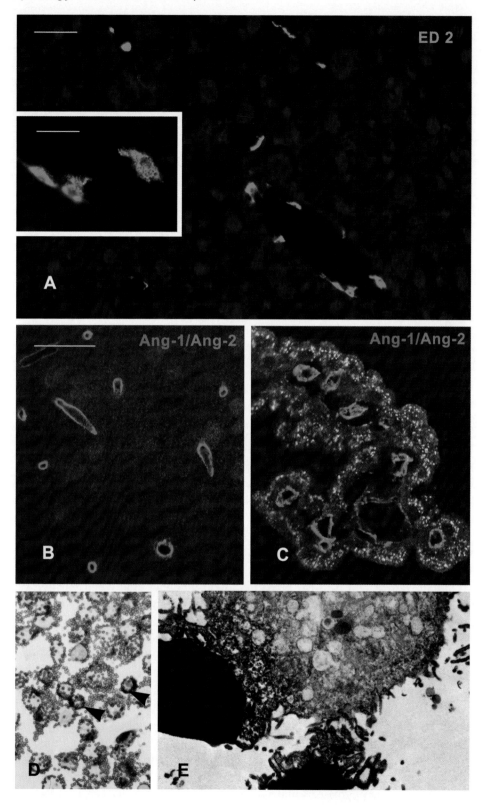

Table 1
Markers for Localization of Normal Cerebral Endothelium

Markers Specific for Cerebral Endothelium	
Glucose transporter-1	Kalaria et al. *(13)*
γ-glutamyl transpeptidase	Albert et al. *(14)*
Neurothelin/HT7 protein (chick)	Risau et al. *(15)*; Schlosshauer and Herzog *(16)*
(human)	Prat et al. *(17)*
OX-47 (Rat homologue of HT7)	Fossum et al. *(18)*
Endothelial barrier antigen (only in rat)	Sternberger & Sternberger *(19)*; Cassella et al. *(20)*
	Jefferies et al. *(21)*
Transferrin receptor	Dermietzel and Krause *(22)*
Tight junction proteins:	Liebner et al. *(23)*
ZO-1	
Occludin	
Markers Common to all Endothelia	
Endoglin (CD 105)	Personal Observation
Factor VIII	Weber et al. *(24)*
Growth Factors:	
VEGF-B	Nag et al. *(25)*
Angiopoietin-1	Nourhaghighi et al. *(26)*
Lectin binding:	
Ulex europaeus agglutinin I	Weber et al. *(24)*
Enzymes: Endothelial nitric oxide synthase	Nag et al. *(27)*
Platelet/endothelial cell adhesion	Plate *(28)*
molecule-1, (PECAM-1, CD31)	

Morphologic studies have shown that brain capillaries are derived from endothelial cells from outside the brain that invade the neuroectoderm and differentiate in response to the neural environment *(28)*. Using chick-quail transplantation experiments, convincing evidence was presented that BBB characteristics could be induced in endothelial cells, which invade brain transplants *(36)*. Conversely, brain capillaries become permeable after invasion of somite transplants *(36)*. These results indicate that organ-specific characteristics of endothelial cells may be induced and maintained by the local environment. Janzer and Raff *(5)* provided direct evidence that when purified type I astrocytes are transplanted into the rat anterior eye chamber or the chick chorioallantoic membrane, the astrocytes induce a permeability barrier in invading endothelial cells.

The specific mechanisms regulating angiogenesis are not fully understood but several potential regulators of this process include fibroblast growth factor, epidermal growth factor, transforming growth factors α and β, platelet-derived growth factor, ephrins, the family of the vascular endothelial growth factors (VEGF) and the angiopoietins (Ang). The VEGF family includes several members; VEGF or VEGF-A is best characterized, and numerous studies indicate the importance of VEGF-A in vasculogenesis and angiogenesis during brain development *(28,33,37)*. During embryonic brain angiogenesis, VEGF-A is expressed in neuroectodermal cells of the subependymal layer correlating with the invasion of endothelial cells from the perineural plexus *(33)*. The high affinity tyrosine kinase receptors that bind VEGF-A, VEGFR-1 (flt-1; *38*) and VEGFR-2 (flk-1; *39,40*) are highly expressed in invading and proliferating endothelial cells during brain development. This suggests that VEGF-A may act as a paracrine angiogenic factor. Both VEGF-A and its receptors are largely switched off in the adult *(37,39)*.

The importance of VEGF-A in cerebral angiogenesis has also been demonstrated in central nervous system (CNS) neoplasia *(28)* and non-neoplastic conditions, such as brain infarction *(41–43)* and brain injury *(25,32)*. Recent studies demonstrate that VEGF-B, another member of the VEGF family, is constitutively expressed in cerebral endothelial cells *(25)*. Angiogenesis following injury is also associated with increased expression of VEGF-B at both the gene and protein level at the injury site during angiogenesis *(25)*.

Angiopoietin-1 and -2 constitute a novel family of endothelial growth factors that function as ligands for the endothelial-specific receptor tyrosine kinase, Tie-2 *(44)*. Angiopoietins are involved in later stages of angiogenesis when vessel remodeling and maturation takes place and have a role in the interaction of endothelial cells with smooth muscle cells/pericytes *(45)*. Recent studies show constitutive expression of angiopoietin-1 protein in endothelium of normal cerebral vessels **(Fig. 2B**; *26,46)*. This protein is not specific for cerebral endothelium because it is also present in endothelium of choroid plexus and pituitary vessels **(Fig. 2C)**. Increased angiopoietin-2 expression at both the gene and protein level occurs during angiogenesis after brain injury *(26,46)*, in cerebral tumors *(47,48)*, and after infarction *(43,49,50)*.

2.2. Properties of Cerebral Endothelial Cells

Features that distinguish cerebral endothelial cells from those of non-neural vessels and form the structural basis of the BBB include the presence of tight junctions between cerebral endothelial cells, reduced endothelial plasmalemmal vesicles or caveolae, and increased numbers of mitochondria.

2.2.1. Endothelial Junctions

2.2.1.1 MORPHOLOGY

Transmission electron microscopy (TEM) shows that the junctions between adjacent cerebral endothelial cells are characterized by fusion of the outer leaflets of adjacent plasma membranes at intervals along the interendothelial space producing a penta-laminar apperarance and forming tight or occluding junctions that prevent paracellular diffusion of solutes via the intercellular route (*see* **Figs. 3** and **4**; *51–53)*. These tight junctions form the most apical element of the junctional complex, which includes both tight and adherens junctions. Subsequent studies using horseradish peroxidase as a tracer suggested that tight junctions extend circumferentially around cerebral endothelial cells; hence, their name zonula occludens *(54,55)*. Permeability of these junctions to protein and protein tracers is further discussed in Chapter 6. Tight junctions are also present between arachnoidal cells located at the outer layers of the dura *(56,57)*, and at the apical ends of choroid plexus epithelial cells *(52,58–60)* and ependymal cells *(61)*.

Certain areas of the brain, most of which are situated close to the ventricle and are therefore called circumventricular organs, have endothelial cells that do not form tight junctions. These areas include the hypothalamic median eminence, pituitary gland, choroid plexus, pineal gland, subfornicial organs, the area postrema, and the organum vasculosum of the lamina terminalis *(60,62–64)*; together they comprise less than 1%

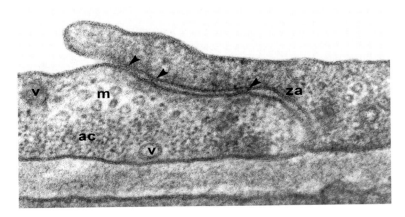

Fig. 3. **(A)** Segment of cortical arteriolar endothelium from a control rat injected intravenously with HRP showing tight junctions (arrowheads) and a zonula adherens (za) junction along the intercellular space between two endothelial cells. Also present are cross sections of actin filaments (ac) and microtubules (m) and two plasmalemmal vesicles (v). The vesicle at the luminal plasma membrane contains HRP. × 132,000.

of the brain. Endothelium in these areas is fenestrated with circular pores having a diameter of 40–60 nm that are covered by diaphragms that are thinner than a plasma membrane and of unknown composition. Fenestrations allow free exchange of molecules between the blood and adjacent neurons. The epithelial cells, which delimit the circumventricular organs, however, impede diffusion into the rest of the brain and the CSF *(65)*. Therefore, substances that have entered these areas do not have unrestricted access to the rest of the brain.

Freeze-fracture studies show that the tight junctions of cerebral endothelium of mammalian species are characterized by the highest complexity of any other body vessels *(66)*. Eight to 12 parallel junctional strands having no discontinuities run in the longitudinal axis of the vessel, with numerous lateral anastomotic strands. This pattern extends into the postcapillary venules, although in a less complex fashion *(66)*. In cerebral arteries, tight junctions consist of simple networks of junctional strands, with occasional discontinuities, whereas collecting veins, of which there are a few, have tight junctional strands that are free-ending and widely discontinuous *(66)*. Another feature of cerebral endothelial tight junctions is the high association with the protoplasmic (P)-face of the membrane leaflet, which is 55% as compared with endothelial cells of non-neural blood vessels, which have a P-face association of only 10% *(67)*. The tight junctions of choroidal epithelium consist of four or more strands or fibrils, arranged in parallel with few interconnections *(68)*. In addition, focal discontinuities have been noted in the junctional strands, which may represent hydrated channels, thus explaining the leakiness of choroid plexus epithelium *(69)*. Further details of freeze fracture studies of endothelial tight junctions are discussed in Chapter 3 and in studies in the literature *(22,23,67,70–72)*.

2.2.1.2. Transendothelial Resistance

The physiologic correlate of tightness in epithelial membranes is transepithelial resistance. Leaky epithelia generally exhibit electrical resistances between 100–200 Ω/cm^2.

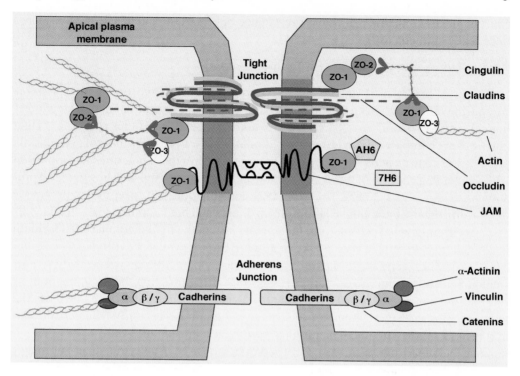

Fig. 4. Proposed locations of the major proteins associated with tight junctions (TJs) at the BBB are shown. The tight junction is embedded in a cholesterol-enriched region of the plasma membrane (shaded). Three integral proteins—claudin 1 and 2, occludin and junctional adhesion molecule (JAM)—form the tight junction. Claudins make up the backbone of the TJ strands forming dimers and bind homotypically to claudins on adjacent cells to produce the primary seal of the TJ. Occludin functions as a dynamic regulatory protein, whose presence in the membrane is correlated with increased electrical resistance across the membrane and decreased paracellular permeability. The tight junction also consists of several accessary proteins, which contribute to its structural support. The zonula occludens proteins (ZO-1 to 3) serve as recognition proteins for tight junctional placement and as a support structure for signal transduction proteins. AF6 is a Ras effector molecule associated with ZO-1. 7H6 antigen is a phosphoprotein found at tight junctions impermeable to ions and molecules. Cingulin is a double-stranded myosin-like protein that binds preferentially to ZO proteins at the globular head and to other cingulin molecules at the globular tail. The primary cytoskeletal protein, actin, has known binding sites on all of the ZO proteins. (Modified from **ref. 93**.)

Cultured brain endothelial cells grown in the absence of astrocytes have an electrical resistance of approximately 90 Ω/cm^2 *(73)*. The latter is 100-fold less than the electrical resistance across the BBB in vivo, which is estimated to be approx 4–8000 Ω/cm^2 *(74,75)*. The electrical resistance of cultured endothelial cells can be increased to 400–1000 by using special substrata such as type IV collagen and fibronectin *(76)*. Co-culture of brain microvascular endothelial with astrocytes increases transendothelial electrical resistance by 71% *(77)* and treatment with glial-derived neurotrophic factor and cAMP increases transendothelial electrical resistance by approx 250%. *(78)*. The resistance of the isolated arachnoid membrane with its tight junctions is less than that of intracerebral vessels being approx 2000 Ω/cm^2 *(79)*, while the junctions of choroid plexus epithelial cells have a resistance of 73 Ω/cm^2 and hence are considered to be leaky epithelia *(80)*.

2.2.1.3. MOLECULAR STRUCTURE OF TIGHT JUNCTIONS

Research in the past decade has provided new information on the proteins composing tight junctions using Madin Darby canine kidney epithelial cells, endothelial cells of non-neural vessels, cerebral endothelial cells, and other cell types. Tight junctions are composed of an intricate combination of transmembrane and cytoplasmic proteins linked to an actin-based cytoskeleton that allows these junctions to form a seal while remaining capable of rapid modulation and regulation **(Fig. 4)**. Three integral proteins—claudin 1 and 2 *(81)*, occludin *(82)* and junction adhesion molecule (JAM) *(83)*—form the tight junction. Claudins form dimers and bind homotypically to claudins on adjacent endothelial cells to form the primary seal of the tight junction *(84)*. Occludin is a regulatory protein, whose presence at the BBB is correlated with increased electrical resistance across the barrier and decreased paracellular permeability *(85)*. Occludin is not present in non-neural vessels thus differentiating the tight junctions of cerebral and non-neural vessels *(85)*. Junctional adhesion molecules are localized at the tight junction and are members of the immunoglobulin superfamily, which can function in association with platelet endothelial cellular adhesion molecule 1 (PECAM) to regulate leukocyte migration *(83)*. Overexpression of JAM in cells that do not normally form tight junctions increases their resistance to the diffusion of soluble tracers, suggesting that JAM functionally contributes to permeability control *(83)*.

Tight junctions are also made up of several accessory proteins that are necessary for structural support such as ZO-1 to 3, AF-6, 7H6 and cingulin. The zonula occludens (ZO) proteins 1–3 *(86,87)* belong to a family of proteins known as membrane-associated guanylate kinase-like proteins *(88)*, a family of multidomain cytoplasmic molecules involved in the coupling of transmembrane proteins to the cytoskeleton. ZO-1 is a component of the human and rat BBB *(89)*. The ALL-1 fusion partner from chromosome 6 (AF-6) is associated with ZO-1 and serves as a scaffolding component of tight junctional complexes by participating in regulation of cell–cell contacts via interaction with ZO-1 at the N terminus Ras-binding domain *(90)*. 7H6 antigen is a phosphoprotein found at tight junctions that are impermeable to ions and macromolecules *(91)*. A recent review suggests that 7H6 is sensitive to the functional state of the tight junction *(92)*. In response to cellular adenosine triphosphate (ATP) depletion, 7H6 reversibly disassociates from the tight junction while ZO-1 remains attached and there is concurrent increase in paracellular permeability *(93)*. Cingulin is a double-stranded myosin-like protein localized at the tight junction and found in endothelial cells as well. Recent in vitro studies have shown that ZO-1, ZO-2, ZO-3, myosin, JAM and A6 interact with cingulin at the N-terminus, while myosin and ZO-3 bind at the C-terminus *(94)*. Thus, cingulin appears to serve as a scaffolding protein that links tight junction accessory proteins to the cytoskeleton.

Availability of antibodies to some of the tight junction proteins has allowed localization of these proteins in cerebral endothelium by immunohistochemistry. ZO-1 immunoreactivity occurs along the entire perimeter of cultured cerebral endothelial cells *(73)* and endothelial cells in cryostat sections of human brain *(95)*. Cryostat sections of adult brain also show anti-ZO-1 immunoreactivity as fibrillar fluorescence along the lateral aspects of brain microvessels *(22)* which constitute interendothelial tight junction domains. The ZO-1 protein occurs in approximately the same quantities (molecules per micron) as the intramembranous particles that constitute the junctional fibrils in freeze-fracture preparations *(96,97)*. ZO-1 immunoreactivity is also observed in brain endothelial cells of chick and rat along with immunoreactivity for occludin, claudin-1 and claudin-5 *(23)*. ZO-1 immunoreactivity is also present at the other known sites of tight junction locations within the CNS such as in the leptomeningeal layer, choroid plexus epithelium, and the ependyma *(22)*.

The primary cytoskeletal protein, actin, has known binding sites on all ZO proteins, and on claudin and occludin *(98)*. Electron microscopy shows microfilaments having the dimensions of actin grouped near the cytoplasmic margins in proximity to cell junctions in cerebral endothelium (*see* **Fig 3**; *10,99,100*) and actin has been localized to the plasma membrane by molecular techniques *(101)*. ZO-1 binds to actin filaments and the C-terminus of occludin *(98)*, which couples the structural and dynamic properties of perijunctional actin to the paracellular barrier.

Tight junctions are localized at cholesterol-enriched regions along the plasma membrane associated with caveolin-1 *(102)*. Caveolin-1 interacts with and regulates the activity of several signal transduction pathways and downstream targets *(103)*. Several cytoplasmic signaling molecules are concentrated at tight junction complexes and are involved in signaling cascades that control assembly and disassembly of tight junctions *(104)*. Regulation of tight junctions is discussed in previous reviews *(67,93,104–106)*.

Adherens junctions are located near the basolateral side of endothelial cells (**Fig. 3**). Adherens junction proteins include the E, P, and N cadherins, which are single-pass transmembrane glycoproteins that interact homotypically in the presence of Ca^{2+} *(107)*. These cadherins are not specific for cerebral endothelial junctions being present in endothelium of non-neural blood vessels as well *(108)*. Cadherins are linked intracellularly to a group of proteins termed catenins *(109)*. α-catenin is a vinculin homolog that binds to β-catenin and probably links cadherins to the actin-based cytoskeleton and to other signaling components. γ-catenin is related to β-catenin and can substitute for it in the cadherin-catenin complex. Catenins are, thereby, part of the system by which adherens and tight junctions communicate. All these molecules are expressed at junctions in brain endothelial cells *(110,111)*. The newer cadherin-associated protein, p120 and a related protein p100 are associated with the cadherin/catenin complex in both epithelial and endothelial cells *(112)*. The interaction of these proteins in junctional permeability has been recently reviewed *(67,106)*.

2.2.2. Endothelial Plasmalemmal Vesicles or Caveolae

2.2.2.1. Morphology

TEM studies show membrane-bound vesicles open to both the luminal and abluminal plasmalemma through a neck 10–40 nm in diameter and also free in the endothelial cytoplasm of vessels of most organs (**Fig. 3**; Chapter 6, Fig. 1). These non-coated structures referred to in the previous literature as pinocytotic vesicles are now generally referred to as plasmalemmal vesicles or caveolae. These vesicles are distinct from clathrin-coated vesicles, which have an electron-dense coat and are involved in receptor-mediated endocytosis. Free cytoplasmic caveolae are spherical structures having a mean diameter of approx 70 nm. Studies of frog mesenteric capillaries suggest that caveolae are part of two racemose systems of invaginations of the luminal and abluminal cell surfaces and not freely moving entities *(113)*. There is considerable heterogeneity in endothelium in different parts of a single organ hence the findings in frog mesenteric capillaries may not necessarily apply to all species or to brain capillaries.

Morphometric studies show that normal cerebral endothelium contains a mean of 5 plasmalemmal vesicles/μm^2 in arteriolar *(114)* and capillary endothelium *(70,115–117)*. Other authors report higher values of 10–14 vesicles/μm^2 in capillary endothelium of rats and mice *(118,119)*, which they attribute to small sample size. As compared with endothelium of non-neural vessels such as myocardial capillaries *(120)*, cerebral endothelium contains 14-fold fewer vesicles. The decreased number of vesicles in cerebral endothelium implies limited transcellular traffic of solutes. In contrast, capillaries in areas where a BBB is absent such as the subfornicial organ and area postrema *(63,118)* and muscle capillaries *(121)* that are highly permeable have significantly higher numbers of endothelial vesicles. Theoretical models of vesicular transport agree in predicting a transport time in the order of seconds *(122)*.

2.2.2.2. FUNCTION

Endothelial caveolae are either endocytic or transcytotic *(123)*. The permeant molecules can either be internalized within endothelial cells by endocytosis or may be translocated across the cell to the interstitial fluid, a process termed transcytosis. Both endocytosis and transcytosis may be receptor-mediated or fluid phase and require ATP and can be inhibited by N-ethylmaleimide (NEM), an inhibitor of membrane fusion *(124)*. There is increasing evidence that in defined vascular beds, receptor-mediated transcytosis of caveoli are involved in transport of low density lipoprotein (LDL), β-very low density lipoprotein (VLDL), transferrin, insulin, albumin, ceruloplasmin, and transcobalamin across the endothelium *(125,126)*. Transcytosis is a known mechanism for passage of plasma solutes and macromolecules across endothelium of non-neural vessels *(127–129)*. Two distinct but not mutually exclusive mechanisms for transendothelial transport by the caveolar system have been proposed and debated without resolution in the past four decades. The oldest or "shuttle" hypothesis suggests that single caveolae at the luminal or abluminal surfaces "bud off" to become free within the cytoplasm carrying their molecular cargo across the cell to fuse with the opposite plasma membrane. Second, caveolae are postulated to fuse to form a transendothelial channel extending from the luminal to the abluminal plasma membrane, which allows passage of substances from the blood to tissues or in the reverse direction *(125,127,130)*. Such channels have been demonstrated in non-neural vessels in steady states *(125,130)* but not in normal cerebral endothelium either by freeze fracture *(71)*, transmission *(131)*, or high voltage electron microscopy *(132)*. The latter technique allows examination of 0.25-0.5 μm thick plastic sections and is further discussed in Chapter 4. Transendothelial channels have been observed in cerebral endothelium following BBB breakdown as discussed in Chapter 6.

In addition to transcytosis of molecules across endothelial cells, caveolae have been implicated in endocytosis, potocytosis, signal transduction, mechano-transduction in endothelial cells and control of cholesterol trafficking *(103,133–140)*.

2.2.2.3. MOLECULAR STRUCTURE OF CAVEOLAE

Studies of non-neural endothelial cells and other cell types have provided information about the molecular structure of caveolae. The specific marker and major component of caveolae is caveolin-1, an integral membrane protein (20–22 kDa) having both amino and carboxyl ends exposed on the cytoplasmic aspect of the membrane *(141)*. Caveolae in most epithelial and endothelial cells contain caveolin-1 *(142)*, which belongs to a multigene family of caveolin-related proteins that show similarities in structure but differ in properties and distribution. Caveolin-2 has a similar distribution with caveolin-1 *(143)* and endothelial cells express only caveolin-1 and -2 *(144)*. Caveolae have a unique lipid composition, of which cholesterol and sphingolipids (sphingomyelin and glycosphingolipid) are the main components. The sphingolipids are substrates for synthesis of a second intracellular messenger, the ceramides *(145)*. Cholesterol may create the frame in which all other caveolar elements are inserted.

Caveolin acts as a multivalent docking site for recruiting and sequestering signaling molecules through the caveolin-scaffolding domain that recognizes a common sequence motif within caveolin-binding signaling molecules *(146)*. Signaling molecules found in caveolar domains and form complexes with caveolin are: heterotrimer G protein, components of the Ras-mitogen-activated protein kinase pathway, Src tyrosine kinase, protein kinase C, H-Ras, and endothelial nitric oxide synthase *(139)*. Receptors having caveolar localization are involved in transport and signaling and include growth factor receptors like epidermal growth factor, insulin, and platelet derived growth factor receptors, G-protein coupled receptors like PAR, mAcR, CCK-, and receptors for endothelin, advanced glycation end products (RAGE), inositol triphosphate, CD36 *(147)*, albumin binding proteins *(148)*, and albondin *(149)*. Other antigens residing in endothelial caveolae are PV-1, and the plasma membrane calcium ATPase *(150–154)*. Other molecules partially localized in caveolae include thrombomodulin, the functional thrombin receptor, GP85/115, heterotrimeric G proteins, and dynamin *(155–159)*.

Caveoli contain the molecular machinery that promote vesicle formation, fission, docking, and fusion with the target membrane. Isolated caveoli from lung capillaries demonstrate vesicle-associated SNAP receptor (vSNARE), vesicle-associated membrane protein-2 (VAMP-2) *(160)*, monomeric and trimeric GTPases, annexins II and IV, N-ethyl maleimide-sensitive fusion factor (NSF), and soluble NSF attachment protein (SNAP) *(161)*. These molecules interact in the stages of transcytosis as follows: Caveoli form at the cell surface through ATP-, guanosine 5′-triphosphate (GTP)- and Mg^{2+}-polymerization of caveolin-1 and 2, a process stabilized by cholesterol *(141)*. Caveolin oligomers may also interact with glycosphingolipids *(162)*; these protein-protein and protein-lipid interactions are thought to be the driving force for caveoli formation *(163)*. A component of the caveolar fission machinery is the large GTPase, dynamin, which oligomerises at the neck of caveolae and probably undergoes hydrolysis for fission and release of caveolae so it becomes free in the cytoplasm *(158,164)*. The transcellular movement of caveolae is facilitated by association with the actin-cytoskeleton related proteins, such as myosin HC, gelsolin, spectrin, and dystrophin *(165)*. Fusion at the abluminal membrane is aided by NSF, which interacts with soluble attachment proteins (SNAPs) that can associate with complementary SNAP receptors to form a functional SNARE fusion complex. Before fusion of the target and vesicle membrane, v-SNARE (VAMP), the targeting receptor located on the vesicles, recognizes and docks with its cognate t-SNARE (syntaxin) on the target membrane *(166,167)*. Specific docking, is aided by endothelial VAMP-2 *(160)*, which is localized in caveolae.

The evidence thus far favors the hypothesis that caveolae are dynamic vesicular carriers budding off from the plasma membrane to form free transport vesicles that fuse with specific target membrane molecules as described in recent reviews *(125,126,168,169)*. Evidence that caveolae can traffic their cargo across cells (transcytosis) has been recently shown *(170)*. In addition, caveolin-1 gene knock-out mice show defects in the uptake and transport of albumin in vivo *(171)*. A new direction is molecular mapping of the endothelium in its native state in tissue in order to identify novel tissue-specific and disease-induced targets on the endothelial-cell surface and within caveolae, which could be targeted by therapeutic strategies such as drug or gene delivery in vivo *(126,172)*.

2.2.2.4. ALBONDIN

Most of the molecular studies on caveolae have been done in non-neural endothelium and other cell types. It would be interesting to know whether the caveolae in brain endothelium are similar or whether tissue-specific differences exist. One difference that has been identified thus far is the low expression of albondin in brain endothelium *(149)*.

Albondin is a 60-kDa albumin-binding sialoglycoprotein that is expressed selectively by vascular endothelium and is present on the luminal surface of continuous endothelium *in situ* and in culture. It binds albumin apparently not only to initiate its transcytosis via caveolae but also to increase capillary permselectivity *(149)*. Microvascular endothelial cells isolated from rat heart, lung and epididymal fat pad express albondin and bind albumin whereas various nonendothelial cells do not. Low expression or lack of expression of albondin in brain-derived microvascular endothelial cells accounts for restricted albumin passage into brain in steady states.

2.2.3. Mitochondria

Most of the mitochondria in cerebral endothelium are located in the vicinity of the nucleus, but occasional mitochondria occur throughout the cytoplasm and these tend to be parallel to the cell surface. Murine cerebral endothelium contains greater numbers of mitochondria *(173)*, with 10% *(174)* and 13.7% *(175)* of the endothelial volume being occupied by mitochondria, which is greater than found in endothelia of other tissues. Increased mitochondria in cerebral endothelium may provide the metabolic work capacity for maintaining the ionic gradient across the BBB. Other studies have obtained a lower value of 2–6% for the proportion of cerebral endothelial cytoplasmic volume occupied by mitochondria in rat *(116)*, chick *(36)*, mouse *(118)*, and human *(117)* cerebral capillaries. The different values obtained by both groups may be related to the different morphometric method used by each group. Endothelial mitochondrial density is lower in capillaries of the subfornicial organ, which lies outside the BBB *(63)*.

2.3. Endothelium in Inflammation

The brain exists in an immunologically privileged environment, however, this does not exclude immune cells from reaching the brain under steady states *(176)*. Many of the same processes that are active in the beneficial monitoring also play key roles in initiating and /or potentiating detrimental inflammatory reactions (*177*). Cerebral endothelial cells, by virtue of their capacity to express adhesion molecules and chemokines and their potential to act as antigen presenting cells and participate in death receptor signaling are intricately involved in inflammatory processes. They also express cytokines and represent a point of passage for immune cells present in the blood into the extracellular environment of the brain as discussed in Chapter 6, Subheading 6.5.

2.3.1. Adhesion Molecule Expression

The interaction of leukocytes with endothelium is a sequential and multistep process that involves leukocyte rolling on the endothelium mediated by adhesion molecules leading to firm adhesion. The latter process is mediated by interaction of the immunoglobulin family (ICAM-1 and VCAM-1) or the selectin family (E-selectin and P-selectin) of cellular adhesion molecules expressed by endothelial cells and their leukocyte ligands. Immunohistochemical studies of frozen normal brain tissue show minimal to nondetectable localization of various integrin adhesion molecules including ICAM-1, VCAM-1 and PECAM-1 in endothelium (*178–180*) and in vitro studies show low E-selectin (*181*) and P-selectin (*182*) localization on human microvascular endothelial cells. Small numbers of immune cells are able to cross into the CNS using the minimal level of these adhesion molecules that exist normally (*177*). The ICAM-1, VCAM-1 and E-selectin levels on brain microvascular endothelial cells are upregulated 3- to 15-fold in response to IFN-γ, TNF-α, and IL-1 treatment (*181,183–186*) while P-selectin levels increase in response to histamine or thrombin (*182*).

2.3.2. Chemokine Production

Chemokines are a subgroup of small cytokines (8–10 kD) that are produced within a target tissue and provide a mechanism by which specific populations of inflammatory cells including Th1 and Th2 cells and monocytes are attracted to the target tissue (*187–191*). Production of chemokines within the CNS is well documented; cellular sources include glial and endothelial cells (*192*). The presence of MCP-1 and MIP-1β-binding sites on human brain microvessels (*193*) suggests that chemokines produced locally by perivascular astrocytes and microglial cells either diffuse or are transported to the endothelial cell surface, where they are immobilized for presentation to leukocytes. Such a process has been demonstrated in the periphery with the chemokine interleukin (IL)-8 (*194*).

MCP-1 immunoreactivity can be detected in brain endothelial cells (*195*) at the onset of inflammation and before clinical expression of disease during experimental allergic encephalomyelitis. Human brain endothelial cells produce and secrete bioactive IL-6 and MCP-1 (*196–198*) and can also express IP-10, MIP1β, and RANTES (*199,200*). Prominent upregulation of these chemokines is observed in response to glial cell-derived pro-inflammatory cytokines (*200,201*).

2.3.3. Antigen Presentation at the BBB

Although both neural antigen-specific and nonspecific T-cells can enter brain, only those that recognize their specific antigen presented by a competant antigen presenting cell would persist and initiate an inflammatory reaction (*176*). Cerebral microvascular endothelial cells are potentially significant antigen presenting cells because of their large cumulative surface area and unique anatomical location between circulating T-cells and the extravascular sites of antigen exposure. Endothelial cells do not constitutively express MHC class II molecules in vivo or in vitro but can be induced to do so *(179,202–204)*. Fc receptor for the constant region of immunoglobulin has also been demonstrated on cerebral endothelial cells **(Fig. 2D,E**; *205,206)*.

2.3.4. Death Receptor Signaling

Recent evidence suggests that endothelial cells can induce immune cell death and downregulate an ongoing immune cell-mediated inflammatory process. Resting human brain endothelial cells produce minimal levels of Fas and TRAIL mRNA *(199)*. However, after stimulation with IFN-γ or IFN-β, up regulation of both Fas and TRAIL mRNA occurs *(199)*. Upon secretion/cleavage from the cell surface, TRAIL binds to death receptors and induces the apoptotic death of the target cell. Conversely, Fas molecules expressed by human brain endothelial cells could be produced as soluble Fas, secreted and act as antagonists to Fas ligand expressed by immune cells, thus inhibiting death mediated signals.

Thus it appears that cerebral endothelium is equipped with the necessary molecules to participate in immune responses. Most of the cited studies are in vitro studies using cultured cerebral endothelial cells. Methods for conducting such studies are given in **part IV** of this book. Further work is necessary to identify the molecular mechanisms occurring in vivo that lead to the inflammatory response in various brain diseases.

3. Basement Membrane

The basement membrane is a specialized, extracellular matrix, which separates endothelial cells and pericytes from the surrounding extracellular space **(Fig. 1)**. The basement membrane develops its "mature" form between postnatal d 21 and 28 by apparent fusion and thickening of the lamellae from both endothelia and astroglia *(207)* resulting in a change in both the quantity of the basement membrane, as measured by the hydroxyproline content *(208)*, and the quality as determined by immunohisto-chemistry *(209,210)*. In adults this membrane is 30–40 nm thick and is synthesized by both astrocytes and endothelial cells which are connected with the basement membrane via fine filaments. Ultrastructural studies show that the basement membrane has an inner electron-dense layer called the lamina densa; and less electron-dense layers called the laminae rarae *(22)*. The basement membrane is composed of laminin *(211)*, collagen IV *(212)*, proteoglycans, notably heparan sulphate *(213,214)*, fibronectins *(215)*, nido-gen *(216)* and entactin *(217)*. The chemical composition of these individual basement membrane components differs among various organs *(218)*.

The subendothelial basal lamina is no impediment to the extracellular flow of tracers such as horseradish peroxidase. The basal lamina of capillaries shows strong labeling with cationic colloidal gold indicating that it forms a negatively charged screen or filter controlling the movement of charged solutes between blood and the brain interstitial fluid *(219)*. Large, charged molecules such as ferritin do not cross the basal lamina *(220)*. The subendothelial basal lamina also serves as a repository for growth factors such as basic fibroblast growth factor and heparin binding proteases and protease inhibitors *(214)*. Therefore regulated release of growth factors and proteases from the basal lamina reservoir could play a role in angiogenesis and the invasion of the interstitium by tumor cells *(214)*.

4. Pericytes

Some authors refer to perivascular cells in relation to all types of vessels whether capillaries, arterioles and venules as pericytes *(221,222)*. In this review pericytes will refer to the cells, which form a discontinuous layer on the outer aspect of capillaries and the small post-capillary venules, both of which are indistinguishable by electron microscopy **(Fig. 1)**. Pericyte coverage of microvessels varies and in rat capillaries coverage is 22–32% *(223)*. In isolated brain capillary preparations, there is approximately one pericyte for every three endothelial cells *(224)*. In the CNS, pericytes have an oval to oblong cell body arranged parallel to the vessel long axis *(225)*. The cell body of the pericyte consists of a prominent nucleus with limited perinuclear cytoplasm from which extend cytoplasmic processes that also run parallel to the long axis of the blood vessel; orthogonally oriented secondary processes arise along the length of the primary process and partially encircle the vascular wall. Pericytes may be "granular" or "agranular," depending on whether cytoplasmic lysosomes are abundant or sparse respectively *(221)*. A quantitative study shows that human cerebral pericytes are exclusively granular *(226)*. Cerebral pericytes are rich in cytoplasmic plasmalemmal vesicles.

Although pericytes are separated from endothelium by the basement membrane, pericyte-endothelial cell gap junctions have been described in human cerebral capillaries *(227)*. Scanning electron microscopic studies show contacts between the pericyte and endothelial cell which is made by cytoplasmic processes of the pericyte indenting the endothelial cell and vice versa, forming so-called "peg-and-socket" contacts *(228,229)*. In brain and retina, the adhesive glycoprotein, fibronectin, has been characterized at junctional sites between pericytes and endothelial cells adjacent to "adhesion plaques" at the pericyte plasma membrane *(230)*. The plaques suggest a mechanical linkage between the two cells, which would permit contractions of the pericyte to be transmitted to the endothelial cell resulting in the reduction of the microvessel diameter.

Pericytes share many markers with smooth muscle cells such as aminopeptidase N, aminopeptidase A and nestin **(Table 2, *230–233*)**. Angiopoietin-2 *(26,44)*, and γ-glutamyl transpeptidase, a frequently used endothelial marker is also present in brain microvessel pericytes *(234)*. Glutamic acid decarboxylase *(235)* and butyrylcholinesterase *(236)* have also been localized in pericytes. Although α-smooth muscle actin protein and mRNA have been demonstrated in isolated cerebral pericytes *(237)*, immunohistochemistry at the light microscopical level does not show this protein in paraffin sections of brain (personal observation) or in isolated bovine brain capillaries *(224)*. However, immunogold labeling of isolated cerebral vessels with anti-α-smooth muscle actin shows smooth muscle cell labeling frequently aligned along the myofilaments and dense labeling of pericytes *(233)*. These authors did not observe luminal/abluminal polarity in the immunogold labeling in either cell.

Three major functional roles have been ascribed to pericytes associated with CNS microvasculature. They include contractility, regulation of endothelial cell activity and a role in inflammation.

4.1. Contractility

Contraction (and reciprocal relaxation) appears to be the way that pericytes regulate microvascular blood flow, similar to the smooth muscle of larger vessels. The demonstration of contractile elements in pericytes support their contractile role. Actin and actin

Table 2
Markers for Cerebral Pericytes, Smooth Muscle Cells
and Perivascular Macrophages

Marker	Pericyte	Smooth muscle cell	Perivascular macrophage
Aminopeptidase A	+	+	NK
Aminopeptidase N	+	+	NK
Nestin	+	+	NK
γ-glutamyl transpeptidase	+	+	+
Angiopoietin-2	+	–	–
α-SM actin	+	+	–
SM myosin	–	+	–
β-actin	+	+	NK
NM myosin	+	+	NK
Vimentin	+	+	+
Calponin	–	+	NK
Desmin	–	+	–
Skeletal/SM	–	+	NK
CD11b	+	–	+
ED-2	±	–	+
CD45	–	–	+
MHC class II	–	–	+
GSA	–	–	+

SM = smooth muscle; NM = nonmuscle; GSA = Griffonia Simplicifolia Agglutinin; NK = not known. References: Alliott et al. *(231)*; Balabanov and Dore-Duffy *(232)*; Bandopadhyay et al. *(233)*.

filaments *(238,239)* have been demonstrated in pericytes in vitro with a portion of this actin corresponding to the α-isoform or muscle-specific form *(233)*. These authors did not observe α-smooth muscle actin in all CNS pericytes suggesting that contraction may not be a universal property of all pericytes. Functional capability of pericyte-derived actin was demonstrated in biochemical assays, where it was shown to activate myosin Mg^{2+}-ATPase *(240)*. Tropomyosin and both muscle-specific and non-muscle isoforms of myosin are present in pericytes *(241,242)* including cerebral pericytes *(243)*, with the muscle-specific myosin being distributed in the same location as α-actin. Further support for the contractile function of pericytes comes from in vitro studies where their contraction has been directly observed *(244)*. A number of agents, which regulate contraction of pericytes such as histamine, serotonin, endothelin-1 and angiotensin II have been reviewed previously *(225)*.

4.2. Pericyte-Endothelial Interactions

Pericytes not only reside in proximity to vascular endothelial cells, but they also have several different types of direct contact with these cells as described above. The combination of intimate cellular contacts and their presence during development suggests that they may regulate endothelia and vascular development *(245,246)*. Pericytes can inhibit endothelial cell proliferation and thereby stabilize developing microvessels *(247)*.

In vitro studies demonstrate inhibition of endothelial cell growth and proliferation by pericytes *(248,249)*, which is mediated by transforming growth factor-β-1 only when there is physical contact between these cells *(45,248)*. On the other hand pericytes also provide factors, which stimulate the growth and development of endothelial cells. Pericytes have been implicated in all stages of angiogenesis including initiation, vascular sprout extension and endothelial migration *(250)*. In vitro studies demonstrate that pericytes produce basic fibroblast growth factor *(45,251)*, VEGF growth factor *(252,253)*, and angiopoietins *(44)*, all of which are involved in neovascularization and angiogenesis. Thus pericytes have both a negative and positive effect on endothelial cells.

Endothelial cells on the other hand have a reciprocal regulatory effect on pericytes. Endothelial cells produce transforming growth factor *(254)*, VEGF *(255)*, endothelin-1 *(256,257)*, and platelet-derived growth factor, all of which have been shown to influence pericyte function. Transforming growth factor inhibits pericyte growth, whereas VEGF stimulates pericyte proliferation in vitro. Endothelin-1 stimulates pericyte contraction and is a pericyte mitogen in vitro. The origin of platelet-derived growth factor from endothelial cells is less certain. However, in the absence of this agent, pericytes do not develop *(258)*, hence it is thought to serve as a migration and/or differentiation signal *(259,260)*.

4.3. Role in Inflammation

Pericytes are thought to serve as macrophages in the brain. Morphologic studies demonstrate numerous cytoplasmic lysosomes that increase in size and number in conditions associated with BBB breakdown to exogenously administered proteins such as horseradish peroxidase *(261–263)*. In this context, pericytes form a second line of defence for the BBB by phagocytosing molecules that pass through endothelium. Pericytes *in situ* contain the components recognized by the macrophage-selective monoclonal antibodies EBM/11 *(264)* and ED2 *(265,266)*, and the macrophage-specific protein class II major histocompatibility complex *(267,268)*. In addition, they exhibit phagocytosis *(265,269)*. Thus there is significant evidence supporting the role of pericytes as macrophages in the CNS.

CNS pericytes may be actively involved in the regulation of leukocyte transmigration, antigen presentation, and T-cell activation *(270–272)*. Pericytes present antigen in vitro and differentially activate Th1 and Th2 CD4+ lymphocytes *(273)*. CNS pericytes produce a number of immunoregulatory cytokines such as interleukin-1β, interleukin-6 and granulocyte-macrophage colony stimulatory factor *(271)*.

For detailed discussions of pericyte functions, readers are referred to reviews in the literature *(221,225,232,274,275)*. Some of the studies described in this review have been done in non-neural pericytes. The availability of a technique to isolate CNS pericytes (*see* Chapter 25) provides a means to study the role of this cell in BBB biology.

5. Perivascular Macrophages

These are a heterogenous population found in the CNS and the peripheral nervous system. Various names have been used to describe these cells, including perivascular cells, perivascular macrophages, perivascular microglia, and fluorescent granular perithelial cells (Mato cells), which reflect heterogeneity within this population as well as different models of neuropathology, anatomic locations, and species studied *(276)*.

There is consensus regarding their location, phenotype, and putative immune functions. Perivascular macrophages are a minor population in the CNS situated adjacent to endothelial cells and are immediately beyond the basement membrane of small arterioles and venules *(Fig. 2 A; 10,267,277)*. Within the perivascular space, these cells abut CNS endothelial cells and can extend long branching processes that enwrap the vessels to which they are apposed *(267,277)*. Under certain pathologic conditions, including autoimmune inflammation and viral encephalitides, perivascular macrophages can accumulate transiently and then disappear *(262,263,278,279)*, or they can remain for long periods *(280)*. Perivascular macrophages are bone marrow derived and continuously replaced by monocytes. Bone marrow chimera studies in rodents *(277,281)* and transplantation studies in humans *(282)* show a steady rate of perivascular macrophage turnover in the normal noninflamed CNS. Approximately 30% of perivascular macrophages are replaced during a 3-mo period in rats *(283)*.

5.1. Immunophenotype of Perivascular Macrophages

ED2 *(Table 2; 266,277,284)* is a universally accepted marker of perivascular cells along with CD45 *(266,278,285)*, CD4 *(286)* and OX-42 *(266,277,278)*, although, populations of these cells are identified that are OX-42 and GSA-I-B4 isolectin negative *(266)*. Fc receptor for the constant region of immunoglobulin is also found on perivascular macrophages *(287–290)*. Rodent perivascular macrophages have constitutive major histocompatibility class II (MHC II) expression that is further augmented after exposure to interferon-γ/tumor necrosis factor-α *(291)*, and in experimental allergic encephalomyelitis *(277,278)*, and neuronal damage *(292,293)*.

Perivascular macrophages in human and nonhuman primates are immunoreactive with CD14, CD45 *(276,289,290,294)* and esterase antibodies *(289,294)*. They also have constitutive MHCII and CD4 expression *(267,295,296)*. Basal levels of MHC II antigens, B7 and CD40, and Fc receptor are increased on perivascular macrophages in inflammatory CNS conditions including multiple sclerosis *(289,290,296–299)*.

Numerous immune functions have been ascribed to perivascular macrophages based solely on the expression of immune molecules. They have been implicated as the resident CNS antigen presenting cells *(277)*, they undergo immune activation in response to inflammation or neuronal injury/death *(292,293,300)* and can perform phagocytosis and pinocytosis or respond to cytokines and LPS in the peripheral blood *(280,301,302)*.

In conclusion, an attempt has been made to describe some of the molecules that have been discovered in cerebral vessels in the past decade. It is hoped that the methods described in this book will aid researchers in the isolation of molecules not described thus far and in increasing our understanding of how these molecules interact at the BBB to maintain cerebral homeostasis as well as the mechanisms that result in BBB breakdown. A greater understanding of the molecular mechanisms occurring at the BBB in diseases is necessary in order to identify substances/molecules that can be targeted for pharmacological manipulation and/or gene therapy.

Acknowledgments

Grant support from the Heart and Stroke Foundation of Ontario for the period 1978 to 2002 is gratefully acknowledged. Thanks are expressed to Drs. David Robertson, Jim Eubanks, and Cynthia Hawkins for their helpful suggestions during the preparation of this manuscript.

References

1. Crone, C. (1971) The blood-brain barrier—facts and questions. In *Ion Homeostasis of the Brain*. (Siesjo, B. K., and Sorensen, S. C., eds.) Munksgaard, Copenhagen, pp. 52–62.

2. Pardridge, W. M., Frank, H. J. L., Cornford, E. M., Braun, L. D., Crane, P. D., and Oldendorf, W. H. (1981) Neuropeptides and the blood-brain barrier. In *Neurosecretion and Brain Peptides*. (Martin, J. B., Reichlin, S., and Bick, K. L., eds.) Raven, New York, pp. 321–328.

3. Hutson, P. H., Sarna, G. S., Katananeni, B. D., and Curzon, G. (1985) Monitoring effects of tryptophan load on endometabolism in freely moving rats by simultaneous cerebrospinal fluid sampling and brain dialysis. *J. Neurochem.* **44,** 1266–1273.

4. DeBault, L., and Cancilla, P. A. (1980) γ-glutamyl transpeptidase in isolated brain endothelial cells: induction by glial cells in vitro. *Science* **207,** 635–645.

5. Janzer, R. C., and Raff, M. C. (1987) Astrocytes induce blood-brain properties in endothelial cells. *Nature* **325,** 253–257.

6. Kacem, K., LaCombe, P., Seylaz, J., and Bonvento, G. (1998) Structural organization of the perivascular astrocyte endfeet and their relationship with the endothelial glucose transporter: a confocal microscopy study. *Glia* **23,** 1–10.

7. Laterra, J., Guerin, C., and Goldstein, G. W. (1990) Astrocytes induce neural microvascular endothelial cells to form capillary-like structures in vitro. *J. Cell Physiol.* **144,** 204–215.

8. Pardridge, W. M. (1995) *The Blood-Brain Barrier. Cellular and Molecular Biology*. Raven, New York, pp. 137–164.

9. Pardridge, W. M. (1998) *Introduction to the Blood-Brain Barrier. Methodology, Biology and Pathology*. Cambridge University Press, Cambridge, UK.

10. Peters, A., Palay, S. L., and DeF Webster, H. (1976) *The Fine Structure of the Nervous System: The Neurons and Supporting Cells*. W.B. Saunders Co., Philadelphia.

11. Raub, T. J., Kuentzel, S. L., and Sawada, G. A. (1992) Permeability of bovine brain microvessel endothelial cells in vitro: barrier tightening by a factor released from astroglioma cells. *Exp. Cell Res.* **199,** 330–340.

12. Crone, C. (1963) The permeability of capillaries in various organs as determined by use of the 'indicator diffusion' method. *Acta Physiol. Scand.* **58,** 292–305.

13. Kalaria, R. N., Gravina, S. A., Schmidley, J. W., Perry, G., and Harik, S. I. (1988) The glucose transporter of the human brain and blood-brain barrier. *Ann. Neurol.* **24,** 757–764.

14. Albert, Z., Orlowski, M., Rzucidlo, Z., and Orlowska, J. (1966) Studies on γ-glutamyl transpeptidase activity and its histochemical localization in the central nervous system of man and different species. *Acta Histochem.* **25,** 312–320.

15. Risau, W., Hallmann, R., Albrecht, U., and Henke-Fahle, S. (1986) Brain induces the expression of an early cell surface marker for blood-brain barrier-specific endothelium. *EMBO J.* **5,** 3179–3183.

16. Schlosshauer, B., and Herzog, K.-H. (1990) Neurothelin: inducible cell surface glycoprotein on blood-brain barrier-specific endothelial cells and distinct neurons. *J. Cell Biol.* **110,** 1261–1274.

17. Prat, A., Biernacki, K., Pouly, S., Nalbantoglu, J., Couture, R., and Antel, J. P. (2000) Kinin B1 receptor expression and function on human brain endothelial cells. *J. Neuropathol. Exp. Neurol.* **59,** 896–906.

18. Fossum, S., Mallett, S., and Barclay, A. N. (1991) The MRC OX-47 antigen is a member of the immunoglobulin superfamily with an unusual transmembrane sequence. *Eur. J. Immunol.* **21,** 671–679.

19. Sternberger, N. H., and Sternberger, L. A. (1987) Blood-brain barrier protein recognized by monoclonal antibody. *Proc. Natl. Acad. Sci. USA* **84,** 8169–8173.

20. Cassella, J. P., Lawrenson, J. G., Lawrence, L., and Firth, J. A. (1997) Differential distribution of an endothelial barrier antigen between the pial and cortical microvessels of the rat. *Brain Res.* **744,** 335–338.

21. Jefferies, W. A., Brandon, M. R., Hunt, S. V., Williams, A. F., Gatter, K. C., and Mason, D. Y. (1984) Transferrin receptor on endothelium of brain capillaries. *Nature* **312,** 162–163.

22. Dermietzel, R., and Krause, D. (1991) Molecular anatomy of the blood-brain barrier as defined by immunocytochemistry. *Int. Rev. Cytol.* **127,** 57–109.

23. Liebner, S., Kniesel, U., Kalbacher, H., and Hartwig, W. (2000) Correlation of tight junction morphology with the expression of tight junction proteins in blood-brain barrier endothelial cells. *Eur. J. Cell Biol.* **79,** 707–717.

24. Weber, T., Seitz, R. J., Liebert, U. G., Gallasch, E., and Wechsler, W. (1995) Affinity cytochemistry of vascular endothelia in brain tumors by biotinylated *Ulex europaeus* type I lectin (UEA I). *Acta Neuropathol. (Berl.)* **67,** 128–135.

25. Nag, S., Eskandarian, M. R., Davis, J., and Stewart, D. J. (2002) Differential expression of vascular endothelial growth factor A and B after brain injury. *J. Neuropathol. Exp. Neurol.* **61,** 778–788.

26. Nourhaghighi, N., Stewart, D. J., and Nag, S. (2000) Immunohistochemical characterization of Angiopoietin-1, Angiopoietin-2, and Tie-2 in an in vivo model of cerebral trauma and angiogenesis. *Brain Pathol.* **10,** 555.

27. Nag, S., Picard, P., and Stewart, D. J. (2001) Expression of nitric oxide synthases and nitrotyrosine during blood-brain barrier breakdown and repair after cold injury. *Lab. Invest.* **81,** 41–49.

28. Plate, K. H. (1999) Mechanisms of angiogenesis in the brain. *J. Neuropathol. Exp. Neurol.* **58,** 313–320.

29. Risau, W. (1997) Mechanisms of angiogenesis. *Nature* **386,** 671–674.

30. Nag, S. (2002) The blood-brain barrier and cerebral angiogenesis: lessons from the cold-injury model. *Trends Mol. Med.* **8,** 38–44.

31. Dvorak, H. F., Nagy, J. A., Feng, D., Brown, L. F., and Dvorak, A. M. (1999) Vascular permeability factor/vascular endothelial growth factor and the significance of microvascular hyperpermeability in angiogenesis. *Curr. Top. Microbiol. Immunol.* **237,** 97–132.

32. Nag, S., Takahashi, J. T., and Kilty, D. (1997) Role of vascular endothelial growth factor in blood-brain barrier breakdown and angiogenesis in brain trauma. *J. Neuropathol. Exp. Neurol.* **56,** 912–921.

33. Risau, W. (1994) Molecular biology of blood-brain barrier ontogenesis and function. *Acta Neurochir.* **(Suppl.) 60,** 109–112.

34. Robertson, P. L., Du Bois, M., Bowman, P. D., and Goldstein, G. W. (1985) Angiogenesis in developing rat brain: an in vivo and in vitro study. *Dev. Brain Res.* **23,** 219–223.

35. Engerman, R. L., Pfaffenbach, D., and Davis, M. D., (1967) Cell turnover of capillaries. *Lab. Invest.* **17,** 738–743.

36. Stewart, P. A., and Wiley, M. J. (1981) Developing nervous tissue induces formation of blood-brain barrier characteristics in invading endothelial cells: a study using quail-chick transplantation chimeras. *Dev. Biol.* **84,** 183–192.

37. Breier, G., Albrecht, U., Sterrer, S., and Risau, W. (1992) Expression of vascular endothelial growth-factor during embryonic angiogenesis and endothelial-cell differentiation. *Development* **114,** 521–532.

38. Devries, C., Escobedo, J. A., Ueno, H., Houck, K., Ferrara, N., and Williams, L. T. (1992) The fms-like tyrosine kinase, a receptor for vascular endothelial growth-factor. *Science* **255,** 989–991.

39. Millauer, B., Wizigmann-Voos, S., Schnürch, H., et al. (1993) High affinity VEGF binding and developmental expression suggests Flk-1 as a major regulator of vasculogenesis and angiogenesis. *Cell* **72,** 835–846.

40. Risau, W., and Wolburg, H. (1991) The importance of the blood-brain barrier in fetuses and embryos—reply. *TINS* **14**, 14–15.
41. Croll, S. D., and Wiegand, S. J. (2001) Vascular growth factors in cerebral ischemia. *Mol. Neurobiol.* **23**, 121–135.
42. Plate, K. H., Beck, H., Danner, S., Allegrini, P. R., and Wiessner, C. (1999) Cell type specific upregulation of vascular endothelial growth factor in a MCA-occlusion model of cerebral infarct. *J. Neuropathol. Exp. Neurol.* **58**, 654–666.
43. Mandriota, S. J., Pyke, C., Di Sanza, C., Quindoz, P., Pittet, B., and Pepper, M. S. (2000) Hypoxia-inducible angiopoietin-2 expression is mimicked by iodonium compounds and occurs in the rat brain and skin in response to systemic hypoxia and tissue ischemia. *Amer. J. Pathol.* **156**, 2077–2089.
44. Maisonpierre, P. C., Suri, C., Jones, P. F., et al. (1997) Angiopoietin-2, a natural antagonist for Tie-2 that disrupts in vivo angiogenesis. *Science* **277**, 55–60.
45. Folkman, J., and D'Amore, P. A. (1996) Blood vessel formation: What is its molecular basis? *Cell* **87**, 1153–1155.
46. Nag, S., Nourhaghighi, N., Teichert-Kuliszewska, K., Davis, J., Papneja, T., and Stewart, D. J. (2002) Altered expression of angiopoietins during blood-brain barrier breakdown and angiogenesis. *J. Neuropathol. Exp. Neurol.* **61**, 453.
47. Stratmann, A., Risau, W., and Plate, K. H. (1998) Cell type-specific expression of angiopoietin-1 and angiopoietin-2 suggests a role in glioblastoma angiogenesis. *Am. J. Pathol.* **153**, 1459–1466.
48. Zagzag, D., Hooper, A., Friedlander, D. R., et al. (1999) In situ expression of angiopoietins in astrocytomas identifies angiopoietin-2 as an early marker of tumor angiogenesis. *Exp. Neurol.* **159**, 391–400.
49. Beck, H., Acker, T., Wiessner, C., Allergrini, P. R., and Plate, K. (2000) Expression of Angiopoietin-1 and Angiopoietin-2, and Tie-2 receptors after middle cerebral artery occlusion in the rat. *Am. J. Pathol.* **157**, 1473–1483.
50. Lin, T. N., Wang, C.-K., Cheung, W.-M., and Hsu, C.-Y. (2000) Induction of angiopoietin and Tie Receptor mRNA expression after cerebral ischemia-reperfusion. *J. Cereb. Blood Flow Metab.* **20**, 387–395.
51. Muir, A. R, and Peters, A. (1962) Quintuple-layered membrane junctions at terminal bars between endothelial cells. *J. Cell Biol.* **12**, 443–447.
52. Brightman, M. W., and Reese, T. S. (1969) Junctions between intimately apposed cell membranes in the vertebrate brain. *J. Cell Biol.* **40**, 648–677.
53. Nag, S., Dinsdale, H. B., and Robertson, D. M. (1977) Cerebral cortical changes in acute experimental hypertension. An ultrastructural study. *Lab. Invest.* **36**, 150–161.
54. Reese, T. S., and Karnovsky, M. J. (1967) Fine structural localization of a blood-brain barrier to exogenous peroxidase. *J. Cell. Biol.* **34**, 207–217.
55. Feder, N., Reese, T. S., and Brightman, M (1969) Microperoxidase, a new tracer of low molecular weight. A study of the interstitial compartments of the mouse brain. *J. Cell Biol.* **43**, 35a–36a.
56. Dermietzel, R. (1975) Junctions in the central nervous system of the cat. V. The junctional complex of the pia-arachnoid membrane. *Cell Tissue Res.* **164**, 309–329.
57. Nabeshima, S., Reese, T. S., Landis, D. M. D., and Brightman, M. W. (1975) Junctions in the meninges and marginal glia. *J. Comp. Neurol.* **164**, 127–169.
58. Dermietzel, R., Meller, N., Tetzlaff, W., and Waelsch, M. (1977) In vivo and in vitro formation of the junctional complex in choroid epithelium. A freeze-etching study. *Cell Tissue Res.* **181**, 427–441.
59. van Deurs, B. (1980) Structural aspects of brain barriers, with special reference to the permeability of the cerebral endothelium and choroidal epithelium. *Int. Rev. Cytol.* **65**, 117–191.

60. Nag, S. (1991) Effect of atrial natriuretic factor on permeability of the blood-cerebrospinal fluid barrier. *Acta Neuropathol. (Berl.)* **82,** 274–279.

61. Brightman, M. W., and Palay, S. L. (1963) The fine structure of ependyma in the brain of the rat. *J. Cell Biol.* **19,** 415–439.

62. Bouchaud, C., and Bosler, O. (1986) The circumventricular organs of the mammalian brain with special reference to monoaminergic innervation. *Int. Rev. Cytol.* **105,** 283–327.

63. Gross, P. M., Sposito, N. M., Pettersen, S. E., and Fenstermacher, J. D. (1986) Differences in function and structure of the capillary endothelium in gray matter, white matter and a circumventricular organ of rat brain. *Blood Vessels* **23,** 261–270.

64. Weindl, A. (1973) Neuroendocrine aspects of circumventricular organs. In *Frontiers in Neuroendocrinology.* (Ganong, W. F., and Martin, L., eds), Oxford University Press, New York, pp. 3–32.

65. Krisch, B., and Leonhardt, H. (1989) Relations between leptomeningeal compartments and the neurohemal regions of circumventricular organs. *Biomed. Res.* **10,** 155–168.

66. Nagy, Z., Peters, H., and Hüttner, I. (1984) Fracture faces of cell junctions in cerebral endothelium during normal and hyperosmotic conditions. *Lab. Invest.* **50,** 313–322.

67. Kniesel, U., and Wolburg, H. (2000) Tight junctions of the blood-brain barrier. *Cell Mol. Neurobiol.* **20,** 57–74.

68. Brightman, M. W., and Tao-Cheng, J. H. (1993) Tight junctions of brain endothelium and epithelium. In *The Blood-Brain Barrier. Cellular and Molecular Biology.* (Pardridge, W. M., ed.), Raven, New York, pp. 107–125.

69. van Deurs, B., and Koehler, J. K. (1979) Tight junctions in the choroid plexus epithelium, a freeze fracture study including complementary replicas. *J. Cell Biol.* **80,** 662–673.

70. Connell, C. J., and Mercer, K. L. (1974) Freeze-fracture appearance of the capillary endothelium in the cerebral cortex of mouse brain. *Am. J. Anat.* **140,** 595–599.

71. Farrell, C. L., and Shivers, R. R. (1984) Capillary junctions of the rat are not affected by osmotic opening of the blood-brain barrier. *Acta Neuropathol. (Berl.)* **63,** 179–189.

72. Shivers, R. R., Edmonds, C. L., and Del Maestro, R. F. (1984) Microvascular permeability in induced astrocytomas and peritumor neuropil of rat brain. A high-voltage electron microscope-protein tracer study. *Acta Neuropathol. (Berl.)* **64,** 192–202.

73. Krause, D., Mischeck, U., Galla, H. J., and Dermietzel, R. (1991) Correlation of zonula occludens ZO-1 antigen and transendothelial resistance in porcine and rat cultured cerebral endothelial cells. *Neurosci. Letts.* **128,** 301–304.

74. Crone, C., and Olesen, S. P. (1982) Electrical resistance of brain microvascular endothelium. *Brain Res.* **241,** 49–55.

75. Smith, Q. R., and Rapoport, S. I. (1986) Cerebrovascular permeability coefficients to sodium, potassium and chloride. *J. Neurochem.* **46,** 1732–1742.

76. Tilling, T., Korte, D., Hoheisel, D., and Galla, H. J. (1998) Basement membrane proteins influence brain capillary endothelial barrier function in vitro. *J. Neurochem.* **71,** 1151–1157.

77. Kondo, T., Kinouchi, H., Kawase, M., and Yoshimoto, T. (1996) Astroglial cells inhibit the increasing permeability of brain endothelial cell monolayer following hypoxia/reoxygenation. *Neurosci. Letts.* **208,** 101–104.

78. Igarashi, Y., Utsumi, H., Chiba, H., et al. (1999) Glial cell line-derived neurotrophic factor induces barrier function of endothelial cells forming the blood-brain barrier. *Biochem. Biophys. Res. Commun.* **261,** 108–112.

79. Perez-Gomez, J., Bindslev, N., Orkand, P. M., and Wright, E. M. (1976) Electrical properties and structure of the frog arachnoid membrane. *J. Neurobiol.* **7,** 259–270.

80. Frömter, E., and Diamond, J. (1972) Route of passive ion permeation in epithelia. *Nature* **235,** 9–13.

81. Furuse, M., Fujita, K., Hiiragi, T., Fujimoto, K., and Tsukita, S. (1998) Claudin-1 and -2: Novel integral membrane proteins localizing at tight junctions. *J. Cell Biol.* **141,** 1539–1550.

82. Furuse, M., Hirase, T., Ito, M., et al. (1993) Occludin: a novel integral membrane protein localizing at tight junctions. *J. Cell Biol.* **123,** 1777–1788.

83. Martìn-Padura, I., Lostaglio, S., Schneemann, M., et al. (1998) Junctional adhesion molecule, a novel member of the immunoglobulin superfamily that distributes at intercellular junctions and modulates monocyte transmigration. *J. Cell Biol.* **142,** 117–127.

84. Furuse, M., Sasaki, H., and Tsukita, S. (1999) Manner of interaction of heterogenous claudin species within and between tight junction strands. *J. Cell Biol.* **147,** 891–903.

85. Hirase, T., Staddon, J. M., Saitou, M., et al. (1997) Occludin as a possible determinant of tight junction permeability in endothelial cells. *J. Cell Sci.* **110,** 1603–1613.

86. Haskins, J., Gu, L., Wittchen, E. S., Hibbard, J., and Stevenson, B. R. (1998) ZO-3, a novel member of the MAGUK protein family found at the tight junction, interacts with ZO-1 and occludin. *J. Cell Biol.* **141,** 199–208.

87. Stevenson, B. R., Siliciano, J. D., Mooseker, M. S., and Goodenough, D. A. (1986) Identification of ZO-1: a high molecular weight polypeptide associated with the tight junction (zonula occludens) in a variety of epithelia. *J. Cell Biol.* **103,** 755–766.

88. Anderson, J. M. (1996) Cell signalling: MAGUK magic. *Curr. Biol.* **6,** 382–384.

89. Watson, P. M. M., Anderson, J., VanItallie C. M., and Doctrow, S. R. (1991) The tight junction-specific protein ZO-1 is a component of the human and rat blood-brain barrier. *Neurosci. Letts.* **129,** 6–10.

90. Yamamoto, T., Harada, N., Kano, K., et al. (1997) The Ras target AF-6 interacts with ZO-1 and serves as a peripheral component of tight junctions in epithelial cells. *J. Cell Biol.* **139,** 785–795.

91. Satoh, H., Zhong, Y., Isomura, H., et al. Saitoh, M., Enomoto, K., Sawada, N., (1996) Localization of 7H6 tight junction-associated antigen along the cell border of vascular endothelial cells correlates with paracellular barrier function against ions, large molecules, and cancer cells. *Exp. Cell Res.* **222,** 269–274.

92. Denker, B. M., and Nigam, S. K. (1998) Molecular structure and assembly of the tight junction. *Am. J. Physiol.* **274,** F1–F9.

93. Huber, J. D., Egleton, R. D., and Davis, T. P. (2001) Molecular physiology and pathophysiology of tight junctions in the blood-brain barrier. *TINS,* **24,** 719–725.

94. Cordenonsi, M., D'Atri, F., Hammar, E., Parry, D. A., Kendrick-Jones, J., Shore, D., and Citi, S. (1999) Cingulin contains globular and coiled-coil domains and interacts with ZO-1, ZO-2, ZO-3, and myosin. *J. Cell Biol.* **147,** 1569–1582.

95. Kuruganti, P. A., Hinojoza, J. R., Eaton, M. J., Ehmann, U. K., and Sobel, R. A. (2002) Interferon-β counteracts inflammatory mediator-induced effects on brain endothelial tight junction molecules—Implications for multiple sclerosis. *J. Neuropathol. Exp. Neurol.* **61,** 710–724.

96. Anderson, J. M., Stevenson, B. R., Jesaitis, L. A., Goodenough, D. A., and Mooseker, M. S. (1988) Characterization of ZO-1, a protein component of the tight junction from mouse liver and Madin-Darby canine kidney cells. *J. Cell Biol.* **106,** 1141–1149.

97. Stevenson, B. R., Anderson, J. M., Goodenough, D. A., and Mooseker, M. S. (1988) Tight junction structure and ZO-1 content are identical in two strains of Madin-Darby canine kidney cells which differ in transepithelial resistance. *J. Cell Biol.* **107,** 2401–2408.

98. Itoh, M., Furuse, M., Morita, K., Kubota, K., Saitou, M., and Tsukita, S. (1999) Direct binding of three tight junction associated MAGUKs, ZO-1, ZO-2, and ZO-3, with the COOH termini of claudins. *J. Cell Biol.* **147,** 1351–1363.

99. Nag, S., Dinsdale, H. B., and Robertson, D. M. (1978) Cytoplasmic filaments in intracerebral cortical vessels. *Ann. Neurol.* **3,** 555–559.

100. Nag, S. (1995) Role of the endothelial cytoskeleton in blood-brain barrier permeability to proteins. *Acta Neuropathol. (Berl.)* **90,** 454–460.
101. Pardridge, W. M., Nowlin, D. M., Choi, T. B., Yang, J., Calaycay, J., and Shively, J. E. (1989) Brain capillary 46,000 Dalton protein is cytoplasmic actin and is localized to endothelial plasma membrane. *J. Cereb. Blood Flow Metab.* **9,** 675–680.
102. Nusrat, A., Parkos, C. A., Verkade, P., et al. (2000) Tight junctions are membrane microdomains. *J. Cell Sci.* **113,** 1771–1781.
103. Schlegel, A., and Lisanti, M. P. (2001) The caveolin triad: caveolae biogenesis, cholesterol trafficking, and signal transduction. *Cytokine Growth Factor Rev.* **12,** 41–51.
104. Madara, J. L., Parkos, C., Colgan, S., Nusrat, A., Atisook, K., and Kaoutzani, P. (1992) The movement of solutes and cells across tight junctions. *Ann. New York Acad. Sci.* **664,** 47–60.
105. Gloor, S. M., Wachtel, M., Bolliger, M. F., Ishihara, H., Landmann, R., and Frei, K. (2001) Molecular and cellular permeability control at the blood-brain barrier. *Brain Res. Rev.* **36,** 258–264.
106. Rubin, L. L., and Staddon, J. M. (1999) The cell biology of the blood-brain barrier. *Annu. Rev. Neurosci.* **22,** 11–28.
107. Takeichi, M. (1995) Morphogenetic roles of classic cadherins. *Curr. Biol.* **7,** 619–627.
108. Lampugnani, M. G., and Dejana, E. (1997) Interendothelial junctions: structure, signalling and functional roles. *Curr. Opin. Cell Biol.* **9,** 674–682.
109. Gumbiner, B. M. (1996) Cell adhesion: the molecular basis of tissue architecture and morphogenesis. *Cell* **84,** 345–357.
110. Staddon, J. M., Herrenknecht, K., Smales, C., and Rubin, L. L. (1995) Evidence that tyrosine phosphorylation may increase tight junction permeability. *J. Cell Sci.* **108,** 606–619.
111. Schultze, C., Smales, C., Rubin, L. L., and Staddon, J. M. (1997) Lysophosphatidic acid increases tight junction permeability in cultured brain endothelial cells. *J. Neurochem.* **86,** 991–1000.
112. Staddon, J. M., Smales, C., Schulze, C., Esch, F. S., and Rubin, L. L. (1995) p120, a p-120 related protein (p100) and the cadherin/catenin complex. *J. Cell Biol.* **130,** 369–381.
113. Frøkjaer-Jensen, J. (1980) Three-dimensional organization of plasmalemmal vesicles in endothelial cells. An analysis by serial sectioning of frog mesenteric capillaries. *J. Ultrastruc. Res.* **73,** 9–20.
114. Nag, S., Robertson, D. M., and Dinsdale, H. B. (1979) Quantitative estimate of pinocytosis in experimental acute hypertension. *Acta Neuropathol. (Berl.)* **46,** 107–116.
115. Dux, E., and Joó, F. (1982) Effects of histamine on brain capillaries: fine structural and immunohistochemical studies after intracarotid infusion. *Exp. Brain Res.* **47,** 252–258.
116. Stewart, P. A., Hayakawa, K., Farrell, C. L., and Del Maestro, R. F. (1985) A quantitative study of blood-brain barrier permeability ultrastructure in a new rat glioma model. *Acta Neuropathol. (Berl.)* **67,** 96–102.
117. Stewart, P. A., Magliocco, M., Hayakawa, K., et al. (1987) A quantitative analysis of blood-brain barrier ultrastructure in the aging human. *Microvasc. Res.* **33,** 270–282.
118. Coomber, B. L., and Stewart, P. A. (1985) Morphometric analysis of CNS microvascular endothelium. *Microvasc. Res.* **30,** 99–115.
119. Stewart, P. A. (2000) Endothelial vesicles in the blood-brain barrier: are they related to permeability? *Cell. Mol. Neurobiol.* **20,** 149–163.
120. Simionescu, N., Simionescu, M., and Palade, G. E. (1974) Morphometric data on the endothelium of blood capillaries. *J. Cell Biol.* **60,** 128–152.
121. Coomber, B. L., Stewart, P. A., Hayakawa, E. M., Farrell, C. L., and Del Maestro, R. F. (1988) A quantitative assessment of microvessel ultrastructure in C6 astrocytoma spheroids transplanted to brain and muscle. *J. Neuropath. Exp. Neurol.* **47,** 29–40.
122. Shea, S. M., and Raskova, J. (1983) Vesicular diffusion and thermal forces. *Fed. Proc.* **42,** 2431–2434.

123. Simionescu, M. (1988) Receptor-mediated transcytosis of plasma molecules by vascular endothelium. In *Endothelial Cell Biology in Health and Disease.* (Simionescu, N., and Simionescu, N., eds.) Plenum, New York,. pp. 69–104.

124. Schnitzer, J. E., Allard, J., and Oh, P. (1995) NEM inhibits transcytosis, endocytosis, and capillary permeability: Implications of caveolae fusion in endothelia. *Am. J. Physiol.* **268,** H48–H55.

125. Simionescu, M., Gafencu, A., and Antohe, F. (2002) Transcytosis of plasma macromolecules in endothelial cells: a cell biological survey. *Microsc. Res. Tech.* **57,** 269–288.

126. Schnitzer, J. E. (2001) Caveolae: from basic trafficking mechanisms to targeting transcytosis for tissue-specific drug and gene delivery in vivo. *Adv. Drug Delivery Rev.* **49,** 265–280.

127. Palade, G. E., Simionescu, M., and Simionescu, N. (1979) Structural aspects of the permeability of the microvascular endothelium. *Acta Physiol. Scand.* **463,** 11–32.

128. Predescu, D., and Palade, G. E. (1993) Plasmalemmal vesicles represent the large pore system of continuous microvascular endothelium. *Am. J. Physiol.* **265,** H725–H733.

129. Predescu, S. A., Predescu, D. N., and Palade, G. E. (1997) Plasmalemmal vesicles function as transcytotic carriers for small proteins in the continuous endothelium. *Am. J. Physiol.* **272,** H937–H949.

130. Simionescu, N., Simionescu, M., and Palade, G. E. (1975) Permeability of muscle capillaries to small hemepeptides. Evidence for the existence of patent transendothelial channels. *J. Cell Biol.* **64,** 585–607.

131. Nag, S. (1998) Blood-brain barrier permeability measured with histochemistry. In *Introduction to the Blood-Brain Barrier. Methodology, Biology and Pathology.* (Pardridge, W. M., ed.) Cambridge University Press, Cambridge, UK, pp. 113–121.

132. Shivers, R. R., and Harris, R. J. (1984) Opening of the blood-brain barrier in Anolis Carolinenis. A high voltage electron microscope protein tracer study. *Neuropathol. Appl. Neurobiol.* **10,** 343–356.

133. Anderson, R. G., Kamen, B. A., Rothberg, K. G., and Lacey, S. W. (1992) Potocytosis: sequestration and transport of small molecules by caveolae. *Science* **255,** 410–411.

134. Anderson, R. G. (1998) The caveolae membrane system. *Annu. Rev. Biochem.* **67,** 199–225.

135. Fielding, C. J. (2001) Caveolae and signaling. *Curr. Opin. Lipidol.* **12,** 281–287.

136. Fujimoto, T., Hagiwara, H., Aoki, T., Kogo, H., and Nomura, R. (1998) Caveolae: from a morphological point of view. *J. Electron Microsc.* **47,** 451–460.

137. Michel, C. C. (1998) Capillaries, caveolae, calcium and cyclic nucleotides: a new look at microvascular permeability. *J. Mol. Cardiol.* **30,** 2541–2546.

138. Mukerjee, S., and Maxfield, F. R. (2000) Role of membrane organization and membrane domains in endocytic lipid trafficking. *Traffic.* **1,** 203–211.

139. Okamoto, T, Schlegel, A., Scherer, P. E., and Lisanti, M. P. (1998) Caveolins, a family of scaffolding proteins for organizing "preassembled signaling complexes" at the plasma membrane. *J. Biol. Chem.* **273,** 5419–5422.

140. Shaul, P. W., and Anderson, R. J. (1998) Role of plasmalemmal caveolae in signal transduction. *Am. J. Physiol.* **275,** L843–L851.

141. Monier, S., Parton, R. G., Vogel, F., Behlke, J., Henske, A., and Kurzchalia, T. (1995) VIP21-caveolin, a membrane protein constituent of the caveolar coat, oligomerises in vivo and in vitro. *Mol. Biol. Cell* **6,** 911–927.

142. Rothberg, K. G., Heuser, J. E., Donzell, W. C., Ying, Y. S., Glenney, J. R., and Anderson, R. G. (1992) Caveolin, a protein component of caveolae membrane coats. *Cell* **68,** 673–682.

143. Scherer, P. E., Okamoto, T., Chun, M., Nishimoto, I., Lodish, H. F., and Lisanti, M. P. (1996) Identification, sequence, and expression of caveolin-2 defines a caveolin gene family. *Proc. Natl. Acad. Sci. U.S.A.* **93,** 131–135.

144. Garcia-Cardena, G., Martasek, P., Masters, B. S., et al. (1997) Dissecting the interaction between nitric oxide synthase (NOS) and caveolin. Functional significance of the nos caveolin binding domain in vivo. *J. Biol. Chem.* **272,** 25,437–25,440.

145. Lui, P., and Anderson, R. G. W. (1995) Compartmentalized production of ceramide at the cell surface. *J. Biol. Chem.* **270,** 27,179–27,185.

146. Li, S., Couet, J., and Lisanti, M. P. (1996) Src tyrosine kinases, Gα subunits and H-ras share a common membrane-anchored scaffolding protein, caveolin: caveolin binding negatively regulates the autoactivation of Src tyrosine kinases. *J. Biol. Chem.* **271,** 29,182–29,190.

147. Couet, J., Li, S., Okamoto, T., Scherer, P. E., and Lisanti, M. P. (1997) Molecular and cellular biology of caveolae paradoxes and elasticities. *Trends Cardiovasc Med.* **7,** 102–110.

148. Antohe, F., Heltianu, C., and Simionescu, M. (1991) Albumin binding proteins of endothelial cells: Immunocytochemical detection of the 18 kDa peptide. *Eur. J. Cell Biol.* **56,** 34–42.

149. Schnitzer, J. E. (1992) gp60 is an albumin-binding glycoprotein expressed by continuous endothelium involved in albumin transcytosis. *Am. J. Physiol.* **262,** H246–H254.

150. Fujimoto, T., Nakade, S., Miyawaki, K., and Ogawa, K. (1992) Localization of 1,4,5-triphosphate receptor-like protein in plasmalemmal caveolae. *J. Cell Biol.* **119,** 1507–1513.

151. Fujimoto, T. (1993) Calcium pump of the plasma membrane is localized in caveolae. *J. Cell Biol.* **120,** 1147–1157.

152. Kurzchalia, T. V., and Parton, R. G. (1999) Membrane microdomains and caveolae. *Curr. Opin. Cell Biol.* **11,** 424–431.

153. Stan, R.-V., Roberts, W. G., Predescu, D., et al. (1997) Immunoisolation and partial characterization of endothelial plasmalemmal vesicles (caveolae). *Mol. Biol. Cell* **8,** 595–605.

154. Stan, R.-V., Ghitescu, L., Jacobson, B. S., and Palade, G. E. (1999) Isolation, cloning, and localization of rat PV-1. A novel endothelial caveolar protein. *J. Cell Biol.* **145,** 1189–1198.

155. Ghitescu, L. D., Crine, P., and Jacobson, B. S. (1997) Antibodies specific to the plasma membrane of rat lung microvascular endothelium. *Exp. Cell Res.* **232,** 47–55.

156. Horvat, R., and Palade, G. E. (1993) Thrombomodulin and thrombin localization on vascular endothelium; their internalization and transcytosis by plasmalemmal vesicles. *Eur. J. Cell Biol.* **61,** 299–313.

157. Huang, C., Hepler, J. R., Chen, L. T., et al. (1997) Organization of G proteins and adenyl cyclase at the plasma membrane. *Mol. Biol. Cell* **8,** 2365–2378.

158. Henley, J. R., Krueger, E. W., Oswald, B. J., and McNiven, M. A. (1998) Dynamin-mediated internalization of caveolae. *J. Cell Biol.* **141,** 85–99.

159. Oh, P., McIntosh, D. P., and Schnitzer, J. E. (1998) Dynamin at the neck of caveolae mediates their budding to form transport vesicles by GTP-driven fission from the plasma membrane of endothelium. *J. Cell Biol.* **141,** 101–114.

160. Shaul, P. W., Smart, E. J., Robinson, L., et al. (1996) Acylation targets endothelial nitric oxide synthase to plasmalemmal caveolae. *J. Biol. Chem.* **271,** 6518–6522.

161. McIntosh, D. P., and Schnitzer, J. E. (1999) Caveolae require intact VAMP for targeted transport in vascular endothelium. *Am. J. Physiol.* **277,** H2222–H2232.

162. Schnitzer, J. E., Liu, J., and Oh, P. (1995) Endothelial caveolae have the molecular transport machinery for vesicle budding, docking, and fusion including VAMP, NSF, SNAP, annexins, and GTPases. *J. Biol. Chem.* **270,** 14,399–14,404.

163. Fra, A. M., Masserini, M., Palestini, P., Sonnino, S., and Simons, K. (1995) A photo-reactive derivative of ganglioside GM1 specifically cross-links VIP21-caveolin on the cell surface. *FEBS Lett.* **375,** 11–14.

164. Sargiacomo, M., Scherer, P. E., Tang, Z., et al. (1995) Oligomeric structure of caveolin: implications for caveolae membrane organization. *Proc. Natl. Acad. Sci. U.S.A.* **92**, 9407–9411.

165. Lisanti, M. P., Scherer, P. E., Vidugiriene, J., et al. (1994) Characterization of caveolin-rich membrane domains isolated from an endothelial-rich source: Implications for human disease. *J. Cell Biol.* **126**, 111–126.

166. Hay, J. C., and Schneller, R. H. (1997) SNAREs and NSF in targeted membrane fusion, *Curr. Opin. Cell Biol.* **9**, 505–512.

167. Rothman, J. E. (1994) Intracellular membrane fusion. *Adv. Second Mess. Phosphoprot. Res.* **29**, 81–96.

168. Couet, J., Belanger, M. M., Roussel, E., and Drolet, M.-C. (2001) Cell biology of caveolae and caveolin. *Advanced Drug Delivery Reviews* **49**, 223–235.

169. Stan, R.-V. (2002) Structure and function of endothelial caveolae. *Microsc. Res. Tech.* **57**, 350–364.

170. McIntosh, D. P., Tan, X.-Y., Oh, P., and Schnitzer, J. E. (2002) Targeting endothelium and its dynamic caveolae for tissue-specific transcytosis in vivo: A pathway to overcome cell barriers to drug and gene delivery. *P.N.A.S.* **99**, 1996–2001.

171. Schubert, W., Frank, P. G., Razani, B., Park, D. S., Chow, C. W., and Lisanti, M. P. (2001) Caveolae-deficient endothelial cells show defects in the uptake and transport of albumin in vivo. *J. Biol. Chem.* **276**, 48,619–48,622.

172. Schnitzer, J. (1998) Vascular targeting as a strategy for cancer therapy. *New Engl. J. Med.* **339**, 472–474.

173. Oldendorf, W. H., and Brown, W. J. (1975) Greater number of capillary endothelial cell mitochondria in brain than muscle. *Proc. Soc. Exp. Biol. Med.* **149**, 736–738.

174. Oldendorf, W. H., Cornford, M. E., and Brown, W. J. (1977) The large apparent work capability of the blood-brain barrier: A study of mitochondrial content of capillary endothelial cells in brain and other tissues of the rat. *Ann Neurol.* **1**, 409–417.

175. Claudio, L., Kress, Y., Norton, W. T., and Brosnan, C. F. (1989) Increased vesicular transport and decreased mitochondrial content in blood-brain barrier endothelial cells during experimental autoimmune encephalomyelitis. *Am. J. Pathol.* **135**, 1157–1168.

176. Hickey, W. F., Hsu, B. L., and Kimura, H. (1991) T-lymphocyte entry into the central nervous system. *J. Neurosci. Res.* **28**, 254–260.

177. Hickey, W. F. (2001) Basic principles of immunological surveillance of the normal central nervous system. *Glia* **36**, 118–124.

178. Lossinsky, A. S., Buttle, K. F., Pluta, R., Mossakowski, M. J., and Wiśniewski, H. M. (1999) Immunoultrastructural expression of intercellular adhesion molecule-1 in endothelial cell vesiculotubular structures and vesiculovacuolar organelles in blood–brain barrier development and injury. *Cell Tissue Res.* **295**, 77–88.

179. Washington, R., Burton, J., Todd, R. F. III, Newman, W., Dragovic, L., and Dore-Duffy, P. (1994) Expression of immunologically relevant endothelial cell activation antigens on isolated central nervous system microvessels from patients with multiple sclerosis. *Ann. Neurol.* **35**, 89–97.

180. Williams, K. C., Zhao, R. W., Ueno, K., and Hickey, W. F. (1996) PECAM-1 (CD31) expression in the central nervous system and its role in experimental allergic encephalomyelitis in the rat. *J. Neurosci. Res.* **45**, 747–757.

181. Wong, D., and Dorovini-Zis, K. (1996) Regulation by cytokines and lipopolysaccharide of E-selectin expression by human brain microvessel endothelial cells in primary culture. *J. Neuropathol. Exp. Neurol.* **55**, 225–235.

182. Easton, A. S., and Dorovini-Zis, K. (2001) The kinetics, function, and regulation of P-selectin expressed by human brain microvessel endothelial cells in primary culture. *Microvasc. Res.* **62,** 335–345.

183. McCarron, R. M., Wang, I., Stanimirovic, D. B., and Spatz, M. (1995) Differential regulation of adhesion molecule expression by human cerebrovascular and umbilical vein endothelial cells. *Endothelium* **2,** 339–346.

184. Prat, A., Becher, B., Blain, M., and Antel, J. P. (1998) Induction of B7.1 and B7.2 co-stimulatory molecules on the surface of human brain endothelial cells. *J. Neuroimmunol.* **90,** 24 (abstract).

185. Wong, D., and Dorovini-Zis, K. (1992) Upregulation of intercellular adhesion molecule-1 (ICAM-1) expression in primary cultures of human brain microvessel endothelial cells by cytokines and liposaccharide. *J. Neuroimmunol.* **39,** 11–21.

186. Wong, D., and Dorovini-Zis, K. (1995) Expression of vascular cell adhesion molecule-1 (VCAM-1) by human brain microvessel endothelial cells in primary culture. *Microvasc. Res.* **49,** 325–339.

187. Bonecchi, R., Bianchi, G., Bordignon, P. P., et al. (1997) Differential expression of chemokine receptors and chemotactic responsiveness of type 1 T helper cells (Th1s) and Th2s. *J. Exp. Med.* **187,** 129–134.

188. Borges, E., Tietz, W., Steegmaier, M., et al. (1997) P-selectin glycoprotein ligand-1 (PSGL-1) on T helper 1 but not on T helper 2 cells binds P-selectin and supports migration into inflamed skin. *J. Exp. Med.* **185,** 573–578.

189. Eng, L. F., Ghirnikar, R. S., and Lee, Y. L. (1996) Inflammation in EAE: role of chemokine/cytokine expression by resident and infiltrating cells. *Neurochem. Res.* **21,** 511–525.

190. Jourdan, P., Abbal, C., Nora, N., et al. (1998) IL-4 induces functional cell-surface expression of CXCR4 on human T cells. *J. Immunol.* **160,** 4153–4157.

191. Meeusen, E. N., Premier, R. R., and Brandon, M. R. (1996) Tissue-specific migration of lymphocytes: a key role for Th1 and Th2 cells? *Immunol. Today.* **17,** 421–424.

192. Zach, O., Bauer, H. C., Richter, K., Webersinke, G., Tontsch, S., and Bauer, H. (1997) Expression of a chemotactic cytokine (MCP-1) in cerebral capillary endothelial cells. *Endothelium* **5,** 143–153.

193. Andjelkovic, A. V., Spencer, D. D., and Pachter, J. S. (1999) Visualization of chemokine binding sites on human brain microvessels. *J. Cell Biol.* **145,** 403–412.

194. Middleton, J., Neil, S., Wintle, J., et al. (1997) Transcytosis and surface presentation of IL-8 by venular endothelial cells. *Cell* **91,** 385–395.

195. Berman, J. W., Guida, M. P., Warren, J., Amat, J., and Brosnan, C. F. (1996) Localization of monocyte chemoattractant peptide-1 expression in the central nervous system in experimental autoimmune encephalomyelitis and trauma in the rat. *J. Immunol.* **156,** 3017–3023.

196. Biernacki, K., Prat, A., Pouly, S., Nalbantoglu, J., Couture, R., and Antel, J. P. (2000) Kinin B1 receptor expression and function on human brain endothelial cells: in vitro blood-brain barrier permeability study. *Neurology* **54,** A167 (abstract).

197. Prat, A., Biernacki, K., Lavoie, J-F., Poirier, J., Duquette, P., and Antel, J. P. (2002) Migration of multiple sclerosis lymphocytes through brain endothelium. *Arch. Neurol.* **59,** 391–397.

198. Zhang, W., Smith, C., Shapiro, A., Monette, R., Hutchison, J., and Stanimirovic, D. (1999) Increased expression of bioactive chemokines in human cerebromicrovascular endothelial cells and astrocytes subjected to simulated ischemia in vitro. *J. Neuroimmunol.* **101,** 148–160.

199. Prat, A., Biernacki, K., Wosik, K., and Antel, J. (2001) Glial cell influence on the human blood-brain barrier. *Glia* **36,** 145–155.

200. Shukaliak, J. A., and Dorovini-Zis, K. (2000) Expression of beta-chemokine RANTES and MIP-1 beta by human brain microvessel endothelial cells in primary culture. *J. Neuropathol. Exp. Neurol.* **59,** 339–352.

201. Biernacki, K., Prat, A., Pouly, S., Nalbantoglu, J., Couture, R., and Antel, J. P. (2000) Kinin B1 receptor expression and function on human brain endothelial cells: in vitro blood-brain barrier permeability study. *Neurology* **54,** A167 (abstract).

202. Jemison, L. M., Williams, S. K., Lublin, F. D., Knobler, R. L., and Korngold, R. (1993) Interferon-gamma-inducible endothelial cell class II major histocompatibility complex expression correlates with strain- and site-specific susceptibility to experimental allergic encephalomyelitis. *J. Neuroimmunol.* **47,** 15–22.

203. Etienne, S., Bourdoulous, S., Strosberg, A. D., and Couraud, P. O. (1999) MHC class II engagement in brain endothelial cells induces protein kinase A-dependent IL-6 secretion and phosphorylation of cAMP response element-binding protein. *J. Immunol.* **163,** 3636–3641.

204. McCarron, R. M., Spatz, M., and Cowan, E. P. (1991) Class II MHC antigen expression by cultured human cerebral vascular endothelial cells. *Brain Res.* **566,** 325–328.

205. Nag, S., and Gupta, S. (1981) Demonstration of Fc receptor on cerebral endothelium. *J. Neuropathol. Exp. Neurol.* **40,** 327.

206. Schlachetzki, F., Zhu, C., and Pardridge, W. M. (2002) Expression of the neonatal Fc receptor (FcRn) at the blood-brain barrier. *J. Neurochem.* **81,** 203–206.

207. Bär, T., and Wolff, J. R. (1972) The formation of capillary basement membranes during internal vascularization of the rat's cerebral cortex. *Z. Zellforsch.* **133,** 231–248.

208. Betz, A. L., and Goldstein, G. W. (1981) Developmental changes in metabolism and transport properties of capillaries isolated from rat brain. *J. Physiol. (Lond.)* **312,** 365–376.

209. Herkin, R., Götz, W., and Thies, M. (1990) Appearance of laminin, heparin sulphate proteoglycan, and collagen type IV during initial stages of vascularization of the neuroepithelium of the mouse embryo. *J. Anat.* **169,** 189–195.

210. Risau, W., and Lemmon, W. (1988) Changes in the vascular extracellular matrix during embryonic vasculogenesis and angiogenesis. *Dev. Biol.* **125,** 441–450.

211. Timpl, R., Rohde, H., Risteli, L., Ott, U., Robey, P. G., and Martin, G. R. (1982) Laminin. *Methods Enzymol.* **82:**Pt A, 831–838.

212. Timpl, R., Wiedemann, H., van Delden, V., Furthmayr, H., and Kuhn, K. (1981) A network model for the organization of type IV collagen molecules in basement membrane. *Eur. J. Biochem.* **120,** 203–211.

213. Linker, A., Hovingh, P., Kanwar, Y. S., and Farquhar, M. G. (1981) Characterization of heparan sulfate isolated from drug glomerular basement membranes. *Lab. Invest.* **44,** 560–565.

214. Timpl, R. (1994) Proteoglycans of basement membranes. *EXS* **70,** 123–144.

215. Hynes, R. O. (1986) Fibronectins. *Sci. Am.* **254,** 42–51.

216. Timpl, R., Dziadek, M., Fujiwara, S., Nowack, H., and Wick, G. (1983) Nidogen: a new, self-aggregating basement protein. *Eur. J. Biochem.* **137,** 455–465.

217. Carlin, B., Jaffe, R., Bender, B., and Chung, A. E. (1981) Entactin, a novel basal lamina-associated sulfated glycoprotein. *J. Biol. Chem.* **256,** 5209–5214.

218. Kefalides, N. A., Alper, R., and Clark, C. C. (1979) Biochemistry and metabolism of basement membranes. *Int. Rev. Cytol.* **61,** 167–228.

219. Vorbrodt, A. W. (1993) Morphological evidence of the functional polarization of brain microvascular endothelium. In *The Blood-Brain Barrier. Cellular and Molecular Biology.* (Pardridge, W. M., ed.), Raven, New York, pp. 137–164.

220. Brightman, M. W., and Kaya, M. (2000) Permeable endothelium and the interstitial space of brain. *Cell. Mol. Neurobiol.* **20,** 111–130.

260. Benjamin, L. E., Hemo, I., and Keshet, E. (1998) A plasticity window for blood vessel remodeling is defined by pericyte coverage of the preformed endothelial network and is regulated by PDGF-B and VEGF. *Development* **125,** 1591–1598.

261. Baker, R. N., Cancilla, P. A., Pollock, P. S., and Frommes, S. P. (1971) The movement of exogenous protein in experimental cerebral edema. An electiron microscopic study after freeze injury. *J. Neuropathol. Exp. Neurol.* **80,** 668–679.

262. Cancilla, P. A., Baker, R. N., Pollock, P. S., and Frommes, B. S. (1972) The reaction of pericytes of the central nervous system to exogenous protein. *Lab. Invest.* **26,** 376–383.

263. van Deurs, B. (1976) Observations on the blood-brain barrier in hypertensive rats, with particular reference to phagocytic pericytes. *J. Ultrastruct. Res.* **56,** 65–77.

264. Esiri, M. M., and McGee, J. (1986) Monoclonal antibody to macrophages (EBM/11) labels macrophages and microglial cells in human brain. *J. Clin. Pathol.* **39,** 615–621.

265. Balabanov, R., Washington, R., Wagnerova, J., and Dore-Duffy, P. (1996) CNS microvascular pericytes express macrophage-like function, cell surface intergrin alphaM, and macrophage marker ED-2. *Microvasc. Res.* **52,** 127–142.

266. Graeber, M. B., Streit, W. J., and Kreutzberg, G. W. (1989) Identity of ED2-positive perivascular cells in rat brain. *J. Neurosci. Res.* **22,** 103–106.

267. Graeber, M. B., Streit, W. J., Buringer, D., Sparks, D. L., and Kreutzberg, G. W. (1992) Ultrastructural location of major histocompatibility complex (MHC) class II positive perivascular cells in histologically normal human brain. *J. Neuropathol. Exp. Neurol.* **51,** 303–311.

268. Pardridge, W. M., Yang, J., Buciak, J., and Tourtellotte, W. W. (1989) Human brain microvascular DR-antigen. *J. Neurosci. Res.* **23,** 337–341.

269. Mato, M., Ookawara, S., Sugamata, M., and Aikawa, E. (1984) Evidence for the possible function of the fluorescent granular perithelial cells in brain as scavengers of high-molecular-weight waste products. *Experientia* **40,** 399–402.

270. Fujikawa, L. S., Reay, C., and Morin, M. E. (1989) Class II antigen on retinal vascular endothelium, pericytes, macrophages and lymphocytes of the rat. *Invest. Ophthamol. Visual Sci.* **30,** 66–73.

271. Fabry, Z., Fitzsimmons, K. M., Herlein, J. A., Moninger, T. O., Dobbs, M. B., and Hart, M. N. (1993) Production of cytokines interleukin 1 and 6 by murine brain microvessel endothelium and smooth muscle pericytes. *J. Neuroimmunol.* **47,** 23–34.

272. Dore-Duffy, P., Balabanov, R., Rafols, J., and Swanborg, R. (1996) The recovery period of acute experimental autoimmune encephalomyelitis in rats corresponds to development of endothelial cell unresponsiveness to interferon gamma activation. *J. Neurosci. Res.* **44,** 223–234.

273. Fabry, Z., Sandor, M., Gajewski, T. F., et al. (1993b) Differential activation of Th1 and Th2 CD4+ cells by murine brain microvessel endothelial cells and smooth muscle/pericytes. *J. Immunol.* **151,** 38–47.

274. Allt, G., and Lawrenson, J. G. (2001) Pericytes: Cell biology and pathology. *Cells Tissues Organs* **169,** 1–11.

275. Herman, I. M. (1995) Microvascular pericytes in development and disease. In *The Blood-Brain Barrier. Cellular and Molecular Biology*. Pardridge, W. M., ed. Raven, New York, pp. 127–135.

276. Williams, K., Alvarez, X., and Lackner, A. A. (2001) Central nervous system perivascular cells are immunoregulatory cells that connect the CNS with the peripheral immune system. *Glia* **36,** 156–164.

277. Hickey, W. F., and Kimura, H. (1988) Perivascular microglial cells of the CNS are bone marrow-derived and present antigen in vivo. *Science* **239,** 290–292.

278. Lassmann, H., Schmied, M., Vass, K., and Hickey, W. F. (1993) Bone marrow derived elements and resident microglia in brain inflammation. *Glia* **7,** 19–24.
279. Bauer, J., Huitinga, I., Zhao, W., Lassmann, H., Hickey, W. F., and Dijkstra, C. D. (1995) The role of macrophages, perivascular cells, and microglial cells in the pathogenesis of experimental autoimmune encephalomyelitis. *Glia* **15,** 437–446.
280. Kida, S., Cteart, P. V., Zhang, E. T., and Weller, R. O. (1993) Perivascular cells act as scavengers in the cerebral perivascular spaces and remain distinct from pericytes, microglia and macrophages. *Acta Neuropathol. (Berl.)* **85,** 646–652.
281. Lassmann, H., Vass, K., Brunner, C., and Wisniewski, H. M. (1986) Peripheral nervous system lesions in experimental allergic encephalomyelitis. Ultrastructural distribution of T cells and Ia-antigen. *Acta Neuropathol. (Berl.)* **69,** 193–204.
282. Unger, E. R., Sung, J. H., Manivel, J. C., Chenggis, M. L., Blazar, B. R., and Krivit, W. (1993) Male donor-derived cells in the brain of female sex-mismatched bone marrow transplant recipients: A Y chromosome specific in situ hybridization study. *J. Neuropathol. Exp. Neurol.* **52,** 460–470.
283. Hickey, W. F., Vass, K., and Lassman, H. (1992) Bone marrow derived elements in the central nervous system: an immunohistochemical and ultrastructural survey of rat chimeras. *J. Neuropathol. Exp. Neurol.* **51,** 246–256.
284. Dijkstra, C. D., Doop, E. A., Joling, P., and Kraal, G. (1985) The heterogeneity of mononuclear phagocytes in lymphoid organs: distinct macrophage subpopulations in the rat recognized by monoclonal antibodies ED1, ED2 and ED3. *Immunology* **54,** 589–599.
285. Ford, A. L., Goodsall, A. L., Hickey, W. F., and Sedgwick, J. D. (1995) Normal adult ramified microglia separated from other central nervous system macrophages by flow cytometric sorting. Phenotypic differences defined and direct ex vivo antigen presentation to myelin basic protein-reactive CD4+ T cells compared. *J. Immunnol.* **154,** 4309–4321.
286. Flaris, N. A., Densmore, T. L., Molleston, M. C., and Hickey, W. F. (1993) Characterization of microglia and macrophages in the central nervous system of rats: definition of the differential expression of molecules using standard and novel monoclonal antibodies in normal CNS and in four models of parenchymal reaction. *Glia* **7,** 34–40.
287. Perry, V. H., Hume, D. A., and Gordon, S. (1985) Immunohistochemical localization of macrophages and microglia in the adult and developing mouse brain. *Neuroscience* **15,** 313–326.
288. Morimura, T., Neuchrist, C., Kitz, K., et al. (1990) Monocyte subpopulations in human gliomas: expression of Fc and complement receptors and correlation with tumor proliferation. *Acta Neuropathol. (Berl.)* **80,** 287–294.
289. Ulvestad, E., Williams, K., Mork, S., Antel, J., and Nyland, H. (1994) Phenotypic differences between human monocyte/macrophages and microglial cells studied in situ and in vitro. *J. Neuropathol. Exp. Neurol.* **53,** 492–501.
290. Ulvestad, E., Williams, K., Vedeler, C, et al. (1994) Reactive microglia in multiple sclerosis lesions have an increased expression of receptors for the Fc part of Ig G. *J. Neurol. Sci.* **121,** 125–131.
291. Lassmann, H., Zimprich, F., Vass, K., and Hickey, W. F. (1991) Microglial cells are a component of the perivascular glia limitans. *J. Neurosci. Res.* **28,** 236–243.
292. Streit, W. J., Graeber, M. B., and Kreutzberg, G. W. (1989) Expression of Ia antigen on perivascular and microglial cells after sublethal and lethal motor neuron injury. *Exp. Neurol.* **105,** 115–126.
293. Streit, W. J., and Graeber, M. B. (1993) Heterogeneity of microglial and perivascular cell populations: insights gained from the facial nucleus paradigm. *Glia* **7,** 68–74.

294. Williams, K., Barr-Or, A., Ulvestad, E., Olivier, A., Antel, J. P., and Yong, W. V. (1992) Biology of adult human microglia in culture: comparisons with peripheral blood monocytes and astrocytes. *J. Neuropathol. Exp. Neurol.* **51,** 538–549.

295. Bö, L., Mork, S., Kong, P. A., Nyland, H., Pardo, C. A., and Trapp, B. D. (1994) Detection of MHC class II-antigens on macrophages and microglia, but not on astrocytes and endothelia in active multiple sclerosis lesions. *J. Neuroimmunol.* **51,** 135–146.

296. Ulvestad, E., Williams, K., Bo, L., Trapp, B., Antel, J., and Mork, S. (1994) HLA class II molecules (HLA-DR, DP, DQ) on cells in the human CNS studied in situ and in vitro. *Immunology* **82,** 535–541.

297. De Simone, R., Giampaolo, A., Giometto, B., et al. (1995) The costimulatory molecule B7 is expressed on human microglia in culture and in multiple sclerosis acute lesions. *J. Neuropathol. Exp. Neurol.* **54,** 175–187.

298. Aloisi, F., Ria, F., and Adorini, L. (2000) Regulation of T-cell responses by CNS antigen-presenting cells: different roles for microglia and astrocytes. *Immunol. Today* **21,** 141–147.

299. Williams, K., Ulvestad, E., and Antel, J. P. (1994) B7/BB-1 expression on adult human microglia studied in vitro and in situ. *Eur. J. Immunol.* **24,** 3031–3037.

300. Streit, W. J., Graeber, M. B., and Kreutzberg, G. W. (1989) Peripheral nerve lesion produces increased levels of major histocompatibility complex antigens in the central nervous system. *J. Neuroimmunol.* **21,** 117–123.

301. Franson, P. (1985) Quantitative electron microscopic observations on the non-neuronal cells and lipid droplets in the posterior funiculus of the cat after dorsal rhizotomy. *J. Comp. Neurol.* **231,** 490–499.

302. Elmquist, J. K., Breder, C. D., Sherin, J. E., et al. (1997) Intravenous lipopolysaccharide induces cyclooxygenase 2-like immunoreactivity in rat brain perivascular microglia and meningeal macrophages. *J. Comp. Neurol.* **381,** 119–129.

2

Studies of Cerebral Vessels by Transmission Electron Microscopy and Morphometry

Sukriti Nag

1. Introduction

Transmission electron microscopy (TEM), a once popular research tool, is used less frequently now. However, the isolation of novel proteins in the past decade has led to renewed use of electron microscopy for the subcellular localization of these proteins. This chapter will describe the standard method for preparation of brain tissue for TEM studies of cerebral vessels.

Optimum tissue fixation is essential to obtain good electron micrographs and in this chapter, primary fixation with an aldehyde mixture containing both glutaraldehyde and paraformaldehyde that crosslink proteins *(1)* will be described. The advantage of an aldehyde mixture is that formaldehyde penetrates cells faster than glutaraldehyde and temporarily stabilizes structures, which are subsequently more permanently stabilized by glutaraldehyde. Structural preservation is superior when the combined aldehyde fixative is used rather than either fixative alone. Paraformaldehyde reacts with proteins, lipids, and nucleic acids. Glutaraldehyde results in the formation of intermolecular and intramolecular links between amino acids, yielding rigid heteropolymers of proteins thus stabilizing cell structures and preventing distortion during processing. It also increases tissue permeability to embedding media. Glutaraldehyde does not stabilize lipids; hence, cell membranes are not visible unless tissues are post-fixed in osmium tetroxide.

Secondary fixation is done using osmium tetroxide, which reacts with unsaturated lipids, proteins, and lipoproteins. It is electron-dense and stains phospholipids in the cell membrane resulting in deposition of lower oxides of osmium, although a small degree of density may be contributed by organically bound but unreduced osmium. Osmium does not react with ribonucleic acid or deoxyribonucleic acid. The next optional step is tertiary fixation of blocks using aqueous uranyl acetate. This increases the overall contrast and further stabilizes membranous and nucleic acid containing structures. However, glycogen is extracted from the tissue. Tissues are dehydrated through ascending concentrations of ethanol, then propylene oxide before embedding in a resin mixture whose major constituent is Epon 812. The latter is the most widely used embedding medium for electron microscopy because it can be easily sectioned and sections can be stained without difficulty. In addition, ultrathin sections of Epon can tolerate the intense

From: *Methods in Molecular Medicine, vol. 89:*
The Blood–Brain Barrier: Biology and Research Protocols
Edited by: S. Nag © Humana Press Inc., Totowa, NJ

heat and strong vacuum in the electron microscope and sections show greater contrast in the electron microscope than do comparable Araldite sections. Since Epon was discontinued in the 1970s, substitutes for Epon 812 became available.

Semithin and especially ultrathin sectioning require patience and practice. The basics of sectioning are included in this chapter as are staining of semithin and ultrathin sections. However, detailed problem solving of sectioning difficulties is beyond the scope of this chapter and the reader is referred to more comprehensive texts for further information *(2,3)*.

2. Materials

Chemicals used should be of high purity and of analytic grade (*see* **Note 1**).

2.1. Primary Fixation

1. Buffer: 0.2 *M* sodium cacodylate buffer containing 8.7% sucrose, pH 7.4. Sodium cacodylate has an osmolality of 400 mOsM while sucrose provides an osmolality of 250 mOsM (*see* **Note 2**). This buffer is diluted in a ratio of 1:1 for use so the final osmolality is approx 300 mOsM.
2. Fixatives: Paraformaldehyde powder and 25% or 70% glutaraldehyde stock solution.
3. 1% Calcium chloride solution.
4. A perfusion apparatus, which consists of a 1L bottle having a rubber stopper through which a glass connector tube and a plastic Y-shaped connector are inserted (**Fig. 1**). The latter is attached to a) tubing that leads to a pressure pump, and b) a segment of tubing at the end of which a Hoffman clamp is placed so that the pressure in the system can be adjusted (*see* **Fig. 1**). The inner end of the glass connector tube in the stopper is attached to tubing, which should be long enough to reach the bottom of the bottle so that the last few drops of fixative can flow into the system. The outer end of the glass connector tube is connected to tubing, which is connected to a plastic Y-shaped connector, which connects to segments of tubing that are attached to a) a manometer, and b) a metal stopcock. The beveled end of a 16-gage needle is filed so the tip has a straight edge. This needle is fitted on the stopcock.

2.2 Secondary Fixation

1. 2% Osmium tetroxide in 0.1 *M* cacodylate buffer is prepared as follows:
   a. Wash the vial containing the osmium tetroxide crystals and score the marked ring on the neck of the vial with a file.
   b. Break off the tip of the vial and drop the vial into an amber colored bottle containing the required amount of distilled water to prepare a 4% solution and replace the lid of the bottle but do not secure tightly.
   c. Stir solution in a fume hood for a few hours until it is clear.
   d. Add an equal volume of 0.2 *M* cacodylate buffer to get a 2% solution of osmium.
   e. Tighten the lid and store at 4°C.
   f. Ensure that the solution is clear before use because this solution precipitates if kept too long.
   g. Osmium tetroxide is a hazardous chemical (*see* **Note 3**).
2. Pasteur pipets, glass tubes 10 × 75 mm or shell vials 12 × 35 mm (Kimble) with cork stoppers, parafilm.
3. Nalgene wash bottles.

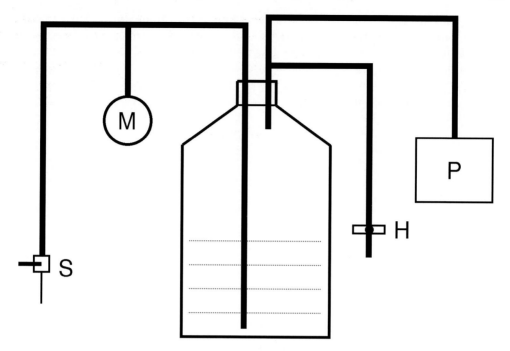

Fig. 1. The diagram shows the apparatus used for vascular perfusion of mice and rats that is described in **Subheading 2.1., step 4**. Abb. S = metal stopcock at the end of which a 16-gage needle is fitted; M = Manometer; H = Hoffman clamp, which can be adjusted to attain the correct perfusion pressure; P = the pressure perfusion pump.

2.3. Tertiary Fixation

1. 0.05 *M* Sodium hydrogen maleate buffer, pH 5.15 for washes and pH 6.0 for the preparation of uranyl acetate stain, is prepared fresh before use as follows:
   a. Prepare a 0.1 *M* sodium hydrogen maleate stock solution containing 1.16% maleic acid and 0.4% NaOH in distilled water.
   b. Prepare a 0.2 *M* NaOH solution
   c. Add the 0.2 *M* NaOH to 50 mL of the sodium hydrogen maleate stock solution until the required pH is reached and bring the total volume to 100 mL with distilled water.
4. 3% Uranyl acetate in 0.05 *M* sodium hydrogen maleate buffer, pH 6.0. Filter using no. 50 Whatman paper (*see* **Note 4**).

2.4. Dehydration and Embedding

1. Ethanol, propylene oxide.
2. Epon mixtures A and B are prepared depending on the weight per epoxide equivalent (WPE) of Epon 812, which is obtained from a table supplied by the manufacturer. If the WPE of Epon 812 is 145, then Mixture A contains 200 g of Jembed 812 resin (J.B. EM Services, Pointe Claire, Quebec) and 254 g of dodecenyl succinic anhydride, and Epon mixture B contains 250 g of Jembed 812 resin and 212 g of nadic methyl anhydride. The Epon mixtures are prepared as follows:
   a. The required amounts of the constituents are added by weight to an amber-colored 1 L bottle placed on a top-loading balance.

  b. Stir using a glass rod.

  c. Tighten lids of bottles and store at 4°C. These stocks are good for several months.

3. The hardness of the final block depends upon the ratio of mixture A and B and an increase in the proportion of mixture B will harden the block. In our laboratory the final resin mixture contains 1 part of Epon mixture A and 4 parts of Epon mixture B to which 1.8% of 2,4,6-tri (dimethylaminomethyl) phenol (DMP-30), an accelerator for epoxy resin is added. This mixture is prepared as follows:

  a. Leave the stock solutions at room temperature for at least 2 h before use.

  b. Pour the required amount of the stock solutions into graduated disposable tri-pour polypropylene beakers.

  c. Place the beaker in a 60°C oven for 10 min and then stir contents for 5 min using a glass rod.

  d. Add the required amount of DMP-30 in a fume hood and stir for at least 10–15 min.

  e. Cover the beaker with a paper lid and let it sit at least 45 min before use so that the air bubbles, which form during stirring, break up.

4. Round mold or flat molds, small paper labels (5 × 7 mm) with case numbers written in pencil.

2.5. Sectioning

1. A knife maker, strips of plate glass (6 mm thick) to make glass knives, masking tape for making a boat at the cutting edge of the glass knife.

2. Dissecting microscope, a diamond knife, an ultramicrotome, tissue sectioner.

3. Small and large plastic petri dishes, gelatin capsules.

4. Segments of hair taped to an applicator stick, platinum wire loop.

5. Copper grids having a 3.05-mm outside diameter and a 300 hexagonal or square mesh. Grids obtained from the manufacturer are cleaned as follows:

  a. Place grids in a glass vial containing 10% HCl and swirl for 2–3 min. 10% HCl is made every week.

  b. Pour out the HCl solution and rinse in several changes of filtered distilled water and then filtered ethanol.

  c. Decant most of the ethanol and pour grids on a filter paper. Separate with forceps and allow them to dry.

  d. The filter paper is placed in a plastic petri dish, which is covered to keep out the dust.

  e. Freshly cleaned grids are required each day ultrathin sectioning is done.

  f. Grids can be cleaned only once otherwise they flake.

2.6. Staining

1. Stainless steel Dumont tweezers, 30-mL capacity amber-colored bottle, 50-mL glass flask with a glass lid.

2. Sodium hydroxide pellets, 5 N sodium hydroxide.

3. Silicone rubber plates with numbered squares on the surface.

4. Stain for semithin sections: Add 1 g of toluidine blue and 1 g of sodium borate to 100 mL of distilled water. Stir for 30 min to dissolve. This stain keeps for long periods when stored at 4°C. It is filtered each time before use using a Whatman no. 1 filter paper or it can be dispensed in a 5-mL syringe fitted with a 0.45-μm filter.

5. Stain for ultrathin sections:

  a. Saturated solution of uranyl acetate in water: Add uranyl acetate to a 30-mL amber-colored bottle containing filtered distilled water and stir for a few hours. A residue remains at the bottom indicating that it is a saturated solution. Store at room temperature in the staining area. When staining, pipette a few drops from just below the surface of the solution taking care not to disturb the solution. This stain is only good for 2 wk (*see* **Note 5**).

b. Lead citrate stain: Boil 100 mL of distilled water in an Erlenmeyer flask to remove carbonates from the water and allow it to cool. Rinse an acid-cleaned 50 mL volumetric flask having a glass stopper with the boiled distilled water and dry. Add 160 mg of lead citrate to 45 mL of the boiled distilled water. Do not use a metal spoon while weighing the lead because it reacts with the metal. Add the glass stopper and shake vigorously for 3–4 min. Add 5 *N* NaOH drop by drop over 15 min until the solution clears. Clean the sidewalls of the flask by pipetting solution down the sidewalls and rinse the glass stopper with carbonate-free water. Allow the solution to settle. It can be used after 2–3 h. This solution is made every week because of the tendency for lead carbonate precipitate to form on being kept. This solution is kept at room temperature (*see* **Note 5**).

3. Methods

The methods described below outline 1) primary fixation, 2) secondary fixation, 3) tertiary fixation, 4) dehydration and embedding, 5) sectioning, 6) staining, and 7) morphometry.

3.1. Primary Fixation

A mixture of paraformaldehyde and glutaraldehyde in cacodylate buffer is used at a pH of 7.4, which is the pH of most animal tissues. Cacodylate buffer is widely used for preparation of tissues for electron microscopy. One of its advantages is that calcium can be added to the fixative solution without the formation of precipitate. It is also resistant to bacterial contamination during specimen storage. Addition of calcium chloride to the fixative solution has many beneficial effects, including 1) decrease in the swelling of cell components, 2) maintenance of cell shape, 3) reduction in the extraction of cellular materials, and 4) membrane and cytoskeletal stabilization.

3.1.1. Paraformaldehyde-Glutaraldehyde Mixture

This fixative contains 1% paraformaldehyde and 1.25% glutaraldehyde in 0.1 *M* cacodylate buffer, pH 7.2 containing 4.35% sucrose and 0.05 % calcium chloride. Preparation of 100 mL of this fixative is done as follows:

1. Heat 35 mL of distilled water to about 60–70°C in an Erlenmeyer flask in a fume hood. Add 2 g of paraformaldehyde and stir for a few minutes using a magnetic stirrer.
2. Add 1–2 drops of 5 *N* NaOH and stir for few minutes until the solution clears.
3. Cool the solution and filter using a Whatman no.1 filter.
4. Add 10 mL of 25% glutaraldehyde, 50 mL of 0.2 *M* cacodylate buffer, and 5 mL of a 1% $CaCl_2$ solution.
5. This solution is used undiluted for immersion fixation. It is diluted in a 1:1 ratio with 0.1 *M* cacodylate buffer for vascular perfusion of experimental animals.
6. The fixative is cooled to 4°C before vascular perfusion. For long-term storage (*see* **Note 6**).

3.1.2. Alternate Fixative

If both electron microscopy and immunohistochemistry have to be performed using the same tissue, then it is best to use a fixative that does not contain glutaraldehyde, which inhibits demonstration of some proteins. In our laboratory, we use 3% paraformaldehyde in 0.1 M phosphate buffer, pH 7.2 for vascular perfusion. Phosphate buffer is one of the most physiologic buffers, because it is found in living systems in the form of inorganic phosphates and phosphate esters. Microorganisms grow readily in this buffer therefore it is advisable to sterilize this solution and store at 4°C. Another disadvantage is that calcium cannot be added to this fixative because a precipitate of calcium phosphate may form which can be visualized by electron microscopy:

1. Prepare 50 mL of 6% paraformaldehyde as described previously (*see* **Subheading 3.1.1**).
2. Add 25 mL of distilled water and 25 mL of 0.4 *M* phosphate buffer, pH 7.2 which is prepared by adding 5 *N* NaOH to a 0.4 *M* solution of KH_2PO_4 until the pH is 7.2.
3. Filter before perfusion using a Whatman no.1 qualitative filter.

3.1.3. Vascular Perfusion of Fixative

This is the best method for obtaining good preservation of animal tissues for electron microscopy. Animals should be perfused in a fume hood to avoid inhaling paraformaldehyde or glutaraldehyde vapours which are irritant to the eyes as well. The method used for perfusing rats in our laboratory is:

1. Load fixative cooled to 4°C into the 1-L bottle of the perfusion apparatus (*see* **Fig. 1**). Switch the pump on and allow pressure to build in the system. Tighten the Hoffman clamp until the manometer reads 110 mm Hg. Open the stopcock above the needle and allow fluid to flow until all air bubbles are expelled. Close the stopcock.
2. Rats are anesthetized using metofane or halothane inhalation. In accordance with animal experimentation regulations, animals must be kept anesthetized throughout the surgery. They do not move when their paws are held tightly and the corneal reflex is absent.
3. The abdominal wall just below the ribcage on the left side is lifted up using toothed forceps and a hole is made. Scissors are then inserted into the hole and a linear cut is made upwards through the diaphragm and the costal cartilages left of the sternum up to the level of the clavicle. A transverse cut is made in the diaphragm.
4. The heart is then grasped using the index and thumb of the left hand or a pair of forceps with blunt ends and the 16-gage needle is inserted through the apex of the heart into the left ventricle. Slide the needle along the interventricular septum of the left ventricle until it enters the aorta.
5. If the heart is pulled down gently the tip of the needle will be seen in the ascending aorta.
6. Clamp the needle in place by placing a pair of curved hemostatic forceps on the heart below the atria.
7. Open the stopcock above the needle and allow fixative to flow into the aorta (*see* **Note 7**).
8. A slit is made in the right atrium for efflux of the perfusate, which initially is bloody and later becomes clear.
9. Perfusion is continued for 10 min or until 500 mL of fixative is perfused (*see* **Note 8**).
10. Perfusion is usually not satisfactory if the lag between death and entry of fixative in the animal exceeds 60–90 s. An adequately fixed brain is firm and yellow in color when an aldehyde mixture is used and white if paraformaldehyde is used alone. A poorly perfused brain showing reddish areas of discoloration on the surface should be discarded since electron microscopy invariably shows poorly preserved tissue.
11. Brains are removed and placed in the same fixative solution for 2 h at room temperature.
12. Our usual practice is to obtain coronal slices of 50–60 µm thickness, from each hemisphere using a tissue sectioner (Sorvall TC-2 Sectioner or a Vibratome).

13. Brain slices should show only few red blood cells in the vasculature on light microscopy if fixation is optimum.
14. Slices are viewed under the stereoscope and rectangles of cortex measuring 2.5 × 4 mm and having central arterioles are cut using a sharp razor blade, which is cleaned with acetone. These blocks are fixed overnight and processed for resin embedding next morning (*see* **Note 9**).

3.2. Secondary Fixation

The minimum processing time for 50–60 µm thick blocks is given. For thicker blocks the time in the different solutions have to be increased. Processing is done in glass tubes or shell vials with cork stoppers and solutions are added or removed using Pasteur pipettes.

1. Rinse blocks in 3 changes of 0.1 *M* cacodylate buffer for 5 min each.
2. Postfix in 2% osmium tetroxide in 0.1 *M* cacodylate buffer for 30 min in a fume hood on ice. The tissue becomes uniformly black at the end of osmication.
3. Rinse blocks in 2 changes of 0.1 *M* cacodylate buffer for 5 min each and then in 3 changes of maleate buffer for 5 min each.

3.3. Tertiary Fixation

1. Perform *en bloc* staining by placing tissues in 3% uranyl acetate in 0.05 *M* maleate buffer for 1 hour in an incubator at 37°C (*see* **Note 10**).
2. Wash in 3 changes of maleate buffer for 5 min each.

3.4. Dehydration and Embedding

These steps remove water from the tissue and replace it with a medium that will withstand the stress of cutting. These steps are also done using a fume hood to avoid inhalation of solvents.

1. The solutions listed below are filled in Nalgene wash bottles for quick dispensing.
   a. 2 changes of 70% ethanol for 7 min each.
   b. 2 changes of 85% ethanol for 7 min each (*see* **Note 11**).
   c. 2 changes of 95% ethanol for 7 min each.
   d. 3 changes of 100% ethanol for 7 min each.
   e. 3 changes of propylene oxide for 5 min each.
2. Infiltration
   Blocks are then placed in the final resin mixture and propylene oxide in a 1 : 1 ratio for at least 2 h and in a 2 : 1 ratio for 36 h. Place cork stoppers on the tubes containing the blocks and wrap parafilm at the junction of the cork and glass tube to prevent evaporation, otherwise the resin mixture may become too viscous (*see* **Note 12**).
3. Embedding
   Fill freshly prepared resin mixture in a 5-mL plastic syringe and fill blank molds. Cubes of brain are embedded in polyethylene BEEM embedding capsules while 50–100 µm thick blocks or vessels are embedded in flat silicone rubber molds.
   a. Pour the contents of the glass tube with the sections onto card paper. Use bamboo splints to transfer the sections into the molds.
   b. Use a stereoscope to position the tissue such that the required surface is at the cutting edge (**Fig. 2A**).

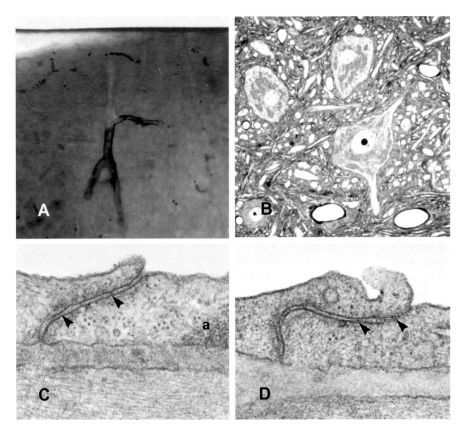

Fig. 2. **(A)** The cutting edge of an epon block prepared using a flat silicone mold, shows a 50-μm thick section of cerebral cortex embedded within the block. The block viewed using a 3.5X objective shows an arteriole permeable to HRP. The vessel details allow the researcher to determine the precise level at which to section this vessel. **(B)** A semithin section stained with toluidine blue provides good visualization of neurons, nerve processes, and blood vessels present in the block. Such a preparation can be used for morphometry to determine the number of neurons or the density of vessels in a particular area. Segments of cerebral endothelium of vessels permeable to horseradish peroxidase processed without **(C)** and with **(D)** *en bloc* stain are shown. Note that unit membranes are better preserved in **(D)** than in **(C)**. The fusion of the outer leaflets of the plasma membranes at tight junctions (**arrowheads**) are clearly seen in **(D)**. Cross sections of actin filaments **(a)** are indicated. **A**, X150; **B**, X185; **C**, X70,000; **D**, X75,000.

    c. Puncture any air bubbles in the resin mixture with a needle and leave molds at room temperature for a few hours. If tissues have moved from the cutting edge reposition them using the bamboo splints or needles.

    d. Place molds in a 60°C oven overnight.

    e. Remove resin blocks from the molds and place in a glass beaker. If blocks are sticky put them back in the oven for a further 2–4 hr (*see* **Note 13**).

    f. Next place blocks in a 90°C oven for 2–3 d to cure.

    g. Residual resin mixture is polymerized in the beaker by placing in the 90°C oven until it hardens. It can then be discarded.

    h. Any Epon spills can be cleaned with ethanol.

3.5. Sectioning

Both semithin and ultrathin sections are cut using an ultramicrotome. Before cutting, wipe the ultramicrotome with a damp Kimwipe or soft cloth to remove dust or debris from previous sectioning. This also reduces static electricity.

3.5.1. Semithin Sectioning

Resin-embedded sections are superior to paraffin sections because tissue shrinkage is less and cellular components are better preserved (**Fig. 2B**). Semithin sections have many uses such as 1) they can be used to determine whether fixation, dehydration, and infiltration are carried out properly; 2) they are suitable for assessment of morphological findings and for morphometric assessments such as number of neurons/mm^2, number of vessels/mm^2, and the diameter of vessels; 3) they are helpful for selection of areas of interest for ultrathin sectioning. Semithin sections are cut using glass knives, which are made using a knife maker as per instructions supplied with the machine. An old diamond knife may also be used to cut these sections. Semithin sections are stained with an alkaline toluidine blue solution (*see* **Fig. 2B**).

1. Trimming of blocks
   a. Place the block in a microtome chuck, which is then mounted on a special base and viewed by transmitted illumination using a dissecting microscope.
   b. Use a fresh single-edge razor blade to trim away the resin around the tissue to form a four-sided pyramid, with the tissue at the top of the pyramid.
   c. The block face is usually rectangular with the top and bottom edges parallel to one another to obtain a ribbon during sectioning. Cut off one corner. This cut end can be seen in sections by light microscopy and helps to orient the block when the size of the block face is cut down further in preparation for ultrathin sectioning.
2. Sectioning
   a. Reset the advance mechanism of the microtome after each block or each day depending on the number of blocks being cut.
   b. Retract both the coarse and fine stage advances.
   c. Mount the chuck containing the block into its holder in the machine and tighten the screw holding it in place. The block is positioned so that its long axis is at right angles to the cutting edge.
   d. Mount the knife and screw tightly into place. Adjust the angle between 2° and 5°.
   e. Cycle the microtome manually and stop when the face of the block is slightly above the height of the cutting edge.
   f. Position the binocular microscope and adjust the angle of illumination system so that some light is reflected on the block face.
   g. Bring the mounted knife forward slowly until the cutting edge and the block face can be viewed through the microscope.
   h. A syringe or micropipette is used to fill filtered distilled water in the trough adjacent to the diamond knife. Sufficient water is added so that the surface of the water is convex in order to wet the knife-edge. Water is withdrawn from the trough using a syringe until the surface is horizontal (*see* **Note 14**).
   i. Adjust the knife holder so that the best portion of the knife is used for sectioning.
   j. Slowly advance the knife toward the block face, by using the coarse adjustment and then the fine adjustment.

k. Cutting Speed: This is the rate at which the specimen block passes the knife during the cutting phase and is expressed in mm/s. Automatic ultramicrotomes maintain a constant cutting speed over a specified distance. As a general rule, sectioning should be performed at a relatively slow speed of 2–3 mm/s. However, the appropriate speed has to be established by trial and error for the tissue being cut (*see* **Note 15**).

l. The ultramicrotome has a control to adjust section thickness. In addition, thickess of sections in a water bath are also judged by the interference colors that are produced when the light reflected from both the upper surfaces of the sections and from the section-water interface move out of phase. A color scale is available, which gives the thickness of sections having a particular color and this is applicable to all embedding materials having a refractive index close to 1.5, including epoxy resins (*4*). Semithin sections are dark gold and have a thickness of approx 900 nm.

m. When floating on the trough liquid, the sections should appear uniformly colored both within each section and from section to section in the ribbon.

n. The surface of sections should be smooth without wrinkles, folds, or knife marks.

o. Make sure that the ribbon is straight.

p. Detach the ribbon from the cutting edge of the knife using a clean hair, which is mounted on wooden handles or cluster sections.

q. Sections are picked up from the water bath and placed on a glass slide using the wire loop.

r. Slides are placed on a hot plate set at 80°C for 10 min to flatten the section, evaporate the water and ensure adhesion of sections to the slides.

3. Staining semithin sections

a. Place a drop of toluidine blue stain on sections and place on the hot plate for a minute or until a dried ring is seen at the edge of the stain.

b. Wash slide with distilled water and dip in acetone to differentiate.

c. Dehydrate in 2 changes of 100% ethanol and then in 2 changes of xylene for 2 min each.

d. Place a drop of Permount on the section and then a cover slip.

3.5.2. Ultrathin Sectioning

Ultrathin sections are cut from a preselected area of the block, usually based on information obtained by viewing the semithin section. Most sections within the silver to pale gold range (60–90 nm) are suitable for normal work, although thinner sections may be required when high resolution is needed. Pale gold sections having a thickness of about 90 nm give well-contrasted, good-quality electron micrographs while micrographs obtained from silver sections usually lack contrast. Ultrathin sections are cut using a diamond knife whose cutting angle is specified by the manufacturer. These sections can also be cut with freshly prepared glass knives. Follow the sectioning method outlined in **Subheading 3.5.1., step 2.**

1. A plexiglass draft protector, which is supplied with the ultramicrotome should be used when ultrathin sections are cut to prevent the sections being blown around in the trough.

2. Collection of Sections

a. Detach the ribbon from the cutting edge of the knife using the clean hair taped to an applicator stick.

b. One edge of a clean copper grid is bent using a pair of tweezers to produce a 90° angle with the rest of the surface. The grid is held at this end and lowered into the trough and positioned under the ribbon such that the rim of the grid is under the edge of the first section. Lift the grid vertically out of the water. Sections are picked up on the dull side of the grid (*see* **Note 16**).

    c. Touch the edge of the grid on the surface of filter paper and place on a filter paper in a petri dish with the section side up and dry grids at least 20–30 min before staining.
3. Usually two to three grids of sections are cut from one block.
4. After sectioning, clean the diamond knife with distilled water contained in a spray bottle and dry it using a container of compressed air. Periodic cleaning in an ultrasonic cleaner is recommended to remove any adherent resin.

3.6. Staining

May be done using an automatic stainer. Manual staining is done in a designated clean area of the laboratory preferably using a bench top hood. The stain solutions are left undisturbed in this hood. The method for manual staining of ultrathin sections is as follows:

1. If *en bloc* staining with uranyl acetate is done proceed to lead citrate staining (**step 4**).
2. Place a drop of uranyl acetate on a silicone rubber plate contained in a petri dish. Float the grid on this drop with the section side down for 20 min. Because the surface of this rubber plate is divided into numbered squares, four to five sections can be stained at a time.
3. Rinse the grid by 10–15 quick dips in a 250-mL beaker containing filtered water. Use a dry forceps to hold the opposite side of the grid and dip again 10–15 times in another beaker containing filtered water (*see* **Note 17**).
4. Touch the edge of the grid to filter paper to remove excess water and place the grid on a drop of lead citrate stain for 1–1.5 min. This stain is placed on the silicone rubber plate, which is surrounded by NaOH pellets in a large petri dish. Keep petri dish covered while staining.
5. Rinse sections as in **step 3**.
6. Place the grid with the section side up on a filter paper and allow it to dry for few minutes and then place the grid in a gelatin capsule, which can be labeled with the case number (*see* **Note 18**).

3.7. Morphometry

Measurements of various parameters of pial and intracerebral cortical vessels such as diameter, cross sectional area of the media and intima and density of cerebral microvessels are available in the literature *(5–7)*. In addition, measurements are also available of lengths of tight junctions, density of fenestrations and organelles in cerebral endothelium such as mitochondria, and endothelial vesicles *(8–10)*. Details on how to perform these measurements are beyond the scope of this chapter because it depends to a certain extent on the type of image analyser being used. An example of a morphometric technique follows.

3.7.1. Determination of the Ratio of the Wall-Lumen Area of Cerebral Cortical Arterioles

1. Areas of the different layers of the vessel wall can only be measured when vessels are fixed after achieving maximal dilatation. This is achieved by perfusing the rat initially with Kreb's solution for 15 min followed by perfusion of a fixative solution.
2. Flat embedded blocks are placed on a glass slide and examined using a 3.5X objective of a light microscope to determine whether the block has an arteriole (*see* **Fig. 2A**). Arterioles having external diameters ranging from 15–25 μm and a single layer of smooth muscle are selected.

3. Sectioning arterioles at a particular depth from the cortical surface is a laborious process. To section arterioles at a depth of 300 μm from the cortical surface:

   a. Place a calibrated micrometer scale in the eyepiece of a light microscope and measure the length of the block using a 3.5X objective. If 1 division of the eyepiece scale is equal to 34.5 μm, then 8.5 divisions of the top of the block have to be cut away to reach a depth of 300 μm from the cortical surface.

   b. The block face is trimmed and sectioned. Once the cortical surface is reached the ultra-microtome is set at a thickness setting of 10,000Å and 300 sections are discarded to reach a depth of 300 μm.

   c. The block can then be re-measured to determine whether the correct depth has been reached.

   d. The block face may have to be trimmed further before thin sections are cut.

4. Overlapping electron micrographs are taken along the circumference of arterioles at a screen magnification of 2700. Only vessels sectioned perpendicular to their long axis are used for photography. These vessels show unit membranes along their entire circumference and the cell wall has a fairly uniform thickness excluding the areas having endothelial nuclei.

5. The magnification of the electron microscope is checked using a carbon diffraction replica.

6. Electron micrographs printed at a constant magnification are taped together to reconstruct the entire vessel. This is placed on an illuminated copy stand (Kaiser). A high resolution CCD Camera is used to transmit the image to an image analyzer.

7. Image analysers such as the Microcomputer Imaging Device system (Imaging Research, St Catherines, On) can receive images from negative of electron micrographs placed on a light box via a CCD camera. The image obtained on the screen can be inverted to obtain a positive image.

8. The image analyser is calibrated for linear measurements in micrometers or area measurements in square-micrometers. The cross-sectional area occupied by the vessel lumen and the total cross sectional area of the vessel (lumen + media and intima) of vessels in the different experimental groups is measured using the image analyzer. Cross sectional area of the arterial wall is calculated by subtracting the lumen area from the total vessel area. The ratio of the wall-lumen area is then calculated.

9. Vessel wall dimensions for the different experimental groups are compared using the unpaired t test.

10. Interpretation: These measurements were used to establish that the ratio of the wall-lumen area of cerebral arterioles is significantly higher ($p > 0.001$) in rats with chronic renal hypertension as compared to arterioles of normotensive rats *(7)*.

Quantitation of endothelial organelles such as plasmalemmal vesicles is given in Chapter 8, Subheading 3.1.2.1.

4. Notes

1. Most chemicals used for electron microscopy such as solvents, resins, and buffers are hazardous and gloves should be worn for all procedures listed in this chapter and is optional only when sectioning.

2. Sodium cacodylate contains approx 30% arsenic by weight and is a health hazard. It should be weighed and dissolved in a fume hood.

3. Osmium tetroxide is hazardous and should be used in a fume hood. It is disposed in a container containing vegetable oil and kitty litter. The Biohazard Department of the Institution has to be contacted to dispose this waste.

4. Uranyl acetate and lead salts used for staining grids are toxic and have to be handled with care. Uranyl acetate is a radiochemical as well. Both these substances are disposed in

different labeled containers, which are disposed by the biohazard department of the institution.

5. Distilled water filtered using a no. 50 Whatman filter is used for the preparation of stains for ultrathin sections and to fill the trough adjacent to knives.

6. Undiluted fixative can be frozen in 250-mL amounts in Nalgene flasks. It is defrosted before use and diluted with an equal amount of 0.1 M cacodylate buffer containing sucrose for perfusion.

7. Some researchers perfuse a buffer or Ringer's solution containing heparin for few minutes to clear the blood followed by the perfusion of the fixative solution. In our studies preservation of endothelium is not as good when buffer followed by fixative is used as compared with fixative alone. Heparin is known to increase cerebral endothelial permeability to protein tracers *(11)* and therefore should not be used when permeability studies are undertaken. Perfusion with Ringers solution is done if vessel morphometry is required to achieve maximal dilatation of vessels before fixation.

8. A perfusion time of 10 min is selected because the results show good preservation of tissue. Shorter periods result in poor preservation and fragile slices.

9. If brain blocks or slices are immersion fixed for longer than 48 h there is extraction of tissue constituents and organelles and nerve processes appear swollen.

10. The advantage of *en bloc* staining is that preservation of membranes is superior (*see* **Figs. 2C** and **D**), and it eliminates having to stain grids with uranyl acetate prior to viewing. The latter technique is more likely to produce precipitates on sections making them unsuitable for photography.

11. Ethanols, which are 85% and higher, are collected in a safety container for disposal by the biohazard department of the institution.

12. Rotary shakers are available for agitating the contents of the tubes to promote infiltration of resin into blocks. These shakers are not suitable for tissue blocks, which are 50 μm thick as the tissue gets caught between the cork and glass tube and disintegrates.

13. If semithin sections are required quickly blocks can be cut after being kept at 60°C overnight. These blocks do have to be placed in the 90°C oven to cure the Epon before ultrathin sections can be cut.

14. A convex meniscus generally tends to overwet the cutting edge resulting in wetting of the block face of the specimen and sometimes the back of the knife face also picks up some trough fluid. Sections cannot be obtained under these conditions. A concave meniscus may not wet the entire knife face, which results in the sections sticking to the knife and crumpling.

15. If the cutting speed is too high, variations in section thickness result. Excessive compression, wrinkles and fine chatter parallel to the cutting edge of the section can also occur. A very slow cutting speed may drag the trough fluid over the back of the knife and changes in temperature and draught during a cycle will cause thermal drift.

16. An alternate method for collecting sections is to place the grid flat on the upper surface of the sections.

17. Static may be a problem during staining especially during the winter months when the air is very dry. This is overcome by running a humidifier near the staining area.

18. An inexpensive way to store grids is to place them in a gelatin capsule, which is numbered with the grid number. Capsules containing grids from one case are placed in a cardboard pill box that is labeled with the case number.

Acknowledgments

This work is funded by the Heart and Stroke Foundation of Ontario. The skilled technical assistance provided by Verna Norkum and Blake Gubbins is gratefully acknowledged.

References

1. Karnovsky, M. J. (1965) A formaldehyde-glutaraldehyde fixative of high osmolality for use in electron microscopy. *J. Cell Biol.* **27,** 137A.
2. Hayat, M. A. (2000) *Principles and Techniques of Electron Microscopy. Biological Applications, 4th ed.* Cambridge University Press, New York.
3. Hunter E. (1993) *Practical Electron Microscopy. A Beginner's Illustrated Guide, 2nd ed.* Cambridge University Press, Cambridge, UK.
4. Peachey, L. D. (1958) Thin sections. I. A study of section thickness and physical distortion produced during microtomy. *J. Biophys. Biochem. Cytol.* **4,** 233–242.
5. Baumbach, G. L., Walmsley, J. G., and Hart, M. N. (1988) Composition and mechanics of cerebral arterioles in hypertensive rats. *Am. J. Pathol.* **133,** 464–471.
6. Gross, P. M., Sposito, N. M., Pettersen, S. E., and Fenstermacher, J. D. (1986) Differences in function and structure of the capillary endothelium in gray matter, white matter and a circumventricular organ of rat brain. *Blood Vessels* **23,** 261–270.
7. Nag, S., and Kilty, D. W. (1997) Cerebrovascular changes in chronic hypertension. Protective effects of enalapril in rats. *Stroke* **28,** 1028–1034.
8. Nag, S. (1995) Role of endothelial cytoskeleton in blood-brain barrier permeability to protein. *Acta Neuropathol. (Berl)* **90,** 454–460.
9. Nag S. (1998) Blood-brain barrier permeability measured with histochemistry. In *Introduction to the Blood-Brain Barrier. Methodology, Biology and Pathology.* Pardridge, W. M. ed. Cambridge University Press, Cambridge, UK, pp 113–121.
10. Stewart, P. A., Hayakawa, K., and Farrell C. (1994) Quantitation of blood-brain barrier ultrastructure. *Microsc. Res. Tech.* **27,** 516–527.
11. Nagy, Z., Peters, H., and Huttner, I. (1983) Charge-related alterations of the cerebral endothelium. *Lab. Invest.* **49,** 662–671.

3

Freeze-Fracture Studies of Cerebral Endothelial Tight Junctions

Hartwig Wolburg, Stefan Liebner, and Andrea Lippoldt

1. Introduction

The tracer experiments of Reese and Karnovsky *(1)* demonstrated that it was the endothelium that formed a permeability barrier because electron-dense tracers such as horseradish peroxidase did not pass from the vessel lumen through the interendothelial cleft. The structure responsible for the lack of permeability of tracers was the tight junction. In endothelial cells, these specialized contact zones were already known from ultrastructural studies *(2)*, and in epithelial cells, their morphology was described in detail by Farquhar and Palade *(3)*.

Around this time, a novel morphologic technique was developed for visualizing the cytoplasmic membranes and intercellular contacts and to corroborate the then-novel fluid mosaic model of membranes *(4)*. This technique was the freeze-fracturing technique, and soon after its introduction, epithelial tight junctions *(5–8)* and also cerebral endothelial tight junctions *(9–12)* were the subject of many freeze-fracture studies.

1.1. Freeze-Fracture Technique

The two main freeze-fracture techniques are freeze-etching and freeze-fracturing. In freeze-etching (**Fig. 1A**), a cooled knife (-196°C) produces a plane below the surface of the specimen, which is etched by holding the knife directly above the plane of the specimen. The depth of etching depends on the difference of temperature between specimen and the etching device, i.e., the knife, and the duration of water sublimation. The disadvantage of this method is that the stabilization of the specimen during sectioning is difficult and only one replica per specimen is produced. In freeze-fracturing (**Fig. 1B**), the specimen is deeply frozen between a double holder device and is then cleaved into two halves by a fracturing device. The etching step is omitted. The advantage is that each half can be shadowed and two complementary replicas per specimen are produced.

An early study of frozen yeast cells suggested that the fracture planes follow the surface of membranes *(13)*. It was then demonstrated that the fracture plane runs through the middle of the membrane, cleaving both lipid layers of the unit membrane since in a vitrified matrix the hydrophobic membrane centre is fragile enough to be easily split

From: *Methods in Molecular Medicine, vol. 89:*
The Blood–Brain Barrier: Biology and Research Protocols
Edited by: S. Nag © Humana Press Inc., Totowa, NJ

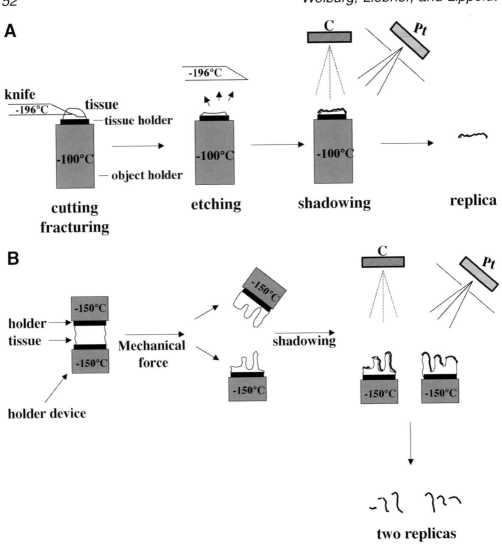

Fig. 1. Highly schematic view of the principle of freeze-etching (**A**) and freeze-fracture (**B**) techniques. In the freeze-etching process, the frozen specimen is cleaved by a cooled knife, etched and shadowed resulting in one platinum-carbon replica. In the freeze-fracturing process, we get two complementary replicas without the etching process.

(14) (**Figs. 1, 2**). The ultimate evidence for membrane cleavage came from labeling the outer surface of red blood cells with ferritin, which was only visualized after etching to show the true surface of the cell, but never without etching proving that the fracture plane had to be run through the middle of the membrane where the ferritin had no access *(15)*. In 1975, an international consortium published the nomenclature and interpretation of fracture faces *(16)* (**Figs. 2, 3**). The reader is referred to a good review of the technology and interpretation of freeze-fracture replicas *(17)*.

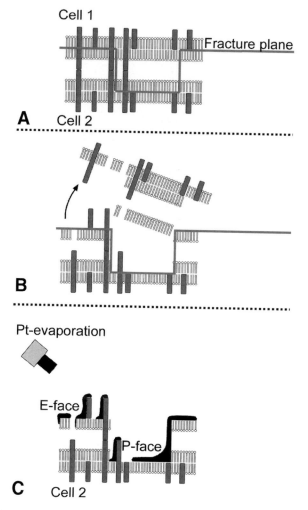

Fig. 2. Highly schematic view of the cleavage process during freeze-fracture. In (**A**), two membranes of adjacent cells are shown, the fracture plane runs through the hydrophobic middle of the membranes. In (**B**), the cleavage process itself is illustrated showing the disconnection of the membranous leaflets, and in (**C**), the Pt-evaporation shadows the exposed membrane fracture faces creating platinum replicas of the extracellular (E-) and protoplasmic (P-) fracture face.

1.2. Fixation

The first step in biologic specimen preparation for electron microscopy is fixation. Fixation methods for ultrathin sectioning are also applicable to the freeze-fracture technique. The aim of fixation is to maintain a "true-to-life" structure of cells and organelles at the time point of immobilization. However, the stabilization of the cellular organization has to be in a manner so that ultrastructural features are very well preserved and that the specimen is stable enough to overcome the necessary steps such as ultrathin sectioning, immunolabeling, or freeze-fracturing. In principle there are three methods

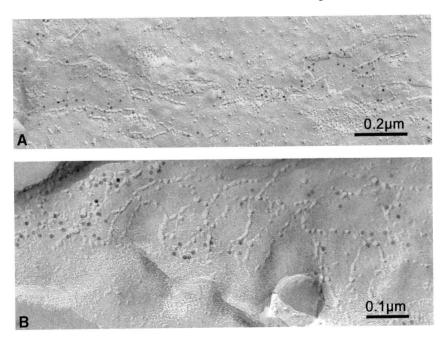

Fig. 4. (**A, B**) Fracture-labeling technique using Madin Darby canine kidney (MDCK) cells for labeling the tight junction protein occludin in the tight junction strands with a polyclonal anti-occludin antibody and a 10-nm gold-conjugated secondary antibody.

7. Cleaning the replicas several times in double distilled water and mounting on formvar-coated copper grids.
8. Analysis of the replicas using a transmission electron microscope (*see* **Notes 1–3**).

3.2. Immunogold Freeze-Fracture Labeling Technique

The limitation of the freeze-fracture technique is that it allows only a morphologic assessment of plasma membranes, providing no direct information on the molecular composition of the structure of interest. To overcome this limitation, freeze- fracture studies have been combined with immunohistochemistry using two different methods. The first method is the label-fracture method, in which the biologic material is prelabeled with the first antibody against the protein of interest and with a gold-conjugated secondary antibody. Afterward, the specimens are fixed, cryo-protected, frozen and fractured *(22,23)*. In contrast, in the fracture-label technique immunohistochemistry is done after the fracture process. The techniques used today, are all based on the freeze-fracture immunocytochemistry methods (**Fig. 4**).

The breakthrough in fracture-labeling that produces a Pt/C-replica was the combination of cryo-immobilization (rapid-/quick-freezing) with subsequent fracturing, and shadowing with Pt/C and sodium dodecyl sulfate (SDS)-digestion *(24,25)*. The SDS-digested freeze-fracture replica labeling (SDS-FRL) method of Fujimoto *(24)* utilizes the modified quick freezing method of Heuser et al. *(26)*, which belongs to the metal mirror quick freezing methods, in which the tissue to be frozen is stamped on a copper bloc cooled by liquid helium. The very quick and firm contact of the tissue with the plane polished (mirror) copper surface leads to the vitrification of several cell layers. The main advantage of the method is the good ultrastructural preservation of the tissue and the high antigenicity. The disadvantage is that only cytoplasmic residues of trans-membrane proteins and, using TritonX-100 for digestion, peripheral membane proteins can be labelled. Another limitation is the time consuming preparation of the copper foil and copper block, needed for the sample sandwich and the freezing process. If many samples have to be frozen, the freezing method of Heuser et al. *(26)* is not suitable.

To partially overcome these methodological problems, the authors used a modification of the technique of Fujimoto et al. *(27)*:

1. Cells were grown according to their requirements, washed twice with PBS and scraped off. Alternatively they were grown on filters, which were cut into appropriate pieces.
2. Cells were mounted between a copper sandwich and frozen quickly in liquid ethan and transferred into the pre-cooled Balzer's freeze-fracture device.
3. The freeze-fracture and shadowing conditions were identical to those described above (*see* **Subheading 3.1.**).
4. After fracturing, the replicas were detached from the copper holder under PBS and washed two more times.
5. For handling the replicas a loop of thin platinum wire and an eyelash were used.
6. Subsequently, the replicas were digested overnight under vigorous stirring with a magnet stirrer (450 rpm) in a 2.5% SDS solution.
7. The next day, approx 20 small pieces of the replica were subjected to a conventional "in solution" immunogold-labeling procedure as follows:
   a. The replicas were washed twice with PBS and blocked in PBS containing 10% BSA for 30 min.
   b. The primary antibody (dilution has to be determined) was incubated either 1 h at room temperature or overnight at 4°C.
   c. The replicas were washed three times for 10 min in PBS, blocked again with PBS/BSA (10%) for 10 min and incubated with the gold-conjugated secondary antibody.
   d. The labeled replicas were washed in PBS several times.
   e. They were then fixed with 2.5% glutaraldehyde in PBS for 10 min at room temperature.
   f. Replicas were washed in distilled water and mounted on pioloform-coated copper grids.
   g. They were analyzed in a Zeiss EM10 transmission electron microscope (*see* **Fig. 4**).
8. Interpretation of the labeling specificity of this method is done by stereo images, which allow visualization of gold particles above and under the replica surface *(25)*, Gold particles above the replica surface are due to nonspecific adsorption to the Platinum (*see* **Note 4**).

3.3. Interpretation of the Freeze-Fracture Replicas

The freeze-fracture method is based on the fact that the fracture plane runs through the middle of the membrane according to the hydrophobicity between the two lipid layers of a membrane (*see* **Figs. 2A, B**). Thus, the membrane is split into two halves: The protoplasmic fracture face (P-face) associated with the protoplasm and the external fracture face (E-face) associated with the extracellular space. The P-face is viewed from outside the cell, and one has to imagine the cytoplasm behind the fracture plane. Correspondingly, the E-face is viewed from inside the cell, and one has to imagine the extracellular space behind the fracture plane (*see* **Figs. 2C** and **3**).

Because of the cleavage of the membrane and the consecutive disconnection of both halves (*see* **Fig. 2B**), one gets two different replicas that have to be examined independently. For this reason, it is extremely difficult to find and to identify the complementary sites of the intact membrane in complementary replicas.

Particles are visible on the surface of the fractured membranes as well as within the tight junction strands. It is believed that most intramembrane protein particles (IMP) are not cleaved by the fracturing process, thus resulting in particles, which are elevated above one or the other plane of the separated membrane halves. A cleavage of proteins in the fracture plane was proposed to be unlikely because it would presume a break of covalent bonds in the transmembrane peptide chain (28).

In general, particles in freeze-fracture replicas are mainly associated with the P-face and less associated with the E-face (*see* **Fig. 3**). This might be caused by the prefixation with aldehydes. Comparing replicas from aldehyde-fixed and rapidly frozen aortic endothelial cells, it appears that after rapid freezing, the IMPs were nearly equally distributed on both fracture faces, whereas chemical prefixation resulted in the clear predominance of P-face associated IMPs (29). This may reflect an aldehyde-strengthened interaction between the cytoskeleton and intramembrane proteins resulting in the anchorage of the proteins in the protoplasmic leaflet. If, however, in series of experiments the cells are consistently fixed with glutaraldehyde before fracturing, the advantage of better reproducibility is considerable, and the disadvantage of an altered particle association is always the same in all experimental systems. Thus, if after glutaraldehyde fixation a given particle or strand is still associated with the E-face (e.g., peripheral non-barrier endothelial cell tight junctions: 30), the linkage to the cytoskeleton must even be weaker than judged from replicas after cryoimmobilization. It is postulated that each particle on one membrane leaflet has to correspond with one pit in the complementary membrane leaflet. In high quality replicas, this is indeed the case. However, in most cases one finds a high number of P-face–associated particles, a low number of E-face–associated particles and a smooth lipid plane at the E-face (*see* **Fig. 3**). Pits are frequently obscured by the evaporated platinum. However, there are some well-known examples of E-face–associated negative structures (pits or grooves) corresponding with P-face associated positive structures: the vertebrate gap junctions (31), the orthogonal arrays of particles in glial cells (32), and the tight junctions (33).

3.4. Morphometric Evaluation of Tight Junctions

3.4.1. Determination of E-Face/P-Face Ratio

The E-face/P-face distribution of tight junction particles is an important feature of tight junctions, which may reflect the functional state of the epithelial or endothelial barrier. The ratio of E-face– or P-face–associated particles to the total length of the tight junctional network is done using morphometry by:

1. Determining the length of the given tight junction structure that is covered with particles.
2. Determining the whole length of the given tight junction structure.
3. Calculation of the ratio of the total amount of particles and the whole length of the tight junction. The ratio is expressed as %PFA or %EFA, respectively.

3.4.2. Quantification of the Tight Junction Complexity

The classical feature of tight junction morphology is the complexity of the network. This can be assessed by different methods:

1. Strand counting (SC): In earlier studies, quantification of tight junction complexity was mainly performed by counting of longitudinal strands by line- or grid-intersection methods (SC), sometimes in combination with the additional evaluation of parameters such as linear junctional density, tight junction depth, meshes, branches, and free ends *(34–36)*.
2. Complexity index (CI): Alternatively, the CI was calculated and defined as the ratio of the number of branch-points in a network to the total of the tight junction length *(37,38)*.
3. Fractal dimension (FD): This provides a more comprehensive description of tight junctional complexity *(39)*. Fractal geometry describes the structural properties of objects that, by definition, show self-similarity, which means that these objects are identical at all magnitudes. Theoretically, this should be true for an infinite range of scales, but in reality objects are self-similar only in a restricted range of magnitudes. Such objects are characterized by a dimension (FD) greater than the corresponding classical topologic dimension *(40,41)*. For tree-like structures with a topologic dimension of 1, FD ranges between 1.0 (for the most simplest objects) and 2.0 (for the most complex structures). FD is a highly integrative parameter, because its value is influenced by a variety of properties such as number, length, tortuosity, and connectivity of elements of a given object. On the other hand, measurement of FD is independent of the orientation of a figure in the plane. These advantageous features of FD have successfully been used to describe the structure of biologic objects or phenomena *(42–45)*, although such structures are self-similar only within a restricted range of scaling levels. Because tight junctions are tree-like structures, they show a certain degree of self-similarity and are therefore a suitable object of fractal analysis *(39)*. For manual evaluation of FD, grids of different scaling levels (grid-sizes) are superimposed to the tight junction image and for each scaling factor the number of boxes containing parts of the tight junction network are counted (N) only once, in repeated measurements. The suitable grid sizes are 0.2, 0.1, 0.05, 0.025, and 0.0125 μm. Because the definition of FD (box counting) is log N / log (l/s), the values obtained for each scaling level are inserted into a log N vs. log (l/s) graph for visualization, and the regression curve is calculated. The slope of the curve can be related to the value for the FD.

The semi-automatized method for FD evaluation includes video-recording, analog to digital conversion with image processing software (*see* **Note 5**). Only the fractal analysis is able to detect the continuous increase of complexity *(39)*.

3.5. Cerebral Endothelial Tight Junctions

In conventional ultrathin sections, tight junctions are considered to form pentalaminar layers resulting from the fusion of the external leaflets of the adjacent cell membranes. Depending on the orientation of the section, the tight junctions mostly appear as a chain of fusion ("kissing") points or as a domain of an occluded intercellular cleft of variable length *(3,46)*. In freeze-fracture replicas, the tight junctions of the blood–brain barrier (BBB) endothelial cells are the most complex in the vascular system *(47)*. The P-face association of particles in BBB endothelial cell tight junctions is approximately 55%, whereas 45% of the IMPs are situated on the E-face *(48)* (**Fig. 5A**). In contrast, epithelial tight junctions are predominantly associated with the P-face forming a network of strands and leaving grooves at the E-face, which are occupied by very few particles *(37,49–51)*.

If tight junction particles occur on the E-face they are frequently arranged in chains; if occurring on the P-face they are, at least in epithelia, formed as smooth continuous cylindrical profiles (*see* **Note 1**). This difference has been explained by the assumption that the discontinuous and irregular appearance of tight junction particles on E-faces

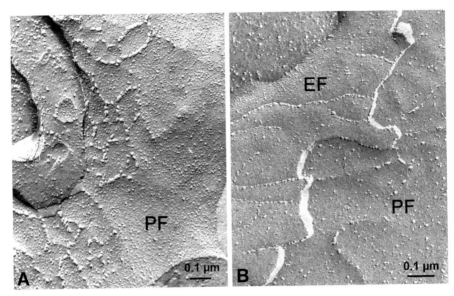

Fig. 5. Freeze-fracture replicas of cerebral endothelial cell tight junctions. **(A)** In capillary preparations, most P-face strands are occupied with particles. **(B)** Cultured brain endothelial cells, show only few tight junction particles in the tight junction ridges of the P-face, but most particles have switched to the E-face.

are due to multiple linkage sites of protein complexes to the cytoskeleton *(52,53)*. Some authors believe that the cylindrical profiles of tight junctions consist of double strands that are arranged in an offset manner so that the fracture plane runs between the partner strands *(53,54)*.

It has been suggested that there is a logrithmic relationship between the number of strands and the transcellular electrical resistance (TER; *34,35)*. It was concluded that the complexity of the network of strands could be used to predict physiologic parameters such as permeability and TER *(55)*. These suggestions are now supported by transfection experiments using DNA constructs of the newly discovered important family of proteins—the claudins, which are responsible for tight junction stability and restrict tight junction permeability *(56–58)*. When transfected with claudin-1 or claudin-3, tight junction-negative L-fibroblasts formed tight junctions associated with the P-face; if transfected with claudin-2 or claudin-5, the cells formed tight junctions associated with the E-face *(59,60)*. Whereas transfected occludin induced short strands the claudin-induced strands are very long and resemble in vivo tight junctions *(56,61)*. Introduction of the claudin-2 cDNA into high-resistant madin-darby canine kidney cells (MDCK-I) that normally express claudin-1 and 4, mimicked both the resistance behaviour and the tight junction morphology of low-resistant MDCK-II cells *(62)*. These experiments support the predicted relationship between tight junction morphology as it appears by freeze-fracture and electrical resistance. Moreover, the combination and stoichiometry of the claudin species directly determines the barrier function and the morphologic outcome of a given tight junction.

In endothelial cells, the claudins detected so far are claudin-1, claudin-5 *(60,63,64)*, and claudin-3 (Engelhardt and Wolburg, unpublished). The high electrical resistance of cerebral endothelial cells is accompanied by an expression of claudin-1/3 and claudin-5 and reflected by P-face/E-face ratio of about 55/45% *(48)*. In contrast, in non-BBB endothelial cells, tight junctions are almost completely associated with the E-face *(30)*, and claudin-1 is rarely or not expressed. BBB endothelial cells cultured in vitro develop tight junctions with particles dramatically switched to the E-face *(38;* **Fig. 5b)** accompanied by a downregulation of claudin-1 *(64)* and by an increased permeability and decreased TER. This means that the morphologic features of tight junctions reflect their physiologic properties.

Thus, morphologic investigations by freeze-fracturing combined with the detection of antigens in freeze-fracture replicas is a powerful method to predict and characterize the physiologic properties of tight junctions (*see* **Note 6**).

4. Notes

1. Experience is necessary to interpret freeze-fracture replicas and to detect tight junctions in them. Therefore, it is advisable to undertake initial studies under the supervision of an experienced scientist in this field. Incorrect interpretations may be made if the investigator is not able to distinguish between leaky (choroid plexus capillaries; large vessels) and tight endothelial (cerebral capillaries) and epithelial (choroid plexus) junctions that occur in brain tissue. A feature of choroid plexus capillary tight junctions is the intercalation of gap junctions, while choroid plexus epithelial tight junctions have a predominant P-face association of particles. Very small tight junction pieces should not be taken for analysis of in vivo material.

2. Detection of tight junctions in fractured brain tissue pieces is very difficult and time-consuming. Therefore, isolated capillary fragments are used for our investigations whenever it is feasible. Cerebral capillaries are isolated according to the method described in Risau et al. *(65)* as follows:
   a. Briefly, the forebrains are carefully dissected out and the meninges and choroid plexuses are removed.
   b. Three to four brains are minced and digested using collagenase (0.75% w/v, Worthington CLS II in 15 mM HEPES) for 80 min at 37°C with gentle agitation.
   c. The capillaries are separated by BSA density gradient centrifugation (25% in PBS; 20 min at about 1000*g* in a Heraeus Centrifuge using a swing-out rotor), washed and fixed according to the protocol (*see* **Subheading 3.1.**).

3. For proper evaluation of freeze-fracture replicas by electron microscopy, a goniometer attached to the microscope is useful.

4. For interpretation of the labeling specificity, controls are done either by omitting the primary antibody or by use of the preimmune serum instead of the primary antibody.

5. For semiquantitative analysis, software from SISanalysis (Münster, Germany) is used, which allows the simultaneous detection of the amount of particles and tight junction length as well as the statistical evaluation.

6. Several authors suggest that the tight junctional strands are lipid in nature *(66,67)*. This model postulated inverse lipid micelles, which should form during a phase transition of the planar lipid bilayer similar to that occurring in models of membrane fusion (e.g., in exocytosis). The important point is that this model involves a continuity of the outer membrane leaflets of two tight junction-connected partner cells (**Fig. 6**). The experiments per-

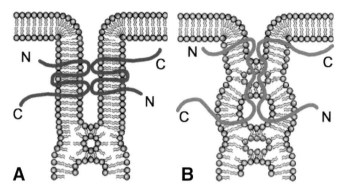

Fig. 6. Diagram depicting the proposed molecular structure of tight junctions. Both inverse micelles and proteins are unequivocally proven (see text). In (**A**), the generally suggested folding mode of the four transmembrane domains proteins such as occludin and the claudins is shown, revealing two extracellular loops and one intracellular loop between transmembrane domain II and III. The inverse micelle depicted below the protein, corresponds to the lipid nature of tight junctions postulated in the literature (see text), but is not stable without stabilizing proteins. Therefore, we propose the working hypothesis of tight junctional structure in (**B**), unifying the protein and lipid hypotheses and showing an alternative folding mechanism of the proteins. The main difference is the translocation of the hydrophilic domain between transmembrane domains II and III from the cytoplasm into the micelle allowing a stabilization of the micelle.

formed by Hein et al. *(68)* and Grebenkämper and Galla *(69)* provided unequivocal evidence for the continuity of outer partner cell leaflets. Accordingly, Kan *(70)* demonstrated the presence of phospholipids in cylindrical tight junction strands using freeze-fracture labeling with gold-conjugated phospholipase A_2. Because the tight junction particles in freeze-fracture replicas have a similar size to the connexins of gap junctions and these structures are constructed from six single units, each particle or strand on the fracture face should consist of more than one transmembrane molecule or represent the backbone of a lipid/protein-suprastructure. However, the manner in which lipids and proteins are interconnected within the tight junction structure is as yet undefined. In fact, reaching high electrical resistances in the order of 10.000 Ωcm^2 seems impossible with proteins only, and because a pure lipid backbone of the tight junction is unstable, stabilizing proteins are required (**Fig. 6A**). Therefore, we propose a unifying model using a special folding scheme of the junctional proteins, which incorporates all known properties of the four transmembrane domains of proteins (**Fig. 6B**). We assume that the hydrophilic loop between transmembrane domain II and III is located in the micelle but not in the cytoplasm. The model allows predictions, which are currently being tested.

References

1. Reese, T. S., and Karnovsky, M. J. (1967) Fine structural localization of a blood-brain barrier to exogenous peroxidase. *J. Cell Biol.* **34**, 207–217.
2. Muir, A. R., and Peters, A. (1962) Quintuple-layered membrane junctions at therminal bars between endothelial cells. *J. Cell Biol.* **12**, 443–448.
3. Farquhar, M. G., and Palade, G. E. (1963) Junctional complexes in various epithelia. *J. Cell Biol.* **17**, 375–412.
4. Singer, S. J., and Nicolson, G. L. (1972) The fluid mosaic model of the structure of cell membranes. *Science* **175**, 720–731.

5. Staehelin, L. A., Mukherjee, T. M., and Williams, A. W. (1969) Freeze-etch appearance of the tight junctions in the epithelium of small and large intestine of mice. *Protoplasma* **67,** 165–187.

6. Staehelin, L. A. (1973) Further observations on the fine structure of freeze-cleaved tight junctions. *J. Cell Sci.* **13,** 763–786.

7. Staehelin, L. A. (1974) Structure and function of intercellular junctions. *Int. Rev. Cytol.* **39,** 191–283.

8. McNutt, N. S., and Weinstein, R. S. (1973) Membrane ultrastructure at mammalian intercellular junctions. *Progr. Biophys. Mol. Biol.* **26,** 45–102.

9. Dermietzel, R. (1975) Junctions in the central nervous system of the cat. IV. Interendothelial junctions of cerebral blood vessels from selected areas of the brain. *Cell Tissue Res.* **164,** 45–62.

10. Mollgard, K., and Saunders, N. R. (1975) Complex tight junctions of epithelial and endothelial cells in early foetal brain. *J. Neurocytol.* **4,** 453–468.

11. Tani, E., Yamagata, S., and Ito, Y. (1977) Freeze-fracture of capillary endothelium in rat brain. *Cell Tissue Res.* **176,** 157–165.

12. Shivers, R. R. (1979) The blood-brain barrier of a reptile, *Anolis carolinensis.* A freeze-fracture study. *Brain Res.* **169,** 221–230.

13. Moor, H., and Mühlethaler, K. (1963) Fine structure of frozen etched yeast cells. *J. Cell Biol.* **17,** 609–628.

14. Branton, D. (1966) Fracture faces of frozen membranes. *Proc. Natl. Acad. Sci. USA* **55,** 1048–1051.

15. Pinto da Silva, P., and Branton, D. (1970) Membrane splitting in freeze-etching: covalently bound ferritin as a membrane marker. *J. Cell Biol.* **45,** 598–605.

16. Branton, D., Bullivant, S., Gilula, N. B., et al. (1975) Freeze-etching nomeclature. *Science* **190,** 54–56.

17. Deamer, D. (1998) Daniel Branton and freeze-fracture analysis of membranes. *Trends Cell Biol.* **8,** 460–462.

18. Menco, B. P. M. (1986) A survey of ultra-rapid cryofixation methods with particular emphasis on applications to freeze-fracturing, freeze-etching, and freeze-substitution. *J. Electr. Micros. Techn.* **4,** 177–240.

19. Glauert, A. M., ed. (1985), *Practical Methods in Electron Microscopy.* Elsevier, Amsterdam, New York, Oxford.

20. Wendt-Gallitelli, M-F., and Wolburg, H. (1984) Rapid freezing, cryosectioning, and X-ray microanalysis on cardiac muscle preparations in defined functional states. *J. Electr. Micros. Techn.* **1,** 151–174.

21. Dahl, R., and Staehelin, L. A. (1989) High-pressure freezing for the preservation of biological structure: theory and practice. *J. Electr. Micros. Techn.* **13,** 165–174.

22. Pinto da Silva, P., and Kan, F. W. (1984) Label-fracture: a method for high resolution labeling of cell surfaces. *J. Cell Biol.* **99,** 1156–1161.

23. Andersson-Forsman, C., and Pinto da Silva (1988) Fracture-flip: new high resolution images of cell surfaces after carbon stabilization of freeze-fractured membranes. *J. Cell Sci.* **90,** 531–541.

24. Fujimoto, K. (1995) Freeze-fracture replica electron microscopy combined with SDS digestion for cytochemical labeling of integral membrane proteins. Application to the immunogold labeling of intercellular junctional complexes. *J. Cell Sci.* **108,** 3443–3449.

25. Fujimoto, K. (1997) SDS-digested freeze-fracture replica labeling electron microscopy to study the two-dimensional distribution of integral membrane proteins and phospholipids in biomembranes: practical procedure, interpretation and application. *Histochem. Cell Biol.* **107,** 87–96.

26. Heuser, J. E., Reese, T. S., Jan, L. Y., Dennis, M. J., and Evans, L. (1979) Synaptic vesicle exocytosis captured by quick-freezing and correlated with quantal transmitter release. *J. Cell Biol.* **81,** 275–300.

27. Fujimoto, K., Noda, T., and Fujimoto, T. (1997) A simple and reliable quick-freezing (freeze-fracturing procedure. *Histochem. Cell Biol.* **107,** 81–84.

28. Quick, D. C., and Letourneau, P. C. (1988) Immuno-labeling for freeze-fracture: application to a cell surface attachment antigen. *Brain Res.* **440,** 243–251.

29. McGuire, P. G., and Twietmeyer, T. A. (1984) Intramembrane particle distribution and junction morphology in rapidly frozen aortic endothelial cells. *Microcirc. Endoth. Lymphatics* **1,** 705–726.

30. Mühleisen, H., Wolburg, H., and Betz, E. (1989) Freeze-fracture analysis of endothelial cell membranes in rabbit carotid arteries subjected to short-term atherogenic stimuli. *Virch. Arch. B Cell Pathol.* **56,** 413–417.

31. Wolburg, H., and Rohlmann, A. (1995) Structure-function relationships in gap junctions. *Int. Rev. Cytol.* **157,** 315–373.

32. Wolburg, H. (1995) Orthogonal arays of intramembranous particles. A review with special reference to astrocytes. *Brain Res.* **36,** 239–258.

33. Wolburg, H., and Risau, W. (1995) *Formation of the blood-brain barrier.* In *Neuroglia.* Kettenmann, H., and Ransom, B., eds. Oxford University Press, pp. 763–776.

34. Claude, P. (1978) Morphologic factors influencing transepithelial permeability. A model for the resistance of the zonula occludens. *J. Membrane Biol.* **39,** 219–232.

35. Claude, P., and Goodenough, D. A. (1973) Fracture faces of zonulae occludentes from "tight" and "leaky" epithelia. *J. Cell Biol.* **58,** 390–400.

36. Bentzel, C. J., Fromm, M., Palant, C. E., and Hegel, U. (1987). Protamine alters structure and conductance of *Necturus* gallbladder tight junctions without major electrical effects on the apical cell membrane. *J. Membrane Biol.* **95,** 9–20.

37. Kniesel, U., and Wolburg, H. (1993) Tight junction complexity in the retinal pigment epithelium of the chicken during development. *Neurosci. Lett.* **149,** 71–74.

38. Wolburg, H., Neuhaus, J., Kniesel, U., et al. (1994) Modulation of tight junction structure in blood-brain barrier endothelial cells. Effects of tissue culture, second messengers and cocultured astrocytes. *J. Cell Sci.* **107,** 1347–1357.

39. Kniesel, U., Reichenbach, A., Risau, W., and Wolburg, H. (1994) Quantification of Tight Junction complexity by means of fractal analysis. *Tissue Cell* **26,** 901–912.

40. Haussdorff, F. (1918) Dimension und äußeres Maß. *Math. Ann.* **79,** 157–179.

41. Mandelbrot, B. B. (1982) *The Fractal Geometry of Nature.* Freeman and Co., New York.

42. Glenny, R. W., Robertson, H. T., Yamashiro, S., and Bassingthwaighte, J. B. (1991) Application of fractal analysis to physiology. *J. Appl. Physiol.* **70,** 2351–2367

43. Smith, T. G., Behar, T. N., Lange, G. D., Marks, W. B., and Sheriff, W. H. (1991) A fractal analysis of cultured rat optic nerve glial growth and differentiation. *Neuroscience* **41,** 159–166.

44. Fielding, A. (1992) Applications of fractal geometry to biology. *Comput. Appl. Biosci.* **8,** 359–366.

45. Soltys, Z., Ziaja, M., Pawlinski, R., Setkowicz, Z., and Janeczko, K. (2001) Morphology of reactive microglia in the injured cerebral cortex. Fractal analysis and complementary methods. *J. Neurosci. Meth.* **63,** 90–97.

46. Brightman, M. W., and Reese, T. S. (1969) Junctions between intimately apposed cell membranes in the vertebrate brain. *J. Cell Biol.* **40,** 648–677.

47. Nagy, Z., Peters, H., and Hüttner, I. (1984) Fracture faces of cell junctions in cerebral endothelium during normal and hyperosmotic conditions. *Lab. Invest.* **50,** 313–322.

48. Kniesel, U., Risau, W., and Wolburg, H. (1996) Development of blood-brain barrier tight junctions in the rat cortex. *Dev. Brain Res.* **96,** 229–240.

49. Griepp, E. B., Dolan, W. J., Robbins, E. S., and Sabatini, D. D. (1983) Participation of plasma membrane proteins in the formation of tight junctions by cultured epithelial cells. *J. Cell Biol.* **96,** 693–702.

50. Madara, J. L., and Dharmsathaphorn, K. (1985) Occluding junction structure-functiuon relationships in a cultured epithelial monolayer. *J. Cell Biol.* **101,** 2124–2133.

51. Noske, W., and Hirsch, M. (1986) Morphology of tight junctions in the ciliary epithelium of rabbits during arachidonic acid-induced breakdown of the blood-aqueous barrier. *Cell Tissue Res.* **245,** 405–412.

52. Suzuki, F., and Nagano, T. (1991) Three-dimensional model of tight junction fibrils based on freeze-fracture images. *Cell Tissue Res.* **264,** 381–384.

53. Lane, N. J., Reese, T. J., and Kachar, B. (1992) Structural domains of the tight junctional intramembrane fibrils. *Tissue Cell* **24,** 291–300.

54. Hirokawa, N. (1982) The intramembrane structure of tight junctions. An experimental analysis of the single-fibril and two-fibrils models using the quick-freeze method. *J. Ultrastr. Res.* **80,** 288–301.

55. Marcial, M. A., Carlson, S. L., and Madara, J. L. (1984) Partitioning of paracellular conductance along the ileal crypt-villus axis: a hypothesis based on structural analysis with detailed consideration of tight junction structure-function relationship. *J. Membrane Biol.* **80,** 59–70.

56. Tsukita, S., and Furuse, M. (1999) Occludin and claudins in tight-junction strands: leading or supporting players? *Trends Cell Biol.* **9,** 268–273.

57. Tsukita, S., and Furuse, M. (2000) Pores in the wall: Claudins constitute tight junction strands containing aqueous pores. *J. Cell Biol.* **149,** 13–16.

58. Tsukita, S., Furuse, M., and Itoh, M. (2001) Multifunctional strands in tight junctions. *Nature Rev. Mol. Cell Biol.* **2,** 285–293.

59. Furuse, M., Fujita, K., Hiiragi, T., Fujimoto, K., and Tsukita, S., (1998) Claudin-1 and -2: novel integral membrane proteins localizing at tight junctions. *J.Cell Biol.* **141,** 1539–1550.

60. Morita, K., Sasaki, H., Furuse, M., and Tsukita, S. (1999) Endothelial claudin: Claudin-5/TMVCF constitutes tight junction strands in endothelial cells. *J. Cell Biol.* **147,** 185–194.

61. Furuse, M., Sasaki, H., and Tsukita, S. (1999) Manner of interaction of heterogenous claudin species within and between tight junction strands. *J. Cell Biol.* **147,** 891–903.

62. Furuse, M, Furuse, K., Sasaki, H., and Tsukita, S. (2001) Conversion of Zonulae occludentes from tight to leaky strand type by introducing claudin-2 into Madin-Darby canine kidney I cells. *J. Cell Biol.* **153,** 263–272.

63. Liebner, S., Fischmann, A., Rascher, G., Duffner, F., Grote, E.-H., and Wolburg, H. (2000) Claudin-1 expression and tight junction morphology are altered in blood vessels of human glioblastoma multiforme. *Acta Neuropathol. (Berl.)* **100,** 323–331.

64. Liebner, S., Kniesel, U., Kalbacher, H., and Wolburg, H. (2000) Correlation of tight junction morphology with the expression of tight junction proteins in blood-brain barrier endothelial cells. *Eur. J. Cell Biol.* **79,** 707–717.

65. Risau,W., Engelhardt, B., and Werkele, H. (1990) Immune function of the blood-brain barrier: Incomplete presentation of protein (auto-)antigens by rat brain microvascular endothelium in vitro. *J. Cell Biol.* **110,** 1757–1766.

66. Kachar, B., and Reese, T. S. (1982) Evidence for the lipid nature of tight junction strands. *Nature* **296,** 464–466.

67. Pinto da Silva, P., and Kachar, B. (1982) On tight-junction structure. *Cell* **28,** 441–450.

68. Hein, M., Madefessel, C., Haag, B., Teichmann, K., Post, A., and Galla, H. (1992) Reversible modulation of transepithelial resistance in high and low resistance MDCK-cells by basic amino acids, Ca^{2+}, protamine and protons. *Chem. Phys. Lipids* **63,** 223–233.

69. Grebenkämper, K., and Galla, H.-J. (1994) Translational diffusion measurements of a fluorescent phospholipid between MDCK-1 cells support the lipid model of the tight junctions. *Chem. Phys. Lipids* **71,** 133–143.
70. Kan, F. W. K. (1993) Cytochemical evidence for the presence of phospholipids in epithelial tight junction strands. *J. Histochem. Cytochem.* **41,** 649–656.

4

Studies of Cerebral Endothelium by Scanning and High-Voltage Electron Microscopy

Albert S. Lossinsky and Richard R. Shivers

1. Introduction

This chapter reviews methods for the preparation of central nervous system (CNS) tissues for scanning (SEM) and high-voltage electron microscopic (HVEM) studies of the cerebral vasculature. The techniques described here have been used by the authors for the past two decades for investigations of pathologic alterations of the blood–brain barrier (BBB) after a variety of experimentally induced injuries in mammalian *(1–4)* and non-mammalian species *(5–7)*. The question of when to use one morphologic application over another becomes important for studies of cerebrovascular endothelial cells (ECs).

The application of the SEM has a distinct advantage for studies related to questions concerning pathologic alterations of the inner (luminal) surfaces of the ECs during BBB injury. Under these conditions, the SEM can provide an enormous amount of information about changes in the topography of large areas of the luminal surface of the vasculature. The SEM has enabled researchers to observe detailed structural changes that manifest on the EC surfaces after BBB injury, including increased cavitations, microvilli, and fronds *(8–10)*.

Surface details of EC microvilli produced by the conventional transmission electron microscope (TEM) in earlier studies, however, were based on compilations of numerous single images collected from serial thin-sections, a mission that is labor-intensive and often impractical. These images are limited to very small areas of the EC surface compared to those that can be examined by the SEM. The HVEM, like the TEM, is applied to answer questions related to internal structural changes within the ECs during BBB injury *(3,4,6,9–12)*, often not possible using the SEM. The major advantage of the HVEM, a high-voltage TEM, is due to the ability of the electron beam to penetrate thicker plastic sections, as much as 100 times thicker than routine thin-sections examined by TEM. Like the SEM, the HVEM can also provide additional information concerning surface structural features and high-resolution images of the three-dimensional structure of thick-sectioned, internal EC organelles.

Together, SEM and HVEM have contributed tremendously to the scientific literature concerning the morphologic changes that occur in cerebral vessels as a consequence of CNS injury.

From: *Methods in Molecular Medicine, vol. 89:*
The Blood–Brain Barrier: Biology and Research Protocols
Edited by: S. Nag © Humana Press Inc., Totowa, NJ

2. Materials

2.1. Experimental Protocol

1. Anesthetic: Sodium pentobarbital (Nembutal, Western Medical Co., Arcadia, CA). The stock solution contains 50 mg/mL in sterile PBS or sterile saline (*see* **Note 1**).
2. Tuberculin syringes with 27-gage needles.
3. Experimental animals
   a. Male or female adult (25 g) BALBc/J, C57BL, SJL/J mouse strains (Harlan, Indianapolis, IN).
   b. Male or female adult (250–400 g) Wistar, Lewis and Sprague Dawley rat strains (Harlan, Indianapolis, IN).
4. Brain tissues
   a. Rat brain C6 astrocytoma cell cultures were obtained from the American Type Culture Collection, Rockville, MD.
   b. Human brain tumors (biopsies of hemangiomas) were obtained from the Center for Research and Clinical Studies, Polish Academy of Sciences, Warsaw, Poland.

2.2. Tissue Fixation and Autopsy

2.2.1. Fixation by Vascular Perfusion

1. Prewash solution: physiologic phosphate-buffered saline (PBS).
2. 0.1 *M* Sodium cacodylate (a toxic, arsenic salt).
3. Storage buffer (S/C buffer): 0.1 *M* sodium cacodylate containing 0.2 *M* sucrose and 0.01% sodium azide (antimycotic agent).
4. Modified (half strength) Karnovsky's Fixative *(13)* (1/2 K fixative): 2% paraformaldehyde, 2.5% glutaraldehyde diluted in 0.1 *M* sodium phosphate or sodium cacodylate buffer (approx 1100 mOsM), adjusted to pH: 7.4 and used at 25°C.
5. Glass or plastic intravenous drip bottles: 2×500 mL or 1000 mL.
6. Two intravenous drip tubing sets with stop/pinch valves.
7. Latex hose (18 inch) with a plastic attachment at tip for a hypodermic needle-cannula.
8. One three-way stopcock.
9. Small pan to hold 2 L of fluid.
10. Instruments:
    a. Small mosquito and large straight hemostats.
    b. Small and large scissors.
    c. Fine toothed forceps.
    d. Scalpel holder with no. 10 blade.
11. Cork or cardboard dissecting board.
12. Rubber bands.
13. Tape.
14. Cannulae: no.18-gage (mice), and no. 12- or 14-gage (rats) (*see* **Note 2**).
15. 30-mL snap-cap or scintillation vials with screw caps for immersion fixation of tissues.

2.2.2. Autopsy and Tissue Collection

1. Dental wax plates.
2. Single edge razor blades.
3. Wheaton vials, 4 mL with rubber gasket.
4. Vibratome.
5. Tissue stubs.

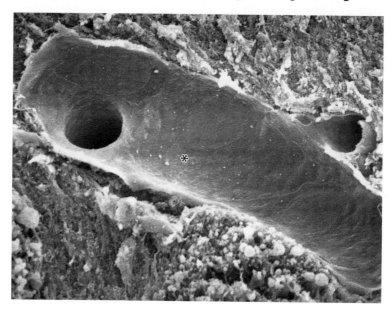

Fig. 1. SEM image of a cerebral cortical blood vessel from a sham-operated control mouse. Note the smooth luminal EC surface (*). Fixation: 2% formalin, 1% glutaraldehyde in 0.1 *M* sodium cacodylate buffer. Original magnification: x2200.

6. Rapid adhesion glue.
7. Smith-Farquhar Tissue Chopper.
8. Agar, 2% aqueous solution.

2.3. Post-Fixation and Dehydration Materials Common to Both SEM and HVEM

1. Ethanol
2. 1-2% osmium tetroxide in 0.1 M sodium cacodylate or S/C buffers.

2.4. Scanning Electron Microscopy

All HVEM and SEM supplies were obtained from Electron Microscopy Sciences, Fort Washington, PA, or Ladd Research Industries. Inc., Burlington, VT or Ted Pella, Redding, CA.

1. Aluminum SEM mounting stubs.
2. Double sticky carbon tape.
3. Silver conductive adhesive.
4. Compressed carbon dioxide gas for specimen critical point drying.
5. Hexamethyldisalizane (HMDS), an optional, rapid chemical drying method.
6. Carbon, gold and platinum metal rods are used for sputter coating.
7. Our studies were conducted using a standard tungsten filament ISI 40 SEM (at the New York State Institute for Basic Research in Developmental Disabilities, Staten Island, NY), which provided excellent visualization of the normal and injured BBB (**Figs.1–4**). We also used environmental and variable pressure (VP) SEMs at the product laboratories of each

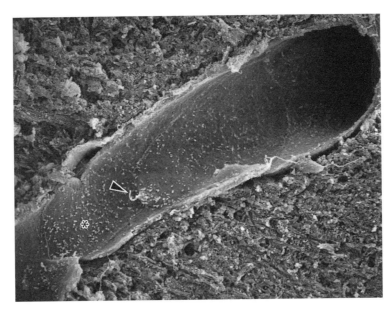

Fig. 2. SEM image of the rat cerebral cortex. This animal was subjected to global ischemia by ligation of the cardiac vessel bundle for 3.5 min, followed by blood recirculation. Note the rough appearance of the EC surface due to increased numbers of microvilli (*) and an elongated frond (**arrowhead**). Fixation: 2% formalin, 1% glutaraldehyde in 0.1 M sodium cacodylate buffer. Original magnification, x2590 (From **ref. 25**, with permission.)

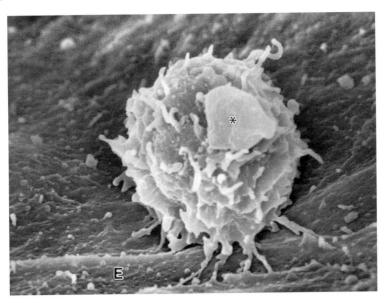

Fig. 3. SEM image of a spinal cord vein from a mouse with chronic relapsing experimental allergic encephalomyelitis (CREAE). Note the leukocyte (*) attached to the rough-surfaced EC (E), presumably seeking an entrance passageway into and across the EC to the CNS side of the BBB. Fixation: 2% formalin, 1% glutaraldehyde in 0.1 M sodium cacodylate buffer. Original magnification, x9600.

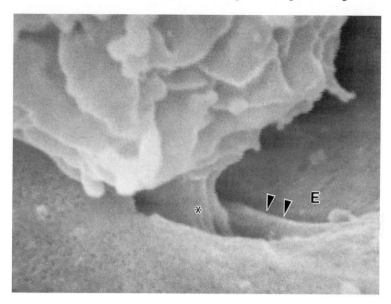

Fig. 4. SEM image of a spinal cord vein from a CREAE mouse. High magnification of a leukocyte (monocyte) inserting a flap-like lamellopodium (*) into the EC (E) at a parajunctional location. The adjacent EC junctional complex is shown (**arrowheads**). Fixation: 2% formalin, 1% glutaraldehyde in 0.1 *M* sodium cacodylate buffer. Original magnification, x32200.

manufacturer (the Hitachi S3500N and Hitachi S4700 Field Emission SEM with a YAG crystal, Mountain View, CA, the JEOL JSM 5900LV, Peabody, MA, the FEI/Philips ESEM XL 30, Wilmington, MA and the LEO/Zeiss 1430 and 1455 VPSE, Thornwood, NY).

2.5. High-Voltage Electron Microscopy

1. Horseradish peroxidase (HRP) Type II (mice), Type VI (rats) (Sigma Chemical, St. Louis, MO).
2. Harvard Infusion Pump (Harvard Apparatus, Holliston, MA).
3. 3, 3'-diaminobenzidine tetrahydrochloride (DAB), a suspected carcinogen (Sigma Chemical, St. Louis, MO).
4. Hypodermic needles: 23-gage and 27-gage.
5. PE 60 or smaller tubing (Clay Adams, Parsippany, NJ).
6. Propylene oxide (flammable) for tissue clearing.
7. Polybed 12 plastic embedding resin.
8. Spurr (liquid plastics are suspected to be toxic and/or carcinogenic) *(14,15)*.
9. BEEM capsules (size 00) are used for preparing the plastic tissues blocks.
10. Formvar plastic to prepare membranes for slot grids *(15)*.
11. 2% uranyl acetate stain.
12. Reynold's lead citrate stain.
13. High-voltage electron microscopy was carried out at 800–1000 kV using an AEI EM7 HVEMs at the HVEM Facility, University of Wisconsin, Madison, WI (RRS), and the AEI EM7 HVEM at 1200 kV at the HVEM Facility in Albany, NY (ASL) (**Figs. 5, 6**).

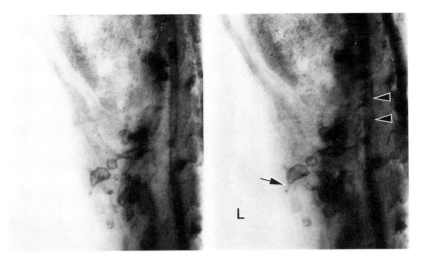

Fig. 5. Stereo-pair HVEM view of a 0.5-µm section of a cerebral cortex capillary from a mouse subjected to stab wound injury. Note the connected series of vesicles filled with HRP reaction product (**arrow**) and the HRP-filled basement membrane (**arrowheads**). Fixation: 2% formalin, 1% glutaraldehyde in 0.1 *M* sodium cacodylate buffer. Vessel lumen = L. Original magnification, x25000. (From **ref. *11***, with permission from Springer Verlag.)

3. Methods
3.1. Experimental Protocol

1. Dilute the stock solution of sodium pentobarbital (Nembutal) (50 mg/mL) with sterile PBS or other sterile physiologic saline in a ratio of 1:6 (1 part Nembutal plus 5 parts saline).
2. Inject mice intraperitoneally (i.p.) with the diluted anesthetic 0.1–0.15 mL, (0.83–1.25 mg) using a Tuberculin syringe with no. 27-gage needle.
3. Inject rats i.p. with 0.1 mL of the stock Nembutal (5 mg/dose).
4. Animals are either placed in restraining holders, or grasped at the nape of the neck and secured by hand to reduce stress during i.p. injection of anesthetics.

3.1.1. Euthanasia

Before vascular perfusion, the animals are given a single i.p. injection of stock Nembutal.

1. Mice are given 0.1 mL (5 mg/dose) (*see* **Note 1**).
2. Rats are given 0.5 mL (25 mg/dose)

3.2. Tissue Fixation and Autopsy

Tissues are either fixed by vascular perfusion or by immersion

3.2.1. Fixation by Vascular Perfusion

1. Use perfusion bottles with hanging plastic drip tubing and a three-way stopcock.
2. Attach the cannulae (no. 18-gage needles for mice, and no. 12- or 14-gage needles for rats) to the end of an 18-inch segment of latex tubing, the other end of which is connected to a three-way stopcock and intravenous (iv) drip tubes and perfusion bottles.
3. Place the deeply anesthetized animals on their backs on a dissecting board with the legs and tail secured to the board with tape (*see* **Note 3**).

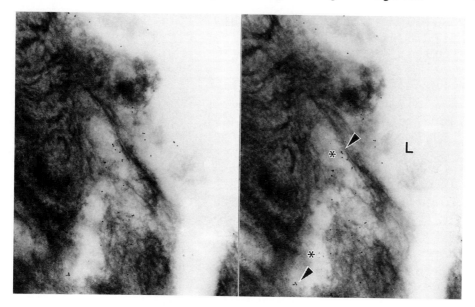

Fig. 6. This HVEM micrograph is a stereo-pair image taken from a 0.5-μm section through a vein from a human brain hemangioma biopsy. Using a pre-embedding technique *(35)*, the tissue was immunoincubated with a 1:100 dilution of mouse anti-human ICAM-1 monoclonal anti-body, 1:50 dilution of goat anti-mouse biotinylated IgG, and then a 1:20 dilution of strepta-vidin gold (15 nm gold particles). HVEM reveals the immunogold particles (**arrowhead**) decorating the internal delimiting membrane surfaces of an elongated vesiculo-vacuolar struc-ture (*). Vessel lumen = L. Fixation: 2% formalin, 0.1 glutaraldehyde in a 0.1 *M* sodium cacody-late buffer. Original magnification, x20000. (From **ref. 3**, with permission from Springer Verlag.)

 4. Grasp the skin over the sternum with forceps and make a midline cut with scissors.
 5. Retract the skin over the chest to the sides by blunt dissection.
 6. Grasp the xiphoid process of the sternum with the large, straight hemostat, and while lifting the hemostat upward, cut the diaphragm transversely under the rib cage.
 7. Cut both sides of the rib cage up to the level of the clavicle.
 8. Lift the entire rib cage over the head of the animal such that it does not interfere with subsequent surgical procedures.
 9. If young adult animals are used, remove the thymus gland and pericardium with care to avoid cutting the underlying large blood vessels of the heart.
10. Grasp the tip of the right atrium using small toothed-forceps and cut it with small dissect-ing scissors (for blood/fluid effluent).
11. Grasp the apex of the heart with a small mosquito hemostat and cut the tip of the apex with the fine scissors.
12. Gently insert the cannula into the slit in the apex and into the ascending aorta, directing the cannula along the left wall of the heart. The end of the cannula can be visualized through the semi-transparent aorta (mice), or by visualizing the movement of the tip of the cannula within the ascending aorta (rats).
13. Place a mosquito hemostat over the apex of the heart and close it to secure the cannula within the ascending aorta.
14. Open the stopcock valve to permit maximum flow of prewash solution (10–20 mL for mice; 20–30 mL rats) to rinse out the blood. The return flow out of the heart usually clears after 20–30 s.

15. Open the stopcock valve to allow the 1/2 K fixative (or other desired fixative) to flow though the animal (mice: 250 mL; rats: 1000 mL). We have successfully used the hydrostatic drip method of Kalimo et al. *(16)* *(see* **Notes 4, 5**).

3.2.2. Fixation by Immersion

1. Place the required piece of fresh tissue (brain cortex after decapitation, tumor biopsy material obtained during surgery, etc.) on a wax plate, a plastic petri dish, or corkboard and dice it into $1 \times 1 \times 4$ mm fragments in several drops of the fixative of choice *(see* **Note 6**).
2. Mince or chop the tissue slices in fixative into smaller 1 mm^3 pieces fixative and allow the tissue to fix in this fixative for either 2 h at 4°C, or the tissue can remain in fixative for longer periods, up to overnight, at 4°C.
3. Place the tissue into S/C buffer, rinse twice, and hold in this buffer at 4°C.

3.2.3. Autopsy and Tissue Selection

This procedure is similar for tissues fixed by vascular perfusion or by immersion *(see* **Notes 6, 7**). For alternate methods of cutting tissue specimens see **Note 8**. After appropriate fixation, the tissue blocks are placed in S/C buffer at 4°C and remain in this buffer until they are post-fixed as described in **Subheading 3.3.**

3.3. Post-Fixation and Dehydration Methods Common to Both SEM and HVEM

1. Fix larger Vibratome slices, chopped tissue slices or smaller hand-minced tissue pieces (for HVEM studies), as well as larger tissue blocks (for SEM studies) for 1–2 h with 1–2% OsO$_4$ at 4°C to stabilize lipid components of the EC membranes. Osmium tetroxide fixative is prepared with 0.1 M buffer including either sodium cacodylate, S/C buffer without sodium azide, or s-collidine according to standard protocols *(15,17)*.
2. Dehydrate the tissue slices, minced pieces or larger blocks in 50%, 70%, 85%, 95% ethanol, 10 min in each concentration, on ice at 4°C. Prepare all dilutions using 95% ethanol.
3. Soak the tissues in 2 changes of 100% (absolute) ethanol for 10 min each at room temperature (25°C).
4. Process the tissue samples further for either SEM *(see* **Subheading 3.4.**) or HVEM analyses *(see* **Subheading 3.5.**).

3.4. SEM

1. Place the specimens into the critical point dryer with about 5 mL 100% ethanol and seal the chamber door.
2. Slowly introduce compressed carbon dioxide into the chamber until the critical point *(21)* is gradually reached at the proper temperature and pressure and preserving the tissue with little structural distortion *(see* **Note 9**).
3. Slowly release the carbon dioxide gas and remove the specimens.
4. Place the dried specimens in a dessicator to prevent rehydration.
5. If a critical point drying apparatus is unavailable, immerse the tissue samples in a 50:50 mixture of 100% ethanol and 100% hexamethyldisalizane (HMDS) *(24)* for 30 min at 25°C.
6. Transfer the tissue to 100% HMDS for 30 min at 25°C *(see end of* **Note 9**).
7. Air-dry the tissue samples on filter paper and maintain the dried specimens in a dessicator to prevent rehydration.
8. Mount the dried specimens on aluminum SEM stubs using either silver conducting paste, double-edged carbon conducting tape, or by adding a small drop of the silver conductive adhesive to the sides of the specimen that overlaps onto the aluminum stub.

9. Sputter coat the dried, mounted specimens with either carbon, gold or platinum in a conventional sputter coater, according to standard protocols *(reviewed in 22)*.
10. The SEM-ready samples are stored for long periods under dessication to prevent rehydration *(see **Note 10**)*.
11. Dried tissues examined by SEM (optional) can be further processed for HVEM so that both SEM/HVEM can be done using the same specimen *(see **Note 11**)*.

3.4.1. SEM Data Interpretation

1. SEM studies of normal mouse and rat brains fixed by perfusion have provided information about the normal appearance of ECs lining the luminal surface of the cerebrovasculature. The normal ECs usually present a smooth surface with short ridges at the boarders of the individual ECs *(see **Fig. 1**)*.
2. After brain and spinal cord injury, however, there is a dramatic change in the surface topography of the ECs including increased microvilli, fronds and surface cavitations *(8,9,25)*. After CNS injury, finger-like microvilli and fronds are thought to upregulate adhesion molecules that facilitate leukocyte-EC adhesion *(26)* and initiates transmigration *(3)* *(see **Fig. 2**)*.
3. The structural interaction between leukocytes and ECs has been studied extensively and the SEM has been an important tool that has contributed to our understanding of how these cells probe the surface of the EC searching for pathways to enter the ECs during inflammatory events of the CNS *(8–10)*. Lymphocytes and mononuclear cells are known to extend thin filipodial *(see **Fig. 3**)*, and lamellipodial extensions of their plasma membranes, and eventually insert them into pores within the EC surface *(see **Fig. 4**)* during the initial phase of their transendothelial passage across the BBB.

3.5. HVEM

3.5.1 Animal Protocol

1. In animals scheduled for BBB permeability studies, HRP, Type II, 15 mg/25 g mice or HRP Type VI, 5 mg/25 g mice, or 50 mg/250 g for larger rats is dissolved in 0.2 mL sterile saline, PBS, or Ringer's solution for administration to mice, or in 1 mL for rats.
2. Under a dissecting microscope, insert a 27-gage needle (attached to a 1 mL tuberculin syringe connected to 12–18 in PE 60 tubing) into either the saphenous or femoral veins in the leg, or into the tail vein.
3. Inject the HRP intravenously slowly either by hand or using an infusion pump.
4. Allow the tracer to circulate within the animal for 5 to 60 min before giving the final dose of nembutal for euthanasia and subsequent vascular perfusion. *(see **Note 12**)*.
5. After fixation, tissue selection and cutting, the 30–40 μm tissue slices are rinsed twice with S/C buffer and then reacted with a filtered solution containing 5 mg DAB dissolved in 10 mL of 0.05 M Tris-HCl buffer containing 0.01% hydrogen peroxide, pH 7.6 for 15–30 min at 25°C *(27,28)*. Negative controls contain all reagents except HRP *(see **Note 13**)*.

3.5.2. Tissue Processing

1. Rinse the sections twice in S/C buffer and hold them in this buffer before post-fixation with OsO_4.
2. Follow **Steps 1–3** given in **Subheading 3.3.**
3. After the second rinse in 100% ethanol, the tissue sections are soaked in 2 changes of propylene oxide (P.O.), for 15 min each.
4. The tissue is then infiltrated with liquid plastic by standard methods using Polybed 812, Spurr etc.) *(14,15)*. Use a 50:50 mixture of P.O. and liquid plastic mixture including accelerator.

5. Transfer the tissue sections to a 100% plastic mixture containing accelerator and allow them to remain overnight in a desiccator under vacuum.
6. Flat-embed the sections in labeled BEEM Capsules in fresh plastic including accelerator.
7. Polymerize the blocks overnight in a 70°C oven.
8. Trim the plastic blocks and cut sections at 0.25, 0.5, 0.75, and 1.0 µm on Leica Ultracut, Sorvall Porter Blum MT-1 and MT-2B or other ultramicrotome.
9. Collect the sections on formvar-coated slot grids. Serial thick sections can also be collected (1–4 sections each/slot grid) to insure accuracy in specimen orientation and to facilitate visualization of sequential three-dimensional stereo-pair images of the internal EC structure in question.
10. Examine the sections either stained or unstained (*see* **Note 14**).

3.5.3. HVEM Data Interpretation

1. Mice subjected to traumatic brain injury and fixed by perfusion have produced useful information concerning the mechanisms of HRP transport across the damaged BBB *(11)*. The advantage of HVEM is that it allows three-dimensional visualization of the relations between internalized EC structures containing the HRP reaction product (*see* **Fig. 5**). In order to acquire this type of data by conventional TEM, numerous serial-thin sections must be stacked up and transformed into three-dimensional images, a technique that is challenging and labor-intensive.
2. Studies of vesiculo-canalicular and vesiculo-vacuolar organelles in thick sections of ECs from human brain tumor biopsy specimens taken during surgery and fixed with weak glutaraldehyde concentrations have also been conducted (**3**). We have studied the nature of the formation of EC vesiculo-canalicular structures after BBB injury using immunocytochemical methods. (*see* **Note 6**) (*see* **Fig.6**). In these studies, immunogold particles labeled the internal delimiting membrane surfaces of vesiculo-canalicular and vesiculo-vacuolar type organelles. These studies indicate that the adhesion molecule-decorated EC vesiculo-canalicular structures may serve as anatomic passageways for invading leukocytes and/or neoplastic cells through the ECs, modulated by upregulated adhesion molecules *(3)*.

These studies and several others provide evidence that the HVEM is, indeed, a unique adjunct tool to the conventional TEM that can be applied to answer specific questions about internal structural details within the ECs related to the pathophysiology of the altered BBB (*see* **Note 15**).

4. Notes

1. Guidelines for animal care and the use of anesthetics are controlled by the Animal Welfare Committee of the University or Institute, in accordance with the guidelines of the National Institutes of Health and the Animal Welfare Act (USA), and the Canadian Council of Animal Care (Canada). Larger mice can be given additional anesthetic doses if they become responsive to toe or tail pinches. In this case, 0.02 – 0.05 mL of the diluted nembutal solution may be given, although one must be aware of the risk of overdose. This amount of nembutal places the animals into a comatose condition prior to opening the chest cavity and perfusion. Surgical opening of the chest for vascular perfusion may begin when the animal no longer responds to toe or tail pinches. An alternative anesthetic that has been used successfully in adult rats is i.p. injections of a cocktail containing 90 mg/kg ketamine and 7 mg/kg xylazine (Western Medical, Aracadia, CA). Animals may be intubated or ventilated via the nose with an oxygen-rich air mixture before thoracotomy and vascular perfusion. This technique is an ideal way to reduce possible hypoxia-related EC structural

changes that may occur at the ultrastructural level if surgical complications prevent a speedy perfusion fixation.

2. Cannulae can be easily prepared by cutting off the beveled ends of 18- and 14-gage hypodermic needles and filing the surfaces to make them smooth. Cannulae may also be purchased from surgical instrument supply companies (e.g., George Tiemann & Co., Hauppauge, NY).

3. A useful technique to ensure good brain fixation is to secure the upper jaw of small- and medium-sized rodents to the cork or cardboard tray using a rubber band. This prevents the animal's head from contorting during perfusion fixation, thus, providing unrestricted flow of fixative through the carotid arteries to the brain.

4. The hydrostatic drip method of Kalimo et al. (16) allows the prewash and fixative solutions to flow into the animals at a pressure of about 150 cm of water, the approximate systolic blood pressure of mice and rats. Using this method, the bottles are positioned approx 5 feet above the animals. Mechanical pumps can also be used for vascular perfusions of mice and rats, but care must be taken to avoid too high an intravascular pressure that can produce hypervolemia and artifactual breach of the EC tight junctional complexes.

5. Caveats about prewash. Prewash solutions are adjusted to pH 7.4. An optional PBS prewash solution may contain heparin, an anticoagulent (final concentration of 10 units/mL) and procaine hydrochloride, a vasodilator (final concentration of 0.01%). Washing periods in excess of 30 s may cause disruption of myelinated fibers and produce perivascular and/or perineuronal halos at the TEM/HVEM level. Once the chest wall is open, one must proceed quicky to avoid blood coagulation within the microvasculature in the event that heparin is not added to the prewash solution.

6. Several excellent fixatives for ultrastructural studies are mentioned in this chapter. Complete discussions concerning the many aspects of tissue fixation can be located in textbooks and review articles (15,17,18). Tissues can be preserved with these fixatives by either perfusion or immersion methods. Although half strength Karnovsky's Fixative is hyperosmolar (ca. 1100 mM), it has been used successfully as a vascular perfusate with excellent preservation of ultrastructural details for SEM and HVEM studies of cerebrovascular ECs. It induces minimal tissue shrinkage without edematous tissue changes. Other excellent fixatives for histologic, TEM, and HVEM studies that approach isotonicity (380 mM) include 2% formalin and 1% glutaraldehyde in 0.1 M sodium phosphate or cacodylate buffers (11). A fixative developed by McDowell and Trump (19) consisting of 4% formalin, 1% glutaraldehyde in a buffer of 176 mM/L was considered to be their choice fixative for optimum ultrastructural results of several different tissue types including the CNS. Fixation of CNS tissues can also be carried out by either perfusion or immersion in one of the following solutions: a) 6% ice-cold glutaraldehyde buffered to pH: 7.35 with 0.1 M sodium cacodylate; b) perfusion of a solution consisting 1% glutaraldehyde, 1% paraformaldehyde containing 4% sucrose in 0. 1 M sodium cacodylate (pH: 7.4) for 5 min, then continue to perfuse for an additional 15 min with a solution of aldehydes increased to 2.5%. For convenience, tissues can be held for extended periods (weeks) in S/C buffer in a refrigerator before postfixation. After fixation, the tissues are changed to S/C buffer and rinsed with 2 changes of this buffer before post-fixation. In general, perfusion fixation preserves the tissue considerably faster compared to immersion fixation. This is because of the fast replacement of blood with fixatives producing rapid penetration of the fixatives across the BBB due to the extensive vascular network in the brain. Thus, assuming that the time required to flush the blood from the animal with the prewash buffer solution is rapid (under 30 s), and hypoxic tissue changes will be avoided, fixation by perfusion is usually considered the optimum method for ultrastructural studies. It is also possible to perform both immunocytochemical and ultrastructural studies in conjunction with SEM, conventional TEM, and HVEM

studies to identify membrane-associated proteins such as adhesion molecules. For these types of studies, one must use fixatives that contain reduced concentrations of glutaraldehyde (e.g., 2–4 % formalin containing 0.1–0.25% glutaraldehyde), which reduces the possibility of denaturing target antigens *(18)*. The concentration of the fixatives, and the time of fixation (not more than 2 h maximum time) may be a critical concern for optimizing the expression of the specific antigens in question *(3)*.

7. Perfusion fixative techniques cannot be applied to preserve human tissues, unless a written consent is obtained from the family. In the state of Maryland, immediate autopsies (with family permission) can be performed on patients with irreversible brain death at the University of Maryland Hospital in Baltimore *(16,20)*. It is customary to immersion-fix human tissues including CNS tumor biopsy materials in either a fixative for ultrastructural preservation, or for good antigen preservation using a more gentle immunofixative, described above in **Note 6**. However, immersion fixation of CNS tissues may have limited value for SEM studies of the microvasculature because the blood vessels may have collapsed before fixation. Thus, in our past experience, observing luminal surfaces of collapsed vessels by SEM has often provided limited results.

8. There are different methods of cutting tissues for peroxidase tracer studies, and other ultrastructural, ultracytochemical, or immunocytochemical studies. Cut the tissues into $2 \times 4 \times 4$ mm thick blocks (SEM studies), or cut the brain pieces into 2×2 mm $\times$ 30–40 μm slices using a Vibratome, chopped at 30–40 μm using a Smith-Farquhar Tissue Chopper, or mince into 1–2 mm^3 pieces on a wax plate in a chosen fixative (HVEM studies). These thick tissue blocks, smaller pieces or thin slices are then placed in 4 mL Wheaton vials in a fixative similar to that which was used for fixation by perfusion or immersion. The total time of fixation can range from 2 h to overnight at 4°C, depending on the goals of the experiment.

9. Critical point drying *(21)* is an excellent method of completing the drying process of the tissue before scanning it in the SEM. This is typically performed after the tissue has been post-fixed with osmium tertroxide and dehydrated with ethanol (*see* **Subheading 3.3.**). The state of critical point within a tissue specimen occurs when the two-phase stage of the majority of volatile liquids (vapor and liquid) disappears from the tissue at this special "critical point," at a specific temperature and pressure. When the critical point is reached, the two phases are in equilibrium and the phase boundary disappears removing any surface tension *(22)*. Thus, critical point drying greatly reduces distortion of the cellular surface topography of the ECs because it avoids the forces of surface tension that occur during evaporation *(23)*. The exchange of liquid/gas and liquid/solid–solid/gas transition using ethanol/carbon dioxide, for example, under high temperature and pressure results in a completely dried specimen. During the final stages of drying, the tissue is exposed to gaseous carbon dioxide, then air. Critical point drying has been considered the optimum technique for specimen preparation for SEM, but some investigators have not found it to be superior to other, more simplified chemical drying methods including HMDS *(24)* (*see* **Subheading 3.4., step 5**).

10. Although coating dried tissue specimens considerably reduces charging artifact in conventional, non-environmental SEMs, environmental SEMs including Hitachi S3500N, JEOL 5910LV, LEO/Zeiss' 1455 VPSE, the FEI/Philips XL30 ESEM, and FEI/Philips Quanta SEMs do not require coating of the samples using variable pressure. These microscopes also have the capacity to examine fixed, totally wet tissue samples. These types of EMs provide excellent tools for cases in forensic medicine, when crime-related samples cannot be dried or coated to avoid altering evidence. Environmental SEMs will hopefully contribute valuable morphologic information from future studies of totally wet tissue samples of both normal ECs, and from ECs after CNS injury.

11. It is also possible to target and evaluate a specific blood vessel combining both SEM and HVEM techniques. The brain tissue samples are initially processed for SEM, including osmication, dehydration, drying, and gold and platinum sputter-coating. These tissue blocks are then scanned in the SEM and images of the targeted blood vessels are captured, either photographically or digitally. Afterward, these same tissue blocks scanned by SEM can be again dehydrated in ethanol and embedded in plastic (*see* **Subheadings 3.3.2., 3.3.3.,** and **3.5.2., steps 3–7**). The plastic blocks can then be trimmed under a dissecting microscope and one can view the reflection of the same, previously SEM-scanned, gold-coated blood vessels through low-power objectives of either dissecting or compound microscopes. Serial thick sections of blood vessels scanned previously by SEM have been successfully examined in the HVEM using this method *(10)*.

12. Horseradish peroxidase has been a valuable tracer to study the ultrastructural characteristics of internalized EC structures involved with increased permeability of the vasculature (*see* **Fig. 5**) and junctional complexes that can open after BBB injury. In our experience Type II HRP (Sigma Chemical) can be tolerated as a vascular-borne tracer in laboratory mice. However, this less purified grade of the peroxidase enzyme is toxic to other experimental animals *(27)*, and this type of HRP has been implicated in histamine-induced opening of the BBB in laboratory rats. Sigma Type VI HRP is the tracer of choice for BBB studies in rats, lizards, cats, and rabbits. Administration of antihistamines can also be used in animals receiving HRP. Forceful, manual injections of HRP may also produce hypervolemia as a result of artifactual opening of the BBB. This potential methodologic pitfall can be alleviated using slow, steady injections with an infusion pump at a rate of approx 1 mL/min using a 27-gage needle attached to polyethylene tubing (PE 60 or smaller).

13. Permeability studies using protein tracers, including HRP, can only be performed in conjunction with the HVEM (and conventional TEM), because the goal using the HVEM is to examine internal structural compartments of the vasculature including EC caveolae, vesiculo-canalicular structures and EC junctional complexes. The cytochemical localization of exogenous HRP is observed after the DAB is reacted with the tissue producing a brown coloration of the extracellular compartments of the CNS including the perivascular basement membranes indicating the presence of a positive exogenous peroxidase activity and increased BBB permeability to this protein tracer. This brown-colored reaction product is subsequently chemically transformed into an electron-dense reaction product (osmium black) after post-fixation with OsO_4. Solutions of DAB should always be filtered before use and the hydrogen peroxide solution should be added to the solution at the last moment. DAB solutions occasionally produce undesired crystal-like precipitates when phosphate buffers are used for perfusion, washes, etc. This precipitate is reduced when the tissue samples are either washed overnight in S/C buffer or when sodium cacodylate buffer is added to the fixative *(19)*.

14. Grids may be either stained with lead citrate *(29)*, stained with uranyl acetate *(30)*, or with alcoholic uranyl magnesium acetate for 3–4 h at 25° C, the latter to provide staining completely through the thickness of the section. A useful method for staining thick sections (0.1–1 µm) using heated alcoholic uranyl acetate has been reported by Locke and Krishnan *(31)*. Other studies have also obtained excellent stain penetration using heated stains including 2% uranyl acetate for 1.5 h at 50°C and 30 min in lead citrate at 25°C *(11)*. Staining of multiple grids for HVEM has always presented problems because the first and last grids may have been stained for different times due to the time required to wash and dry each grid. Thus, the last few grids are often overstained. To rectify this problem, one can use devices that facilitate the simultaneous staining of numerous grids, resulting in equal staining time/grid. We have used a modification of the Hiraoka Staining Kit (Electron Microscopy Sciences, Fort Washington, PA) *(32)*. This technique was developed by

Giammara *(33)*, improved by K. F. Buttle, Albany, NY, and is also described by Hayat *(34)*. The Hiraoka staining pad consists of a plastic disk with 4 rows of 10 slots into which grids are inserted and held in position as the entire disk is submerged into staining baths and distilled water rinses. If one cuts the plastic disk into individual strips, these 0.5×4 cm strips will accommodate up to 10 grids/strip. A single strip securing 10 grids is inserted into a modified plastic bulb-type pipet after the tip has been cut away. The staining fluids (uranyl acetate, lead citrate) can be drawn into the pipet. By capillary attraction, the fluid is held within the pipet as the entire pipet assembly is inserted into a small vial filled with the staining fluid or distilled water for rinsing. Thus, the pipet containing the plastic strip with grids can be stained (and washed) either at room temperature or in a 50°C oven.

15. The main advantage of ultrastructural studies of brain microvessel ECs (the anatomic basis of the BBB) offered by HVEM is the ability to see cross sectional views within thick sections of plastic-embedded CNS tissue. In standard TEMs, the accelerating voltage is usually less than 100 kV. Consequently, the penetration of the specimen by the electron beam is much reduced, thereby limiting resolution of components of the cells in the tissue. To increase the penetration of electrons through thick specimens and retain a resolution that approximates its theoretical resolution of less than 1 Å, a higher accelerating voltage must be used. The million-volt electron microscope provides the necessary beam strength and resolving capabilities to render the contents of thick plastic sections of tissue (0.25–1 um thick) visible. Advantages of the HVEM include recording of three-dimensional architectural relationships of subcellular structures, and even rotational manipulation of the specimen to reveal subtle details of spatial interrelationships. HVEM specimens in thick sections can be examined and photographed at various tilt angles to permit stereo-pair views of the tissues. Examination of thick plastic sections employing standard TEMs before studies with HVEMs permitted screening of samples for technical artifacts and other problems, but provided limited results. Early examination of thick plastic tissue sections in a Philips 201 electron microscope operating at 100 kV served as an efficient method for selecting optimal specimens for HVEM. In cases where access to HVEM facilities is limited or impossible, very acceptable results can be obtained from 0.1–0.5 μm sections and an electron microscope operating at voltages of 250 to 500 kV. Intermediate-range HVEMs equipped with tilt-stage goniometers can reveal valuable information about biologic material in semi-thin sections *(36)*. Stereo-pair electron micrographs can be viewed with either a stereo-viewer lens, or by simply holding the side-by-side paired micrographs at approx 10 inches from the eyes.

Acknowledgments

The authors thank Drs. William F. Agnew, Douglas McCreery (HMRI), and Ms. Karolyn F. Buttle, for their valuable editorial comments and assistance in preparation of this chapter. These studies have been supported in part by the New York State Office of Mental Retardation and Developmental Disabilities, NYS Institute for Basic Research in Developmental Disabilities, Staten Island, NY, the Huntington Medical Research Institutes, Pasadena, CA (ASL), and by the Natural Sciences and Engineering Research Council of Canada, University of Western Ontario, Ontario, Canada (RRS).

References

1. Lossinsky, A. S., Garcia, J. H., Iwanowski, L., and Lightfoote, W. E. Jr. (1979) New ultrastructural evidence for a protein transport system in endothelial cells of gerbil brains. *Acta Neuropathol. (Berl.)* **47,** 105–110.
2. Farrell, C. L., and Shivers, R. R. (1984) Capillary junctions of the rat are not affected by osmotic opening of the blood-brain barrier. *Acta Neuropathol. (Berl.)* **63,** 179–189.

3. Lossinsky, A. S., Buttle, Ǩ. F., Pluta, R., Mossakowski, M. J., and Wisniewski, H. M. (1999) Immunoultrastructural expression of intercellular adhesion molecule-1 (ICAM-1/CD54) in endothelial cell vesiculo-tubular structures (VTS) and vesiculo-vacuolar organelles in blood-brain barrier development and injury. *Cell Tiss. Res.* **295,** 77–88.

4. Shivers, R. R., Edmonds, C. L., and Del Maestro, R. F. (1984) Microvascular permeability in induced astrocytoma and peritumoral neuropil of rat brain. A high-voltage electron microscope-protein tracer study. *Acta Neuropathol.* **64,** 192–202.

5. Shivers, R. R. (1979) The effect of hyperglycemia on brain capillary permeability in the lizard, *Anolis carolinensis.* A freeze-fracture analysis of blood-brain barrier pathology. *Brain Res.* **170,** 509–522.

6. Shivers, R. R. (1980) The blood-brain barrier of *Anolis carolinensis.* A high voltage EM-protein tracer study. *Anat. Rec.* **196,** 172A.

7. Kenny, T. P., and Shivers, R. R. (1974) The blood-brain barrier in a reptile, *Anolis carolinensis. Tiss. Cell* **6,** 319–333

8. Lossinsky, A. S., Badmajew, V., Robson, J., Moretz, R. C., and Wisniewski, H. M. (1989) Sites of egress of inflammatory cells and horseradish peroxidase transport across the blood-brain barrier in a murine model of chronic relapsing experimental allergic encephalomyelitis. *Acta Neuropathol.* **78,** 359–371.

9. Lossinsky, A. S., Pluta, R., Song, M. J., Badmajew, V., Moretz, R. C., and Wisniewski, H. M. (1991) Mechanisms of inflammatory cell attachment in chronic relapsing experimental allergic encephalomyelitis. A scanning and high-voltage electron microscopic study of the injured mouse blood-brain barrier. *Microvasc. Res.* **41,** 299–310.

10. Lossinsky, A. S., Song, M. J., Pluta, R., Moretz, R. C., and Wisniewski, H. M. (1990) Combined conventional transmission, scanning, and high-voltage electron microscopy of the same blood vessel for the study of targeted inflammatory cells in blood-brain barrier inflammation. *Microvasc. Res.* **40,** 427–438.

11. Lossinsky, A. S., Song, M. J., and Wisniewski, H. M. (1989) High-voltage electron microscopic studies of endothelial cell tubular structures in the mouse blood-brain barrier following brain trauma. *Acta Neuropathol.* **77,** 480–488.

12. Shivers, R. R., and Harris, R. J. (1984) Opening of the blood-brain barrier in *Anolis carolinensis.* A high voltage electron microscope protein tracer study. *Neuropathol. Appl. Neurobiol.* **10,** 343–356.

13. Karnovsky, M. J. (1965) A formaldehyde-glutaraldehyde fixative of high osmolarity for use in electron microscopy. *J. Cell Biol.* **27,** 137A.

14. Spurr, A. R. (1969) A low-viscosity epoxy resin embedding medium for electron microscopy. *J. Ultrastruct. Res.* **26,** 31–43.

15. Hayat, M. A., ed. (1970) *Principles and Techniques of Electron. Biological Applications.,* vol 1. Van Nostrand Reinhold, New York.

16. Kalimo, H., Garcia, J. H., Kamijyo, Y., et al. (1974) Cellular and subcellular alterations of human CNS. Studies utilizing in situ perfusion fixation at immediate autopsy. *Arch. Pathol.* **97,** 352–359.

17. Hayat, M. A., ed. (1981) *Fixation for Electron Microscopy.* Academic Press, New York.

18. Ribeiro-Da-Silva, A., Priestley J. V., and Cuello, A. C. (1993) Pre-embedding ultrastructural immunocytochemistry. In *Immunohistochemistry.* Cuello, A. C., ed. John Wiley & Sons, NY, pp. 181–227.

19. McDowell, E. M., and Trump, B. F. (1976) Histologic fixatives suitable for diagnostic light and electron microscopy. *Arch. Pathol. Lab. Med.* **100,** 405–414.

20. Mergner, W. J., Sutherland, J. C., Tigertt, W. D., and Trump, B. F. (1980) To answer questions. A review of an autopsy service. *Arch. Pathol. Lab. Med.* **104,** 167–170.

21. Anderson, T. F. (1951) Technique for the preservation of three-dimensional structure in preparing specimens for the electron microscope. *Trans. NY. Acad. Sci.* **13,** 130–134.

22. Goldstein, J. I., Newbury, D. E., Echlin, P., et al., eds. (1992) *Scanning Electron Microscopy and X-Ray Microanalysis. A Text for Biologists, Materials Scientists, and Geologists*, 2nd ed. Plenum Press, New York.

23. Hayat, M. A., and Zirkin, B. R. (1973) Critical point drying method. In *Principles and Techniques of Electron Microscopy*, Vol. 3. Hayat, M. A., ed. Van Nostrand Reinhold, New York, pp. 297–313.

24. Bray, D. F., Bagu, J., and Koegler, P. (1993) Comparison of hexamethyldisalizane (HMDS), peldri II, and critical-point drying methods for scanning electron microscopy of biological specimens. *Micros. Res. Techn.* **26**, 489–495.

25. Pluta, R., Lossinsky, A. S., Mossakowski, M. J., Faso, L., and Wisniewski, H. M. (1991) Reassessment of a new model of complete cerebral ischemia in rats. Method of induction of clinical death, pathophysiology and cerebrovascular pathology. *Acta Neuropathol.* **83**, 1–11.

26. Wong, D., and Dorovini-Zis, K. (1995) Expression of vascular cell adhesion molecule-1 (VCAM-1) by human brain microvessel endothelial cells in primary culture. *Microvasc. Res.* **49**, 325–339.

27. Graham, R. C., and Karnovsky, M. J. (1966) The early stages of absorption of injected horse-radish peroxidase in the proximal tubules of mouse kidney. Ultrastructural correlates by a new technique. *J. Histochem. Cytochem.* **14**, 291–302.

28. Reese, T. S, and Karnovsky, M. J. (1967) Fine structural localization of a blood-brain barrier to exogenous peroxidase. *J. Cell Biol.* **34**, 207–217.

29. Reynolds, E. S. (1963) The use of lead citrate at high pH as an electron opaque stain in electron microscopy. *J. Cell Biol.* **17**, 208–215.

30. Heuser, J. E., and Reese, T. S. (1973) Evidence for recycling of synaptic vesicle membrane during transmitter release at the frog neuromuscular junction. *J. Cell Biol.* **57**, 315–344.

31. Locke, M., and Krishnan, N. (1971) Hot alcoholic phosphotungstic acid and uranyl acetate as routine stains for thick and thin sections. *J. Cell Biol.* **50**, 550–557.

32. Hiraoka, J. I. (1972) A holder for mass treatment of grids, adapted especially to electron staining and autoradiography. *Stain Technol.* **47**, 297–301.

33. Giammara, B. L. (1981) The Grid-All: a multiple grid handling, staining, and storage device for use in electron microscopy. *Proc. Annu. Meet. Elect. Microsc. Soc. Am.* **39**, 552.

34. Hayat, M. A., ed. (1986) *Basic Techniques for Transmission Electron Microscopy.* Academic Press, San Diego, pp. 217–222.

35. Lossinsky, A. S., Wisniewski, H. M., Dambska, M., and Mossakowski, M. J. (1997) Ultrastructural studies of PECAM-1/CD31 expression in the developing mouse blood-brain barrier with the application of a pre-embedding technique. *Fol. Neuropathol.* **35**, 163–170.

36. Davis, E. C., and Shivers, R. R. (1987) Membrane-associated dense plaques in smooth muscle cells of the common slug, *Limax maximus:* possible sites of transmembrane interaction of filaments. *J. Submicrosc. Cytol.* **19**, 537–544.

5

Detection of Endothelial/Lymphocyte Interaction in Spinal Cord Microvasculature by Intravital Videomicroscopy

Britta Engelhardt, Peter Vajkoczy, and Melanie Laschinger

1. Introduction

The central nervous system (CNS) is considered to be an immune-privileged site. Entry of lymphocytes into the CNS is tightly controlled by endothelium, which is an important component of the blood–brain barrier (BBB). Under physiologic conditions, lymphocyte traffic into the CNS is low, whereas during inflammatory diseases of the CNS such as multiple sclerosis (MS) or in the animal model experimental autoimmune encephalomyelitis (EAE) large numbers of circulating lymphocytes readily gain access to the CNS later during disease facilitated by the loss of BBB integrity. Thus, the interaction of circulating inflammatory cells with endothelium is a critical step in the pathogenesis of EAE.

In general, lymphocyte recruitment across the vascular wall is regulated by the sequential interaction of different adhesion or signaling molecules on lymphocytes and endothelial cells lining the vessel wall. An initial transient contact of the circulating leukocyte with the vascular endothelium, generally mediated by adhesion molecules of the selectin family and their respective carbohydrate ligands, slows down the leukocyte in the bloodstream. Subsequently, the leukocyte rolls along the vascular wall with greatly reduced velocity. The rolling leukocyte can receive endothelial signals resulting in its firm adhesion to the endothelial surface. Such signals can be derived from chemokines, the activity of which is transduced via G-protein–coupled receptors on the leukocyte surface. Binding of a chemokine to its receptor results in a pertussis toxin–sensitive activation of integrins on the leukocyte surface. Only activated integrins mediate the firm adhesion of the leukocytes to the vascular endothelium by binding to their endothelial ligands, which belong to the immunoglobulin (Ig)-superfamily. This ultimately leads to the extravasation of the leukocyte. Successful recruitment of circulating leukocytes into the tissue depends on the leukocyte/endothelial interaction during each of these sequential steps (1).

Investigation of the molecular mechanism involved in lymphocyte/endothelial interaction in vivo can only be achieved by direct observation, i.e., by intravital fluorescence microscopy. Observation of the CNS microcirculation by intravital fluorescence

From: *Methods in Molecular Medicine, vol. 89:*
The Blood–Brain Barrier: Biology and Research Protocols
Edited by: S. Nag © Humana Press Inc., Totowa, NJ

microscopy is, however, hampered by the protected localization of the brain and spinal cord within the skull and the spinal column, respectively. To visualize the CNS microcirculation in vivo, acute and chronic cranial window preparations are available for rodents, which allow observation of the pial and superficial cerebral cortical microcirculations, respectively (2,3). In EAE, however, inflammation is located preferentially in the spinal cord white matter. Thus, to investigate the physiologically relevant interaction of autoaggressive T-lymphocytes with endothelium, we developed a novel spinal cord window preparation that allows direct visualization of CNS white matter microcirculation by intravital fluorescence videomicroscopy. This chapter describes the methods that allow visualization of endothelial/lymphocyte interaction in spinal cord vessels by intravital videomicroscopy.

Using this model, we assessed the interaction of encephalitogenic T-cell blasts with the endothelium of white matter vessels in SJL/N mice. Our study provides, to our knowledge, the first in vivo evidence that α-integrin mediates the G-protein–independent capture and subsequently the G-protein–dependent adhesion strengthening of encephalitogenic T-cell blasts to microvascular VCAM-1 in spinal cord white matter microvasculature (4).

2. Materials
2.1. Labeling of Encephalitogenic T-Cell Blasts with Fluorescent Dye

1. Tissue culture centrifuge.
2. Fluorescent microscope (Axiophot, Zeiss) and/or FACScan (BD Biosciences).
3. Pipetboy (Integra Biosciences) and sterile (serologic disposable or glass) pipets: 5 mL, 10 mL.
4. Micropipets and sterile microtips.
5. Sterile 15-mL and 50-mL Falcon tubes.
6. Tissue culture Petri dishes (10 cm).
7. RPMI 1640 medium.
8. Phosphate-buffered saline (PBS) without Ca^{2+}/Mg^{2+} (Biochrom KG, Berlin, Germany).
9. Wash buffer: Hank's Balanced Salt Solution (HBSS) supplemented with 10% fetal calf serum (FCS) and 25 mM HEPES.
10. Isotonic Nycodenz solution (density 1.09 g/mL at 20°C, Nycomed, Norway): to 17 g Nycodenz add 26 mL of RPMI 1640 and 10 mL of FCS. Add sterile distilled water to a total of 100 mL. Nycodenz solution can be stored at 4°C in the dark for several months.
11. Cell Tracker™ orange (CMTMR) (Molecular Probes, Eugene, Oregon, US). Stock solution: 10 µM in water-free dimethyl sulfoxide (DMSO) stored at –20°C.

2.2. Preparation of the Spinal Cord Window

1. Female SJL/N mice susceptible to EAE (from Bomholdgård Breeding, Ry, Denmark) were used for the experiments described here. In general, mice should be 17 to 19 wk or weigh at least 18 g. Spinal cord window preparations have also successfully been performed using other mouse strains such as BALB/c.
2. Operating stereomicroscope with up to 40-fold magnification.
3. Heating pad.
4. Stereotactic rodent head holder.
5. Ketamine and xylazine.

6. Sterile 1-mL disposable syringes.
7. Instruments: Fine Adson forceps and scissors, jewelry forceps, iris scissors, needle holder.
8. 4-0 silk suture.
9. Polyethylene catheter (PE-10).
10. Impermeable transparent membrane (Saran wrap).

2.3. Intravital Fluorescence Videomicroscopy

1. Axiotech vario microscope with a 100-W mercury lamp emitting an adjustable light intensity (Attoarc; Zeiss, Germany).
2. Combined blue (450–490 nm) and green (520–570 nm) filter block (Zeiss, Germany).
3. X3.2 long distance, X10 long distance, and X20 water immersion working objectives (Zeiss, Germany).
4. Low-light level CCD-video camera with an optional image intensifier for weak fluorescence (Kappa, Germany).
5. S-VHS videosystem (Panasonic, Germany) and video cassettes for on-line and off-line evaluation.
6. Microphone (optional).
7. Sterile disposable 1 mL syringes.
8. 2% FITC-conjugated Dextran (0.1 mL FITC-Dextran$_{150}$ i.v.; MW = 150,000; Sigma Chemical, St. Louis, MO).

2.4. Blocking of Cell Adhesion Molecules on Endothelium or T-Lymphoblasts

1. Tissue culture centrifuge.
2. Pipetboy with 5-mL and 10-mL sterile pipets (either serologic disposable or glass).
3. Micropipets with sterile microtips.
4. Sterile 15-mL Falcon tubes.
5. PBS.
6. Sterile, purified, endotoxin-free blocking antibodies directed against CAMs.

2.5. Detection of Extravasated Cell Tracker Orange-Labeled T-Lymphoblasts in the Spinal Cord Parenchyma by Immunofluorescence

1. Cryostat (Microm GmbH, Walldorf, Germany).
2. Fluorescence microscope (Axiophot, Zeiss, Germany).
3. Pipetboy and 5-mL and 10-mL pipettes.
4. Micropipettes and microtips.
5. Silanized glass slides.
6. Glass coverslips.
7. Dewar vessel.
8. Dry ice.
9. 1% Formaldehyde in PBS.
10. Tissue Tek (OCT; Miles Inc.,Vogel, Germany).
11. 2-Methylbutane (Merck, Darmstadt, Germany).
12. Acetone.
13. Antibodies: Rat anti-mouse endoglin; MJ7/18, FITC-conjugated goat-anti-rat IgG.
14. Bovine serum albumin (BSA).
15. Normal mouse serum.
16. Mowiol 4-88 (Calbiochem-Novbiochem GmbH, Bad Soden, Germany).

3. Methods

3.1. Labeling of Encephalitogenic T-Cell Blasts with Fluorescent Dye

Establishment and culture of the CD4[+] MHC class II restricted protein lipid protein (PLP)–specific T-cell lines derived from SJL/N mice, induction and treatment of EAE are not the topic of this chapter but have been described elsewhere in great detail (*4–6*). Cell Tracker orange labeling of T-lymphocytes was chosen because it does not alter the physiologic characteristics of the T-lymphocytes, i.e., cell surface expression of adhesion molecules or encephalitogenicity (*4*) and reproducibly produces a 100% viable and brightly stained T-lymphoblast population stably fluorescing for several days. Note, the staining intensity of Cell Tracker dye remaines unchanged in non-dividing cells, but decreases nonlinearly with cell division. In our hands, Cell Tracker orange labeling in T-lymphoblasts is detectable up to 6 d after labeling in vitro and up to 4 d after labeling in vivo. The labeling procedure is performed according to the protocol of Hamann and Jonas (*7*), whereby the concentration of the Cell Tracker orange dye has to be optimized for every individual lymphocyte population.

1. For labeling 5×10^6 T-lymphoblasts, the cells are incubated in 20 nM Cell Tracker orange in 1 mL RPMI/10% FCS in an open 50-mL tube for 45 min at 37°C and 7% CO_2.
2. Wash the labeled T-cells with 30 mL wash buffer and pellet by centrifuging for 10 min at 300g.
3. Excess dye is removed by placing 5×10^6-labeled T cells in 10 mL RPMI/10% FCS in a petri dish for 30 min at 37°C and 7% CO_2.
4. Subsequent removal of dead cells and debris is achieved by pelleting the labeled T-cells for 10 min at 300g, resuspending them in 1 mL RPMI 1640 and placing the cell suspension on top of a gradient of 3-mL Nycodenz solution overlaid with 5-mL FCS in a 15-mL tube. The gradient is run for 10 min at 250g without brake.
5. Live cells are harvested from the interphase between the Nycodenz solution and FCS, washed three times with 30 mL of wash buffer and the intensity of fluorescence staining is controlled either under the fluorescence microscope or by using a FACScan.
6. Cell Tracker orange-labeled T-cells can be used immediately for intravital microscopic studies or alternatively be stored for up to 6 h in RPMI/10%FCS/25 mM HEPES on ice before use.

3.2. Preparation of the Spinal Cord Window

1. Animals are anesthetized by subcutaneous injection of 7.5 mg/kg ketamine-hydrochloride and 1 mg xylazine per 100 g body weight and placed on a heating pad (37°C) with continuous control of the body temperature (*see* **Note 1**).
2. A polyethylene catheter is inserted into the right common carotid artery for systemic administration of fluorescent markers and injection of cells.
3. Next, the animal is turned to the prone position and the head fixed in a stereotactic rodent head holder.
4. Using the Adson forceps and scissors a 3–4 cm midline incision is made on the skin of the neck.
5. Using low magnification, the paravertebral musculature is detached from the cervical spinous processes and retracted laterally exposing the vertebral laminae. Retraction is achieved by the use of 4-0 threads.
6. Following exposure of the vertebral laminae, a C1 to C7 laminectomy is performed. Care has to be taken not to injure the underlying spinal cord.

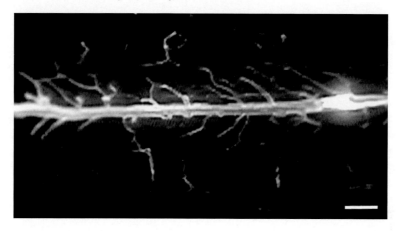

Fig. 1. Spinal cord window preparation for direct intravital microscopic assessment of white matter microcirculation. Intravital fluorescence micoscopic view of a spinal cord window preparation exposing the dorsal spinal cord after a C1 to C7 laminectomy. A large collecting venule is present in the midline of the dorsal medulla, which drains blood caudally. The function of this venule is analogous to that of the superior sagittal sinus in the cerebral microcirculation. The bar represents 500 μm.

7. Using high magnification, the dura is opened over the dorsal spinal cord with the jewelry forceps and the Iris scissors, avoiding trauma to the underlying parenchyma and the spinal microvasculature.
8. The preparation is covered with the impermeable transparent membrane to prevent dehydration of the preparation and the influence of ambient oxygen.
9. Preparations that are traumatized during surgery or reveal signs of acute inflammation (distorted vessels, hyperemia, stagnant blood flow) must be excluded from experiments (*see* **Note 2**).

3.3. Intravital Fluorescence Videomicroscopy

1. The mouse with its head in the stereotactic head holder is transfered to the microscope stage.
2. Intravital fluorescence videomicroscopy is performed by epi-illumination techniques using a modified Axiotech vario microscope with a 100-W mercury lamp emitting an adjustable light intensity, which is attached to a combined blue (450–490 nm) and a green (520–570 nm) filter block. The x3.2 long distance, x10 long distance, and x20 water immersion working objectives result in magnifications of x50, x200, and x400, respectively.
3. The microscopic images are recorded using a low-light level CCD-video camera. To visualize even weak flourescence, the camera should be equipped with an optional image intensifier (Kappa, Germany).
4. For later off-line evaluation the images are recorded using a S-VHS videosystem.
5. To visualize the spinal cord microvasculature 0.1 mL of the fluorescent plasma marker 2% FITC-conjugated Dextran is injected via the arterial catheter and blue light epi-illumination is applied.
6. The microhemodynamics of the vasculature are assessed by the behavior of the negatively contrasted red blood cells (**Fig. 1**).
7. The quality of the preparation is evaluated to exclude the microvascular beds with surgical trauma that are activated inadvertently. This decision is based on the number of leukocytes sticking to the vessel wall (should be less then 5 cells/100 μm vessel length in

postcapillary venules) and the extent of extravasation of the fluorescent marker FITC-Dextran, which reveals BBB breakdown.

8. To study the interaction between the T-lymphoblasts and the white matter endothelium, 3×10^6 Cell Tracker orange-labeled T-cell blasts are injected in aliquots of 100 μL, each containing 1×10^6 T-cells, via the arterial catheter and are visualized within the spinal cord microcirculation using the green light epi-illumination.

9. This combination of two fluorescent markers with distinct excitation wavelengths allows for localization of Cell Tracker orange-labeled T-cell blasts within the FITC-stained vessel lumina (**Fig. 2**).

10. It is essential to use the lowest possible light intensity and the highest sensitivity of the CCD camera otherwise phototoxic damage may develop within seconds and result in activation of the endothelium or BBB breakdown.

11. Due to the angioarchitecture of the spinal cord microvasculature, T-cell blast/endothelium interaction can only be assessed within the spinal capillary bed and postcapillary venules (20–60 μm), but not within precapillary arterioles, that are located within the spinal cord parenchyma, and can therefore not be visualized by intravital fluorescence videomicroscopy.

3.4. Blocking of Cell Adhesion Molecules on Endothelium or T-Lymphoblasts

1. For blocking cell adhesion molecules (CAMs) on the surface of the lymphocytes, 3×10^6 Cell Tracker orange-labeled T-cells are preincubated with 90 μg monoclonal antibody (mAb) in 300 μL PBS for 20 min at 20°C (*see* **Note 3**).

2. For blocking endothelial CAMs, mice are injected with 90 μg mAb in 150 μL PBS 20 min before infusion of T-lymphoblasts.

3.5. Intravital Microscopic Image Analysis

Quantitative analysis of the spinal cord microcirculation may include vessel densities, vessel diameter, microvascular permeability, and the velocity of circulating, fluorescently labeled cells (**Fig. 3**). Any computer-assisted image analysis system such as NIH Image may be used.

1. The highest cell velocity per vessel segment, v_{max}, is used to calculate the mean blood flow velocity, as previously described (*8*):

   $v_{mean} = v_{max} / (2 - \varepsilon^2)$ (μm/s) where ε is the ratio of the cell diameter to vessel diameter (d).

2. From v_{mean} and d, the wall shear rate (γ) is calculated according to Hagen Poiseulle's Law as $\gamma = 8 \times v_{mean} / d$ (s^{-1}) (*8*).

Fig. 2. (*see facing page*) T-lymphoblast/endothelium-interaction within spinal cord white matter postcapillary venules during cell infusion. **A**, The white matter vasculature after contrast enhancement of spinal cord microvasculature using FITC-Dextran$_{150,000}$ is shown after the first and prior to the secondary cell infusion. **B–D**, Intravital microscopic sequence of Cell Tracker orange-labeled T-lymphocytes over time in seconds within the same postcapillary segments as indicated in **A**. Also present in **A** and **B** are examples of a firmly adherent T-lymphoblast from the first injection (*arrows*), and a firmly adherent T-lymphoblast (*arrowhead*) outside of the present focus of the microscope. Intravital microscopic sequence of two Cell Tracker orange-labeled T-lymphocytes 1 and 2 over time in seconds can be observed within the same postcapillary segment as indicated in **A** (*dashed vascular segment*). Cells either lack interaction with endothelium (*2*) or are captured to endothelium without prior rolling (*1*). Bar represents 100 μm.

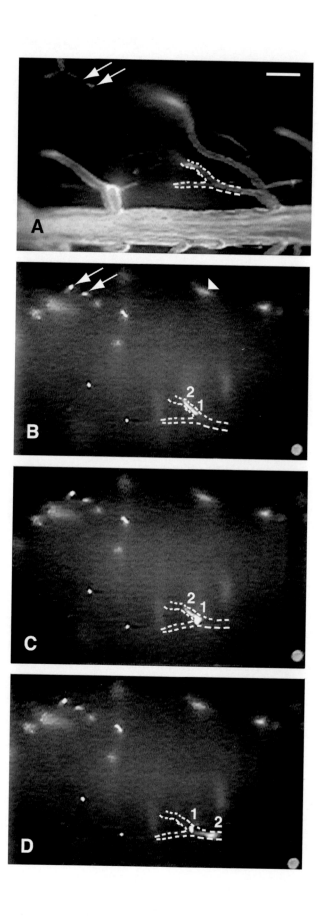

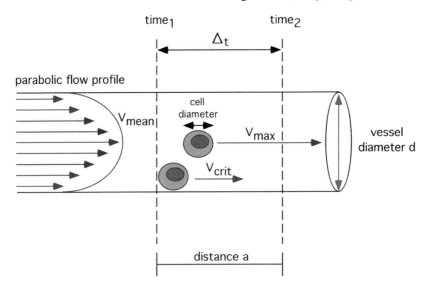

Fig. 3. Intravital microscopic image analysis. Schematic drawing of the hemodynamic parameters applied to image analysis following intravital microscopy. Using a computer assisted image analysis system vessel diameters can be measured. Next the time Δt required by a cell to travel a certain distance between two lines drawn arbitrarily across the vessel diameter is measured for approx 50 cells per vessel. From these measurements the microhemodynamic parameters can be calculated as described in the text.

3. The shear stress (τ) is approximated by $\tau = \gamma \times$ blood viscosity (= 0.025 poise) (dyn/cm^2) (**9**).

4. Typical microhemodynamic parameters as measured in SJL/N spinal postcapillary venules are as follows: vessel diameter: 20 – 60 µm; mean velocities: 500 – 1000 µm/s; wall shear rates: 100 – 350 s^{-1}; wall shear stress: 2.5 – 9.5 dynes/cm^2.

5. In postcapillary venules permanently adherent T-lymphoblasts can be identified as cells that bind to the vessel wall without moving or detaching from the endothelium within an observation period of at least 20 s.

6. Non-permanently adherent T-lymphoblasts are further categorized using a velocity criterion derived from the assumption of a parabolic velocity profile in the microvessel (**Fig. 3**; **8**). Therefore, v_{crit}, the velocity of an idealized cell traveling along, but not interacting with the vessel wall, is calculated as $v_{crit} = v_{mean} \times \varepsilon \times (2 - \varepsilon)$. Consequently, any cell traveling below v_{crit} is regarded as a cell interacting with and therefore rolling along the vessel wall, any cell traveling above v_{crit} is regarded as a non-interacting cell (**8,10**).

7. Within the capillary network, plugging T-lymphoblasts are defined as cells that do not move and block the capillary lumen, as evident by induction of blood flow stasis in the corresponding vascular segment (**Fig. 4**). Permanent T-lymphoblast adhesion at 10 min, 1 h, and 2 h after cell injection is expressed as the number of both adherent and plugging T-lymphoblasts per region of interest (mm^{-2}).

3.6. Detection of Extravasated Cell Tracker Orange-Labeled T-Lymphoblasts in the Spinal Cord Parenchyma by Immunofluorescence

1. To investigate extravasation of Cell Tracker orange-labeled T-lymphoblasts in the spinal cord after intravital microscopy, the mice are placed on a heating pad for thermocontrol and can be kept under anesthesia for up to 9 h after preparation of the spinal cord window.

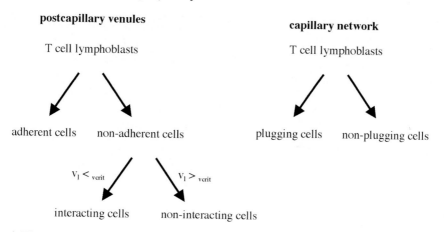

Fig. 4. Flow chart of T-cell interaction with the microvascular endothelium. The chart shows how T-cell/endothelial interactions can be distinguished within capillaries and postcapillary venules of the spinal cord vasculature.

2. At defined time points after injection of Cell Tracker orange-labeled T-lymphoblasts, mice are perfused with 15 mL 1% PFA in PBS via the left ventricle of the heart.
3. Spinal cords are carefully removed, embedded in Tissue-Tek, and snap frozen in a 2-methylbutane bath cooled with dry ice to –80°C in a dewar vessel.
4. Serial 6-μm thick cryosections are cut at a temperature of –15°C, placed on silanized glass slides, dried overnight and acetone fixed for 10 min at –20°C.
5. Fixed and air-dried spinal cord cryosections are visually scanned for Cell Tracker orange-labeled T-lymphoblasts using fluorescence microscopy. We recommend the use of Fluor-objectives by Zeiss.
6. Judgement of the localization of Cell Tracker orange-labeled T-lymphoblasts can only be achieved after visualization of vessels by immunofluorescence with an endothelial cell specific antibody.
   a. Sections containing Cell Tracker orange-labeled T-lymphoblasts are incubated with a rat-anti-mouse endoglin mAb MJ7/18 (BD Biosciences; 10 μg/mL in PBS/0.1% BSA) for 30 min and washed with PBS.
   b. Secondary antibody (i.e., FITC-conjugated goat anti-rat IgG at 10 μg/mL in PBS/10% normal mouse serum) is added for another 30 min.
   c. After washing, slides are coverslipped with Mowiol and immediately analyzed by fluorescence microscopy.
   d. The number of cells within and outside of vessels are counted manually (**Fig. 5**).
   e. To judge if a Cell Tracker orange-labeled T-lymphoblast is located within a spinal cord microvessel or has already migrated into the parenchyma, serial sections have to be analyzed.

3.7. Application

We have used this technique successfully to demonstrate that encephalitogenic T-cell blasts interact with the healthy spinal cord white matter microvasculature without rolling and that α4-integrin mediates the G-protein–independent capture and subsequently the G-protein dependent adhesion strengthening of T-cell blasts to microvascular VCAM-1. This novel spinal cord window model will allow the characterization of molecules involved in the multistep interaction of circulating leukocyte subpopulations with the CNS microvascular endothelium during health and disease in vivo.

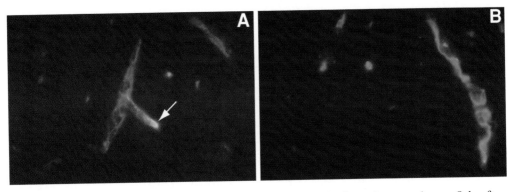

Fig. 5. Localization of T-lymphoblasts within the spinal cord parenchyma 8 h after infusion. Merged photomicrographs showing Cell Tracker orange-labeled T-lymphoblasts (red fluorescence—round dots) and spinal cord microvasculature (green fluorescence—*elongated structures*). One T-cell blast is attached within a spinal cord venule (**A**, *arrow*, yellow fluorescence) and one T-cell blast is present outside the vessel within the spinal cord parenchyma (**B**, red fluorescence—*round dot*). Magnification, x460.

4. Notes

1. Mice have to be carefully observed during the entire time because it might be necessary to re-inject anesthetics. For reinjection half the dose of the primary dose is used.
2. In our experience successful preparation of the spinal cord window in mice requires either experience in microsurgery or alternatively extensive training. Good results are only achieved if the window preparation is completed within 30 min or less.
3. A complete record of the results can be kept by audiovideotaping the entire experiment.
4. Working with whole antibodies in vivo requires stringent controls. Include a control nonblocking antibody of the same isotype as your blocking reagent to rule out any nonspecific effects mediated by the Fc-portions of the antibodies. Also, instead of using irrelevant control immunoglobulins we suggest using control antibodies, which also specifically bind to molecules on either T-lymphocytes or endothelial cells that are not involved in lymphocyte trafficking. We have used successfully anti-endoglin antibodies (clone MJ7/18).

References

1. Butcher, E. C., Williams, M., Youngman, K., Rott, L., and Briskin, M. (1999) Lymphocyte trafficking and regional immunity. *Adv. Immunol.* **72,** 209–253.
2. Uhl, E., Pickelmann, S., Rohrich, F., Baethmann, A., and Schuerer, L. (1999) Influence of platelet-activating factor on cerebral microcirculation in rats: part 2. Local application. *Stroke.* **30,** 880–886.
3. Vajkoczy, P., Ullrich, A., and Menger, M. D. (2000) Intravital fluorescence videomicroscopy to study tumor angiogenesis and microcirculation. *Neoplasia (New York).* **2,** 53–61.
4. Vajkoczy, P., Laschinger, M., and Engelhardt, B. (2001) α4-integrin-VCAM-1 binding mediates G-protein independent capture of encephalitogenic T cell blasts in CNS white matter microvessels. *J Clin Invest.* **108,** 557–565.
5. Engelhardt, B., Laschinger, M., Schulz, M., Samulowitz, U., Vestweber, D., and Hoch, G. (1998) The development of experimental autoimmune encephalomyelitis in the mouse requires alpha4-integrin but not alpha4beta7-integrin. *J. Clin. Invest.* **102,** 2096–2105.
6. Engelhardt, B., Vestweber, D., Hallmann, R., and Schulz, M. (1997) E- and P-selectin are not involved in the recruitment of inflammatory cells across the blood-brain barrier in experimental autoimmune encephalomyelitis. *Blood.* **90,** 4459–4472.

7. Hamann, A., and Jonas, P. (1997) Lymphocyte migration in vivo: the mouse model. In *Immunology Methods Manual.* Lefkovits, I., ed. Academic Press, London, p 1333–1341.

8. Ley, K., and Gaehtgens, P. (1991) Endothelial, not hemodynamic, differences are responsible for preferential leukocyte rolling in rat mesenteric venules. *Circ. Res.* **69,** 1034–1041.

9. Von Andrian, U. H., Hansell, P., Chambers, J. D., et al. (1992) L-selectin function is required for beta 2-integrin-mediated neutrophil adhesion at physiological shear rates in vivo. *Am. J. Physiol.* **263,** H1034–H1044.

10. Robert, C., Fuhlbrigge, R. C., Kieffer, J. D., et al. (1999) Interaction of dendritic cells with skin endothelium: A new perspective on immunosurveillance. *J. Exp. Med.* **189,** 627–636.

II

BLOOD–BRAIN BARRIER PERMEABILITY TECHNIQUES

6

Pathophysiology of Blood–Brain Barrier Breakdown

Sukriti Nag

1. Historical Perspective

The innovative experiments of Paul Ehrlich (*1*) more than a century ago were the first to demonstrate that the permeability properties of cerebral vessels were different from those of non-neural vessels. He injected the aniline dye coerulean-S intravenously into rats and found that all body organs turned blue but the brain did not. These findings were later confirmed by Bouffard (*2*) and Goldmann (*3*) using intravenously injected trypan blue. These early studies led to the recognition of a barrier between blood and brain, which was not passable by trypan blue and which also led to the hypothesis at that time that the barrier was at the endothelium of intracerebral vessels (*4*). Although the term "blood–brain barrier" (BBB) was introduced to describe the absolute restriction of penetration of certain molecules into the brain, it now includes a variety of mechanisms, which maintain cerebral homeostasis.

Tracers such as trypan blue and Evans blue were used extensively in the early part of the 20th century and because they were detectable on visual examination, they established that certain areas of the brain, which collectively are known as the circumventricular organs, were normally permeable to protein. In addition, use of these tracers established that BBB breakdown to protein was present in a variety of experimental conditions such as infection, inflammation, infarction, brain injury, tumors, and hypertensive encephalopathy. In the cold injury model, extensive BBB breakdown was observed at the lesion site with spread of tracer into the white matter in the ipsilateral and contralateral hemisphere (*5*) providing information for the first time on the extent of edema associated with focal brain lesions.

Introduction of electron microscopy (EM) unfolded the ultrastructure of cerebral endothelium as reviewed in Chapter 1, Subheading 2.2. A major advance occurred when the plant enzyme horseradish peroxidase (HRP) became available and was used as a tracer to study the permeability properties of cerebral vessels. Reese and Karnovsky (*6*) demonstrated that HRP injected intravenously into mice was unable to reach the vascular subendothelium despite a circulation time of 1 h. This suggested that the tight junctions of cerebral endothelium were circumferential; hence, their name zonula occludens as compared with junctions of the endothelium of non-neural vessels, which are spotlike tight junctions and named macula occludens. However, longer circulation times of HRP showed this tracer in endothelial plasmalemmal vesicles and in the extracellular

From: *Methods in Molecular Medicine, vol. 89:*
The Blood–Brain Barrier: Biology and Research Protocols
Edited by: S. Nag © Humana Press Inc., Totowa, NJ

space and cerebrospinal fluid (CSF) (7). This may be due to passage of tracer via the fenestrated capillaries at the root of the choroid plexuses to reach the intercellular clefts of the ependyma and thence the CSF (8). Arterioles in certain regions of the brain show passage of HRP via plasmalemmal vesicles in steady states. These areas include the arterioles below the entorhinal sulcus in rat brains (9) and pial vessels in the anterior part of the dorsal sagittal fissure, the cerebellar sulci, and in the sulcus between the olfactory bulb and the cerebral hemisphere of mice (10). These authors also noted increased permeability of intrinsic vessels in the ventral part of the diencephalon and midbrain. These findings emphasize the heterogeneity that is present in endothelium in various parts of the brain.

Normal cerebral vessels are also not permeable to smaller tracers such as ruthenium red (10) or ionic lanthanum (see Chapter 1, Fig. 1; Chapter 8, Fig. 1; 11–13).

2. BBB Development

The permeability of cerebral microvessels to exogenous protein tracers decreases during development (14–16). This decrease in protein permeability appears to be a gradual process and occurs at different times in different locations of the brain. In the mouse, the barrier to protein forms last in the telencephalon where an ependymal-cortical surface gradient of barrier differentiation occurs, with injected protein being still visible in the subependymal layer at embryonic day 16 (15). Intravascular injection of fluorescent-labeled albumin is not detectable in rat brain parenchyma at embryonic day 15 and older (17) while in sheep, a significant barrier to albumin and IgG is present at the early gestational age of embryonic day 60 (18). Decrease in permeability to protein is highly correlated with loss of fenestrations, which occurs by embryonic day 13 in rat intraparenchymal vessels and embryonic day 17 in rat pial vessels due to change in conformation of tight junctions between endothelial cells lining the brain vessels (19). The characteristic appearance of tight junctions in human cerebral endothelial cells occurs as early as week 7 of gestation (20).

Protein permeability studies suggest that although barriers to large molecules are formed, junctions appear to be more permeable to low molecular weight lipid-insoluble compounds such as inulin or sucrose (21,22). The route for this greater permeability cannot be determined since these molecules, unlike proteins, are not visualized at the EM level. It is possible that the progressive changes documented in tight junctions during brain development (19,23) are correlated with the documented decline in passive permeability to molecules, such as sucrose (18,22).

Expression of occludin, a tight junction protein (see Chapter 1, Subheading 2.2.1) in brain endothelial cells is developmentally regulated, being low in rat brain endothelial cells at postnatal day 8 but clearly detectable at postnatal day 70 (24).

3. Pial Vessel Permeability

Pial vessels are more accessible than intracerebral vessels and are used extensively for electrophysiologic and pharmacologic studies of BBB characteristics. However, they are not totally representative of intracerebral vessels, as there are differences in the structure and some molecular characteristics between the endothelia of pial and intracerebral vessels (25). Pial vessels are surrounded by a layer of loose connective tissue and not glial end feet as are intracerebral vessels. The tight junctions of intracerebral vessels have been described previously (see Chapter 1, Subheading 2.2.1). Endothelium of pial vessels show two types of tight junctions. Approximately 25% of the tight junctions are similar to those of intracerebral vessels *(26)*; however, 75% of junctions show a gap of approx 2.8 nm *(27)*. All recorded cerebral endothelial resistance measurements, which are a direct measurement of ion permeability, have been done using endothelia of pial vessels. Values vary from 1870 *(28)* to 6000 $\Omega.cm^2$ *(29)*, supporting the ultrastructural finding of the two types of junctions in pial endothelium.

Studies of normal pial vessel permeability to electron dense tracers, such as ferritin *(10)*, HRP *(8,10)*, microperoxidase *(30)*, and ionic lanthanum *(8)*, give similar results in that tracer is only able to penetrate the first junction at the interendothelial space. Tracer is never seen to form a continuous column from the luminal end of the interendothelial space to the abluminal end. Therefore these tracer studies are at variance with the observation that 2.8-nm gaps are present in majority of pial tight junctions. A possible explanation offered by these authors is that lanthanum, which has a diameter of 2.78 nm in its hydrated state, is cationic and binds to the surface membranes obstructing narrow gaps to prevent further passage of tracer *(25)*.

Comparison of the immunolocalization of OX-47, and Glut-1 shows no differences between the endothelium of pial and intracerebral vessels *(31)*. However, endothelial barrier antigen immunoreactivity is uniform in endothelium of intracerebral vessels but is irregular in endothelia of pial vessels and in these vessels only the endothelial cells in contact with the glia limitans show localization of endothelial barrier antigen *(32)*.

Despite these differences, the major contribution of pial vessel studies is that direct observation of these vessels is possible during the application of various vasoactive agents such as histamine *(33)*, bradykinin *(34,35)*, phenylephrine *(36)*, and angiotensin II amide *(37)* allowing the identification of which vascular segments leak protein. In addition, such studies confirm the findings in non-neural vessels that it is the dilated vascular segments that leak protein *(37)*. Details of the methods used to study pial vessel permeability to tracers using cranial windows are given in Chapter 7.

4. BBB Breakdown in Pathologic States Using HRP as a Tracer

How plasma molecules traverse cerebral endothelium has been studied for a few decades using HRP as a tracer in diverse experimental models *(38–40)*. Methods for studying BBB permeability to tracers are given in Chapters 8 and 11.

The three principal routes for tracer passage have been described previously (*see* Chapter 1, Subheading 2.2). Briefly they are by: 1) transcytosis of plasmalemmal vesicles or caveolae containing plasma macromolecules, apparently by shuttling their contents adsorbed from blood from the luminal to the antiluminal aspect of endothelium; 2) passage via transendothelial channels, which may form transiently by the fusion of two or more plasmalemmal vesicles or caveolae, and provide a direct conduit for the exchange of both small and large plasma molecules; 3) passage via intercellular junctions that form the paracellular pathway for the passive, pressure-driven filtration of water and small solutes.

4.1. Transcytosis

Multifocal areas of HRP extravasation from cerebral vessels occur in diverse models, such as hypertension *(41,42)*; spinal cord injury *(43)*; seizures *(44,45)*; experimental autoimmune encephalomyelitis *(46)*; excitotoxic brain damage *(47)*; brain trauma *(48)*; and BBB breakdown induced by bradykinin *(49,50)*, histamine *(51)*, leukotriene C4

measurements to assess tight junction opening, namely the cleft index and cleft area index *(50)*. In this study only the increase in endothelial vesicles over controls was statistically significant.

An attempt was made to identify drugs that could decrease endothelial vesicles and thus attenuate BBB breakdown, and a number drugs having these properties have been identified such as indomethacin, trifluoperazine, or imidazole, which reduce the number of vessels showing increased permeability following an intracarotid infusion of bradykinin *(49)*. In the case of imidazole, although the number of vessels showing permeability change is markedly reduced, the magnitude of vesicular increase in the involved vessels is not altered. It is of interest that many of these drugs are used in the treatment of psychiatric disorders of humans. Dexamethasone, a drug widely used to reduce increased intracranial pressure resulting from cerebral edema associated with brain tumors and pseudotumors, decreases the number of endothelial vesicles in normal mouse brain vessels *(60)*.

4.2. Transendothelial Channels

Plasmalemmal vesicles are known to fuse to form transendothelial channels in normal non-neural vessels (*see* Chapter 1, Subheading 2.2.2.2.). Although these channels do not occur in normal cerebral endothelium, they have been demonstrated in cerebral endothelium by high-voltage EM, following osmotic opening of the BBB *(61)*, hyperglycaemia *(62)*, and brain trauma *(63)*. In addition, transendothelial channels occur in endothelium of vessels permeable to HRP in acute *(64)* and chronic hypertension (**Fig. 2A,B**; *65*). Transendothelial channels are not easily demonstrated because they do not run perpendicular to the luminal surface. Ultrathin sections cut perpendicular to the length of the vessel as is done in transmission EM studies usually demonstrate incomplete channels (**Fig. 1C**; *55,66*) and a careful search is required along with sectioning at multiple levels of the permeable vessels to demonstrate these channels. It must be emphasized that transendothelial channel formation is a transient phenomenon and can only be demonstrated if EM studies are done within minutes after the onset of the pathologic process.

Despite the accumulated evidence from many models, the role of noncoated plasmalemmal vesicles in the transcytosis of HRP from the circulating blood across the endothelium to the interstitium in pathologic states remains a controversial issue with some opposed to this mechanism based on their studies of cerebral endothelial reactivity in steady states *(67,68)*. The major advances made in the molecular biology of endothelial vesicles in the past decade and its role in transcytosis in non-neural vessels (*see* Chapter 1, Subheading 2.2.2.3), should renew interest in this mechanism of BBB breakdown.

4.3. Tight Junctions

The morphology and molecular composition of tight junctions have been reviewed previously (*see* Chapter 1, Subheading 2.2.1). In steady states these junctions prevent the passage of HRP via the intercellular space. Most of the short-term studies of BBB permeability in diverse experimental models failed to demonstrate passage of protein tracers such as HRP via tight junctions. In the case of osmotic opening of the BBB by hypertonic solutions such as mannitol and urea infused via the carotid artery, HRP was observed in two or more of the interjunctional pools between intact tight junctions and

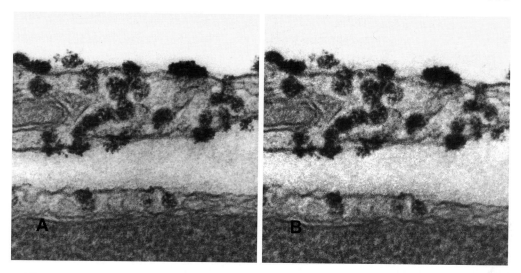

Fig. 2. A segment of arteriolar endothelium from a rat with chronic renal hypertension photographed using a goniometer at 0 degrees (**A**) and + 6 degrees tilt (**B**). No tracer was administered to this rat, which showed extravasation of endogenous serum proteins from cortical vessels (not shown). Note the presence of increased plasmalemmal vesicles in endothelium and the presence of a transendothelial channel, which is formed by fusion of adjacent vesicles. The vesicles are accentuated by the ultracytochemical localization of $Ca^{2+}ATPase$, an enzyme known to be present in these vesicles. **A, B**; x90,000.

this finding along with lack of increase in plasmalemmal vesicles suggested that passage of tracer occurred via opening of tight junctions (*69*). The authors comment "however, no visible cleft separated the outer leaflets of contiguous cell membranes at the junctions nor did the endothelial cells appear shrunken in the brains receiving 3 M urea" (*69*). Other studies of osmotic BBB opening failed to demonstrate HRP passage through tight junctions and reported that passage of tracer through endothelium occurred by enhanced endothelial vesicles (*61,70–72*).

In summary, in vivo studies of BBB breakdown using HRP as a tracer in pathologic states demonstrate passage of this tracer via transcytosis and transendothelial channels suggesting that these are the major routes for passage of plasma proteins in diseases associated with vasogenic edema. Passage of protein across cerebral endothelium most likely occurs by fluid phase transcytosis rather than by receptor-mediated transcytosis as suggested previously (*73*). In favor of fluid-phase transcytosis, is the finding by immunohistochemistry in serial sections of an area of BBB breakdown, of the presence of a variety of plasma proteins such as albumin, fibronectin, fibrinogen, factor VIII, and the various immunoglobulins. Thus, if this breakdown is a consequence of receptor-mediated transcytosis, there would have to be concurrent upregulation of a large number of receptors in a cell already stressed by the pathological state. In addition, passage of protein by transcytosis is calculated to be in the order of seconds, possibly milliseconds. Therefore, there is probably insufficient time for the cell to produce the large variety of receptors required for receptor-mediated transcytosis.

Passage of HRP via interendothelial junctions did not occur in these studies and this may be due to several reasons. First, the junctional alterations may be of a minor degree therefore they are not detectable by HRP, which is a relatively large tracer (MW 40,000). Second, it is possible that routine transmission electron microscopy is not the proper technique to detect alterations of tight junctions because only a small part of the junction is visualized in ultrathin sections. However, tight junction alterations are also not reported in studies using high-voltage EM, which allows thicker sections and larger areas of the junction to be examined.

5. Tight Junction Alterations During BBB Breakdown

5.1. Osmotic BBB Opening

Smaller molecular markers such as ionic lanthanum (MW 138.9 Da) and [^{14}C] sucrose (MW 342 Da) demonstrate BBB breakdown following intracarotid administration of hyperosmotic agents. Ionic lanthanum labeling occurs along the entire length of the interendothelial space, which is widened (12). This finding is not frequent and the author states "a diligent search is usually required to find this pattern" (38). Transient BBB breakdown to [^{14}C] sucrose occurs in the ipsilateral hemisphere following osmotic BBB opening and this is associated with 1–1.5% increase in brain water (74). By relating the cerebrovascular permeability to sucrose, after barrier opening, with its aqueous diffusion co-efficient and the length of the diffusion path across the endothelium, it is estimated that only 0.001% of the endothelial surface need become patent to account for the increased permeability to macromolecules (12). These calculations support passage of tracer either through tight junctions or via plasmalemmal vesicles.

Analysis of water composition and volumes of the intracranial compartment (cerebrospinal fluid, brain tissue, and blood) in dogs (75) and dynamic measurements in animals and humans using positron emission tomography (76) or magnetic resonance imaging (77) confirm that acute hypertonic exposure of brain increases cerebral blood volume. The resulting vasodilatation is postulated to stretch cerebrovascular endothelium to mediate tight-junctional opening (78). Increased BBB permeability after intracarotid hypertonic infusion is essentially reversed within 10 min in rats (74) as well as in monkeys (79).

Regardless of the mechanism of BBB breakdown, clinical application of osmotic modification of the BBB began in 1979 to increase the delivery of chemotherapeutic agents for the treatment of brain tumors in humans. Intracarotid infusion of a 1.4 M mannitol solution, in conjunction with intracarotid methotrexate and intravenous procarbazine and cyclophosphamide infusion prolongs survival of patients with primary central nervous system lymphomas or high-grade gliomas (80,81). However, this therapy is not without its side effects (82).

5.2. Alterations in Tight Junction Proteins and Electrical Resistance

Decreased immunolocalization of zonula occludens (ZO)-1 and occludin occur in areas known to have BBB breakdown to HRP following intracerebral injection of interleukin (83) or BBB breakdown to serum proteins in human cerebral malaria (84). Dual-labeled immunofluorescence confocal microscopy allows the detection of alterations in tight junction proteins in vessels with BBB breakdown. This technique demonstrates altered occludin localization in vessels leaking serum fibrinogen in acute multiple sclerosis lesions (85) and human immunodeficiency virus-1 encephalitis (86). The effect of adenosine triphosphate (ATP) depletion and ischemia on tight junctional proteins has been studied in the Madin-Darby canine kidney cell model as reviewed previously *(87,88)*.

Hypoxic conditions result in a significant decrease by 24.8% in transendothelial electrical resistance after 4 h and a 95% decrease by 8 h *(89)* in rat brain endothelial cells co-cultured with astroglial cells, thereby indicating increased paracellular flux between endothelial cells.

6. Factors Affecting BBB Permeability

Only selected factors affecting BBB permeability will be discussed in this review. Although individual groups study the effect of one or few factors affecting BBB permeability it should be noted that increased permeability is associated with the comcomitant alteration of several factors at a structural and molecular level and the upregulation of several vasoactive factors which act at different times following injury. As an example, BBB breakdown in the cold-injury model is associated with an increase in several mediators such as bradykinin *(90)*, polyamines *(91)*, oxygen radicals *(92)*, nitric oxide *(93)*, and VEGF-A *(94,95)*.

6.1. Endothelial Surface Properties and BBB Breakdown

The luminal plasma membrane of normal cerebral endothelium has a net negative charge *(96)*, which is greater on the luminal than abluminal plasma membrane *(97)* (*see* Chapter 9, Subheading 1.1.). This net negative charge is essential for maintenance of the BBB to protein since its neutralization by intracarotid infusion of positively charged substances such as polycationic protamine or poly-L-lysine *(98,99)* results in BBB breakdown to tracers in the ipsilateral hemisphere. In experimental conditions associated with BBB breakdown such as acute hypertension *(96)*, and cold injury *(97)* vessels with BBB breakdown to HRP show marked reduction or loss of surface charge on endothelium indicating that endothelial injury results in loss of the surface charge on endothelium and contributes to BBB breakdown.

Loss of endothelial surface charge during BBB breakdown suggests that sialyl and other oligosaccharide residues on the luminal plasma membrane might be altered during BBB breakdown. The oligosaccharide residues localized on cerebral endothelium are given in Chapter 9. The only lectin that does not bind to normal cerebral endothelium is peanut agglutinin *(100)*. However, it does bind to endothelium of arterioles permeable to HRP in acute hypertension *(55)* while 10 min after the onset of hypertension when the luminal plasma membrane of endothelium regains its net negative charge peanut agglutinin binding is no longer demonstrable *(55)*. Loss of the terminal sialic acid groups on endothelium in hypertension results in loss of the net negative charge on endothelium and exposes the β-D-gal-(1-3)-D-gal N-acetyl groups that normally are not accessible to peanut agglutinin and binding of this lectin occurs. Supporting this hypothesis is the finding that exposure of cerebral endothelium to neuraminidase, which is known to cleave the terminal sialic acid groups on endothelium, results in loss of the net negative charge on endothelium and allows peanut agglutinin binding *(55,100,101)*. Thus these studies demonstrate that in acute hypertension alterations of the charge and the oligosaccharide residues on the luminal plasma membrane of endothelium allows HRP to be taken up by enhanced transcytosis. These changes are transient and not demonstrable when the BBB is restored.

The surface negative charge on cerebral endothelium is being exploited to introduce chimeric peptides across the BBB. These chimeric peptides are formed when a non-transportable peptide therapeutic is coupled to a BBB drug transport vector such as cationized albumin as described in a recent review *(102)*.

6.2. The Endothelial Cytoskeleton and BBB Breakdown

6.2.1. Actin

Endothelial cells, like most eukaryotic cells, have a cytoskeleton consisting of micro-filaments, intermediate filaments and microtubules. All these filaments have been described in endothelium of normal cerebral vessels *(59,103)*. Transmission EM studies show that microfilament bundles having the dimension of actin are present in proximity to cell junctions and actin has been localized to the endothelial plasma membrane by molecular techniques *(104)*. Integrity of the endothelial microfilaments is necessary for maintenance of the BBB to protein in steady states because increased permeability occurs in the presence of cytochalasin B, an actin-disrupting agent *(59)*. In this study, HRP was used as a marker of BBB permeability alterations and quantitative studies showed a significant increase in endothelial plasmalemmal vesicles in permeable vessels and lack of tracer passage via interendothelial junctions.

Structural organization of actin is necessary for maintenance of tight junction integrity as well. Actin-disrupting substances, such as cytochalasin D, cytokines, and phalloidin, disrupt tight junction structure and function *(105,106)*. A 24-h hypoxic insult results in an approx 2.5-fold increase in paracellular sucrose flux in bovine brain microvascular endothelial cells *(107)*. Increased junctional permeability is associated with increased expression of actin and a redistribution of actin and the tight junctional proteins, occludin, ZO-1 or 2 from the plasma membrane to a cytosolic location.

6.2.2. Microtubules

Rats pretreated with colchicine, an agent that disrupts microtubules, fails to result in BBB breakdown to HRP in steady states suggesting that the microtubular network has no demonstrable role in vascular homeostasis. However, pretreatment with colchicine attenuates the BBB breakdown that is known to occur in acute hypertension *(59)*. Microtubules are known to form intracellular pathways along which protein-bearing vesicles pass in certain cell types *(108,109)*. A possible explanation may be that disruption of microtubules impairs the passage of endothelial vesicles from the luminal to the abluminal plasma membrane thus attenuating BBB breakdown. A study of bovine aortic endothelial cells demonstrates transport of fluorescein isothiocyanate dextran from the luminal to the abluminal side by chains of vesicles *(110)*. This activity was reduced by colchicine supporting the role of microtubules in vesicle passage through endothelium.

6.3. Calcium and BBB Breakdown

Calcium is implicated in both transcytosis and the opening of tight junctions.

6.3.1. Calcium and Transcytosis

The role of calcium (Ca^{2+}) in BBB breakdown to protein was studied in the acute hypertension model in which enhanced numbers of endothelial vesicles and transcytosis of vesicles was demonstrated to lead to BBB breakdown to HRP *(40,111)*. It was hypothesized that transient fluxes of increased intra-endothelial calcium could mediate enhanced permeability by transcytosis *(111)*. This could result by the following mechanisms. Inhibition of the endothelial plasma membrane Na^+, K^+-adenosine triphosphatase (ATPase) by hypertension may result in sodium accumulation within the cell. This would decrease the efficiency of the Na/Ca exchanger leading to intracellular calcium accumulation. Inhibition of the plasma membrane Ca^{2+}-ATPase could contribute to further increases of intra-endothelial Ca^{2+}, which could mediate increased endothelial permeability by enhanced transcytosis. The role of increased intra-endothelial calcium in BBB breakdown in hypertension is supported firstly by ultracytochemical studies that demonstrate reduced localization of both Ca^{2+}-ATPase and Na^+, K^+-ATPase only in the endothelium of arterioles permeable to HRP in hypertension *(64,111)*. Subsequent studies demonstrate that reduced activity of the endothelial ATPases precedes BBB breakdown in hypertension *(112)*. Pretreatment of acutely hypertensive rats with the calcium entry blocker Flunarizine decreases the number and size of areas of Evans blue extravasation and results in a significant decrease in protein transfer of ^{125}I-labeled human serum albumin in total brain as well as in individual brain areas *(113)*.

6.3.2. Calcium and Tight Junctions

Calcium acts both extracellularly and intracellularly to regulate tight junction activity. Removal of extracellular Ca^{2+} results in a concurrent decrease in electrical resistance across the membrane and an increase in permeability *(106)*, which involves heterotrimeric G protein and protein kinase C signaling pathways. Intracellular Ca^{2+} plays a role in cell-cell contact *(114)*, increased electrical resistance *(115)*, ZO-1 migration from intracellular sites to the plasma membrane *(116)*, and tight junction assembly *(117)*. Studies of cultured brain endothelial cells demonstrate that vasoactive agents such as histamine, bradykinin, endothelin, and the nucleotides ATP, adenosine diphosphate, and uridine triphosphate, which are known to increase BBB permeability, cause activation of phospholipase C and elevation of intracellular Ca^{2+} *(118,119)*. Although it was suggested that the data were consistent with a Ca^{2+}-dependent contractile mechanism for opening of tight junctions, no proof of this was provided. Ca^{2+} modulation of tight junction function is discussed in reviews *(88,120)*.

Calcium flux is implicated in structural and functional variations in cultured cerebral endothelial cells during ischemic stress. Bovine brain microvascular endothelial cells exposed to hypoxia-aglycemia for 6 h and hypoxia for 48 h show increased permeability to $[^{14}C]$ sucrose, which is reversed by treatment with the L-type calcium channel blocker nifedipine *(121)*. Of interest is the finding that 48 h of hypoxia is associated with alteration of the f-actin cytoskeleton of endothelial cells. These authors suggest that endothelial cell calcium flux may be responsible for the observed structural and functional variations.

6.4. BBB Breakdown in Inflammation

Proinflammatory cytokines produced and secreted by brain microvascular endothelial cells include interleukin (IL)-α and β, IL-6, and GM-CSF *(122–125)*. In many cases these cytokines have an important role in initiating change in adhesion molecules on endothelial cells such as PECAM-1, E-selectin, and ICAM-1. This allows immune cells to infiltrate the CNS and cause BBB breakdown. Injection of IL-1β into the brain results in rapid induction of the neutrophil chemoattractants MIP-2 and CINC-1. Administration of an antibody neutralizing the activity of CINC-1 suppresses both the number of neutrophils in the brain parenchyma as well as BBB breakdown *(126)*. Endogenous agents, including many cytokines regulate tight junctions in vitro. Several studies have reported marked permeability increases in cultured endothelial cells after exposure to vasoactive cytokines, such as tumor necrosis factor-α (TNF-α), IL-1β, interleukin 6, interferon γ, and histamine *(119,127)*.

Brain diseases such as multiple sclerosis, human immunodeficiency virus encephalitis, Alzheimer's disease, and stroke show BBB breakdown associated with leukocyte migration *(83,128,129)*. Ultrastructural studies show lymphocytes and monocytes penetrate the endothelial cell adjacent to the junctional area in autoimmune demyelination *(130)* and in chronic relapsing experimental encephalomyelitis *(131)*. Migration of activated rat T cells through porcine brain endothelial monolayers at parajunctional areas occurs within 20–40 min of co-culture and is accompanied by a decay of endothelial resistance *(132)*. However, other studies show that T-lymphocytes penetrate human brain endothelial cells by movement across the cytoplasm and through tight junctions between these cells without apparent disruption of the integrity or solute permeability of the monolayer *(133)*. Increased leukocyte migration alters the molecular organization of the tight junction complex, including breakdown of occludin and ZO-1 *(83)*.

An interesting observation in IL-1–treated human umbilical vein endothelial cells cultured in astrocyte conditioned medium is that leukocyte migration occurs at tricellular corners where the borders of three endothelial cells meet and where tight and adherens junctions are discontinuous and immunostaining for occludin, ZO-1, cadherin, and β-catenin are not present *(134)*. Supporting this finding is the observation by these authors that neutrophil migration does not result in widespread proteolytic loss of the tight junction proteins ZO-1, ZO-2, and occludin from endothelial borders and that transendothelial electrical resistance is unaffected *(135)*. The transendothelial electrical resistance of endothelial cells grown in the astrocyte conditioned medium in this study is reported to be approx 12,000 Ω; hence, these findings may have relevance to cerebral endothelial cells.

Endothelial cells can also produce T-cell growth factors β1 and β2, which are considered to downregulate adhesion molecules and markedly diminish leukocyte migration across CNS endothelial cells *(136,137)*.

6.5. Angiogenic Factors and BBB Breakdown

The endothelial-specific angiogenic factors vascular endothelial growth factor (VEGF)-A and VEGF-B and the angiopoietins (Ang) described previously (*see* Chapter 1, Subheading 2.1.) are known to affect BBB permeability as reviewed recently *(73,138)*.

6.5.1. VEGF-A

Although numerous studies show upregulation of VEGF-A gene in models associated with cerebral angiogenesis, very few have related the increased expression to BBB breakdown *(139–142)*. One of the earliest studies relating VEGF-A with BBB breakdown used serial sections to demonstrate VEGF-A protein in the endothelium of vessels showing BBB breakdown to fibronectin at the site of a cortical cold injury *(94)*.

This was later confirmed by confocal microscopy of tissues dual labeled for VEGF-A and fibronectin *(95)* (*see* Chapter 33, Figs. 2E and F). Other studies have correlated the increased expression of VEGF-A at the messenger ribonucleic acid (mRNA) and protein level with edema formation in human meningiomas *(143,144)*. A single intracortical injection of VEGF-A in mice *(145)* or chronic administration of VEGF-A to rat brains via miniosmotic pumps *(146)* causes BBB breakdown at the lesion site. BBB breakdown and edema associated with cerebral infarcts and the size of infarcts can be decreased by pretreatment of mice with a soluble VEGF receptor chimeric protein (Flt-[1-3]-IgG), which inactivates endogenous VEGF *(147)*. Thus VEGF is partly responsible for the cortical edema formation in this model. However, administration of VEGF to ischemic rats one hr after onset of ischemia exacerbates BBB leakage in the ischemic hemisphere but not in the nonischemic hemisphere *(148)*. These studies suggest that acute inhibition of VEGF-A may have therapeutic potential in BBB leakage.

Activation by VEGF-A of its receptors leads to a multifaceted activation of downstream signaling pathways, some of which are involved in increasing permeability as reviewed recently *(149)*. This results in structural alterations in endothelium, which leads to BBB breakdown. Ultrastructural studies following intracortical injection of VEGF-A show interendothelial gaps and formation of segmental fenestrae-like narrowings in cortical vessels permeable to endogenous albumin *(145)*. In contrast, VEGF-A–induced hyperpermeability of the blood-retinal barrier endothelium is associated predominantly with enhanced numbers of endothelial vesicles while alterations of tight junctions and fenestrations in endothelium are not observed *(150)*. VEGF-A is also known to induce changes in tight junction proteins. VEGF-A increases occludin as well as its phosphorylation and that of tyrosine phosphorylation of ZO-1 within 15 min in cultured retinal endothelial cells as well as when injected into the vitreous cavity of rat eyes *(151)*. Greater than 3-h exposures of VEGF-A with retinal *(152)* or brain *(153)* endothelial cells reduces occludin expression coinciding with changes in permeability.

The fact that high-grade gliomas show abnormalities of tight junction structure has been known since the 1970s *(154)* when EM studies demonstrated that there was lack of fusion of adjacent plasma membranes at the tight junctions leaving a space. Molecular studies show an almost 50-fold upregulation of VEGF-A in glioblastomas *(155)*. VEGF-A is known to produce structural alterations in endothelium *(145,150)*, including both an increase in endothelial vesicles and altered structure of tight junctions; therefore, both factors contribute to increased permeability in high-grade gliomas. In addition, VEGF-A can alter junctional proteins *(153,156)* leading to increased junctional permeability and cerebral edema in high-grade gliomas.

VEGF-B, on the other hand, is expressed constitutively in endothelium of cerebral vessels *(95)*. The finding that BBB breakdown to fibronectin in the cold injury model shows loss of VEGF-B in the endothelium of the permeable vessels suggests that VEGF-B, unlike VEGF-A, may be an essential factor in maintenance of endothelial homeostasis *(95)*.

6.5.2. Angiopoietin-1

Angiopoietin-1 (Ang-1) protein is constitutively expressed in endothelium of non-neural and cerebral vessels, consistent with it having a constitutive stabilizing effect by maintaining cell-to-cell and cell-matrix interactions *(157)* (Chapter 1, Fig. 2B).

18. Dzieglielewska, K. M., Evans, C. A. N., Malinowska, D. H., Mollgard, K., Reynolds, J. M., Reynolds, M. L., and Saunders, N. R. (1979) Studies of the development of the brain-barrier systems to lipid insoluble molecules in fetal sheep. *J. Physiol.* **292,** 207–231.

19. Stewart, P. A., and Hayakawa, E. M. (1994) Early ultrastructural changes in blood-brain barrier vessels of the rat embryo. *Dev. Brain Res.* **78,** 25–34.

20. Møllgård, K., and Saunders, N. R. (1986) The development of the human blood-brain and blood-CSF barriers. *Neuropathol. Appl Neurobiol.* **12,** 337–358.

21. Saunders, N. R., and Dziegielewska, K. M. (1997) Barriers in the developing brain. *News Physiol. Sci.* **12,** 21–31.

22. Habgood, M. D., Knott, G. W., Dziegielewska, K. M., and Saunders, N. R. (1993) The nature of the decrease in cerebrospinal fluid barrier exchange during postnatal brain development in the rat. *J. Physiol.* **468,** 73–83.

23. Kniesel, U., Risau, W., and Wolburg, H. (1996) Development of blood-brain barrier tight junctions in the rat cortex. *Dev. Brain Res.* **96,** 229–240.

24. Hirase, T., Staddon, J. M., Saitou, M., Ando-Akatsuka, Y., Itoh, M., Furuse, M., Fujimoto, K., Tsukita, S., and Rubin, L. L. (1997) Occludin as a possible determinant of tight junction permeability in endothelial cells. *J. Cell Sci.* **110,** 1603–1613.

25. Allt, G. and Lawrenson, J. G. (1997) Is the pial microvessel a good model for blood-brain barrier studies ? *Brain Res. Rev.* **24,** 67–76.

26. Schulze, C., and Firth, A. J. (1992) Interendothelial junctions during blood-brain barrier development in the rat: morphological changes at the level of individual tight junctional contacts. *Dev. Brain Res.* **69,** 85–95.

27. Cassella, J. P., Lawrenson, J. G., and Firth, J. A. (1996) Rat pial microvasculature: the incomplete expression of a blood-brain barrier ? *Ann. Anat.* **178** (Suppl), 54–55.

28. Crone, C., and Olesen, S. P. (1982) Electrical resistance of brain microvascular endothelium. *Brain Res.* **241,** 49–55.

29. Butt, A. M., Jones, H. C., and Abbott, N. J. (1990) Electrical resistance across the blood-brain barrier in anesthetized rats: a developmental study. *J. Physiol.* **429,** 47–62.

30. Westergaard, E. (1980) Transport of microperoxidase across segments of cerebral arterioles under normal conditions. *Neuropathol. Appl. Neurobiol.* **6,** 267–277.

31. Cassella, J. P., Lawrenson, J. G., Allt, G., and Firth, J. A. (1996b) Ontogeny of four blood-brain barrier markers: An immunocytochemical comparison of pial and cerebral cortical microvessels. *J. Anat.* **189,** 407–415.

32. Cassella, J. P., Lawrenson, J. G., Lawrence, L. and Firth, J. A. (1997) Differential distribution of an endothelial barrier antigen between the pial and cortical microvessels of the rat. *Brain Res.* **744,** 335–338.

33. Wahl, M., and Kuschinsky, W. (1979) The dilating effect of histamine on pial arteries of cats and its mediation by H_2 receptors. *Circ. Res.* **44,** 161–165.

34. Unterberg, A., Wahl, M., and Baethmann, A. Effects of bradykinin on permeability and diameter of pial vessels in vivo. *J. Cereb. Blood Flow Metab.* **4,** 574–585.

35. Wahl, M., Unterberg, A., and Baethman, A. (1985) Intravital fluorescence microscopy for the study of blood-brain barrier function. *Int. J. Microcirc. Clin. Exp.* **4,** 3–18.

36. Mayhan, W. G. (1991) Disruption of the blood-brain barrier in open and closed cranial window preparations in rats. *Stroke* **22,** 1059–1063.

37. Farrar, J. K., Jones, J. V., Graham, D. I., Strandgaard, S., and MacKenzie, E. T. (1976) Evidence against cerebral vasospasm during acutely induced hypertension. *Brain Res.* **104,** 176–180.

38. Brightman, M. W., Zis, K., and Anders, J. (1983) Morphology of cerebral endothelium and astrocytes as determinants of the neuronal microenvironment. *Acta Neuropathol. (Berl.)* **Suppl VIII,** 21–33.

39. Cervos-Navarro, J., Artigas J., and Mrsulja, B. J. (1983). Morphofunctional aspects of the normal and pathological blood-brain barrier. *Acta Neuropathol. (Berl.)* **Suppl VIII,** 1–19.

40. Nag S. (1998) Blood-brain barrier permeability measured with histochemistry. In *Introduction to the Blood-Brain Barrier. Methodology, Biology and Pathology.* Pardridge, W. M., ed., Cambridge University Press, Cambridge, UK, pp. 113–121.

41. Nag, S., Robertson, D. M., and Dinsdale, H. B. (1977) Cerebral cortical changes in acute experimental hypertension. An ultrastructural study. *Lab. Invest.* **33,** 150–171.

42. Westergaard, E., Van Deurs, B., and Brondsted, H. E. (1977). Increased vesicular transfer of peroxidase across cerebral endothelium, evoked by acute hypertension. *Acta Neuropathol. (Berl.)* **37,** 141–152.

43. Beggs, J. L., and Waggener, J. D. (1976) Transendothelial vesicular transport of protein following compression injury to the spinal cord. *Lab. Invest.* **34,** 428–439.

44. Hedley-Whyte, E. T., Lorenzo, A. V., and Hsu, D. W. (1977) Protein transport across cerebral vessels during metrazole-induced convulsions. *Am. J. Physiol.* **233,** C74–C85.

45. Nitsch, C., Goping, G., Laursen, H., and Klatzo, I. (1986) The blood-brain barrier to horseradish peroxidase at the onset of bicuculline-induced seizures in hypothalamus, pallidum, hippocampus, and other selected regions of the rabbit. *Acta Neuropathol. (Berl.)* **69,** 1–16.

46. Claudio, L. (1996) Ultrastructural features of the blood-brain barrier in biopsy tissue from Alzheimer's patients. *Acta Neuropathol. (Berl.)* **91,** 6–14.

47. Nag, S. (1992) Vascular changes in the spinal cord in N-methyl-D-aspartate-induced excitotoxicity: morphological and permeability studies. *Acta Neuropathol. (Berl.)* **84,** 471–477.

48. Povlishock, J. T., Becker, D. P., Sullivan, H. G., and Miller, J. D. (1978) Vascular permeability alterations to horseradish peroxidase in experimental brain injury. *Brain Res.* **153,** 223–239.

49. Raymond, J. J., Robertson, D. M., and Dinsdale, H. B. (1986). Pharmacological modification of bradykinin induced breakdown of the blood-brain barrier. *Can. J. Neurol. Sci.* **13,** 214–220.

50. Hashizume, K., and Black, K. L. (2002) Increased endothelial vesicular transport correlates with increased blood-tumor barrier permeability induced by bradykinin and leukotriene C4. *J. Neuropathol. Exp. Neurol.* **61,** 725–735.

51. Dux, E., and Joó, F. (1982) Effects of histamine on brain capillaries: Fine structural and immunohistochemical studies after intracarotid infusion. *Exp. Brain Res.* **47,** 252–258.

52. Brightman, M. W., Klatzo, I., Olsson, Y., and Reese, T. S. (1970) The blood-brain barrier to proteins under normal and pathological conditions. *J. Neurol. Sci.* **10,** 215–239.

53. Nag, S., Robertson, D. M., and Dinsdale, H. B. (1980) Morphological changes in spontaneously hypertensive rats. *Acta Neuropathol. (Berl.)* **52,** 27–34.

54. Nag, S. (1984) Cerebral changes in chronic hypertension, combined permeability and immunohistochemical studies. *Acta Neuropathol. (Berl.)* **62,** 178–184.

55. Nag, S. (1986) Cerebral endothelial plasma membrane alterations in acute hypertension. *Acta Neuropathol. (Berl.)* **70,** 38–43.

56. Nag, S., and Harik, S. I. (1997) Cerebrovascular permeability to horseradish peroxidase in hypertensive rats: effects of unilateral locus ceruleus lesion. *Acta Neuropathol. (Berl.)* **73,** 247–253.

57. Nag, S., Robertson, D. M., and Dinsdale, H. B. (1981) Cerebrovascular permeability in mechanically induced hypertension. *Can. J. Neurol. Sci.* **8,** 215–220.

58. Petito, C. K., and Levy, D. E. (1980) The importance of cerebral arterioles in alterations of the blood-brain barrier. *Lab. Invest.* **43,** 262–268.

59. Nag, S. (1995). Role of the endothelial cytoskeleton in blood-brain barrier permeability to proteins. *Acta Neuropathol. (Berl.)* **90,** 454–460.

60. Hedley-Whyte, E. T., and Hsu, D. W. (1986) Effect of dexamethasone on blood-brain barrier in the normal mouse. *Ann. Neurol.* **19,** 373–377.

61. Farrell, C. L. and Shivers, R. R. (1984) Capillary junctions of the rat are not affected by osmotic opening of the blood-brain barrier. *Acta Neuropathol. (Berl.)* **63,** 179–189.

62. Shivers, R. R., and Harris, R. J. (1984) Opening of the blood-brain barrier in *Anolis carolinenis.* A high voltage electron microscope protein tracer study. *Neuropathol. Appl. Neurobiol.* **10,** 343–356.

63. Lossinsky, A. S., Song, M. J., and Wisniewski, H. M. (1989) High voltage electron microscopic studies of endothelial cell tubular structures in the mouse blood-brain barrier following brain trauma. *Acta Neuropathol. (Berl.)* **77,** 480–488.

64. Nag, S. (1988) Localisation of calcium-activated adenosine-triphosphatase (Ca2+-ATPase) in intracerebral arterioles in acute hypertension. *Acta Neuropathol. (Berl.)* **75,** 547–553.

65. Nag, S. (1990) Presence of transendothelial channels in cerebral endothelium in chronic hypertension. *Acta Neurochir.* **51,** 335–337.

66. Castejón, O. J. (1984) Formation of transendothelial channels in traumatic human brain edema. *Path. Res. Pract.* **179,** 7–12.

67. Broadwell, R. D. (1989) Transcytosis of macromolecules through the blood-brain barrier: a cell biological perspective and critical appraisal. *Acta Neuropathol. (Berl.)* **79,** 117–128.

68. Stewart, P. A. (2000) Endothelial vesicles in the blood-brain barrier: Are they related to permeability ? *Cell. Mol. Neurobiol.* **20,** 149–163.

69. Brightman, M. W., Hori, M., Rapoport, S. I., Reese, T. S., and Westergaard, E. (1973) Osmotic opening of tight junctions in cerebral endothelium. *J. Comp. Neurol.* **152,** 317–325.

70. Hansson, H. A. and Johansson, B. B. (1980) Induction of pinocytosis in cerebral vessels by acute hypertension and hyperosmolar solutions. *J. Neurosci. Res.* **5,** 183–190.

71. Houthoff, H. J., Go, K. G., and Gerrits, P. O. (1982) The mechanisms of blood-brain barrier impairment by hyperosmolar perfusion. *Acta Neuropathol. (Berl.)* **56,** 99–112.

72. Nagy, Z., Peters, H., and Hüttner, I. (1984) Fracture faces of cell junctions in cerebral endothelium during normal and hyperosmotic conditions. *Lab. Invest.* **50,** 313–322.

73. Nag, S. (2002) The blood-brain barrier and cerebral angiogenesis: lessons from the cold-injury model. *Trends Mol. Med.* **8,** 38–44.

74. Rapoport, S. I., Fredericks, W. R., Ohno, K., and Pettigrew, K. D. (1980) Quantitative aspects of reversible osmotic opening of the blood-brain barrier. *Am J. Physiol.* **238,** R421–R431.

75. Rosomoff, H. L. (1962) Distribution of intracranial contents after hypertonic urea. *J. Neurosurg.* **19,** 859–864.

76. Ravussin, P., Archer, D. P., Tyler, J. L., et al. (1986) Effects of rapid mannitol infusion on cerebral blood volume. A positron emission tomographic study in dogs and man. *J. Neurosurg.* **64,** 104–113.

77. Lin, W., Paczynski, R. P., Kuppusamy, K., Hsu, C. Y., and Haacke, E. M. (1997) Quantitative measurements of regional cerebral blood volume using MRI in rats: Effects of arterial carbon dioxide tension and mannitol. *Magn. Reson. Med.* **38,** 420–428.

78. Rapoport, S. I. (2000) Osmotic opening of the blood-brain barrier: principles, mechanism, and therapeutic applications. *Cell. Mol. Neurobiol.* **20,** 217–230.

79. Kessler, R. M., Goble, J. C., Bird, J. H., et al. (1984) Measurement of blood-brain barrier permeability with positron emission tomography and [^{68}Ga]EDTA. *J. Cereb. Blood Flow Metab.* **4,** 323–328.

80. Dahlborg, S. A., Henner, W. D., Crossen, J. R., et al. (1996) Non-AIDS primary CNS lymphoma: The first example of a durable response in primary brain tumor using enhanced chemotherapy delivery without cognitive loss and without radiotherapy. *Cancer J. Sci. Am.* **2,** 166–174.

81. Gumerlock, M. K., York, D., and Durkis, D. (1994) Osmotic blood-brain barrier disruption and chemotherapy in the treatment of high grade malignant glioma: Patient series and literature review. *J Neurooncol.* **12,** 33–46.

82. Roman-Goldstein, S., Mitchell, P., Crossen, J. R., Williams, P. C., Tindall, A., and Neuwelt, E. A. (1995) MR and cognitive testing of patients undergoing osmotic blood-brain barrier disruption with intra-arterial chemotherapy. *Am. J. Neuroradiol.* **16,** 543–553.

83. Bolton, S. J., Anthony, D. C., and Perry, V. H. (1998) Loss of the tight junction proteins occludin and zonula occludens-1 from cerebral vascular endothelium during neutrophil-induced blood-brain barrier breakdown *in vivo. Neurosci.* **86,** 1245–1257.

84. Brown, H., Hien, T., Day, N., et al. (1999) Evidence of blood-brain barrier dysfunction in human cerebral malaria. *Neuropathol. Appl. Neurobiol.* **25,** 331–340.

85. Plumb, J., McQuaid, S., Mirakhur, M., and Kirk, J. (2002) Abnormal endothelial tight junctions in active lesions and normal-appearing white matter in multiple sclerosis. *Brain Pathol.* **12,** 154–169.

86. Dallasta, L. M., Pisaaarov, L. A., Esplen, J. E., et al. (1999) Blood-brain barrier tight junction disruption in human immunodeficiency virus-1 encephalitis. *Am. J. Pathol.* **155,** 1915–1927.

87. Tsukamoto, T., and Nigam, S. K. (1997) Tight junction proteins form large complexes and associate with the cytoskeleton in an ATP depletion model for reversible junction assembly. *J. Biol. Chem.* **272,** 16133–16139.

88. Denker, B. M. and Nigam, S. K. (1998) Molecular structure and assembly of the tight junction. *Am. J. Physiol.* **274,** F1–F9.

89. Kondo, T., Kinouchi, H., Kawase, M., and Yoshimoto, T. (1996) Astroglial cells inhibit the increasing permeability of brain endothelial cell monolayer following hypoxia/reoxygenation. *Neurosci. Lett.* **208,** 101–104.

90. Maier-Hauff, K., Baethmann, A. J., Lange, M., Schürer, L., and Unterberg, A. (1984) The kallikrein-kinin system as mediator in vasogenic brain edema. Part 2: Studies on kinin formation in focal and perifocal brain tissue. *J. Neurosurg.* **61,** 97–106.

91. Trout, J. J., Koenig, H., Goldstone, A. D. and Lu, C. Y. Blood-brain barrier breakdown by cold injury. Polyamine signals mediate acute stimulation of endocytosis, vesicular transport, and microvillus formation in rat cerebral capillaries. *Lab. Invest.* **55,** 622–631.

92. Murakami, K., Kondo, T., Yang, G., Chen, S. F., Morita-Fujimura, Y., and Chan P. H. (1999) Cold-injury in mice: A model to study mechanisms of brain edema and neuronal apoptosis. *Prog. Neurobiol.* **57,** 289–299.

93. Nag, S., Picard, P., and Stewart, D. J. (2001) Expression of nitric oxide synthases and nitrotyrosine during blood-brain barrier breakdown and repair following cold-injury. *Lab. Invest.* **81,** 41–49.

94. Nag, S., Takahashi, J. T., and Kilty, D. (1997) Role of vascular endothelial growth factor in blood-brain barrier breakdown and angiogenesis in brain trauma. *J. Neuropathol. Exp. Neurol.* **56,** 912–921.

95. Nag, S., Eskandarian, M. R., Davis, J., and Stewart, D. J. (2002) Differential expression of vascular endothelial growth factor A and B after brain injury. *J. Neuropathol. Exp. Neurol.* **61,** 778–788.

96. Nag, S. (1984) Cerebral endothelial surface charge in hypertension. *Acta Neuropathol. (Berl.)* **63,** 276–281.

97. Vorbrodt, A. W. (1993) Morphological evidence of the functional polarization of brain microvascular endothelium. In *The Blood-Brain Barrier. Cellular and Molecular Biology.* Pardridge, W. M., ed., Raven Press, New York, pp. 137–164.

98. Hardebo, J. E., and Kåhrström, J. Endothelial negative charge areas and blood-brain barrier function. *Acta Physiol. Scand.* **125,** 495–499.

99. Nagy Z., Peters H., and Hüttner I. (1983) Charge-related alterations of the cerebral endothelium. *Lab. Invest.* **49,** 662–671.

100. Nag, S. (1985) Ultrastructural localization of lectin receptors on cerebral endothelium. *Acta Neuropathol. (Berl.)* **66,** 105–110.

101. Vorbrodt, A. W. (1986) Changes in the distribution of endothelial surface glycoconjugates associated with altered permeability of brain micro-blood vessels. *Acta Neuropathol. (Berl.)* **70,** 103–111.

102. Bickel, U., Yoshikawa, T., and Pardridge, W. M. (2001) Delivery of peptides and proteins through the blood-brain barrier. *Adv. Drug Delivery Rev.* **46,** 247–279.

103. Nag S., Robertson, D. M., and Dinsdale, H. B. (1978) Cytoplasmic filaments in intracerebral cortical vessels. *Ann. Neurol.* **3,** 555–559.

104. Pardridge, W. M., Nowlin, D. M., Choi, T. B., Yang, J., Calaycay, J and Shively, J. E. (1989) Brain capillary 46,000 dalton protein is cytoplasmic actin and is localized to endothelial plasma membrane. *J. Cereb. Blood Flow Metab.* **9,** 675–680.

105. Bentzel, C. J., Hainau, B., Edelman, A., Anagnodtopoulos, T., and Benedetti, E. L. (1976) Effect of plant cytokinins on microfilaments and tight junction permeability. *Nature* **264,** 666–668.

106. Stevenson, B. R., and Beggs, D. A. (1994) Concentration-dependent effects of cytochalasin D on tight junctions and actin filaments in MDCK epithelial cells. *J Cell Sci.* **107,** 367–375.

107. Mark, K. S., and Davis, T. P. (2002) Cerebral microvascular changes in permeability and tight junctions induced by hypoxia-reoxygenation. *Am J. Physiol. Heart Circ. Physiol.* **282,** H1485–H1494.

108. Allen, R. D., Weiss, P. S., Hayden, J. H., Brown, D. T., Fijiwake, H., and Simpson, M. (1985) Gliding movement of a bidirectional organelle transport along single native microtubules from squid axoplasm: Evidence for an active role of microtubules in cytoplasmic transport. *J. Cell Biol.* **100,** 1736–1752.

109. Schnapp, B. J., Vale, R. D., Sheetz, N. P., and Reese, T. S. (1985) Single microtubules from squid axon support bidirectional movement of organelles. *Cell* **40,** 455–462.

110. Lui, S. M., Magnusson, K-E, and Sundqvist, T. (1993) Microtubules are involved in transport of macromolecules by vesicles in cultured bovine aortic endothelial cells. *J. Physiol.* **156,** 311–316.

111. Nag, S. (1990). Ultracytochemical localisation of Na$^+$, K$^+$-ATPase in cerebral endothelium in acute hypertension. *Acta Neuropathol. (Berl.)* **80,** 7–11.

112. Nag, S. (1993) Cerebral endothelial mechanisms in increased permeability in chronic hypertension. *Adv. Exp. Med. Biol.* **331,** 263–266.

113. Nag, S. (1991) Protective effect of flunarizine on blood-brain barrier permeability alterations in acutely hypertensive rats. *Stroke* **22,** 1265–1269.

114. Gumbiner, B. M. (1996) Cell adhesion: The molecular basis of tissue architecture and morphogenesis. *Cell* **84,** 345–357.

115. Nigam, S. K., Rodriguez-Boulan, E., and Silver, R. B. (1992) Changes in intracellular calcium during the development of epithelial polarity and junctions. *Proc. Natl. Acad. Sci. U.S.A.* **89,** 6162–6166.

116. Stuart, R. O., Sun, A., Panichas, M., Hebert, S. C., Brenner, B. M., and Nigam, S. K. (1994) Critical role for intracellular calcium in tight junction biogenesis. *J. Cell Physiol.* **159,** 423–433.

117. Stuart, R. O., Sun, A., Bush, K. T., and Nigam, S. K. (1996) Dependence of epithelial intercellular junction biogenesis on thapsigargin-sensitive intracellular calcium stores. *J. Biol. Chem.* **271,** 13636–13641.

118. Abbott, N. J. (1998) Role of intracellular calcium in regulation of brain endothelial permeability. In: *Introduction to the Blood-Brain Barrier. Methodology, Biology and*

Pathology. Pardridge, W. M., ed. Cambridge University Press, Cambridge, UK. pp. 345–351.

119. Abbott, N. J. (2000) Inflammatory mediators and modulation of blood-brain barrier permeability. *Cell. Mol. Neurobiol.* **20,** 131–147.

120. Brown, R. C., and Davis, T. P. (2002) Calcium modulation of adherens and tight junction function. *Stroke* **33,** 1706–1711.

121. Abbruscato, T. J., and Davis, T. P. (1999) Combination of hypoxia/aglycemia compromises in vitro blood-brain barrier integrity. *J. Pharmacol. Exp. Ther.* **289,** 668–675.

122. Fabry, Z., Fitzsimmons, K. M., Herlein, J. A., Moninger, T. O., Dobbs, M. B., and Hart, M. N. (1993) Production of cytokines interleukin 1 and 6 by murine brain microvessel endothelium and smooth muscle pericytes. *J. Neuroimmunol.* **47,** 23–34.

123. Frigerio, S., Gelati, M., Ciusani, E., Corsini, E., Dufour, A., Massa, G., and Salmaggi, A. (1998) Immunocompetance of human microvascular brain endothelial cells: cytokine regulation of IL-1beta, MCP-1, IL-10, sICAM-1, and sVCAM-1. *J. Neurol.* **245,** 727–730.

124. Zhang, W., Smith, C., Howlett, C., and Stanimirovic, A. (2000) Inflammatory activation of human brain endothelial cells by hypoxic astrocytes in vitro is mediated by IL-1beta. *J. Cereb. Blood Flow Metab.* **20,** 967–978.

125. Stanimirovic, D., and Satoh, K. (2000) Inflammatory mediators of cerebral endothelium: a role in ischemic brain inflammation. *Brain Pathol.* **10,** 113–126.

126. Anthony, D. C., Miller, K. M., Fearn, S., et al. (1998) Matrix metalloproteinase expression in an experimentally-induced DTH model of multiple sclerosis in the rat CNS. *J. Neuroimmunol.* **87,** 62–72.

127. De Vries, H. E., Blom-Roosemalen, M. C. M., Van Oosten, M., et al. (1996) The influence of cytokines on the integrity of the blood-brain barrier in vitro. *J. Neuroimmunol.* **64,** 37–43.

128. Claudio, L., Kress, Y., Norton, W. T., and Brosnan, C. F. (1989) Increased vesicular transport and decreased mitochondrial content in blood-brain barrier endothelial cells during experimental autoimmune encephalomyelitis. *Am. J. Pathol.* **135,** 1157–1168.

129. Lou, J., Chofflon, M., Juillard, C., et al. (1997) Brain microvascular endothelial cells and leukocytes derived from patients with multiple sclerosis exhibit increased adhesion capacity. *Neuroreport* **8,** 629–633.

130. Raine, C. S., Lee, S. C., Scheinberg, L. C., Duijvestijn, A. M., and Cross, A. H. (1990) Short analytical review: adhesion molecules on endothelial cells in the central nervous system. An emerging area in the neuroimmunology of multiple sclerosis. *Clin. Immunol. Immunopathol.* **57,** 173–187.

131. Wisniewski, H. M., and Lossinsky, A. S. (1991) Structural and functional aspects of the interaction of inflammatory cells with the blood-brain barrier in experimental brain inflammation. *Brain Pathol.* **1,** 89–96.

132. Wekerle, H., Engelhardt, B., Risau, W., and Meyerman, R. (1991) Interaction of T lymphocytes with cerebral endothelial cells in vitro. *Brain Pathol.* **1,** 107–114.

133. Wong, D., Prameya, R., and Dorovini-Zis, K. (1999) In vitro adhesion and migration of T lymphocytes across monolayers of human brain microvessel endothelial cells: regulation by ICAM-1, VCAM-1, E-selectin and PECAM-1. *J. Neuropathol. Exp. Neurol.* **58,** 138–152.

134. Burns, A. R., Walker, D. C., Brown, E. S., et al. (1997) Neutrophil transendothelial migration is independent of tight junctions and occurs preferentially at tricellular corners. *J. Immunol.* **159,** 2893–2903.

135. Burns, A. R., Bowden, R. A., MacDonell, S. D., et al. (2000) Analysis of tight junctions during neutrophil transendothelial migration. *J. Cell Sci.* **113,** 45–57.

136. Fabry, Z., Topham, D. J., Fee, D., et al. (1995) TGF-beta 2 decreases migration of lymphocytes in vitro and homing of cells into the central nervous system in vivo. *J. Immunol.* **155,** 325–332.

137. Prat, A., Biernacki, K., Wosik, K., and Antel, J. (2001) Glial cell influence on the human blood-brain barrier. *Glia* **36,** 145–155.
138. Zhang, Z., and Chopp, M. (2002) Vascular endothelial growth factor and angiopoietins in focal cerebral ischemia. *Trends Cardiovasc. Med.* **12,** 62–66.
139. Beck, H., Acker, T., Wiessner, C., Allergrini, P. R., and Plate, K. (2000) Expression of Angiopoietin-1 and Angiopoietin-2, and Tie-2 receptors after middle cerebral artery occlusion in the rat. *Am. J. Pathol.* **157,** 1473–1483.
140. Hayashi T., Abe, K., Suzuki, H., and Itoyama, Y. (1997) Rapid induction of vascular endothelial growth factor gene expression after transient middle cerebral artery occlusion in rats. *Stroke* **28,** 2039–2044.
141. Mandriota, S. J., Pyke, C., Di Sanza, C., Quindoz, P., Pittet, B., and Pepper, M. S. (2000) Hypoxia-inducible angiopoietin-2 expression is mimicked by iodonium compounds and occurs in the rat brain and skin in response to systemic hypoxia and tissue ischemia. *Am. J. Pathol.* **156,** 2077–2089.
142. Pichiule, P., Chávez, J. C., Xu, K., and LaManna, J. C. (1999) Vascular endothelial growth factor upregulation in transient global ischemia induced by cardiac arrest and resuscitation in rat brain. *Mol. Brain Res.* **74,** 83–90.
143. Provias, J. P., Claffey, K., delAguila, L., Lau, N., Feldkamp, M., and Guha, A. (1997) Meningiomas: Role of vascular endothelial growth factor/vascular permeability factor in angiogenesis and peritumoral edema. *Neurosurgery* **40,** 1016–1026.
144. Goldman, C. K., Bharara, S., Palmer, C. A., Vitek, J., Tsai, J.-C., Weiss, H. L., and Gillespie, G. Y. (1997) Brain edema in meningiomas is associated with increased vascular endothelial growth factor expression. *Neurosurgery* **40,** 1269–1277.
145. Dobrogowska, D. H., Lossinsky, A. S., Tarnawski, M., and Vorbrodt, A. W. (1998) Increased blood-brain barrier permeability abnormalities induced by vascular endothelial growth factor. *J. Neurocytol.* **27,** 163–173.
146. Proescholdt, M. A., Heiss, J. D., Walbridge, S., et al. (1999) Vascular endothelial growth factor (VEGF) modulates vascular permeability and inflammation in rat brain. *J. Neuropathol. Exp. Neurol.* **58,** 613–627.
147. van Bruggen, N., Thibodeaux, H., Palmer, J. T., et al. (1997) VEGF antagonism reduces edema formation and tissue damage after ischemia/reperfusion injury in the mouse brain. *J. Clin. Invest.* **104,** 1613–1620.
148. Zhang, Z. G., Zhang, L., Jiang, Q., et al. (2000) VEGF enhances angiogenesis and promotes BBB breakdown leakage in the ischemic brain. *J. Clin. Invest.* **106,** 829–838.
149. Bates, D. O., Hillman, N. J., Williams, B., Neal, C. R., and Pocock, T. M. (2002) Regulation of microvascular permeability by vascular endothelial growth factors. *J. Anat.* **200,** 581–597.
150. Hofman, P., Blaauwgeers, G. T., Tolentino, M. J., et al. (2000) VEGF-A induced hyperpermeability of blood-retinal barrier endothelium in vivo is predominantly associated with pinocytotic vesicular transport and not with formation of fenestrations. *Curr. Eye Res.* **21,** 637–645.
151. Antonetti, D. A., Barber, A. J., Hollinger, L. A., Wolpert, E. B., and Gardner, T. W. (1999) Vascular endothelial growth factor induces rapid phosphorylation of tight junction proteins occludin and zonula occludins 1. A potential mechanism for vascular permeability in diabetic retinopathy and tumors. *J. Biol. Chem.* **274,** 23463–23467.
152. Antonetti, D. A., Barber, A. J., Khin, S., Lieth, E., Tarbell, J. M., and Gardner, T. W. (1998) Vascular permeability in experimental diabetes is associated with reduced endothelial occludin content: Vascular endothelial growth factor decreases occludin in retinal endothelial cells. Penn State Retina Research group. *Diabetes* **47,** 1953–1959.

153. Wang, W., Dentler, W. L., and Borchardt, R. T. (2001) VEGF increases BMEC monolayer permeability by affecting occludin expression and tight junction assembly. *Am. J. Physiol. Heart Circ. Physiol.* **280,** H434–H440.

154. Long, D. M. (1970) Capillary ultrastructure and the blood-brain barrier in human malignant brain tumors. *J. Neurosurg.* **32,** 127–144.

155. Weindel, K., Moringlane, J. R., Marmé, D., and Weich, H. A. (1994) Detection and quantification of vascular endothelial growth factor/vascular permeability factor in brain tumor tissue and cyst fluid. *Neurosurgery* **35,** 439–449.

156. Davies, D. C. (2002) Blood-brain barrier breakdown in septic encephalopathy and brain tumours. *J. Anat.* **200,** 639–646.

157. Maisonpierre, P. C., Suri, C., Jones, P. F., et al. (1997) Angiopoietin-2, a natural antagonist for Tie-2 that disrupts in vivo angiogenesis. *Science* **277,** 55–60.

158. Thurston, G., Suri, C., Smith, K., et al. (1999) Leakage resistant blood vessels in mice transgenically over-expressing angiopoietin-1. *Science* **286,** 2511–2514.

159. Nag, S., Nourhaghighi, N., Teichert-Kuliszewska, K., Davis, J., Papneja, T., and Stewart, D. J. (2002c) Altered expression of angiopoietins during blood-brain barrier breakdown and angiogenesis. *J. Neuropathol. Exp. Neurol.* **61,** 453.

160. Nourhaghighi, N., Stewart, D. J., and Nag, S. (2000) Immunohistochemical characterization of Angiopoietin-1, Angiopoietin-2, and Tie-2 in an in vivo model of cerebral trauma and angiogenesis. *Brain Pathol.* **10,** 555.

161. Zhang, Z. G., Zhang, L., Croll, S. D., and Chopp, M. (2002) Angiopoietin-1 reduces cerebral blood vessel leakage and ischemic lesion volume after focal cerebral embolic ischemia in mice. *Neuroscience* **113,** 683–687.

162. Gamble, J., Drew, J., Trezise, L., et al. (2000) Angiopoietin-1 is an antipermeability and anti-inflammatory agent in vitro and targets cell junctions. *Circ. Res.* **87,** 603–607.

7

Pial Vessel Permeability to Tracers Using Cranial Windows

William G. Mayhan

1. Introduction

Many investigators, including numerous studies from our laboratory, have used fluorescence microscopy to directly examine changes in permeability of pial vessels in rats (1–4), cats (5–7), mice (8,9), and hamsters (10–12) during basal and stimulated states. The use of cranial windows and fluorescence microscopy to examine the cerebral microcirculation has several advantages. First, the techniques allow the investigator to directly visualize cerebral microvessels. Thus, the investigator can visually examine changes in permeability of pial vessels during various stimuli and can visualize changes in reactivity of cerebral blood vessels during alterations in permeability of the blood–brain barrier (BBB). This allows the investigator to correlate changes in permeability of pial vessels with local cerebral hemodynamics. Second, the techniques allow the investigator to examine changes in permeability of pial vessels to differently sized and charged molecules. Thus, one can examine the effect of a specific stimuli on the selectivity of pial vessels. Third, the techniques allow the investigator to determine the precise sites, i.e., arteries, arterioles, capillaries, venules and/or veins, of disruption of the BBB during pathophysiologic conditions. This information may be of potential importance in the delivery of therapeutic agents to brain tissue during disease states. Fourth, the techniques allow the investigator to not only qualitatively, but quantitatively evaluate the permeability of pial vessels under basal conditions and during various conditions. With the calculation of clearance, the investigator can evaluate basal permeability of pial vessels and can determine the magnitude of disruption of the BBB in response to various stimuli. This chapter will focus on the materials and methods required to view and measure permeability of the pial microcirculation using fluorescent tracers.

2. Materials

1. Artificial cerebrospinal fluid (CSF): Potassium chloride (0.22 g/L), magnesium chloride (0.132 g/L), calcium chloride (0.22 g/L), sodium chloride (7.7 g/L), sodium bicarbonate (2.06 g/L), and glucose (0.55 g/L). The artificial cerebral spinal fluid is prepared in volume (20 L) on a weekly basis using milli-Q water.

From: *Methods in Molecular Medicine, vol. 89:*
The Blood–Brain Barrier: Biology and Research Protocols
Edited by: S. Nag © Humana Press Inc., Totowa, NJ

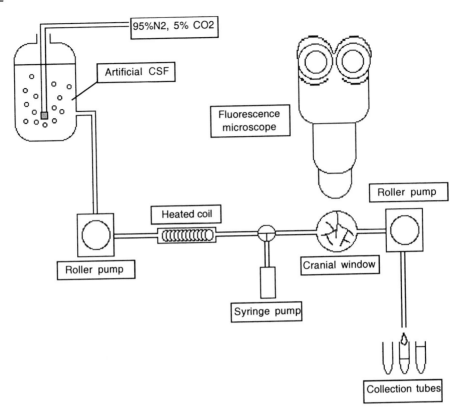

Fig. 1. Schematic of the artificial cerebrospinal fluid (CSF) suffusion apparatus. The artificial CSF is bubbled with 95% nitrogen and 5% carbon dioxide. The suffusate is then pumped through a heating coil and into the cranial window. The presence of a three-way valve also allows for the infusion of agents into the CSF and into the cranial window. A second roller pump is used to withdraw CSF from the cranial window and into collection tubes for later determination of fluorescence intensity.

2. The suffusion apparatus (**Fig. 1**) for artificial CSF consists of:
   a. An aspirator bottle (2 L; Fisher Scientific; Pittsburgh, PA) connected via polyethylene tubing to a custom made glass-jacketed heating coil (manufactured in the Department of Chemistry; University of Iowa). This glass-jacketed coil is connected to a heated circulating water bath (Fisher Scientific; Pittsburgh, PA).
   b. The outflow of suffusate from the heating coil is connected via polyethylene tubing to a variable speed roller pump (Mini Plus 2; Gilson Medical Electronics, Inc.; Middleton, WI).
   c. The polyethylene tubing from the roller pump is then connected to a three-way valve (model HV 3-3; Hamilton Co., Reno, NV).
   d. The outflow from the three-way valve is PE-90 tubing that is connected to the cranial window via a 20-gage needle and to a syringe pump that allows for the infusion of agents into the artificial CSF, and thus into the cranial window.
   e. The total volume of artificial CSF contained within the suffusion apparatus (independent of the aspirator bottle) is approx 1–2 mL. This small volume ensures a rapid turnover of artificial CSF in the cranial window and is important for maintaining gases and temperature constant within the cranial window.
   f. The outflow from the cranial window is connected, via a 20-gage needle and PE-90 tubing, to a variable speed roller pump (Cole-Parmer Inc.; Chicago, IL). The outflow tubing can be connected to a fractionator for the constant collection of artificial CSF.

3. Thiobutabarbital sodium (Inactin; Research Biochemical International [RBI]; Natick, MA).
4. Wistar-Furth rats (*see* **Note 1**; 200–400 g; Harlan, Inc.; Indianapolis, IN).
5. Polyethylene tubing (Becton Dickinson and Co.; Franklin Lakes, NJ): PE-260 to make the trachea cannula, and PE-90, and PE-50 tubing for arterial and venous cannulas.
6. Rat ventilator (Harvard Apparatus; Holliston, MA).
7. Body temperature control unit (Cole-Parmer; Chicago, IL).
8. Rat head holder (David Kopf, Inst.; Tujunga, CA).
9. Dremel motor tool (Dremel; Racine, WI).
10. High-speed carbide dental burs (Biomedical Research Inst.; Bethesda, MD).
11. Micro-dissecting scissors (Biomedical Research Inst.; Bethesda, MD).
12. Micro-dissecting forceps (Biomedical Research Inst.; Bethesda, MD).
13. Dura hook (30-gage needle; Becton Dickinson and Co.; Franklin Lakes, NJ).
14. Fluorescein isothiocyanate-dextran (FITC-dextran) of various molecular weights: FITC-dextran-70K, -40K, -20K, -10K, -4K (Sigma Chemical Co.; St. Louis, MO), FITC-albumin and sodium fluorescein (Sigma Chemical, St. Louis, MO). FITC-dextrans and FITC-albumin are dissolved in saline (40 mg/mL) and are made up on a weekly basis. Sodium fluorescein is prepared as a 1% solution in saline and is made up on a weekly basis. The fluorescent tracers are light and temperature sensitive, and thus are refrigerated in amber bottles that are covered with aluminum foil.
15. Variable speed syringe pump (Sage syringe pump; Orion Research, Inc.; Boston, MA) for the intravenous infusion of fluorescent tracers.
16. Fluorescence microscope (Leitz stereo-zoom microscope with a 75 W xenon vertical illuminator (North Central Instr.; Minneapolis, MN).
17. PowerLab (ADInstruments).
18. Fraction collector (Gilson Medical Electronics, Inc.; Middleton, WI) to collect suffusate samples.
19. Spectrophotofluorometer (LS-30; Perkin Elmer; Norwalk, CT) to measure suffusate and plasma concentration of the fluorescent tracers.

3. Methods

The methods described below outline 1) the suffusion of the cranial window preparation with artificial CSF, 2) the preparation of the cranial window to view the pial microcirculation, and 3) the use of fluorescence microscopy to evaluate the permeability characteristics of pial vessels.

3.1. Artificial CSF Suffusion

1. On the day of the experiment, sodium bicarbonate (2.06 g/L) and glucose (0.55 g/L) are added to 2 L of artificial CSF contained within an aspirator bottle.
2. The aspirator bottle is attached to the suffusion apparatus and the artificial CSF is bubbled continuously with a mixture of nitrogen (95%) and carbon dioxide (5%) at least 30–60 min before preparation of the cranial window (*see* **Fig. 1**). We have shown previously *(2)*, that this procedure maintains gases (PO_2 and PCO_2) and pH of the artificial CSF within physiologic limits.
3. Once the craniotomy is prepared, the artificial CSF is suffused across the pial microcirculation at a rate of approx 2 mL/min, which maintains the temperature of the suffusate in the cranial window preparation at $37 \pm 1°$ C.

3.2. Preparation of the Animal

1. Rats are anesthetized with Inactin (thiobutabarbital sodium; 100 mg/kg intraperitoneally). Inactin produces sustained anesthesia in rats *(13)*, and thus systemic and cerebral vascular hemodynamics remain constant during the experimental protocol.

2. After the animal is anesthetized, it is placed on a heated stage to maintain body temperature within physiologic limits.
3. A tracheotomy is performed (PE-260 tubing), and the rat is mechanically ventilated (tidal volume of 1 mL/100 g body weight and ventilatory rate of 45–50 breaths/min) with room air and supplemental oxygen to maintain arterial PO_2, PCO_2, and pH within normal limits.
4. A cut-down is performed over the left femoral area, and the femoral artery and vein are exposed and isolated using 3-0 suture. A modest amount of tension is placed on the femoral artery and vein using the sutures.
5. A small incision is then made in the femoral artery and vein using the microdissecting scissors. A cannula consisting of PE-90 tubing with a tip of PE-50 tubing (approx 4 cm in length) is filled with saline, and the PE-50 tubing is inserted into the femoral artery or vein with the aid of microdissecting forceps.
6. The femoral artery cannula is connected to a pressure transducer for the measurement of systemic blood pressure, mean arterial pressure and heart rate using a PowerLab and for obtaining blood samples for the measurement of plasma concentration of fluorescent tracer.
7. The femoral vein cannula is used for injection of the fluorescent tracer.
8. Following these initial procedures, the rat is then positioned in a head holder in preparation for the craniotomy.

3.2.1. Cranial Window

1. To visualize the pial microcirculation, a cranial window is prepared, as we have described previously (2). The cranial window is positioned over the parietal cortex and is enclosed by the sagittal, coronal, and interparietal suture lines. The constant position of the window in this area assures that we are examining the permeability of pial vessels and reactivity of pial arterioles in similar regions in all experiments.
2. The preparation of the cranial window requires several steps. First, a small incision is made in the skin with a no. 10 scalpel blade to expose the skull.
3. Second, the skin is retracted with 3-0 sutures and allows the skin to serve as a "well" for the suffusion fluid (**Fig. 2**).
4. Third, an inlet port and an outlet port (20-gage needle hubs) are made at opposite ends of the cranial window (**Fig. 3**). These ports allow for the constant flow of suffusate across the pial microcirculation.
5. Fourth, a craniotomy is performed using a dental burr attached to an air-cooled dremel tool (10–12,000 rpm) (**Fig. 4**). To minimize the transfer of heat from the dental burr to the skull/brain, extreme care is taken so that the dental burr does not remain in contact with one specific area of the skull for longer than 1–3 s (*see* **Note 2**). In addition, to prevent excessive bleeding, care is taken so that no drilling of the skull is performed over the suture lines. Drilling continues until small cracks are observed in the skull. At this point, the cerebral microcirculation can be viewed through the translucent bone.
6. Then, the piece of skull is carefully removed using microforceps, exposing the dura with the underlying pial microcirculation.
7. The suffusion of artificial CSF over the cranial window preparation is now started.
8. Twenty to thirty minutes after starting the suffusion of artificial CSF, an incision is made in the dura using a dura hook. This hook is fashioned from a 30-gage needle that is bent at an angle of approx 90° using a needle holder.
9. After an incision is made in the dura, the dura is opened using microdissecting scissors to expose the pial microcirculation (**Fig. 5**).
10. Sections of the dura are then cut away for better viewing of the pial microcirculation (*see* **Fig. 5**). To quantitate the permeability of pial vessels using fluorescent tracers, it is critical that there be no bleeding from either the bone or dura. Bleeding from the skull is usu-

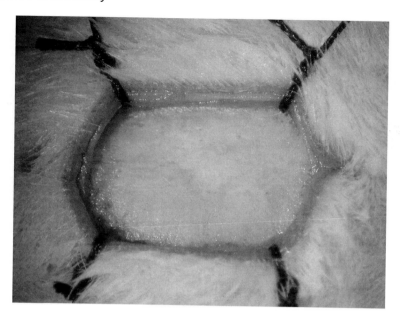

Fig. 2. Photograph of the cranial window. Sutures placed in the skin allow for the formation of a "well" that contains the cerebrospinal fluid.

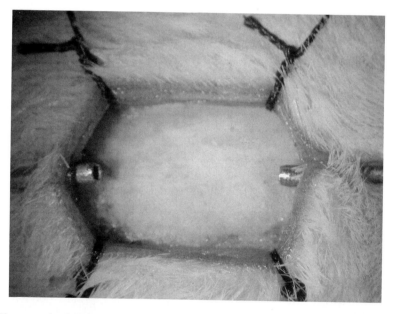

Fig. 3. Photograph of the cranial window after insertion of inflow and outflow ports. These ports allow for the continuous infusion of cerebrospinal fluid across the cerebral microcirculation.

ally not a major concern because drilling rarely produces bleeding. However, bleeding from the dura can be, at times, difficult to control. Normally, bleeding from the dura can be stopped by grasping the dura with microdissecting forceps and applying a "pinching" pressure for several seconds (*see* **Note 3**). Care must be taken so that the dissecting forceps do not touch the exposed brain surface.

2. The suffusate fluid is collected in glass test tubes at predetermined intervals (usually every 4–5 min) with the aid of a fraction collector.

3. Arterial blood samples (approximately 60 µL/sample) also are drawn at various times (usually every 20–30 min) throughout the experimental period.

4. Although there are no "standard" scenarios for protocols used to evaluate the permeability characteristics of pial vessels, there are several key ingredients.
   a. Several control samples of artificial CSF must be collected before injection of fluorescent tracer. These control samples serve as background and are subtracted from all subsequent samples.
   b. An adequate number of samples must be collected before stimulation of the BBB. This gives the investigator an index of "basal" levels of permeability before stimulation.
   c. Adequate samples must be collected in order to observe a profile for the effects of a stimulus on the permeability of pial vessels.
   d. Time control experiments must be completed to evaluate a temporal relationship in changes in permeability of pial vessels. With these considerations in mind, one can then design appropriate protocols to meaningfully evaluate changes in permeability of pial vessels during various conditions.

5. Once the suffusate and plasma samples are collected, the investigator can quantitate the concentration of fluorescent tracer in these samples using a spectrophotofluorometer.

6. First, a standard curve for concentration of fluorescent tracer vs percent transmission is obtained with the aid of a spectrophotofluorometer. The standards are prepared on a weight-per-volume basis, i.e., 1, 5, 10, 50, and 100 ng/mL. A standard curve is generated for each experiment and each curve is subjected to linear regression analysis.

7. The percent transmission for unknown samples (suffusate and plasma) are then measured on the spectrophotofluorometer and the concentration is calculated from the standard curve. The suffusate samples usually are in the ng/mL range and do not require dilution. However, because plasma samples are in the mg/mL range they must be diluted before the measurement of fluorescent intensity. By using linear regression analysis, one can calculate the plasma concentration of fluorescent tracer at each interval. Finally, clearance of fluorescent tracer at every time interval is calculated by multiplying the ratio of suffusate-to-plasma concentration by the suffusate flow rate that is kept constant during the experimental period (2 mL/min) (*2,17,18*).

8. The calculation of clearance is given as follows:

$$CLR = ([FT]s \times [FT]p) \times Qs$$

where CLR is clearance of fluorescent tracer (mL/s $\times$ 10^{-6}), [FT]s is the fluorescent tracer concentration in the suffusate (ng/mL), [FT]p is the fluorescent tracer concentration in the plasma (mg/mL), and Qs is the suffusate flow rate (mL/min). Because the concentration of fluorescent tracer in the suffusate ([FT]s) is in ng/mL and the concentration of fluorescent tracer in the plasma ([FT]p) is in mg/mL, the product of this ratio is standardized to ng/mL. Given that the suffusate flow rate (Qs) is given as mL/min, the units of clearance (CLR) must be divided by 60 s/min to report CLR as mL/s $\times$ 10^{-6}.

Thus, using these methods, an investigator can visualize and quantitate the permeability characteristics of pial vessels in a variety of animal models under numerous physiologic and pathophysiologic conditions.

4. Notes

1. Rats: To examine the permeability of pial vessels using FITC-dextrans, we initially used Sprague-Dawley rats (*2*). However, because these rats are anaphylactic to dextran, they had

to be injected with antihistamines before the infusion of FITC-dextran. Because the cerebral circulation contains many mast cells and the degranulation of these structures may influence permeability of pial vessels during various stimuli, we elected to switch our model to the Wistar-Furth rat. These rats respond similarly to vasoactive agents as other strains of rats, but do not exhibit anaphylaxis to dextran and thus are not influenced by the potential effects of antihistamines.

2. Damage during drilling: Because the brain is in proximity to the skull, it is possible that the drilling process could produce damage to the brain. To prevent this damage, one could use controlled hemorrhage to decrease mean arterial pressure, and thus intracranial pressure. This would produce an increase in the space separating the brain and the skull. We have used controlled hemorrhage (withdrawal of approx 0.5–1 mL/100 g body weight of blood) to prevent damage to the brain during the drilling procedure. In addition, the use of controlled hemorrhage aids in decreasing the amount of bleeding from the dura following the craniotomy. The decrease in blood pressure observed during controlled hemorrhage (approx 20–30 mm Hg) usually recovers within 5–15 min.

3. Bleeding: The skin is very vascular and bleeds following the initial incision to expose the skull. Because we do not want any fluorescent tracer to leak from the skin and contaminate the readings obtained from the pial microcirculation, it is important to stop all bleeding. This bleeding can be stopped by using ferric chloride. Thus, after the initial skin incision, ferric chloride is painted onto the skin and skull to stop all bleeding. Once the "well" is formed by the skin, the area is washed extensively with artificial CSF for at least 30 min to remove all ferric chloride from this area before making the cranial window.

4. In some instances, bleeding from the dura can not be alleviated by a "pinching" pressure applied by the microdissecting forceps. Over the years, we have used many procedures to attempt to stop bleeding from the dura, including vasoconstrictor agents (thromboxane, endothelin, norepinephrine, etc.), a coagulating agent (thrombin), and electrocauterizing forceps. Vasoconstrictor agents are not useful. These agents fail to adequately control bleeding from the dura and they have direct effects on pial blood vessels, which may be detrimental to the experimental protocol. The use of thrombin has been modestly successful in stopping bleeding from the dura. However, if thrombin is left on the dura/brain for an extended period, it damages the pial microcirculation by coagulating blood within pial vessels. In addition, thrombin has been shown to increase the permeability of the BBB, and thus its use is a problem. The use of electrocauterization can produce damage to the pial microcirculation. Because the dura is very thin, the electrocautery can actually cut the dura and produce more bleeding. In addition, the electrocautery can produce irreversible dilation of pial arterioles, presumably via heat generated and/or the formation of reactive oxygen radicals. Thus, these methods have not been successful in aiding in the control of bleeding from the dura.

Careful use of very small amounts of ferric chloride can prevent bleeding from the dura. Once the dura has been opened and we have "pinched" bleeders with the microdissecting forceps, we stop the flow of suffusate from the pial microcirculation and carefully apply a small, but concentrated solution of ferric chloride to specific sites on the dura from which bleeding has occurred. Thus, neither the dura nor the pial microcirculation is bathed with ferric chloride, but rather very specific sites of the dura are exposed to this agent. The ferric chloride is applied to the dura via a finely drawn tip of a kim wipe. This method allows the investigator to apply ferric chloride to very precise areas on the dura and prevents the ferric chloride from coming into contact with the brain. Once the dura has been treated with ferric chloride, it is carefully removed and the flow of suffusate is re-started. During this procedure, the brain is only devoid of suffusate for less than 30 s. This procedure can be repeated

8

Blood–Brain Barrier Permeability Using Tracers and Immunohistochemistry

Sukriti Nag

1. Introduction

The first report of blood–brain barrier (BBB) permeability by Paul Ehrlich *(1)* involved the use of the exogenous tracer Coerulean-S as described in Chapter 6. Over the years, tracers of different sizes were introduced to study the permeability properties of normal cerebral vessels in physiologic and pathologic states (**Table 1**). Tracers provide information about the permeability status of vessels immediately before sacrifice. The disadvantage of exogenous tracers is that there are side effects associated with the administration of some tracers in live animals. The properties and methods by which some of these tracers are used to study BBB permeability to proteins and ions in pathologic states will be described.

Endogenous protein extravasation into brain can be detected by immunohistochemistry. Antibodies to circulating proteins such as albumin, serum proteins, fibrinogen, or fibronectin have been used to detect BBB permeability *(2,3)*. The advantage of immunohistochemistry is that it not only provides information about the status of vessels at the time of sacrifice *(2,3)* but it also detects areas of previous protein extravasation in brain and the associated inflammatory reaction can be used to date these areas of protein extravasation *(2)*. Dual labeling can also be done to identify a protein of interest in a vessel with BBB breakdown to circulating proteins *(4–6)*.

2. Materials

1. PE50 polyethylene tubing, syringes: 1 mL, 5 mL
2. 250 units of heparin per 100 mL of sterile 0.9% sodium chloride for syringes and polyethylene cannulas.
3. Tracers:
   a. 2% Evans Blue dissolved in sterile 0.9% sodium chloride.
   b. Horseradish peroxidase, type II (Sigma Chem. Co., St Louis, MO) dissolved in sterile 0.9% sodium chloride. (*See* **Note 1.**)
   c. Microperoxidase (Sigma Chemical, St Louis, MO).
   d. Iodine-125–labeled serum albumin (RISA) (ICN Biomedicals, Canada Ltd., Montreal, Quebec).
   e. 5 m*M* Lanthanum chloride (Sigma Chemical, St Louis, MO) dissolved in sterile 0.9% sodium chloride.

From: *Methods in Molecular Medicine, vol. 89:*
The Blood–Brain Barrier: Biology and Research Protocols
Edited by: S. Nag © Humana Press Inc., Totowa, NJ

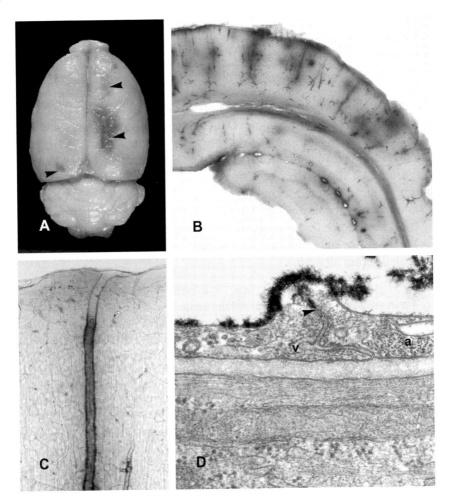

Fig. 1. (A), Brain of a rat with chronic hypertension showing areas of Evans blue extravasation in the boundary zone areas (*arrowheads*). (B), Fixed slice of rat brain showing multifocal areas of HRP extravasation following intracarotid administration of cytochalasin B. (C), High magnification shows an arteriole with BBB breakdown to HRP in norepinephrine-induced acute hypertension. (D), Segment of arteriolar wall from an intracerebral vessel showing lack of passage of ionic lanthanum through the tight junction (*arrowhead*). A plasmalemmal vesicle is seen to open into an interjunctional space (*v*). Note the cross sections of actin filaments (*a*) occupying the entire width of the endothelium and along the intercellular space. **A**, x3; **B**, x15; **C**, x100; **D**, x60,000.

ride and injected intravenously via a PE50 polyethylene cannula inserted into the femoral vein. A slow injection administered over 1–2 min prevents fluctuations in the blood pressure of rats (*see* **Note 5**).

4. Mice are injected with 10–20 mg of HRP dissolved in 0.1–0.4 mL of 0.9% sodium chloride via the tail vein.

5. The circulation period of HRP depends on the biologic process being studied. To detect how tracer reaches the vascular subendothelium very short circulation times such as 90 s may be sufficient (*17*). To determine how HRP is cleared from the extracellular spaces, longer circulation times are required.

6. Experiments are teminated by vascular perfusion of fixative into anesthetized rats as described in Chapter 2, Subheadings 3.1.1. and 3.1.3. (*see* **Notes 6–8**).
7. Brains are removed and placed in the same fixative for 2 h (*see* **Note 9**).
8. Brain blocks of the required areas are sectioned at 50-μm intervals using a tissue sectioner (Sorvall MT-2C or Vibratome). *See* **Note 10**.
9. Slices are collected in 0.1 *M* cacodylate buffer in a petri dish.
10. Sections are then processed for the demonstration of HRP reaction product as follows:
    a. Rinse slices in 1 change of Tris-HCl, which is added to the petri dish containing the slices and removed by pipetting.
    b. The substrate solution containing 0.06 % DAB in 0.06 *M* Tris-HCl, pH 8.0 is then added for 20 min and replaced by fresh substrate containing 0.02% H_2O_2 for 20 min. The DAB is discarded into a large beaker (*see* **Notes 11** and **12**).
    c. Slices are then rinsed with cacodylate buffer and transferred to fresh petri dishes containing the same buffer using bamboo splints.
    d. Large areas of HRP extravasation are visible as brown spots in slices. Examine slices under a stereoscope or light microscope for the presence of HRP reaction product in individual vessels (**Figs. 1B,C**).
    e. Selected slices are mounted in 60% glycerine. The edges of the coverslip are sealed with nail polish (*see* **Note 13**). These sections are ideal for photography.
    f. Positive controls: Residual red blood cells in the vascular system serve as a positive control as they turn brown due to their high peroxidase content. Vessels with endothelium having pores and fenestrae located in regions such as area postrema, the median eminence, and choroid plexuses, also stain brown. Pial arterioles located in the cerebellar sulci, superior saggital fissure and the sulcus between the olfactory bulb and cerebral hemisphere and few intracerebral vessels located in the entorhinal cortex are normally permeable to HRP *(17,18)*. Intracerebral cortical vessels are not permeable to HRP even after a circulation time of 60 min *(13)* nor are the apical tight junctions between normal choroidal epithelial cells *(19)*.
    g. Negative controls: Delete H_2O_2 or DAB from the substrate solution.
9. Appropriate sized blocks containing vessels permeable to HRP and nonpermeable vessels are then processed for electron microscopy as given in Chapter 2, Subheadings 3.2–3.6.
10. Electron micrographs are then obtained to assess findings.

3.2.1. Assessment of Results

1. Light microscopy: The number of permeable vessels and areas showing BBB breakdown in brain slices of test and control rats can be counted to determine the magnitude of BBB breakdown in test and control rats *(20)*.
2. Electron microscopy: Qualitative studies can be done to determine how tracer traverses endothelium to reach the endothelial basement membrane. Studies done in our laboratory have demonstrated an increase in endothelial plasmalemmal vesicles in arterioles of test rats showing BBB breakdown to HRP as compared to the nonpermeable vessels of control rats *(17,21)*. The method for quantitation of endothelial vesicles is as follows:
    a. An overlapping series of electron micrographs are taken along the circumference of nonpermeable arterioles or microvessels and those permeable to HRP at a screen magnification of x22,000.
    b. In case of arterioles, five vessels may be used for quantitation per group while in the case of microvessels a larger number are measured.
    c. Images from the negatives of electron micrographs placed on a light box are transmitted to an image analyser (Microcomputer Imaging Device, Imaging Research Inc., St. Catherines, On) by a high-resolution CCD camera.

d. The image analyzer is calibrated for area measurements in square micrometers. Because plasmalemmal vesicles have an uneven distribution along the circumference of endothelium we suggest obtaining the total area of endothelium excluding the area occupied by the nucleus.

e. The number of vesicles/μm^2 of endothelium for the vessel is then obtained. Set criteria for identification of endothelial plasmalemmal vesicles. These are roughly spherical structures measuring 75–80 nm in diameter (*see* Chapter 6, Fig. 2D). They are present in the endothelial cytoplasm singly or in small groups or chains. At the luminal aspect, only those vesicles with a neck in the process of being pinched off from the surface are counted. Within the endothelium only vesicles with a full profile and definite unit membranes are counted.

f. The mean vesicles/μm^2 of endothelium for vessels in the different experimental groups are compared simultaneously by the one way analysis of variance with the Tukey post hoc test (SPSS 11.0).

The mean vesicles/μm^2 of arteriolar endothelium in normal rat brain cortex is 5 ± 1 *(21)* and for microvessels is 10–14 *(22)*. The tracer HRP by itself does not alter the number of pinocytotic vesicles in arteriolar endothelium. Breakdown of the BBB to HRP in diverse experimental states is associated with a significant increase ($p < 0.001$) in the mean vesicles/μm^2 of arteriolar endothelium to 10 with 65% of vesicles containing tracer *(21)*.

3. Quantitative estimate of HRP: When HRP was first studied it was measured in tissue fractions using calorimetric techniques. Modification of a prevalent calorimetric method has also been used to determine HRP activity in brains of test and control rats *(23)*.

3.3. Microperoxidase

This tracer, prepared from cytochrome c, has a molecular weight of 1900 daltons and molecular diameter of 2 nm. It is localized in tissues at an ultrastructural level at the same dose and by the same methods as used for HRP with minor modifications. Despite its smaller molecular weight than HRP, it never gained popular usage for studies of BBB permeability perhaps due to its cost and because it did not provide any additional information as compared with HRP.

1. The tracer solution contains 0.5% microperoxidase in 0.154 *M* NaCl adjusted to pH 7.0 with 0.1 *N* NaOH. The tracer solution is prewarmed to 38°C immediately before use. It is injected in a dose of 1 mL/100 g body weight at a rate of 0.1 mL/sec.

2. 10–20 mg of microperoxidase has also been used in mice.

3. At the termination of the experiment, animals are fixed by vascular perfusion of fixative as described in Chapter 2, Subheadings 3.1.1–3.1.3.

4. The method for demonstration of microperoxidase in brain slices is the same as described for HRP in **Subheading 3.1.2.** with a minor modification.

5. The substrate contains 0.05% DAB and 0.1 *M* imidazole in 0.05 *M* Tris-HCl, pH 7.6. The final pH is adjusted to 8.8 using NaOH. Hydrogen peroxide 0.02% is added before use. Imidazole increases the activity of the peroxidase.

6. Interpretation: Normal cerebral cortical vessels are impermeable to microperoxidase *(24)*.

3.4. Combined Permeability and Ultracytochemical Studies

In an area of BBB breakdown, particularly in the acute hypertension model, not every vessel shows a permeability change to HRP. Therefore to determine whether vessels with

BBB breakdown show alterations in specific vascular enzymes, permeability studies can be combined with histochemistry for the specific enzyme in the same brain slices *(25,26)*. In these experiments rats are injected with HRP as a tracer of BBB breakdown. The fixative used must be compatible for the histochemical reactions of both HRP and the enzyme of interest. The brain is processed for the demonstration of HRP reaction product first as described in **Subheading 3.1.2.** and then slices are reacted for the demonstration of the enzyme of interest as described in the next chapter (*see* Chapter 9).

3.5. Radio-Iodinated Serum Albumin

The principal advantage of this method is that a quantitative estimate of BBB permeability to protein can be obtained using iodine-125–labeled serum albumin (RISA) *(10,27)*. A disadvantage of this method is that precautions have to be used throughout the experiment because of the use of a radioisotope. This tracer cannot be detected by visual examination; therefore, to identify areas with BBB breakdown before selection of areas for measurement of radioactivity, this tracer has to be used in combination with Evans blue. This method provides no information about how BBB breakdown occurs at the tissue level.

1. Gloves and plastic aprons are worn throughout this procedure and the work area is covered with plastic coated bench paper (*see* **Note 14**).
2. Polyethylene cannulas have to be inserted one in the femoral artery for the withdrawal of blood for total plasma radioactivity measurements and the other in the femoral vein for administration of test substances.
3. Rats (200 g) are injected with 10 µCi RISA intravenously.
4. Blood (0.3 mL) is collected at regular intervals throughout the duration of the experiment after the administration of RISA.
5. Evans blue is injected intravenously 10 min before the termination of the experiment.
6. The experiment is terminated by an intravenous bolus of sodium pentobarbital (37.5 mg).
7. A thoracotomy is then performed and 0.9% saline is perfused for 10 min via a cannula in the ascending aorta to remove RISA from the lumen of vessels.
8. The brain is rapidly removed and divided into the regions of interest, which are dropped into pre-weighed scintillation vials that are re-weighed to determine the weight of the blood and the brain areas.
9. Radioactivity of the samples is determined using a gamma counter.
10. The sum of the radioactivity of all brain regions is used to calculate the leakage of RISA in the whole brain.
11. The amount of leakage of RISA into brain is expressed as a percentage of plasma radioactivity using the following formula:

$$\frac{Cpm/mg\ brain}{Cpm/\mu l\ plasma} \times 100\% = \%\ protein\ transfer$$

12. Values of the mean percent protein transfer in the whole brain, in individual brain regions of rats in the different experimental groups are compared simultaneously by the one-way analysis of variance and the Tukey post-hoc test (SPSS 11.0).

This method was used to obtain a quantitative estimate of BBB permeability in acutely hypertensive rats treated with the calcium entry blocker, flunarizine. The latter results in decrease in protein transfer in total brain and individual brain regions of hypertensive rats showing increased permeability to Evans blue as compared with untreated hypertensive rats suggesting that calcium plays a role in increased endothelial permeability in hypertension (10).

3.6. Lanthanum

One of the smallest electron-dense tracers available is ionic lanthanum, which belongs to the lanthanide series of rare earth metals. Ionic lanthanum has a molecular mass of 138.9 daltons and a radius of 0.12 nm; hence, it appears to be a marker of ion permeability. Small concentrations of ionic lanthanum can be infused intravenously in living rodents without apparent ill effects. This tracer has been used to study permeability of tight junctions of normal cerebral vessels and following injury (28,29). Ionic lanthanum is not visible by light microscopy making selection of abnormal vessels for electron microcopy difficult. This is overcome by using both HRP and ionic lanthanum as tracers in the same animal.

1. Rats are injected with HRP intravenously and simultaneously infused with a 5 mM lanthanum chloride solution at a rate of 0.34 mL/min via the other femoral vein.
2. The circulation time of lanthanum depends on the experimental design.
3. Animals are fixed by vascular perfusion of a paraformaldehyde-glutaraldehyde mixture for 5 min as given in Chapter 2, Subheading 3.1.3. This is followed by perfusion of the same fixative containing 5 mM lanthanum chloride for 10 min.
4. Brains are then processed for demonstration of HRP reaction product as outlined in **Subheading 3.1.2**. Five millimolar lanthanum chloride is added to the cacodylate buffer to prevent the lanthanum from being washed out of the tissues.
5. Vessels showing permeability to HRP and nonpermeable vessels are selected for processing for electron microscopy as described in Chapter 2, Subheadings 3.2–3.6.
6. Interpretation:
   a. Lanthanum appears as an electron-dense precipitate on electron microscopy (**Fig. 1D**). Normal intracerebral vessels are not permeable to ionic lanthanum infused for 6 min prior to fixation (28,29).
   b. Following intracarotid administration of a hypertonic solution of arabinose, ionic lanthanum can penetrate cerebral endothelial tight junctions and extend into continuity in the basal lamina of vessels (28).
   c. A 6-min circulation of ionic lanthanum allows passage of this tracer through the choroid plexus vessels but not through the apical tight junctions between the choroidal epithelial cells (19). Longer circulation of lanthanum for 30 min (30) or administration of ionic lanthanum via the carotid (31) does result in leakage of this tracer through the apical tight junctions between choroidal epithalial cells with extravasation of this tracer into the ventricular cavity.

3.7. Detection of BBB Breakdown by Immunohistochemistry

3.7.1. Single Labeling

Immunohistochemistry at the light microscopical level can detect areas having vessels that are normally permeable to protein such as the area postrema and the median eminence (**Figs. 2A,B**). This technique has also been used to detect BBB breakdown to endogenous circulating proteins such as serum proteins (**Fig. 2C**), fibrinogen and fibronectin using specific antibodies to these proteins (2,3). These proteins can be detected in paraffin-embedded animal or human brain sections by the standard protocol given in Chapter 33.

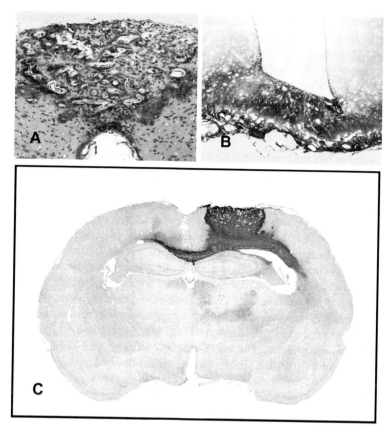

Fig. 2. Immunohistochemical demonstration of serum protein extravasation in paraffin sections of rat brain are shown. Areas with vessels having no BBB such as the area postrema (**A**) and the median eminence (**B**) show extravasation of serum proteins from vessel walls into the surrounding neuropil. (**C**), Rat brain showing BBB breakdown with serum protein extravasation at the cortical cold injury site. Note extravasation of serum proteins in the ipsilateral and contralateral white matter. **A, B**, x100; **C**, x50.

3.7.2. Dual Labeling for Permeability and a Specific Endothelial Protein

Protocols for dual labeling using immunoperoxidase or immunofluorescence are given in Chapter 33. These techniques allow the detection of vessels showing BBB breakdown by using an antibody to circulating proteins such as serum proteins, fibrinogen or fibronectin. Another protein related to permeability can be demonstrated in the endothelium of the vessel showing BBB breakdown. Double labeling for VEGF-A in a vessel with BBB breakdown to fibronectin is shown in Chapter 33; Figs. 2E,F.

In areas of BBB breakdown not every vessel is permeable. Therefore studies documenting alterations in tight junction proteins such as zonula occludens-1 and occludin *(32)* should be combined with evidence of protein extravasation from vessels to determine whether the alterations in the tight junctional proteins are related to BBB breakdown in the same vessel as demonstrated by others *(4,5)*.

In conclusion, immunohistochemical techniques have largely replaced tracer techniques in studies of BBB permeability in pathological states. Immunohistochemical

20. Nag, S. (1995) Role of endothelial cytoskeleton in blood-brain barrier permeability to protein. *Acta Neuropathol. (Berl.)* **90**, 454–460.

21. Nag, S. (1998) Blood-brain barrier permeability measured with histochemistry. In *Introduction to the Blood-Brain Barrier. Methodology, Biology and Pathology.* Pardridge, W. M., ed. Cambridge University Press, Cambridge, UK, pp. 113–121.

22. Stewart, P. A. (2000) Endothelial vesicles in the blood-brain barrier: are they related to permeability ? *Cell. Mol. Neurobiol.* **20**, 149–163.

23. Raymond, J. J., Robertson, D. M., Dinsdale, H. B., Nag, S. (1884) Pharmacological modification of blood-brain barrier permeability following a cold lesion. *Can. J. Neurol. Sci.* **11**, 447–451.

24. Reese T. S., Feder N., and Brightman, M. W. (1971) Electron microscopic study of the blood-brain and blood-cerebrospinal fluid barriers with microperoxidase. *J. Neuropathol. Exp. Neurol.* **30**, 137–138.

25. Nag, S. (1988) Calcium-activated adenosine-triphosphatase in intracerebral arterioles in acute hypertension. *Acta Neuropathol. (Berl.)* **75**, 547–553.

26. Nag, S. (1990) Ultrastructural localization of Na^+, K^+-ATPase in cerebral endothelium in acute hypertension. *Acta Neuropathol. (Berl.)* **80**, 7–11.

27. Harik, S. I., McGunigal, T. Jr. (1984) The protective influence of the locus ceruleus on the blood-brain barrier. *Ann. Neurol.* **15**, 568–574.

28. Dorovini-Zis, K., Sato, M., Goping, G., Rapoport, S., and Brightman, M. (1983) Ionic lanthanum passage across cerebral endothelium exposed to hyperosmotic arabinose. *Acta Neuropathol. (Berl.)* **60**, 49–60.

29. Nag, S., and Pang, S. C. (1989) Effect of atrial natriuretic factor on blood-brain barrier permeability. *Can. J. Physiol. Pharmacol.* **67**, 637–640.

30. Bouldin, T. W., and Krigman, M. R. (1975) Differential permeability of cerebral capillary and choroid plexus to lanthanum ions. *Brain Res.* **99**, 444–448.

31. Castel, M., Sahar, A., and Erlij, D. (1974) The movement of lanthanum across diffusion barriers in the choroid plexus of the cat. *Brain Res.* **67**, 178–184.

32. Bolton, S. J., Anthony, D. C., and Perry, V. H. (1998) Loss of the tight junction proteins occludin and zonula occludens-1 from cerebral vascular endothelium during neutrophil-induced blood-brain barrier breakdown in vivo. *Neuroscience* **86**, 1245–1257.

9

Ultracytochemical Studies of the Compromised Blood–Brain Barrier

Sukriti Nag

1. Introduction

Several cerebral endothelial properties can be studied by electron microscopy such as endothelial surface charge, the composition of monosaccharide residues on the endothelial plasma membrane and enzyme localization in intracerebal vessels. All these factors play a role in blood–brain barrier (BBB) breakdown to protein in pathological states as discussed in Chapter 6.

1.1. Endothelial Surface Charge

The surface of all cells have a net negative charge which is mainly due to sialyl groups (1) along with carboxyl groups of proteins, phosphates of phospholipids, amines of proteins, and proteoglycans (2,3). The surface charge on endothelium can be determined at the ultrastructural level by binding of agents such as cationized ferritin (4), colloidal iron (5), cationized colloidal gold (6), or poly-L-lysine-gold complexes (7). These techniques demonstrate a net negative charge on cerebral endothelium as well (5,8–10).

1.2. Monosaccharide Residues on the Endothelial Luminal Plasma Membrane

Specific oligosaccharide residues were localized on the endothelial plasma membrane of non-neural vessels by light microscopy using lectins conjugated with fluorescent dyes (11) or antibodies to lectins (12) or by electron microscopy of tissues incubated with lectins followed by an horseradish peroxidase (HRP)-gold conjugate, or mucin-gold complex or lactosaminated albumin-gold complex (13). These methods were adapted for electron microcopic studies of monosaccharide residues on the surface of cerebral endothelium (7,14). Most of the lectins studied (**Table 1**) were localized on the surface of cerebral endothelium (7,14) with the exception of peanut agglutinin (14,15). There was no difference in lectin localization in vessels of areas having a BBB such as the cerebral cortex and areas having no BBB such as the pituitary (16).

1.3. Ultracytochemistry of Enzymes in Cerebral Vessels

Early enzyme localizations in cerebral vessels used histochemical techniques on paraffin sections (17), imprints and squash preparations of brain (18) or biochemical

From: *Methods in Molecular Medicine, vol. 89:*
The Blood–Brain Barrier: Biology and Research Protocols
Edited by: S. Nag © Humana Press Inc., Totowa, NJ

Table 1
Biotinylated Lectins with Their Major Sugar Specificites and the Binding Inhibitor Used to Block Lectin Binding

Biotinylated lectins	Major sugar specificities	Binding inhibitor used	Binding in normal vessels	Binding in vessels permeable to HRP
Concanavlin A (Con A)	α-D-mannosyl, α-D-glucosyl-	O-methyl-α-D-mannopyranoside	+	+
Dolichos biflorus agglutinin (DBA)	α-N-acetyl-D-galactosaminyl	N-acetylgalactosamine	+	+
Peanut agglutinin (PNA)	β-D-gal-(1-3)-D-gal N-acetyl, β-D-galactosyl-		–	+
Ricinus communis I agglutinin (RCA I)	β-D-galactosyl	B-methyl-D-galactoside	+	+
Soybean agglutinin (SBA)	α-N-acetylgalactosaminyl	N-acetylgalactosamine	+	+
Ulex europaeus I agglutinin (UEA I)	α-L-fucosyl	L-fucose	+	+
Wheat germ agglutinin (WGA)	β-N-acetylglucosaminyl, N-acetylneuraminic (sialic) acid	N-acetylglucosamine	+	+

All these lectins bind to the luminal plasma membrane of normal cerebral endothelial cells except peanut agglutinin, which only binds to the luminal plasma membrane of vessels permeable to horseradish peroxidase (HRP).

techniques using brain homogenates *(19,20)*. Enzyme localizations in cerebral vessels by light microscopy demonstrated many interesting findings *(9,10,21–23)*. Some enzymes such as alkaline phosphatase and the transport adenosine triphosphatases (ATPases) though ubiquitous being present in both non-neural and cerebral capillaries are highly enriched in cerebral capillaries *(24,25)*. Other enzymes such as γ-glutamyl transpeptidase *(19)* and butrylcholinesterase *(20,26,27)* are present only in endothelium of vessels in areas having a BBB while the enzymes dopa decarboxylase and monoamine oxidase are located in the cytoplasm of cerebral endothelial cells but not in endothelium of non-neural vessels *(28,29)*.

Histochemistry done at the light microscopic level is sometimes difficult to interpret because it is difficult to be certain which component of the vessel wall contains the enzyme of interest and also which organelles contain the enzyme. This is overcome by localizing enzymes at the ultrastructural level, which is done by two principal methods. One method is based on a histochemical reaction, which results in the formation of an end product, which is electron-dense and can be localized by electron microscopy. The latter not only provides information about the subcellular localization of enzymes but also gives an indication of the intensity of the catalytic activity. The other ultracytochemical method for enzyme localization is immunohistochemistry using antibodies to specific enzymes, which are conjugated with HRP or gold particles. Details of the latter method are given in the next chapter (Chapter 10) and studies in the literature *(9,10)*. Some of the enzymes localized in cerebral vessels at the ultrastructural level are shown (**Table 2**).

In some ultracytochemical studies, different amounts of reaction product were observed on the luminal and abluminal plasma membranes of cerebral endothelial cells leading to the conclusion that both membranes are functionally different and gave rise to the concept of polarity, which should permit active solute transport across brain capillary endothelial cells *(9,42)*. These authors found that Na^+,K^+-ATPase is located predominantly on the abluminal plasma membrane by histochemistry and quantitation and proposed that it acts as an ion pump, removing potassium from the brain to the blood against the concentration gradient and removing sodium from the cytoplasm into the brain interstitium, also against the concentration gradient. An early attempt to correlate enzyme localization with vessel function demonstrated BBB breakdown to trypan blue following inhibition of butyrylcholinesterase and acetyl cholinesterase *(30)*. Ultracytochemistry of cerebral vessels in disease states such as scrapie, vessel injury and BBB breakdown during hypertension (*see* **Table 2**) show alterations in enzyme localization or activity.

In this chapter, the methods for ultrastructural localization of the transport adenosine triphosphatases, Na^+,K^+-ATPase, and Ca^{2+}-ATPase in cerebral vessels will be described in steady states and in vessels with BBB breakdown in acute hypertension. In these studies enzyme localization is combined with the use of the tracer HRP to localize vessels with BBB breakdown to detect alterations in enzyme localization in these vessels. It is possible to perform two histochemical reactions using the same tissue since the lead phosphate formed during the ATPase reactions is jet-black and easily differentiated from HRP reaction product, which is less electron-dense.

Table 2
Some of the Enzymes Localized by Ultracytochemistry in Intracerebral Vessels in Normal and Disease States

Histochemical Demonstration of Enzymes

Enzyme	State	Reference
Acetylcholinesterase	Normal vessels and post-injury	Joo and Varkonyi (*30*)
	Normal vessels	Flumerfelt et al. (*26*)
Adenylate cyclase	Normal vessels	Vorbrodt et al. (*31*)
Acid phosphatase	Normal vessels	Vorbrodt (*10*)
	Chronic hypertension	Yamada et al. (*32*)
Alkaline phosphatase	Scrapie	Vorbrodt et al. (*33*)
	Experimental Allergic Encephalomyelitis	Kato and Nakamura (*34*)
Butyryl cholinesterase	Normal vessels and post-injury	Joo and Varkonyi (*30*)
	Normal vessels	Flumerfelt et al. (*26*)
Calcium-dependent adenosine triphosphatase	Normal vessels	Inomata et al. (*35*) Nag (*36*)
	Acute Hypertension	Nag (*37*)
	Brain edema	Kawai et al. (*38*)
Glucose-6-Phosphatase	Normal vessels	Broadwell et al. (*39*)
Guanylate cyclase	Normal capillaries	Karnushina et al. (*40*)
Magnesium-dependent adenosine triphosphatase	Normal vessels	Inomata et al. (*35*) Vorbrodt (*10*)
Monamine oxidase	Normal vessels	Fujimoto et al. (*41*)
	Normal vessels	Inomata et al. (*35*)
5'-Nucleotidase	Normal vessels	Inomata et al. (*35*)
	Scrapie	Vorbrodt et al. (*33*)
Nucleoside diphosphatase	Scrapie	Vorbrodt et al. (*33*)
Ouabain-sensitive and potassium-dependent p-nitrophenyl phosphatase	Normal vessels	Betz et al. (*42*), Inomata et al. (*35*), Nag (*43*), Vorbrodt et al. (*44*)
	Acute Hypertension	Nag (*43*)
	Brain Injury	Vorbrodt et al. (*44*)
	Experimental Allergic Encephalomyelitis	Kato and Nakamura (*45*)
Thiamine monophosphatase	Normal vessels	Inomata et al. (*35*)
Thiamine pyrophosphatase	Normal vessels	Inomata et al. (*35*)
	Scrapie	Vorbrodt et al. (*33*)
Thiamine triphosphatase	Normal vessels	Inomata et al. (*35*)

Immunocytochemical Demonstration of Enzymes

Enzyme	State	Reference
γ-glutamyl transpeptidase	Normal vessels	Ghandour et al. (*46*)

2. Materials
2.1 Experimental Model

1. Male Wistar-Furth rats weighing 200–220 g.
2. Metofane (Janssen-Ortho Inc., Toronto, Canada).
3. Two 20 cm segments of PE-50 polyethylene tubing (Clay Adams, Parsippany, NJ) attached to 10-mL plastic syringes by a 23-gage needle. The syringes are filled with heparinized saline (2.5 U heparin/mL saline) for femoral artery and vein cannulations.
4. Pressure transducer (NARCO Biosystems, Austin, TX) connected to a polygraph (Grass Instruments, Quincy, MA).
5. Angiotensin amide (Ciba).
6. Horseradish Peroxidase, Type II (Sigma Chem. Co., St Louis, MO).

2.2. Endothelial Surface Charge

1. Fixative: 1% paraformaldehyde and 1.25% glutaraldehyde in 0.1 M cacodylate buffer (pH 7.2) containing 4.35% sucrose and 0.05% calcium chloride.
2. Dulbecco's PBS (DPBS), GIBCO-BRL.
3. DPBS containing 0.1 M Glycine.
4. Phosphate-buffered saline (PBS).
5. Cationized ferritin solution 0.5 mg/mL, pH 8.5 (Miles-Yeda, Rehovat, Israel) and native ferritin, pH 4.5 (horse spleen, Calbiochem-Behring, San Diego, CA).
6. Neuraminidase, type X (Sigma), 4 U/mL in 0.1 M sodium acetate buffer, pH 5.4. Add 9 mg NaCl and 1 mg $CaCl_2$/mL of buffer.
7. Substrate for demonstration of HRP reaction product: 0.06 % 3,3'-diaminobenzidine tetrahydrochloride (Sigma) in 0.06 M Tris-HCl buffer, pH 8.0.
8. 1% hydrogen peroxide solution.
9. TC-2 Tissue sectioner (Sorvall) or Vibratome.

2.3. Monosaccharide Residues on the Endothelial Luminal Plasma Membrane

1. Biotinylated lectins (*see* **Table 1**) and avidin D-HRP (Vector Laboratories, Burlingame, CA).
2. Binding inhibitors (Sigma): O-methyl-α-D-mannopyranoside, N-acetylgalactosamine, B-methyl-D-galactoside, N-acetylgalactosamine, L-fucose, and N-acetylglucosamine.

2.4. Ultracytochemistry of Enzymes in Cerebral Vessels
2.4.1. Localization of Na^+,K^+-ATPase

1. Fixative: 2% paraformaldehyde and 0.2 % glutaraldehyde in 0.1 M cacodylate buffer (pH 7.2) containing 4.35 % sucrose (*see* **Notes 1** and **2**).
2. 1% Lead citrate in 50 mM KOH solution.
3. 1 M glycine-KOH buffer, pH 9.0.
4. Substrate: 1% lead citrate in KOH, 8 mL; glycine KOH buffer, 5 mL; Levamisole 12 mg (Sigma); 0.1 M p-nitrophenylphosphate (Mg salt, Sigma); and 2 mL of distilled water. Stir to dissolve, and add 5 mL of dimethyl sulfoxide (DMSO, Sigma). Adjust the pH of this solution to 9.0 (*see* **Note 3**). Levamisole inhibits nonspecific alkaline phosphatase activity while DMSO enhances K^+-NPPase activity.
5. 10 mM ouabain solution.

2.4.2. Localization of Ca^{2+}-ATPase

1. 2 mM Lead citrate solution: Add the required amount of lead to distilled water and stir. 1 M NaOH is added drop by drop until the solution clears.

2. Substrate: 2 mM lead citrate containing 250 mM glycine, 10 mM calcium chloride, and 10 mM levamisole. Adjust the pH of this solution to 9.0 and add 2 mM Na$_2$ATP (Sigma) (*see* **Note 3**).
3. Chemicals (Sigma): adenosine monophosphate (AMP), 10 mM p-chloromercuric benzoate, 0.1 mM quercetin, 0.1 mM oligomycin, and 10 mM MgCl$_2$.

3. Methods

The ultrastructural methods outline how to study (1) endothelial surface charge, (2) monosaccharide residues on the endothelial luminal plasma membrane and (3) Na$^+$, K$^+$-ATPase and Ca^{2+}-ATPase localization in normal intracerebral cortical vessels and vessels with BBB breakdown to HRP in acute hypertension (*see* **Note 4**).

3.1. Endothelial Surface Charge

3.1.1. Experimental Model

The methods described in this chapter utilize the acute hypertension model in which sudden rapid elevations of blood pressure are accompanied by BBB breakdown of cerebral cortical vessels in multifocal areas of the cortex, particularly in the temporo-parietal lobes *(8,37,43,47)*. This protocol will now be described.

1. Male Wistar-Furth rats weighing 200–220 g are anesthetized using methoxyflurane inhalation.
2. A polyethylene catheter (PE-50) inserted into the left femoral artery is connected to a pressure transducer for continuous monitoring of the blood pressure and sampling of blood for gas and pH determinations.
3. Another polyethylene catheter (PE-50) is inserted into the femoral vein for infusion of test substances.
4. Rats are injected intravenously with 250 mg/kg HRP, type II.
5. One minute later, hypertension is induced by a 2-min intravenous infusion of angiotensin II (24 ug).
6. Control rats are injected with saline instead of angiotensin.
7. Thirty seconds after the termination of the angiotensin or saline infusion, a thoracotomy is performed and rats are perfused with fixative via the ascending aorta at a pressure of 120 mm Hg. Details of perfusion fixation are given in Chapter 2, Subheading 3.1.3.
8. Brains are removed and placed in the same fixative for 1 h at room temperature or as stated in the specific method.
9. The temporo-parietal lobes are sliced coronally at a thickness of 50 μ using a tissue sectioner and sections are collected in cacodylate buffer.
10. The cortex is cut into rectangular blocks (*see* **Note 5**).

3.1.2. Demonstration of Endothelial Surface Charge

1. Wash brain blocks of test and control rats with 3 changes of DPBS for 5 min each and then place blocks in DPBS containing 0.1 M glycine for 30 min. to quench the unbound aldehyde groups and prevent nonspecific adsorption of ferritin on the surface of membranes.
2. Place blocks in DPBS containing cationized ferritin 0.5 mg/mL for 20–30 min at room temperature (*see* **Notes 6** and **7**).
3. Place brain blocks of control rats in DPBS containing native ferritin at the same concentration and for the same time as the cationized ferritin.
4. Another set of brain blocks from control rats are treated with 4 U/mL of neuraminidase for 1 h before incubation in the cationized ferritin solution.

5. Wash blocks with DPBS twice for 5 min each.
6. Wash blocks once with 0.06 *M* Tris-HCl buffer, pH 8.0 and then perform the histochemistry for the demonstration of HRP reaction product by placing blocks in the substrate for 20 min (*see* **Subheading 2.2.**).
7. Blocks are then placed in the same substrate containing 2 mL of 1% hydrogen peroxide for 20 min. Details of this procedure and the precautions to be observed are given in the previous chapter (*see* Chapter 8, Subheading 3.1.2.). Vessels and areas with BBB breakdown to HRP appear brown and this can be detected visually or by light microscopy.
8. Leave blocks overnight at 4°C in the same fixative as used for the perfusion.
9. Process blocks for electron microscopy the next morning as described in Chapter 2, Subheadings 3.2.–3.6.
10. Ultrastructural Findings
    a. Intracerebral cortical arterioles of control rats incubated with native ferritin do not show ferritin particles on the endothelial surface.
    b. Cortical arterioles from control rats incubated with cationized ferritin show uniform binding of this agent on the luminal plasma membrane of endothelium (**Fig. 1A**), *(8)* (*see* **Note 8**).
    c. Cortical vessels of control rats incubated with neuraminidase prior to cationized ferritin show marked reduction of cationized ferritin binding on endothelium due to loss of the sialyl groups *(15)*.
    d. Arterioles of hypertensive rats with BBB breakdown to HRP show tracer in all layers of the vessel wall with extravasation into gap junctions between astrocytic foot processes. The endothelium and occasionally the smooth muscle cells show increased numbers of plasmalemmal vesicles or caveolae many of which contain tracer (**Fig. 1B**).
    e. Arterioles of hypertensive rats with BBB breakdown to HRP show loss of CF binding on endothelium therefore increased permeability is associated with loss of the net negative charge on endothelium (**Fig. 1B**) *(8)*.
11. Limitations

This technique only allows localization of anionic groups on the luminal plasma membrane of live or fixed endothelium in brain slices. To determine the charge on both the luminal and abluminal plasma membrane, techniques such as cationic colloidal gold *(6)* or of poly-L-lysine-gold complex *(7)* binding to ultrathin sections of brain can be used. Using these techniques the binding of cationic colloidal gold was observed to be more regular and more intense on the luminal than the abluminal plasma membrane suggesting a polar distribution of anionic sites on cerebral endothelium *(9)*.

3.2. Monosaccharide Residues on the Endothelial Luminal Plasma Membrane

1. The experimental model is the same as described in **Subheading 3.1.1.** and the fixative is the same as described in **Subheading 2.2.**
2. Incubate fixed brain blocks of test and control rats in the biotinylated lectins listed in **Table 1**, at concentrations of 0.5–1 mg/mL except for wheat germ agglutinin, which is used at a concentration of 0.05 mg/mL. The blocks are incubated for 1.5 h with continuous gentle agitation (*see* **Note 7**).
3. Specificity of lectin binding is checked by competitive inhibition in which 0.5 *M* of the monosaccharide to which each lectin binds is added to the incubation medium.
4. Incubate brain blocks of control rats with biotinylated lectin solutions with non-competing sugars or with avidin-HRP alone.

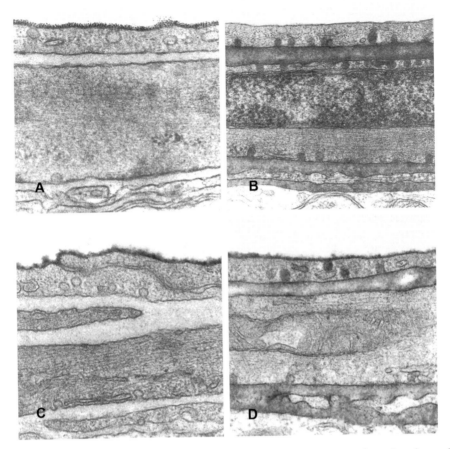

Fig. 1. (A), Segment of an intracerebral cortical arteriole from a control rat showing uniform binding of cationized ferritin on the luminal plasma membrane of endothelium. (B), Arteriolar segment from an acutely hypertensive rat showing increased permeability to HRP which is present in the basement membrane of the endothelium, smooth muscle cell layer and adventitia. An increased number of endothelial vesicles are present and many of the vesicles contain HRP. Note lack of cationized ferritin binding on the luminal plasma membrane. (C), Segment of arteriolar wall from a control rat showing linear electron-dense reaction product on the luminal plasma membrane of endothelium indicating the binding of biotinylated ricinus communis lectin. (D), Permeable arteriolar segment of a hypertensive rat showing HRP in the basement membranes of the vessel wall. Note the luminal binding of peanut agglutinin. A, D, x60,000; B, x40,000; C, x52,000.

5. In a separate series, brain blocks of control rats are pretreated with 4 U/mL of Neuraminidase, for 1 hr at 37°C before the demonstration of lectin binding.
6. Wash blocks three times with PBS for 5 min each.
7. Place blocks in Avidin D-HRP (15 µg/mL) for 30 min at room temperature.
8. Repeat **step 6** and then incubate the blocks in the substrate for demonstration of HRP reaction product as described in **steps 6–8**.
9. Process blocks for electron microscopy as described in **Subheading 3.1.2., step 9** (*see* **Notes 9** and **10**).

10. Ultrastructural Findings

    a. Lectin binding is demonstrated as an uniform, linear, electron-dense deposit on the luminal plasma membrane of endothelium of intracerebral arterioles and microvessels (**Fig. 1C**). Lectin binding is present in plasmalemmal vesicles at the luminal plasma membrane and at the luminal end of the interendothelial space but no binding is present on plasma membranes beyond the first tight junction. The same pattern of lectin binding is present for Con A, SBA, WGA, and RCA I *(14,15)*. Staining intensity is less for UEA I and DBA even when higher concentrations of lectin are used. Biotinylated PNA does not bind to the endothelial luminal plasma membrane of normal cerebral vessels.

    b. Lectin binding does not occur when blocks are treated as given in **step 3** or with Avidin-HRP alone.

    c. Blocks treated as given in **step 5** show a similar pattern of lectin binding as observed previously. In addition there is binding of PNA to the luminal plasma membrane of endothelium and accentuated staining for SBA *(15)*. Thus neuraminidase removes the terminal sialyl groups exposing the subterminal β-D-gal-(1-3)-D-gal *N*-acetyl groups to which PNA binds.

    d. Arterioles of hypertensive rats with BBB breakdown to HRP show similar findings as given in **Subheading 3.1.2., step 10d**.

    e. Vessels permeable to HRP in angiotensin-induced acute hypertension show similar results as control vessels with regard to localization of all the lectins studied except for PNA. Permeable vessels show PNA binding on endothelium (**Fig. 1D**) *(47)*. The significance of this finding is discussed in Chapter 6.

11. Limitation

    This technique allows localization of monosaccharide residues on the luminal plasma membrane only. In order to study both the luminal and abluminal plasma membrane it is necessary to use a technique which can be applied to ultrathin sections of brain embedded in a hydrophilic media such as Lowicryl K4M and use lectin-gold complexes *(10)*. These authors demonstrate differences in the binding of some of the lectins studied to the luminal and abluminal plasma membrane of normal cerebral endothelium, reinforcing the concept of polarity and that both membranes have different properties *(9)*.

3.3. Ultracytochemistry of Enzymes in Cerebral Vessels

3.3.1. Na$^+$,K$^+$-ATPase and HRP Histochemistry

The Na$^+$, K$^+$ pump operates as an antiport, actively pumping Na$^+$ out of the cell and K$^+$ in against their concentration gradients. Energy for these processes are derived from the breakdown of adenosine triphosphate to adenosine diphosphate and phosphate by the Na$^+$,K$^+$-ATPase enzyme. Localization of Na$^+$,K$^+$-ATPase at the ultrastructural level by the one-step method *(48)* gives reliable results and this technique is used to localize the ouabain-sensitive, K$^+$-dependent p-nitrophenylphosphatase (K$^+$-NPPase) activity of the Na$^+$, K$^+$-ATPase complex in normal intracerebral cortical vessels and vessels of rats with angiotensin-induced acute hypertension.

1. The experimental protocol is the same as given in **Subheading 3.1.1.**
2. Brains are fixed by vascular perfusion of fixative for 8 min only (*see* **Subheading 2.4.1.** and **Note 2**).
3. Remove brains and place in the same fixative as used for perfusion for 1 h at 4°C.
4. Transfer brains to 0.1 *M* cacodylate buffer containing 10% DMSO and slice at 50-μm intervals using a tissue sectioner.

membranes of all types of vessels (**Fig. 2D**). In the case of arterioles, reaction product is also present on the plasma membranes of the smooth muscle and adventitial cells (*see* **Note 13**) *(37)*.

b. Multifocal deposits of reaction product indicative of Ca^{2+}-ATPase activity are present in association with endothelial actin filaments both those present in the cytoplasm and those along the interendothelial junctions.

c. A striking finding is the localization of Ca^{2+}-ATPase in plasmalemmal vesicles of the endothelial and smooth muscle cells. Invaginating vesicles at both the luminal and abluminal plasma membrane also show reaction product on the outer plasma membrane.

d. Occasional red blood cells present in the vascular lumina show reaction product on the outer plasma membrane serving as a good positive control for this histochemical reaction.

e. The controls treated as given in **steps 5a–d** of this section fail to show reaction product in cerebral vessels.

f. The controls treated as given in **step 5e** show marked reduction of reaction product in endothelial, smooth muscle and adventitial plasma membranes while no effect is observed on Ca^{2+}-ATPase localization in association with actin filaments or plasmalemmal vesicles.

g. Localization of HRP is similar to the findings described previously in **Subheading 3.1.2., step 10d**.

h. Vessels permeable to HRP in acute hypertension show marked reduction of Ca^{2+}-ATPase localization in vessel walls (**Fig. 2E**).

3.4. Concluding Remarks

The methods listed in this chapter demonstrate that ultracytochemistry can be used to detect alterations in enzyme distribution and activity in cerebral vessels with a compromised BBB. Reduced enzyme localization of Na^+, K^+-ATPase and Ca^{2+}-ATPase in cerebral vessels with BBB breakdown to protein has also been observed in rats with chronic renal hypertension *(50)* and the significance of these findings in the pathogenesis of BBB breakdown in hypertension has been discussed previously *(51)*.

The techniques listed in this chapter can be applied to cultured cells. A recent study describes the use of high-voltage microscopy to obtain three-dimensional views of organelles of cells cultured on grids and stained by histochemical procedures for the demonstration of various enzymes *(52)*.

The availability of antibodies to specific enzymes has resulted in histochemical techniques being replaced by immunohistochemical techniques to demonstrate enzymes of interest using the immunogold technique, which is described in the next chapter (Chapter 10). Ultracytochemistry can also be combined with the immunogold technique in the same tissue to answer specific questions as done previously *(53)*. In such cases, cerium should be used as a capture agent for phosphate instead of lead as cerium phosphate precipitates are more stable than lead phosphate during postembedding immunocytochemistry *(54)*.

4. Notes

1. In the ultracytochemical methods listed in this chapter cacodylate buffer was used with satisfactory results. Other authors suggest that Pipes buffer is more preferable than cacodylate to preserve both enzyme activity and ultrastructure *(54)*.

2. Concentration of glutaraldehyde greater than 0.2 % inhibits the K^+-NPPase activity.

3. The substrates are freshly prepared and should be clear otherwise they have to be filtered before use. Cloudy substrates may result in a nonspecific electron-dense particulate precipitate throughout the tissue.

4. It is essential to treat tissues from the control and test rats in an identical manner. The total time the brain is in fixative, the time when the histochemical reaction is done after fixation and the time when tissues are processed for electron microscopy must be kept constant. Minor variations in these times may alter the final reaction product obtained and preclude comparison of results of control and test rats. For this reason, studies in the literature pertaining to localization of a single enzyme cannot be compared unless identical experimental protocols are used.

5. Using a stereoscope the cortex is cut into rectangular blocks and only blocks containing penetrating arterioles are processed for electron microscopy. The pial surface and the superficial part of the molecular layer are cut away so that solutions can access the vessel lumina for methods given in **Subheadings 3.1** and **3.2**.

6. Histochemical reactions are done using fixed brain slices. Perfusing reagents such as buffers, cationized ferritin and substrate intra-arterially as done previously *(42,55)* damages the endothelial luminal plasma membrane and precludes demonstration of the luminal plasma membrane properties. This may be the reason why these authors did not observe any enzyme in the luminal plasma membrane of endothelium.

7. Incubations are done by placing brain blocks in shell vials with microstir bars and placing the vials on a magnetic stirrer set at a low speed so that the solution is gently agitated and the tissue does not fragment.

8. This method is not reliable for the demonstration of charge on endothelium of microvessels since lack of cationized ferritin binding could result from inadequate penetration of cationized ferritin into the lumen of microvessels.

9. Following all histochemical procedures the edges of the blocks may show nonspecific positive staining for HRP. These edges are cut away before processing the tissue for electron microscopy or when the blocks are trimmed for ultrathin sectioning.

10. HRP reaction product indicative of lectin binding on the luminal plasma membrane can be seen in semithin unstained sections. To save time, examine these sections to ensure that HRP reaction product is present before proceeding to ultrathin sectioning.

11. Histochemical reactions are done about 90 min after perfusion fixation. Longer fixation results in marked decrease in the amount of reaction product indicating decrease in enzyme activity.

12. When the histochemical reaction for demonstration of HRP reaction product is done after the histochemistry for demonstration of Ca^{2+}-ATPase, reaction product of the latter is very much reduced. Therefore in our studies HRP reaction product is demonstrated first followed by the demonstration of Ca^{2+}-ATPase.

13. In this study histochemical reactions for demonstration of both Ca^{2+}-ATPase and HRP was done in control rats therefore the reaction product representing Ca^{2+}-ATPase activity was less than controls in which only Ca^{2+}-ATPase was demonstrated *(36)*.

Acknowledgments

Thanks are expressed to Verna Norkum and Blake Gubbins for their skilled technical assistance. This work is supported by the Heart and Stroke Foundation of Ontario.

References

1. Danon, D., and Skutelsky, E. (1976) Endothelial surface charge and its possible relationship to thrombogenesis. *Ann. NY Acad. Sci.* **275**, 47–63.

42. Betz, A. L., Firth, J. A., and Goldstein, G. W. (1980) Polarity of the blood-brain barrier: distribution of enzymes between luminal and antiluminal membranes of brain capillary endothelial cells. *Brain Res.* **192,** 17–28.

43. Nag, S. (1990) Ultrastructural localization of Na⁺, K⁺-ATPase in cerebral endothelium in acute hypertension. *Acta Neuropathol. (Berl.)* **80,** 7–11.

44. Vorbrodt, A. W., Lossinsky, A. S., and Wisniewski, H. M. (1982) Cytochemical localization of ouabain-sensitive, K+-dependent p-nitro-phenylphosphatase (transport ATPase) in mouse central and peripheral nervous systems. *Brain Res.* **243,** 225–234.

45. Ghandour, M. S., Langley, O. K., and Varga, V. (1980) Immunohistological localization of γ-glutamyl transpeptidase in cerebellum at light and electron microscope level. *Neurosci. Lett.* **20,** 125–129.

46. Kato, S., and Nakamura, H. (1989) Ultrastructural and ultracytochemical studies on blood-brain barrier in chronic relapsing experimental allergic encephalomyelitis. *Acta Neuropathol. (Berl.)* **77,** 455–464.

47. Nag, S. (1986) Cerebral endothelial plasma membrane alterations in acute hypertension. *Acta Neuropathol. (Berl.)* **70,** 38–43.

48. Mayahara, H., Fujimoto, K., Ando, T., and Ogawa, K. (1980) A new one-step method for the cytochemical localization of ouabain-sensitive potassium-dependent p-nitrophenylphosphatase activity. *Histochemistry* **67,** 125–138.

49. Ando, T., Fujimoto, K., Mayahara, H., Miyajima, H., and Ogawa, K. (1981) A new one-step method for the histochemistry and cytochemistry of Ca²⁺-ATPase activity. *Acta Histochem. Cytochem.* **14,** 705–726.

50. Nag, S. (1993) Cerebral endothelial mechanisms in increased permeability in chronic hypertension. *Adv. Exp. Med. Biol.* **331,** 263–266.

51. Nag, S. (1998) Blood-brain barrier permeability measured with histochemistry. In *Introduction to the Blood-Brain Barrier. Methodology, Biology and Pathology.* Pardridge, W. M., ed. Cambridge University Press, Cambridge, UK, pp. 113–121.

52. Nagata, T. Three-dimensional high voltage electron microscopy of thick biological specimens. *Micron* **32,** 387–404.

53. Araki, N., Yokota, S., Takashima, Y., and Ogawa, K. (1995) The distribution of cathepsin D in two types of lysosomal or endosomal profiles of rat hepatocytes as revealed by combined immunocytochemistry and acid phosphatase enzyme cytochemistry. *Exp. Cell Res.* **217,** 469–476.

54. Araki, N., and Hatae, T. (1999) Electron microscopic enzyme cytochemistry. *Methods Mol. Biol.* **117,** 159–165.

55. Firth, J. A. (1977) Cytochemical localization of the K⁺ regulation interface between blood and brain. *Experientia (Basel)* **33,** 1093–1094.

10

Immunogold Detection of Microvascular Proteins in the Compromised Blood–Brain Barrier

Eain M. Cornford, Shigeyo Hyman, and Marcia E. Cornford

1. Introduction

Few techniques have approached the high resolution afforded by immunoelectron microscopy using gold markers for the detection of specific cellular proteins and other molecules (*1*) or by post-embedding procedures (*2*). In principal, a primary antibody to a particular protein, which has been fixed, embedded, and placed on a grid, is identified by a host-specific secondary antibody conjugated to a gold particle of defined size. The gold particle identifying the targeted protein is then detected by electron microscope observation. This method has undergone a steady development over the past few years, because it uniquely meets the need to precisely assign macromolecules to specific locations and domains within both tissues and cells. It has also been used to reveal antigens that may be present in low or trace amounts and thus, has contributed to a greater understanding of functional specialization domains within cells and tissues. Its advantages over light microscopic immunocytochemistry and confocal immunofluorescence localization studies are that it can be carried out on very minute specimens and that it can provide a permanent record for quantitative analyses of multiple domains. Its disadvantage is that, because of the small tissue size, more sampling is needed and more expertise is required in handling, thus requiring more overall experimental time. Additionally, the operation of an electron microscope (EM) can result in prohibitive costs.

Specialized methods have been developed for virus-infected cells (*3*), neurotransmitters (*4*), glutamate receptors (*5*), and ganglioside antigens (*6*), indicating that many ligand-specific techniques are available to the neuroscientist. Blood–brain barrier (BBB)-specific proteins (*7*) including monocarboxylic acid transporters (*8,9*), ZO-1 tight junctional protein (*10*), endothelin (*11*), and p-selectin (*12*) have also been identified with EM immunogold. In our laboratory, we routinely process adjacent tissue samples for EM, and also fix samples in buffered formalin for paraffin-embedded light microscopic immunocytochemistry. The parallel light microscopy provides orientation and some assurance that the EM sample prepared does not represent an aberrant or nontypical area of the cells under study. This chapter describes techniques and methods of analysis, and will introduce the application of these methods by presenting selected examples of the contributions quantitative EM immunogold techniques have made to our understanding of the central nervous system microvasculature.

From: *Methods in Molecular Medicine, vol. 89:*
The Blood–Brain Barrier: Biology and Research Protocols
Edited by: S. Nag © Humana Press Inc., Totowa, NJ

2. Materials

1. 0.1 M Sorensen's buffer: 0.1 M KH_2PO_4/Na_2HPO_4, pH 7.4.
2. EM fixative: 0.05% glutaraldehyde , 2% paraformaldehyde, 2% sucrose, 0.15 M NaCl, 0.15 mg/mL $CaCl_2$, in 0.01 M Sorensen's buffer.
3. LM fixative: 10% buffered formalin.
4. LR white resin, medium grade (Ted Pella, Redding, CA).
5. Absolute ethanol (Gold Shield Chemical, Hayward, CA).
6. Gelatin capsules, size 4 (Polysciences, Warrington, PA).
7. Nickel slot grids (Sjostrand type) and storage boxes (Ted Pella, Redding, CA).
8. Ultramicrotome (LKB).
9. Primary antibodies:
   a. Rabbit anti-GLUT1 (Chemicon, Temecula CA).
   b. Mouse anti GFAP (Glial fibrillary acidic protein; DAKO, Carpinteria, CA).
   c. Rabbit /mouse anti-human serum albumin (DAKO).
10. Gold-conjugated secondary antibodies
    a. Goat anti-rabbit, 5 nm, 10 nm gold (Amersham, Piscataway, NJ).
    b. Goat anti-mouse, 10 nm, 20 nm gold (Ted Pella).
11. Buffer A:1% bovine serum albumin, 0.1% fish gelatin, 0.15 M NaCl, .05% Tween-20, 0.01 M sodium phosphate pH 7.4.
12. Blocking Buffer: 2.5% (wt/vol) ovalbumin, 2.5% (vol/vol) normal goat serum in Buffer A.
13. PBS.
14. Staining microwell mold (Ted Pella).
15. Parafilm.
16. Whatman no. 1 filter paper.
17. Osmium tetroxide, 4% aqueous (Ted Pella).
18. Uranyl acetate (Polysciences, Warrington, PA).
19. 0.22 um syringe filter units (4, 13, 25 mm; Millipore Corp, Bedford, MA).
20. Non-magnetic no. 7 curved forceps (Ted Pella).
21. 0.54% Formvar (Ted Pella) in ethylene dichloride (Mallinckrodt, St Louis, MO).
22. Photographic negatives, enlarger, paper, and developing supplies (Kodak).
23. Electron microscope (JEOL).
24. Digitizing tablet (Numonics, Montgomeryville, PA).
25. SigmaScan /Image Measurement Software (Jandel Scientific, Corte Madera, CA).

3. Immunogold Electron Microscopic Methods

3.1. EM Tissue Fixation

   Traditionally, glutaraldehyde concentrations of approx 1–2% have been used for EM tissue fixation of antigens, and alteration of fixation concentration and time may be the first step to optimizing any procedure *(13)*. Too little fixative compromises ultrastructural detail in tissue preservation, whereas too much fixative results in excessive crosslinking, and the antigenic epitopes of interest are then not recognized by the primary antiserum (*see* **Note 1**).

1. Brain tissues are immersion-fixed as rapidly as possible, to minimize autolysis.
2. Using a binocular stereomicroscope, shavings of brain tissue are cut into small (approx 1 mm³) pieces while immersed in the EM fixation buffer (30–45 min at 4°C).
3. The fixative is replaced by Sorensen's buffer for at least 15 min, then dehydrated in ethanol and embedded in resin.
4. Fixed tissue may be stored in refrigerated buffer overnight if necessary, without compromising the procedure.

3.2. Dehydration and Embedding

Once dehydration begins, the tissue is processed through to resin polymerization, without interruptions.

1. Dehydration involves 10-min changes in 50%, 70%, 95%, and 100% ethanol before resin infiltration. Care is taken to ensure that the tissue samples are not allowed to dry out during the graded alcohol solution changes.
2. Embed in LR-White resin (*see* **Note 2**), using 15-min changes of 100% ethanol : resin 2 : 1, 1 : 1, 1 : 2, and pure resin (3x).
3. Sample glass vials are supported on a slowly rotating, tilted platform to promote mixing of the viscous resin.
4. Tissue samples are then placed in individual gelatin capsules and polymerized overnight at $50 \pm 2^\circ$ C.

3.3. Grid Preparation

1. Nickel slot grids are used for immunogold EM.
2. To coat grids with a formvar support, a solution of 0.54% formvar in ethylene dichloride is prepared and filtered.
3. A cleaned glass microscope slide is dipped in the formvar and then scored with a sharp knife.
4. The slide is gently inserted in a bath of distilled water at an angle, and the formvar layer allowed to float free on the water surface.
5. Grids are then placed on the formvar layer to coat, then picked up with tweezers, dried, and placed in grid storing boxes for later use.
6. Head-band, flip-up magnifying lenses (2–3x) facilitate working with grids for extended periods.
7. Thin sections (70–80 nm) of brain tissue are cut on an ultramicrotome with a diamond knife, then mounted on the formvar-coated slot grids.

3.4. Immunostaining

3.4.1 Single Labeling

To avoid deposition of debris on the grids, all chemicals are EM grade and all solutions are filtered using 0.22-μm filters before use.

1. Incubate in (goat) blocking buffer for 30 min at room temperature. A drop (approximately 80–100 uL) of blocking buffer is pipetted onto clean parafilm, and the grid placed vertically against the side of the drop.
2. Incubate in rabbit primary antibody (30 uL in microwell staining mold) overnight at 4°C (*see* **Notes 3,4**).
3. After a gentle syringe "jet" wash, grids are washed in 6 drops of Buffer A for 10 min each. The drops are placed on large sheets of parafilm as described above for the blocking procedure. When transferring the grid to the next drop, drain off excess fluid by holding the edge of the grid with forceps and touching the contact point with a triangular piece of Whatman no. 1 filter paper. Fluids drain by capillary action from the grid to the paper.
4. Place grids in drops of 10 nm-gold conjugated goat antirabbit serum on the parafilm for 2 h at room temperature (*see* **Note 5**).
5. Wash six times with PBS for 10 min each.
6. Fix for 5 min in 1% glutaraldehyde to immobilize the immunogold, then wash with distilled water for 10 min.

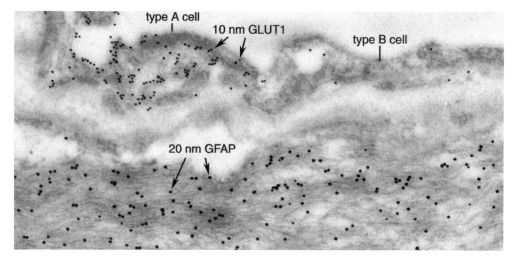

Fig. 1. Double-label immunogold EM demonstration of endothelial (GLUT1) glucose transporter protein with 10-nm gold particles, and perivascular glial fibrillary acidic protein (GFAP) with 20-nm gold particles. The human brain tissue was resected from an actively spiking lesion of a patient with complex partial seizures. Note the type A endothelial cell (in the upper left) exhibits abundant GLUT1 expression, in contrast to the reduced GLUT1 expression in the abutting type B endothelial cell (*upper right*). Glial scarring is seen immediately below the pericapillary basal lamina. Over-expression of GFAP epitopes is apparent along the many fibrils (seen in the lower half of the image) within the glial foot process.

7. Treat with 2% osmium tetroxide for 10 min and wash with distilled water for 10 min.
8. Stain for 30 min with 2% uranyl acetate, and wash twice with distilled water for 10 min each (*see* **Note 6**).
9. Transfer to grid boxes for drying, storing, and transporting.

3.4.2. Double Labeling

We typically localize two different BBB epitopes simultaneously, using different-sized gold particle markers: 10 nm gold to identify Glut1 and 20 nm gold for GFAP (**Fig. 1**). The procedure is the same as for single labeling except that a mixture of polyclonal Glut1 (1 : 800) and monoclonal GFAP (1 : 400) are present in **Subheading 3.4.1., step 2**, and a mixture of Goat anti-rabbit 10nm gold (1 : 75) plus Goat anti-mouse 20nm gold (1 : 75) are present in **Subheading 3.4.1., step 4**.

3.5. Quantitative Analyses of Epitope Densities

Newer electron microscopic systems capable of supporting digital image analyses permit direct immunogold quantification *(13)*, but our experience lies with human discrimination. (In dual label studies, it is important to confirm that if digital software recognition programs are used, that they can accurately identify membrane-associated particles, as well as discriminating two attached or adjacent 5-nm gold particles from a single 10-nm particle.)

1. Grids containing one section are scanned to record all capillary profiles at ×1900 in an electron microscope operating at 80 kV, and endothelial cell profiles photographed for glucose transporter, GFAP, serum albumin, and other epitope analyses.

2. In traditional electron microscopy, there is an emphasis on selecting and photographing artistic images. The textbook image of a BBB capillary is a perfectly circular profile, from a perfusion-fixed preparation.
3. In contrast, with quantitative electron microscopy, the emphasis is on complete (and therefore random) sampling. Every available capillary profile on each grid is photographed, regardless of its shape or aesthetic appeal, at magnifications of x7–x14,000 to ensure no bias in sample analyses.
4. Montages of longitudinal capillary sections are prepared when necessary for quantification and digital analyses of the immunogold particles.
5. The negatives are photographically enlarged to x19–x38,000 for detailed examination of the small gold particles, and the luminal and abluminal endothelial cell membrane lengths measured *(24)*.
6. A population of 15–25 capillaries is sufficient to characterize the tissue block studied.

3.5.1. Microvascular Parameters

1. The initial measuring procedure involves inspection of capillary photographic enlargements under x3–x5 magnification, and the weakly contrasted (no lead staining) abluminal endothelial and pericyte membranes along with the basal laminae, termed BM-1 and BM-3 *(25)*, are manually traced out.
2. The luminal membrane is typically more sharply contrasted against the lumen, and does not require such definition.
3. Gold particles are manually counted and particles are assumed to be membrane-associated when found within ±20 nm of the membranes.
4. The application of this limit is supported by the study of Takata et al. *(26)*, who indicated that 83% of the immunolocalizing gold particles identifying the sodium-dependent glucose transporter protein (SGLT1) were found within a 40-nm zone around the membrane.
5. Cell nuclei are osmiophilic and typically contain condensed electron dense chromatin material within the nuclear envelope. Immunogold particles tend to adhere nonspecifically at these sites *(27,28)* when rinsing procedures are insufficient. It has often been useful to compare gold particle counts at the nuclear border, and within the lumen of a capillary that is devoid of plasma proteins or blood cells. This analysis indicates the effectiveness of the immunostaining and wash procedures, as well as suggesting the background density of gold particles in a staining preparation of experimental and control slides.
6. Areas of pericytes, basal lamina, nuclei and vacuoles are also identified for digitization, and a wide variety of capillary and perivascular structures are systematically measured from the photographs *(29)*.
7. The digital analyses of the membrane lengths and cytoplasmic areas are performed using a digitizing tablet and Sigma Scan image analyzer, which is calibrated for the specific magnification of the photograph.
8. For every capillary profile examined, the following parameters are determined: luminal membrane length, abluminal membrane length, net endothelial cytoplasmic area, capillary lumen area, mean endothelial cell thickness, and (pericyte + basal lamina) area.
9. Nuclear areas of endothelia and pericytes are measured when present, as are the areas of any cytoplasmic vesicles or vacuoles. The sum of these components is subtracted from the total cytoplasmic area to estimate net cytoplasmic areas, which are remarkably constant in cross-sectional profiles (with or without a nucleus).
10. GLUT1 immunogold particles are totaled for every capillary profile location, and the number per micrometer of luminal and abluminal circumference determined along with GLUT1 density per square micrometer of endothelial cytoplasm, and pericyte/basal lamina areas.

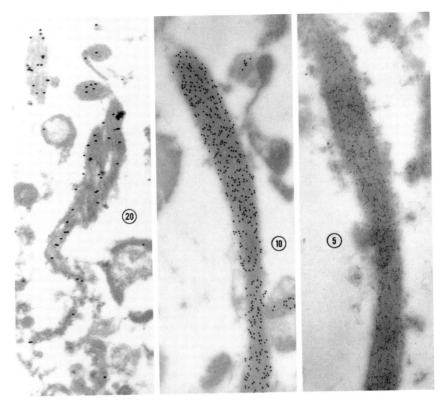

Fig. 2. Comparison of GFAP epitope identification and gold particle size in pericapillary fib-rils from a human brain resection. Sequential sections from the same tissue block were mounted on separate grids, to compare epitope densities with either 20-nm gold particles (*left panel*), 10-nm gold (*center panel*), or 5-nm gold particles (*right panel*). In both the left and right panels, a mouse monoclonal antiserum to human GFAP was employed, together with 5-nm or 20-nm gold–labeled goat antimouse sera. In the center panel, a rabbit polyclonal antibody to bovine GFAP was used in conjunction with 10-nm gold-labeled goat antirabbit serum. Note that when smaller sized gold particles are used, relatively more antigenic sites are apparent.

11. Reduced data may be presented in figures and tables as (length or area) means ± S.D.
12. Considerable differences in epitope densities may be seen in comparing different brain regions (*18,19*). For example, brain capillary membrane GLUT1 transporter densities vary by a factor of two- to threefold in a comparison of cerebellum, hippocampus and frontal cortex microvasculature (*30*).

3.5.2. Influence of Gold Particle Size on Epitope Density

Smaller (1 nm) gold particles may be difficult to see, requiring silver enhancement (*7*). The size of gold particle preparations from different manufacturers may also be quite variable. Differentiation of 10- and 5-nm particles in some instances is easily executed, but in other cases the same two sizes (from different manufacturers) may be so similar that they are almost impossible to distinguish with certainty. Thus, new supplies or lot numbers should be checked before use in double labeling studies.

Another caveat of quantitative immunogold work is that apparent epitope density is influenced by gold particle size, as illustrated in **Fig. 2**. Pericapillary GFAP epitopes

Table 1
Alteration in Hippocampal Capillaries in Pentylene Tetrazole-Induced Seizures

Capillary origin	Luminal area μm^2	Luminal circumference (μm)	Mean cytoplasmic thickness (nm)
Naive ($n = 44$)	5.7 ± 2.3	11.4 ± 2.4	213 ± 46
Kindled ($n = 28$)	7.5 ± 3.6	11.0 ± 2.5	188 ± 47
PTZ-Seizure ($n = 29$)	9.7 ± 3.4	12.5 ± 2.3	176 ± 43

Animals were either seizure-free (naive), or kindled rats had experienced at least one seizure for 10 consecutive days after injection of subconvulsant doses of pentylene tetrazole (PTZ). Capillaries from convulsant-treated (PTZ) rats were obtained immediately after previously naive animals experienced a single seizure.

(which are present in fibrils of uniform assembly of subunits) were immunogold-stained in three sections from the same tissue block with either 5-nm, 10-nm, or 20-nm gold. Note that the most epitopes are apparent using 5-nm gold, and slightly less with 10-nm particles. Significantly fewer 20-nm gold particles are apparent, suggesting that antigen-antibody binding and/or the wash procedure results in retention of reduced quantities of bound 20-nm particles. A quantitative analysis of GFAP epitopes per unit area of fibril further confirms that binding is relatively reduced, as a function of increasing particle size. The number of GFAP epitopes binding the one antibody per micron of fibril was 688 ± 184 for 5-nm gold; 463 ± 107 for 10-nm gold particles; and 64 ± 23 for the 20-nm gold particles. In the illustration used here, a rabbit polyclonal antisera to bovine GFAP was the primary antibody in the 10-nm gold series, followed by goat antirabbit 10-nm gold secondary. In the 5-nm and 20-nm gold particle–labeled material, a mouse monoclonal human GFAP primary antiserum was used; with goat antimouse, 5-nm gold or goat antimouse 20-nm gold–labeled antisera was used.

3.5.3. Microvascular Morphometry

Comparison of the total numbers of abluminal and luminal capillary membrane gold particles in two parallel preparations requires analyses of morphometric parameters to conclude that epitope expression is altered. One cannot assume that capillary endothelial cells from different species, brain regions, or clinical conditions, are inherently morphologically identical *(31)*. For example, retinoic acid and phorbol ester treatments profoundly affect BBB morphology in converting some, but not all, brain endothelia into fenestrated microvessels *(32,33)*. Rabbit brain capillaries have greater cytoplasmic thickness, and membrane circumferences *(29)* than rats and mice. In rat models, a single seizure causes significant size changes in hippocampal capillaries compared to naive animals; rats which have been kindled (experienced multiple seizures before examination) are, surprisingly, morphologically intermediate (**Table 1**). Similarly, capillary size and thickness is significantly increased in both RG2 and C6 tumor microvasculature compared to non-implanted or contralateral controls *(34)*.

Probably the most compelling evidence supporting the need for detailed morphometry in conjunction with quantitative EM immunogold comes from the definition of high GLUT1-expressing type A and low GLUT1-expressing type B capillaries. High expression of GLUT1 glucose transporter occurred coincidentally with a 30–40% size increase in the type A capillaries. This observation may be related to the reports that glucose transporters (of the GLUT supergene family) in addition to regulating glucose entry, co-transport water *(35,36)* in the absence of an osmotic gradient *(37)*, perhaps via a gated channel mechanism *(38)*. Thus membrane transporters are multifunctional proteins, capable of translocating more than one substrate. If the brain microvascular GLUT1 protein also couples water and glucose flux in vivo, then altered GLUT1 expression will not only affect cerebral glycolysis by substrate limitation, but will have an additionally important impact on capillary water permeability and resultant CNS edema.

3.6. Application of EM immunogold to Functional-Structural Analysis

3.6.1. Asymmetric Distribution of GLUT-1 in Cerebral Endothelium

The *polarity* of the BBB is an unusual feature of the cerebral microvasculature. The concept is that luminal and abluminal membranes of the brain capillaries are not identical, but instead exhibit asymmetric differences characterized by either qualitative or quantitative differences in expression of selected membrane components. For example, from in vivo studies of glucose transport, it was generally accepted that blood-to-brain influx, and brain-to-blood efflux of glucose were equal (or symmetrical) *(39–41)*. Elegant studies considered the possibility that BBB glucose transport could be kinetically analyzed in terms of translocation through a single membrane, or alternatively via individually (asymmetric) luminal and abluminal membranes. From these double-membrane analyses of rat brain glucose transport, asymmetric distributions of capillary glucose transporter were predicted *(42–44)*.

Immunogold EM studies, using antisera to the GLUT1 showed an asymmetrical distribution (wherein three- to fourfold more epitopes were seen on the abluminal membrane) in rat brain capillaries *(16)*. GLUT1 asymmetry was confirmed in developing rabbits *(29)*, rats *(45)*, and mice *(19)*. This glucose transporter asymmetry between luminal and abluminal membranes appears to be a unique feature of BBB endothelia, characterized by species-specific functional patterns in canines *(46)*, humans and other non-human primates *(30)*.

3.6.2. GLUT-1 Capillary Subtypes in Human Brain

Two distinct capillary types were noted in the expression of human BBB GLUT1. In vivo changes in the regional distribution of GLUT1 glucose transporters were evident in capillaries of resected seizure foci. Two configurations of endothelial cell GLUT1 distribution were characterized by a bimodal GLUT1 distribution in which the thinner (type B) endothelial cells displayed low GLUT1 immunoreactivity (on both luminal and abluminal membranes. Adjacent, larger type A endothelial cells showed a five- to tenfold greater expression of membrane GLUT1 transporter protein in immunogold epitope analyses *(24)*. Erythrocytes seen in the lumens of both type A and type B capillaries exhibited identical concentrations of membrane GLUT1, emphasizing that the phenomenon was not an artifact of processing or immunogold staining technique (**Fig. 3**). Furthermore, when GLUT1 percentage was compared within endothelial cell domains, type A capillaries exhibited relatively more **luminal** membrane GLUT1, whereas **abluminal** membrane GLUT1 was predominant in type B capillaries (**Fig. 4**). This analysis indicates that there was not only a marked change in glucose transporter

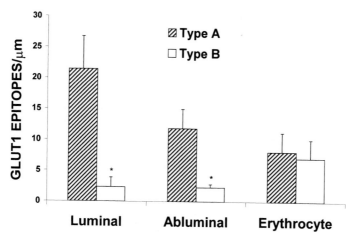

Fig. 3. GLUT1 glucose transporter densities seen with quantitative EM immunogold in human brain capillary endothelial cell membranes, and in capillary luminal erythrocytes. Note that two configurations of endothelial GLUT1 expression were observed in tissues resected from patients undergoing surgical treatment for complex partial seizure disorders. Type B endothelial cells exhibit similar levels of transporter expression as seen in normal animal brains. In contrast, type A endothelial cells exhibit approx tenfold higher GLUT1 expression. Red cell membranes, regardless of whether they are from lumens of type A or type B capillary profiles, exhibit a common GLUT1 expression density, that is intermediate between the type A and type B endothelial membrane density.

expression, but also a concomitant shift in GLUT1 polarity, related to abnormal physiologic conditions (8). The same bimodal pattern was seen in focal human brain injury (47). Brain regional GLUT1 differences, as well as shifts in GLUT1 polarity, have also been defined in the diabetic eye (48) and mouse models of disease (19–21). The conclusion we would emphasize from these studies is that **quantitative** immunogold electron microscopy has already substantially enhanced our understanding of brain capillary biology, and has the potential to provide significant contributions in many similar applications.

3.7. Extravasation of Plasma Protein Indicates Compromised Barrier Function

Vorbodt and colleagues (49,50) pioneered the use of quantitative albumin immunocytochemistry to assess pathophysiologic changes in BBB function. In further studies of human brain injury (47) and complex partial seizures (24) it has been shown that immunogold-labeled human serum albumin (HSA) can define localized regions where extravasation of serum proteins has occurred. In seizure patients, intraoperative EEG monitoring has provided the opportunity to compare tissue from the more actively spiking and less actively spiking regions. Morphologic examinations suggest these capillary profiles are intact, without albumin leakage. However, the HSA analyses indicate that in the capillaries from areas of more active epileptiform activity, there is increased extravasation of HSA into the endothelial cells, to the pericyte/basal lamina, and neuropil domains (24). The endothelial cells represent the first and primary barrier

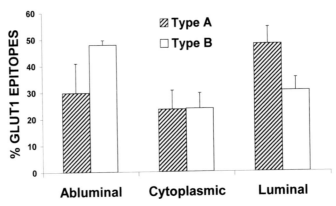

Fig. 4. Polarity changes in GLUT1 domain expression seen with quantitative immunogold EM in comparing (high-GLUT1 density) type A and low GLUT1-density type B endothelial cells. Note than when GLUT1 is highly expressed in the (type A) endothelial cell, the highest percentage of GLUT1 proteins (approx 45%) are found in the luminal membrane domain, with about 30% in the abluminal membranes, and approx 25% within the cytoplasmic domain. In contrast, within type B endothelial cells the highest percentage (almost 45%) of the GLUT1 protein are found at the abluminal membrane, with less than 30% at the luminal membrane and 25% in the cytoplasmic compartment.

to extravasation, and the basal lamina (+pericyte) is a second impediment, before albumin gains access to brain tissue. Quantitative analyses of serum protein extravasation identify regions of compromised function even when the anatomic integrity of the capillary unit is apparently normal. Furthermore, the same capillary-basal lamina sequential restriction pattern is apparent when extravasated serum IgG epitopes are identified with immunogold methods (**Fig. 5**).

In our experience, capillary profiles on the same grid sometimes exhibit quite different degrees of extravasation, suggesting that these changes may be highly localized in pathological situations. In different pathologies, for example, we have observed increasing degrees of serum albumin extravasation in resections from seizures < grade II astrocytomas < traumatic injury. However, a preliminary comparison of HSA extravasation in type A capillary profiles with high glucose transporter expression, and type B low-glucose transporter-expressing capillaries, showed HSA extravasation was essentially identical in the type A and type B capillary profiles (**Fig. 6**).

4. Notes

1. Some workers report that chemical etching or dissolving resin on sections fixed with higher (<2%) glutaraldehyde concentrations exposes the antibody binding sites (*4*), but we have not been able to successfully expose brain capillary epitopes by etching. Microwave fixation has also been used to enhance penetration of fixative into the tissue (*14*). It reportedly increases immunogold densities and shortens processing times, without compromising hepatic fine structure or labeling specificity (*15*). Picric acid fixation is another alternative to glutaraldehyde, but low glutaraldehyde (0.1–0.2%) levels may be optimal (*16*), with the needed concentration depending on the specific epitope studied (*17*).

2. LR-white resin is generally preferred over epoxy resins (*18*), but Lowicryl is also popular (*19–21*) as the entire process (including ultraviolet polymerization) may be completed at

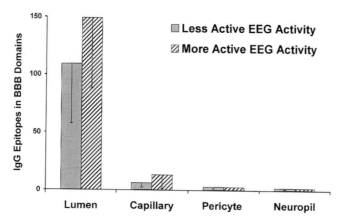

Fig. 5. Extravasation of plasma immunoglobulin G in resected brain from a patient with complex partial seizures. Two samples were resected, one the more actively spiking, and the other from the less actively spiking zones of a temporal lobe resections show a similar pattern to that seen where human serum albumin epitopes were identified *(24)*. Intraoperative electroencephalograph monitoring identified more active and less actively spiking sites. Note that luminal IgG density is about tenfold higher than in the capillary endothelial cell. A slightly reduced density is seen in the pericyte + basal lamina, with even less in the neuropil domain. These studies suggest that the capillary endothelial cell, and then the pericyte-basal lamina domain, form a sequential barrier(s) to plasma protein extravasation in the central nervous system.

   low temperatures. The use of LR gold resin has been reviewed elsewhere *(22)*. Resins are irritants, and adequate ventilation (chemical hood) and gloves are essential. Specialized 4H gloves (Ted Pella) are required if Lowicryl resin is used.
3. For each new antiserum studied, the immunoreactivity is examined by testing the primary antiserum at concentrations of 1:10, 1:100, and 1:500 on a standard specimen control. After examining gold particle distribution, another two concentrations might be tested (e.g., 1:250, 1:1000) and in this manner the optimal concentration is experimentally determined. Multiple incubations in the primary antibody have also been suggested as a signal amplification strategy *(2)*.
4. In our experience, antisera to complete proteins or larger peptides seem in general to provide better immunogold labeling than antisera to short fragments. However, in at least one study, no differences in labeling patterns were observed when full-length GLUT1 antiserum was compared with antiserum to the C-terminus of the glucose transporter *(16)*. Many commercial antibodies that are prepared by immunization with an amino acid sequence linked to keyhole limpet hemocyanin (KLH) protein, bind nonspecifically to brain and red cells. Pre-absorption of these antisera with KLH, or staining in the presence or absence of excess KLH protein, may be essential when used in EM immunogold preparations to reduce nonspecific background.
5. It may also be appropriate to optimize the level of immunogold in a series of dilutions, as there may be commercial lot variation. In our studies, the optimal concentration of gold-labeled anti-rabbit IgG antibody (with minimal nonspecific localization) was found to be at concentrations between 1:50 and 1:125. We have seen an inverse relationship between optimal concentration and gold particle size; higher dilutions are used with the smaller-sized gold particles.
6. We do not usually counterstain immunogold preparations so that gold particles are easily contrasted, but routine lead citrate *(23)* can be applied, if desired.

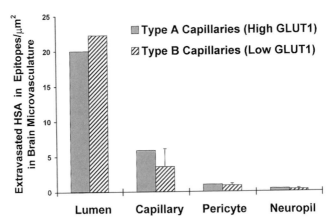

Fig. 6. Extravasation of plasma human serum albumin (HSA) compared in perivasculature of high GLUT1 glucose transporter-expressing (type A) capillaries, and low (type B) GLUT1 glucose transporter-expressing capillaries. The human brain tissue (a temporal lobe resection) came from a patient with complex partial seizures. Dual label EM immunogold studies were performed with 5-nm gold labeled GLUT1 and 10-nm gold–labeled HSA. We observed no significant differences in mean HSA density in endothelium, pericyte + basal lamina, and neuropil domains despite the significant differences in GLUT1 glucose transporter expression observed in the capillary profiles.

Acknowledgments

We thank Mrs. Birgitta Sjostrand for helpful suggestions, and Dr. William M. Pardridge for critically reading the manuscript. These studies were supported by NIH grant NS 37360.

References

1. Verkleij A., and Leunissen, J. L. M. (1989) *Immuno-Gold Labelling in Cell Biology*. CRC Press, Boca Raton, FL.
2. Mayer, G., and Bendayan, M. (2001) Amplification methods for the imunolocalization of rare molecules in cells and tissues. *Prog. Histochem. Cytochem.* **36,** 3–85.
3. Jensen, H. L., and Norrild, B. (1999) Easy and reliable double-immunogold labelling of herpes simplex virus type-1 infected cells using primary monoclonal antibodies and studied by cryosection electron microscopy. *Histochem. J.* **31,** 523–533.
4. Renno, W. M. (2001) Post-embedding double-gold labeling immunoelectron microscopic co-localization of neurotransmitters in the rat brain. *Med. Sci. Monit.* **7,** 188–200.
5. Wang, X. S., Ong, W. Y., Lee, H. K., and Huganir, R. L. (2000) A light and electron microscopic study of glutamate receptors in the monkey subthalamic nucleus. *J. Neurocytol.* **29,** 743–754.
6. Cahill, C. J., and Nayak, R. C. (2000) Immunoelectron microscopic detection of tissue ganglioside antigens. *J. Immunol. Methods* **238,** 45–53.
7. Farrell, C. L., and Pardridge, W. M. (1991) Ultrastructural localization of blood-brain barrier specific antibodies using immunogold-silver enhancement techniques. *J. Neurosci. Methods* **37,** 103–110.
8. Cornford, E. M., and Hyman, S. (1999) Blood-brain barrier permeability to small and large molecules. *Adv. Drug Delivery Rev.* **36,** 145–163.

9. Leino, R. L., Gerhart, D. Z., and Drewes, L. R. (1999) Monocarboxylic acid transporter (MCT1) abundance in brains of suckling and adult rats: a quantitative electron microscopic immunogold study. *Brain Res. Dev. Brain Res.* **113,** 47–54.

10. Nico, B., Quondamatteo, F., Herken, R., et al. (1999) Developmental expression of ZO-1 antigen in the mouse blood-brain barrier. *Brain Res. Dev. Brain Res.* **114,** 161–169.

11. Gajkowska, B., and Mossakowski, M. J. (1997) Endothelin-loke immunoreactivitiy in hippocampus following transient global cerebral ischemia. II. The blood-brain interphase. *Folia Neuropathol.* **35,** 49–59.

12. Easton, A. S., and Dorovini-Zis, K. (2001) The kinetics, function, and regulation of p-selectin expressed by human brain microvessel endothelial cells in primary culture. *Microvasc. Res.* **62,** 335–345.

13. Sierralta, W. D. (2001) Immunoelectron microscopy in embryos. *Methods* **24,** 61–69.

14. Paupard, M. C., Miller, A., Grant, B., Hirsh, D., and Hall, D. H. Immuno-EM localization of GFP-tagged yolk proteins in C. elegans using microwave fixation. *J. Histochem Cytochem.* **49,** 949–956.

15. Rangell, L. K., and Keller, G. A. (2000) Application of microwave technology to the processing and immunolabeling of plastic-embedded and cryosections. *J. Histochem. Cytochem.* **48,** 1153–1159.

16. Farrell, C. L., and Pardridge, W. M. (1991) Blood-brain barrier glucose transporter is asymmetrically distributed on brain capillary endothelial lumenal and ablumenal membranes: An electron microscopic immunogold study. *Proc. Natl. Acad. Sci. USA* **88,** 5779–5783.

17. Ramandeep, Dikshit, K. L., and Raje, M. (2001) Optimization of immunogold labeling TEM. An ELISA-based method for rapid and convenient simulation of processing conditions for quantitative detection of antigen. *J. Histochem. Cytochem.* **49,** 355–368.

18. Brorson, S. H. (1998) Comparison of the immunogold labeling of single light chains and whole immunoglobulins with anti-kappa on LR-white and epoxy sections. *Micron* **29,** 439–443.

19. Vorbrodt, A. W., Dobrogowska, D. H., Meeker, H. C., and Carp, R. I. (1999) Immunogold study of regional differences in the distribution of glucose transporter (GLUT-1) in mouse brain associated with physiological and accelerated aging and scrapie infection. *J. Neurocytol.* **28,** 711–719.

20. Vorbrodt, A. W., Dobrogowska, D. H., Kozlowski, P., Tarnawski, M., Dumas, R., and Rabe, R. (2001) Effect of a single embryonic exposure to alcohol on glucose transporter (GLUT-1) distribution in brain vessels of aged mouse. *J. Neurocytol.* **30,** 167–174.

21. Vorbrodt, A. W., Dobrogowska, D. H., Tarnawski, M., Meeker, H. C., and Carp, R. I. (2001b) Quantitative immunogold study of glucose transporter (GLUT-1) in five brain regions of scrapie-infected mice showing obesity and reduced glucose tolerance. *Acta Neuropathol. (Berl.)* **102,** 278–284.

22. Thorpe, J. R. (1999) The application of LR gold resin for immunogold labelling. *Methods Mol. Biol.* **117,** 99–110.

23. Reynolds, E. S. (1963) The use of lead citrate at high pH as an electron opaque stain in electron microscopy. *J. Cell Biol.* **17,** 208–213.

24. Cornford, E. M., Hyman, S., Cornford, M. E., Landaw, E. M., and Delgado-Escueta, A. V. (1998) Interictal seizure resections show two configurations of endothelial Glut1 glucose transporter in the human blood-brain barrier. *J. Cereb. Blood Flow Metab.* **18,** 26–42.

25. Liwnicz, B. H., Leach, J. L., Yeh, M. S., and Privatera, M. (1990) Pericyte degeneration and thickening of basement membranes of cerebral microvessels in complex partial seizures: electron microscopic study of surgically removed tissue. *Neurosurgery* **26,** 409–420.

26. Takata, K., Kasahara, T., Kasahara, M., Ezaki, O., and Hirano, H. (1991) Localization of Na^+-dependent active type and erythrocyte/HepG2-type glucose transporters in rat kidney: immunofluorescence and immunogold study. *J. Histochem. Cytochem.* **39,** 287–298.

27. Bendayan, M., Roth., J., Perrelet, A., and Orci, L. (1980) Quantitative immunocytochemical localisation of pancreatic secretory proteins of the rat acinar cell. *J. Histochem. Cytochem.* **28,** 149–160.

28. Craig, S. and Goodchild, D. J. (1982) Postembedding immunolabelling. Some effects of tissue preparation on he antigenicity of plant proteins. *Eur. J. Cell Biol.* **28,** 251–256.

29. Cornford, E. M., Hyman, S., and Pardridge, W. M. (1993) An electron microscopic immunogold analysis of developmental upregulation of the blood-brain barrier GLUT1 glucose transporter. *J. Cereb. Blood Flow Metab.* **663,** 7–18.

30. Cornford, E. M., Hyman, S., Cornford, M. E., Damian, R. T., and Raliegh, M. J. (1998) A single glucose transporter configuration in normal primate brain endothelium: Comparison with resected human brain. *J. Neuropath. Expl. Neurol.* **57,** 699–713.

31. Wolff, J. R., and Bar, T. (1972) "Seamless" endothelia in brain capillaries during development of the rat's cerebral cortex. *Brain Res.* **41,** 17–24.

32. Brightman, M.W., and Kaya, M. (2000) Permeable endothelium and the interstitial space of brain *Cell. Mol. Neurobiol.* **20,** 111–130.

33. Kaya, M., Chang, L., Truong, A., and Brightman, M. W. (1996) Chemical induction of fenestrae in vessels of the blood brain barrier. *Exp. Neurol.* **142,** 6–13.

34. Hashizume, K., and Black, K. L. (2002) Increased endothelial vesicular transport correlates with increased blood-tumor barrier permeability induced by bradykinin and leukotriene C_4. *J. Neuropath. Expl. Neurol.* **61,** 725–735.

35. Fischbarg, J., Kuang, K. Y., Hirsch, J., Lecuona, S., Rogozuiaski, L., and Silverstein, S. C. (1989) Evidence that the glucose transporter serves as a water channel. *Proc. Natl. Acad. Sci. USA* **86,** 8397–8401.

36. Fischbarg, J., Kuang, K. Y., Vera, J. C., Arant, S., Silverstein, S. C., Loike, J., and Rosen, O. M. (1990) Glucose transporters serve as water channels. *Proc. Natl. Acad. Sci. USA* **87,** 3244–3247.

37. Loike, J. D., Cao, L., Kuang, K., Vera, J. C., Silverstein, S. C., and Fischbarg, J. (1993) Role of facilitative glucose transporters in diffusional water permeability through J744 cells. *J. Gen. Physiol.* **102,** 897–906.

38. Fischbarg, J., and Vera, J. C. (1995) Multifunctional transporter models: Lesson from the transport of water sugars and ring compounds by GLUTs. *Amer. J. Physiol.* **268,** C1077–C1089.

39. Pappenheimer, J. R., and Setchell, B. P. (1973) Cerebral glucose transport and oxygen consumption in sheep and rabbits. *J. Physiol. London* **233,** 529–551.

40. Pardridge, W. M., and Oldendorf, W. H. (1975) Kinetics of blood-brain barrier transport of hexoses. *Biochim. Biophys. Acta* **382,** 377–392.

41. Gjedde, A., and Christensen, O. (1984) Estimates of Michaelis-Menten constants for the two membranes of the brain endothelium. *J. Cereb. Blood Flow Metab.* **4,** 241–249.

42. Cunningham, V. J., Hargreaves, R. J., Pelling, D., and Moorhouse, S. R, (1986) Regional blood-brain glucose transfer in the rat: A novel double-membrane kinetic analysis. *J. Cereb. Blood Flow Metab.* **6,** 305–314.

43. Hargreaves, R. J., Planas, A. M., Cremer, J. E., and Cunningham, V. J. (1986) Studies on the relationship between cerebral glucose transport and phosphorylation using 2-deoxyglucose. *J. Cereb. Blood Flow Metabol.* **6,** 708–716.

44. Cremer, J. E., Seville, M. P., and Cunningham, V. J. (1988) Tracer 2-deoxyglucose kinetics in brain regions of rats given kainic acid. *J. Cereb. Blood Flow Metabol.* **8,** 244–253.

45. Bolz, S., Farrell, C. L., Dietz, K., and Wolburg, H. (1996) Subcellular distribution of glucose transporter (GLUT1) during development of the blood brain barrier in rats. *Cell Tissue Res.* **284,** 355–365.

46. Gerhart, D. Z., LeVasseur, R. J., Broderius, M. A., and Drewes, L. R. (1989) Glucose transporter localization in brain using light and electron immunocytochemistry. *J. Neurosci. Res.* **22,** 464–472.

47. Cornford, E. M., Hyman, S., Cornford, M. E., and Caron, M. J. (1996) Glut1 glucose transporter activity in human brain injury. *J. Neurotrauma* **13,** 523–536.

48. Kumagai, A. K., Vinores, S. A., and Pardridge, W. M. (1996) Pathological upregulation of inner blood-retinal barrier Glut1 glucose transporter expression in diabetes mellitus. *Brain Res.* **706,** 313–317.

49. Vorbrodt, A. W., Dobrogowska, D. H., Ueno, M., and Tarnawski, M. (1995) A quantitative immunocytochemical study of blood-brain barrier to endogenous albumin in cerebral cortex and hippocampus of senescence-accelerated mice (SAM). *Folia Histochem. Cytobiol.* **33,** 229–237.

50. Vorbrodt, A. W., Dobrogowska, D. H., Tarnawski, M., and Lossinski, A. S. (1994) A quantitative immunocytochemical study of the osmotic opening of the blood-brain barrier to endogenous albumin. *J. Neurocytol.* **23,** 772–800.

11

Measurement of Blood–Brain and Blood–Tumor Barrier Permeabilities with [^{14}C]-Labeled Tracers

Kamlesh Asotra, Nagendra Ningaraj, and Keith L. Black

1. Introduction

Quantitative autoradiographic (QAR) method allows precise measurement of the regional permeabilities of the blood–brain barrier (BBB) and the blood–tumor barrier (BTB) for evaluation of the efficacy of selective drug delivery strategies. In particular, this method has been critical for validation of a number of vasoactive agents that increase the BTB permeability without affecting the BBB permeability to enhance selective delivery of various sized molecules to the brain tumor (1–11; reviewed in 12–14) and ischemic brain regions (15–17). QAR has been used to measure the unidirectional transfer constant (Ki) for radiotracers such as [^{14}C]-α-aminoisobutyric acid (AIB; MW 103 Da), [^{14}C]-sucrose (MW 342 Da), [^{14}C]-methotrexate (MW 455 Da), [^{14}C]-inulin (MW 5 kDa), and [^{14}C]-dextran (MW 70 kDa) following intracarotid or intravenous infusion of a variety of vasoactive substances in glioma-bearing and ischemia-reperfusion injury models (1–11, 15–17). This technique has been pivotal to the development and improvement of noninvasive strategies for transient or sustained increases in BTB permeability. QAR method has allowed us to identify novel target molecules and mechanistic steps for controlled biochemical modulation of BTB permeability that would not have been possible by any other technique (18–21).

Future preclinical testing of new candidate targets for BBB and BTB permeability modulation for enhanced and selective delivery of CNS therapeutic agents noninvasively will critically depend upon QAR method. This chapter provides a detailed description of preparation of the RG2 glioma-bearing rat model and application of QAR method for measurement of BBB and BTB permeabilities following bradykinin infusion and use of [^{14}C]-labeled AIB in this model.

2. Materials

2.1. Culture and Preparation of Glioma Cells for Tumor Implantation

1. Glioma implantation medium: Dissolve 1.2 g of methylcellulose (Sigma Chemical, St. Louis, MO) in 100 mL of sterile Ham's F-12 medium (Gibco BRL Life Technologies, Grand Island, NY) in a capped, sterile glass bottle by gentle magnetic stirring for approx 3 h on a hot plate set at 40°C. Stored at 4°C, this medium can be used for several months.

From: *Methods in Molecular Medicine, vol. 89:*
The Blood–Brain Barrier: Biology and Research Protocols
Edited by: S. Nag © Humana Press Inc., Totowa, NJ

2. Rat RG2 glioma, C6 glioma, or 9L gliosarcoma cell lines (available from American Type Culture Collection, Manassas, VA).
3. Ham's F-12 medium (Gibco-BRL Life Technologies, Grand Island, NY), containing 10% calf serum.
4. Trypsin (Gibco-BRL Life Technologies, Grand Island, NY), 0.5% with ethylene diamine tetraacetic acid, 10X. Dilute to 1X with phosphate-buffered saline (PBS), pH 7.4 to trypsinize confluent cell cultures.
5. Microtubes 1.5 mL.

2.2. Animal Surgery

1. Anesthetic: ketamine-xylazine (Loyd Labs, Sherandough, IA).
2. Bradykinin (Sigma Chemical), 10 µg/kg.
3. Female Wistar or Fisher 344 rats, 160–200 g body weight for intracranial implantation of glioma cells.
4. Bone wax (Eticone, Somerville, NJ).
5. Hydrogen peroxide, 30% solution (Sigma Chemical).
6. 2-Propanol.
7. Tissue solubilizer (Fisher Scientific, Fair Lawn, NJ).
8. Scintillation fluid
9. Betadine (povidone-iodine, 10%; Purdue Frederick, Norwalk, CT)
10. Microsurgery instruments and supplies: fine scissors, forceps, artery clamps, wound clips, surgical blades, surgical nylon (Roboz Instruments and Harvard Apparatus).
11. Equipment:
    a. Stereotaxy frame for intracranial tumor implantation (Harvard Instruments).
    b. Surgical microscope (Seiler Instruments, model MC-M900).
    c. Dentist's drill with 2 mm drill bit (Dremel, model 232-S).
    d. Monopolar/Bipolar coagulator (Martin model ME 102 with blue foot switches, Harvard Apparatus).
    e. Novaflex high intensity illuminator (World Precision Instruments).
    f. Syringe perfusion pumps (Model 22, Harvard Apparatus).
    g. Peristaltic pump (Harvard Apparatus).
    h. Blood pH/gas/hematocrit Analyzer (Model 348, Bayor Corporation).
    i. Scintillation counter.
    j. Coulter particle counter.
    k. Sterilizer for surgical instruments (Inotech Biosystems, Lansing, MI).

2.3. Catheterization of External Carotid Artery and Femoral Vessels

1. 1% Evans blue dye (Sigma) in PBS.
2. 1% Heparin, sodium salt (100,000 units, Sigma Chemical)
3. Disposable 1 mL syringes with:
    a. 30-gage needles for attachment of PE-10 tubing (Intramedic Clay Adams, Parsippany, NJ).
    b. 22-gage needles for attachment of PE-50 tubing.
    c. 20-gage needles for attachment of PE-90 tubing.
4. Radiotracers:
    a. [^{14}C]-α-aminoisobutyric acid (AIB, NEC 212).
    b. [^{14}C]-dextran (NEC 218D), [^{14}C]-iodoantipyrine (NEC 712).
    c. [^{14}C]-inulin (NEC 164A).
    d. [^{14}C]-sucrose (NEC 100) (NEN Life Science Products, Boston, MA).
5. Homeothermic blankets (Harvard Apparatus).
6. Invasive blood pressure monitor (Digi-Med, Model BPA 4000).

2.4. Tissue Harvesting and Sectioning

1. Guillotine (Harvard Instruments).
2. M-1 embedding matrix (Lipshaw, Pittsburgh, PA).
3. Glass slides (25 × 75 × 1 mm) and glass cover slips (24 × 50 mm, no.1).
4. Cryostat (Model HM505E, Mikron Instruments).
5. Microscope slide warming table (Model 285, Lipshaw Mfg.).
6. Harris Modified Hematoxylin stain with acetic acid (Fisher Scientific, Fair Lawn, NJ).

2.5. Quantitative Autoradiography and Data Acquisition

1. [^{14}C] Autoradiographic micro-scales, 120-μm multi-level reference strips, 20 × 2 nCi, cat. RPA 504L (Amersham, Arlington Heights, IL).
2. Kodak XAR-5 photographic film (8 × 10 inch) and intensifying cassettes (Kodak).
3. Apple G3 or G4 computer and film scanner (Epson Scan II) for digital image acquisition from QAR films with PhotoShop, NIH Image, and Microsoft Excel software.

3. Methods

The QAR method for measurement of BBB and BTB permeabilities in glioma-bearing rats following intracarotid infusion of saline and bradykinin, using [^{14}C]-α-AIB is described. This method can also be adapted to study the efficacy of intravenously infused vasoactive substances on BTB permeability.

The methods described below outline 1) culture and preparation of glioma cells for tumor implantation, 2) stereotaxic intracranial implantation procedure, 3) catheterization of external carotid artery and femoral vessels, 4) intracarotid infusion of vasoactive substances and tracers, 5) tissue harvest and sectioning, 6) quantitative autoradiography (QAR) and data acquisition from QAR films, 7) calculation of regional BBB and BTB permeability constant (Ki), and 8) results.

3.1. Culture and Preparation of Glioma Cells for Implantation

1. Culture the RG2, C6 or 9L glioma cells in 75 cm^2 flasks for 2–3 d in F12 medium containing 10% calf serum.
2. Trypsinize glioma cells for 5–10 min at 37°C, collect cells in 10 mL of serum-free F12 medium, prepare a single-cell suspension by repeated trituration, and count cells in an aliquot using a Coulter particle counter.
3. Remove a specific volume to obtain 1.5 × 10^6 cells in a 1.5-mL Eppendorf microfuge tube, centrifuge to obtain a cell pellet, remove supernatant, add 75 μL of serum-free glioma implantation medium. Vortex the tube for approx 1 min, and keep the tube on ice. Intracranial tumor implantation in each rat will require 100,000 cells in 5 μL volume (*see* **Note 1**).

3.2. Animal Surgeries

Animal protocols involving surgeries and use of vasomodulators and radioisotopes in animals have been approved by the Institutional Animal Care and Use Committee and the Radiation Safety Committee of Cedars-Sinai Medical Center. Animal surgeries are performed by aseptic techniques using gloves, respiratory masks, and aprons throughout the experiments.

3.2.1. Stereotaxic Intracranial Tumor Implantation

The implantation surgery takes approx 15 min.

3.2.1.1. ANESTHESIA

1. Check the general level of motor activity, hair condition, and physical features of each rat before any surgical procedures. Animals showing any abnormalities should be excluded.
2. Anesthetize rats using ketamine (80 mg/kg)/xylazine (10 mg/kg) intraperitoneally (i.p) in a single dose.
3. Shave hair from the frontal surface of scalp with electric clippers, and apply ophthalmic ointment to the rat's eyes.

3.2.1.2. CRANIOTOMY

1. Place the rat in a stereotaxic frame, immobilize the head by inserting the side bits into the ear canals (**Fig. 1A**).
2. Apply betadine to the shaved scalp, and make a 5- to 7-mm long incision in the skin on the right side of the head with a sterile scalpel, to expose the underlying skull bones.
3. A hole is drilled on the right coronal suture using a dentist's drill with a 2-mm drill bit, without perforating the dura, at a distance of 3 mm (for 140–150 g rat) or 4–5 mm (for 200 g rats) from the bregma (*see* **Note 2**).

3.2.1.3. TUMOR IMPLANTATION

1. To draw glioma cells, pull the plunger to 5 µL mark on the Hamilton syringe with the needle tip immersed in the viscous cell suspension, and allow the syringe to stand upright for 3–5 min to permit the viscous solution to rise up into the barrel of the syringe.
2. Wipe the outer surface of the needle with tissue paper to remove any adherent cells.
3. The needle of the Hamilton syringe is inserted into the drilled hole to a depth of 5 mm, and then the needle is pulled back by 0.5 mm (i.e., to sterotaxic depth of 4.5 mm). This gap of 0.5 mm creates sufficient space intracranially in the cerebral cortex to accommodate 5 µL volume and to allow for the injected cells to settle down.
4. The cells are injected slowly (2–3 min), and the needle is allowed to stay at that position for approx 5 min (**Fig. 1B**). Then the needle is slowly removed.
5. Outside surface of the skull surrounding the hole is swiped with alcohol.
6. The skin incision is closed with wound clips. Usually, the procedure is bloodless. If there is clear evidence of bleeding, seal the craniotomy area with bone wax.

3.3. Animal Care After Tumor Implantation

1. Postoperative care of rats requires maintenance of body temperature to ensure proper recovery. Keep animals on a thermal blanket immediately after surgery, and monitor body temperature via a thermometer inserted in the rat's rectum.
2. To prevent dehydration, administer subcutaneously 1 mL of warm lactated Ringer's solution per 100 g body weight. Administer buprenorphine (analgesia) subcutaneously to all rats postoperatively at a dose of 0.5–1.0 mg/kg body weight before recovery from general anesthesia.
3. Place the rat back in its cage to recover from anesthesia, and monitor mobility, heart rate, and respiratory patterns approximately every 10 min. After animals have fully recovered, return them to the vivarium. Incision sites must be regularly inspected for signs of infection.
4. Allow the tumor to grow for 7–9 d before conducting experiments for measurement of BBB and BTB permeabilities (*see* **Note 3**).

3.4. Catheterization of External Carotid Artery and Femoral Vessels

1. Permeability studies are performed 7–9 d after tumor implantation.
2. These experiments require experience in microsurgical techniques for vessel catheterization, as well as exercise of precautions necessary for use of radioactive materials.

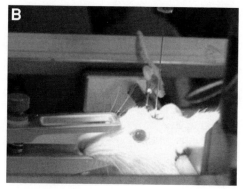

Fig. 1. Immobilization of a rat in the stereotactic frame for craniotomy and intracranial implantation of tumor cells is shown. The head of an anesthetized rat is immobilized by placing the side bits into the ear canals and clamping the incisors (**A**). Once a small hole is drilled at predetermined coordinates on the right side of the bregma, the needle of the Hamilton syringe is inserted to a depth of 5 mm and retracted by 0.5 mm before injection of glioma cells (**B**).

3. It is advisable to gain proficiency in these techniques and in initial experiments, Evans blue dye (1% in PBS) can be used instead of radiotracers.

3.4.1. Catheters

For every rat, prepare four catheters, each attached to a 1-mL disposable syringe containing 1% heparin (in PBS) as follows:

1. A 22–25 cm length of PE-10 tubing connected to a 30-gage needle for the external carotid artery.
2. A 22–25 cm length of PE-50 tubing connected to a 22-gage needle for the right femoral artery catheter. This catheter will later be connected to the blood pressure monitor.
3. A 22–25 cm length of PE-50 tubing connected to a 22-gage needle for the left femoral artery. After inserting one end of this catheter into the left femoral artery, the distal end of the tubing will be coupled to 1-m long P-90 tubing connected to the reverse peristaltic pump for continuous withdrawal and collection of arterial blood upon intravenous bolus injection of the radiotracer.
4. A 22–25 cm length of PE-50 tubing connected to a 22-gage needle for the right femoral vein. After catheterization of the femoral vein, the needle end of the tubing is connected to a 1-mL disposable syringe containing 1 mL of a [^{14}C]-labeled radiotracer (20 µCi/rat) or 1% Evans blue dye solution.

3.4.2. Preparation of Animal Surgery and Infusion Areas

1. Use fresh, disposable absorbant pads to cover surfaces at the microsurgical station, drug and radiotracer infusion area, and a bench near a sink where rats will be decapitated at the end of infusions for harvest of brain. These areas should be properly labeled as "Radioactive Work Area."
2. Cover knobs, switches, and door handles of necessary instruments and equipment with plastic wrap material to prevent potential contamination with radioactivity. After each day's work, these surfaces should be monitored for and cleaned of radioactivity.
3. Ensure availability of appropriate bags and containers for disposal and storage of liquid and dry radioactive waste, including sharps (needles, syringes, blades, glass slides, glass cover slips), animal carcasses, and harvested brain tissue. Radioactive waste is segregated

by individual nuclides (^{14}C waste separated from ^{3}H waste, for example) in appropriately labeled containers as dry, liquid, sharps, and animal carcasses (stored at –20°C), and handed over to the Radiation Safety Office for disposal (*see* **Note 4**).

4. Switch on the thermal blankets (set at 35°C) on which rats will be placed for catheterization procedures and infusions.
5. In an ice bucket, crush dry ice to fine powder, for rapid freezing of harvested brains.
6. Keep several fresh scintillation vials ready to collect arterial blood samples withdrawn immediately upon and after 10 min of intravenous bolus injection of [^{14}C]-labeled tracers.
7. Calibrate the blood pressure monitor, and connect with one PE-50 catheter (intended for the right femoral artery).

3.4.3. Catheterization of Femoral Vessels

1. Anesthetize rats using ketamine (80 mg/kg)/xylazine (10 mg/kg) i.p. at a single dose.
2. Shave hair from the neck and groin regions of the anesthetized rat and apply betadine to skin surfaces. Place ophthalmic ointment in the rat's eyes.
3. Make approx 1 cm oblique skin incisions in the left and right groin areas of the rat.
4. Expose the left and right femoral arteries and the right femoral vein.
5. One by one, tightly clamp the distal and proximal points (approx 1 cm apart) of each femoral vessel, partially cut each (approx one-half to two-third diameter), and insert PE-50 catheter into each, and tightly ligate the vessels (slightly distal to the point of catheterization) to hold the catheters in place (**Fig. 2A**). Remove the distal clamps from the femoral vessels.
6. At this time, collect approx 300 µL of arterial blood for measuring hematocrit.
7. In a typical intracarotid infusion setup (**Fig. 2C**), the right femoral artery catheter is connected to the blood pressure monitor, the left femoral artery catheter is connected to a reverse-flow peristaltic pump for continuous withdrawal of arterial blood samples for 10 min after injection of the tracer, and the right femoral vein is connected to 1 mL disposable syringe containing 20 µCi of [^{14}C]-labeled tracer (800 µL; stock diluted in PBS). Alternatively, if intravenous administration of vasomodulators is desired, the right femoral vein can be used instead of the external carotid artery (ECA).

3.4.4. Catherization of the ECA for Intracarotid Infusions

Procedural steps are depicted in **Fig. 3**, and their sequence is described below.

1. Make a 1-cm skin incision on the right side of the ventral aspect of the neck region, and expose the common carotid artery (CCA) and the ECA.
2. Using fine forceps, carefully remove the connective tissue surrounding the CCA and ECA.
3. Locate, coagulate and cut the occipital artery, which is a small branch of the ECA immediately distal to the emerging ICA (**Fig. 3A**).
4. Follow the ICA to locate the pterigopalatine artery. Care should be taken not to damage the vagus nerve situated close to the bifurcation of ICA.
5. Coagulate but do not cut the pterigopalatine artery (**Fig. 3A**).
6. Expose the ECA as distal as possible and tightly ligate the distal part of the ECA. Then, clamp the proximal part of the CCA and the distal part of the ICA (**Fig. 3B**).
7. Using micro-scissors, partially cut the ECA (approx one-half to two-third diameter of the ECA), blot away blood, and insert retrogradely one end of the PE-10 catheter into the ECA toward the CCA until just before the bifurcation point of the CCA but not beyond this point (**Fig. 3C**). Ligate the ECA tightly to hold the catheter in place.
8. Next, remove the clamp from ICA and ensure that there is backflow of blood into the PE-10 catheter, which would indicate that cannulation is correct.
9. Now release the CCA clamp and check for arterial pulsation.

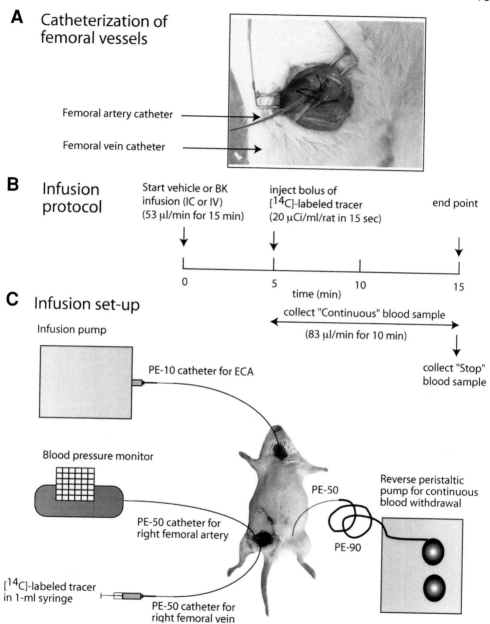

Fig. 2. Diagrammatic representation of the infusion setup and infusion protocol for measurement of blood–brain barrier and blood–tumor barrier permeabilities is shown. The right femoral vessels (**A**) and the left femoral artery are cannulated with PE-50 tubing. (**B**) The sequential steps of the infusion and collection protocols. (**C**) A typical infusion setup for intracarotid infusion of vehicle or test substances coupled with intravenous bolus injection of radiotracer.

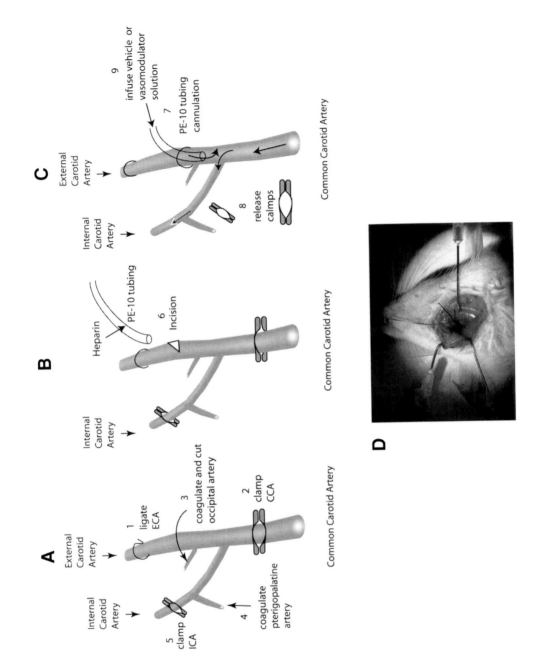

A

Internal Carotid Artery →

External Carotid Artery →

1 ligate ECA

3 coagulate and cut occipital artery

2 clamp CCA

5 clamp ICA

4 coagulate pterigopalatine artery

Common Carotid Artery

B

Internal Carotid Artery →

Heparin

PE-10 tubing

6 Incision

Common Carotid Artery

C

External Carotid Artery →

Internal Carotid Artery →

9 infuse vehicle or vasomodulator solution

7 PE-10 tubing cannulation

8 release calmps

Common Carotid Artery

D

184

10. Slowly infuse heparin from the PE-10 catheter syringe, and retrieve some blood to ensure that the blood flow through the arteries is smooth. **Fig. 3D** shows the catheterized ECA in a rat, ready for intracarotid infusion.

3.5. Infusion of Vasoactive Substances and Tracers

1. At time zero, infusion of the vehicle or a test substance (vasomodulator, agonist, or antagonist) solution is initiated through the cannula placed in the ECA (IC infusion) or the left femoral vein (intravenous infusion) and allowed to continue at a rate of 53.3 µL/min for 15 min (**Fig. 2B**).
2. At 5 min of IC infusion, a bolus of [^{14}C]-labeled tracer is injected through the right femoral vein in 15 s, and IC infusion of the vehicle or test substance continued for another 10 min.
3. Immediately upon injection of the radiotracer, blood samples are continuously collected from the left femoral artery at the rate of 0.083 mL/min for 10 min. At the end of 15 min infusion, both the pumps are stopped.
4. Cut PE-50 tubing connector of the catheter from the left femoral artery, and collect blood sample at the "Stop Time" in one scintillation vial. The continuously collected blood in the PE-90 catheter (for 10 min after tracer injection) is flushed into another vial with an air-filled syringe ("Continuous Sample").
5. Duplicate aliquots (20 µL) of both the "Stop" and "Continuous" blood samples are added to vials containing 250 µL of tissue solubilizer solution and 500 µL of 2-propanol, followed by 500 µL of 30% hydrogen peroxide and 5 mL of scintillation fluid. Scintillation counting of these samples provides initial and final radioactivity concentration in the plasma (dpm/µL).

3.6. Tissue Harvest and Sectioning

1. While still under deep anesthesia, the rat is decapitated using a guillotine, and the brain is rapidly removed and frozen by covering under finely powdered dry ice.
2. Mount the frozen brains onto pedestals with M-1 embedding matrix and cut 20-µm thick coronal sections with a cryostat, and thaw-mount the sections onto square cover slips (one section/cover slip) or glass slides (six sections/slide).
3. Attach the glass cover slips or glass slides, with brain sections on top, on 8 × 11-inch paper sheet, using rubber cement. Pertinent experimental information, such as the date, tumor type, vasoactive substance and radiotracer used, can be recorded on this sheet.
4. Stain the serial sections with hematoxylin to correlate histologically verifiable tumor areas with corresponding autoradiograms.

3.7. Quantitative Autoradiography and Data Acquisition from QAR Films

1. Generate QAR films by co-exposing the sections for two weeks on Kodak XAR-5 film with tissue-calibrated [^{14}C] autoradiographic microscale strips. Tissue-calibrated [^{14}C] autora-

---

Fig. 3. *(see opposite page)* Diagrammatic representation of sequential microsurgical steps for cannulation of the external carotid artery. **(A)** The distal portion of the external carotid artery (ECA) is ligated (*1*), the common carotid artery (CCA) is clamped (*2*), the occipital artery is coagulated and cut (*3*), the pterigopalatine artery is coagulated (*4*), and the internal carotid artery (ICA) is clamped (*5*). **(B)** An oblique incision is made approximately two thirds the diameter of the ECA (*6*). **(C)** The PE-10 tubing connected to a syringe containing heparin solution is inserted retrogradely to a point immediately before, but not past, the CCA bifurcation and securely ligated (*7*), the clamps from the ICA and CCA are removed (*8*), heparin is infused, and a small volume of blood aspirated to ensure patent blood flow before infusions (*9*). The direction of flow of blood and infused substances along the ICA is indicated by *arrows*. **(D)** Ventral aspect of the neck region of the rat after ECA cannulation.

diographic microscales serve as standards to calculate the concentration of [^{14}C]-AIB within regions of interest in the brain sections used for QAR. Once QAR films of satisfactory quality have been obtained, glass slides (or glass cover slips) with [^{14}C]-containing brain sections can be stored in dedicated containers for 6 mo, and then handed over to the radiation safety personnel for appropriate disposal.

2. Quantitative analysis of the regional radioactivity is performed using a Macintosh G3 computer and the NIH Image 1.61 software. Several spots within specific regions of the digitized QAR image of the coronal sections of tumor-bearing brain are sampled and by comparison with the signal intensities of the [^{14}C]-standards, the corresponding regional radioactivity is measured in the tumor, BST (areas within 2 mm of the tumor margin), left (contralateral) and right (ipsilateral) cortex, left and right white matter, left and right basal ganglia (**Fig. 4A**).

3. Representative QAR films from a study using intracarotid and intravenous infusion of vehicle and bradykinin are shown (**Fig. 4B**). Typically, data are acquired from 6 coronal brain sections of four to seven glioma-bearing rats infused with vehicle and a vasomodulator.

3.8. Calculation of Ki for Regional Permeabilities

Following QAR analysis, the regional permeabilities in the normal brain and tumor tissues are expressed by the unidirectional transfer constant, K_i (µl/g/min). The initial rate for blood-to-brain transfer is calculated by using the following equation *(22)*:

$$Ki = \frac{Cbr - V_0 Cbl}{\int_0^T Cpl \cdot dt}$$

where *Cbr* is the brain or tumor concentration (dpm/g) of the tracer at the end of the experiment, *Cbl* is the blood concentration (dpm/g) of the tracer at the end of experiment, *T* (min) is the duration of the experiment, and *Cpl* the arterial plasma concentration of radiotracer (dpm/ml). V_0 is the regional cerebral blood volume in the tissue (µL/g), and *dt* denotes integration over time. Blood volumes (V_0) of RG2, C6, and 9L tumors are derived from previously published studies *(2,3,17)*. Macros for calculating Ki values using an Apple computer and procedural details are available from authors, on request.

3.9 Results

The effect of bradykinin infusion (both intracarotid and intravenous) on the BBB and BTB permeabilities, as compared with vehicle infusion is shown in **Table 1**. Mean Ki values for the BBB permeability modulation are derived from the sampled contralateral brain regions, and those for BTB permeability are derived from ipsilateral regions of the same brain section. The results show that intravascular infusion of bradykinin selectively increases the regional BTB permeability for [^{14}C]-AIB in the tumor center and tumor periphery but does not affect BBB permeability measured in brain surrounding tumor and selected brain regions, both ipsilateral and contralateral to the brain tumor.

4. Notes

1. The tumor cell suspension in 1.2% methylcellulose is kept on ice for 2–3 h for the entire time it takes to work with 10 rats. The cells remain viable and yield reproducible size of tumors. Although only 50 µL of this cell suspension is needed for 10 rats, allow the other 25 µL for losses when the syringe needle is repeatedly dipped in to the tube containing the injectable cells. Tap the tube each time to ensure homogeneous distribution of cells before withdrawing them for injection.

A

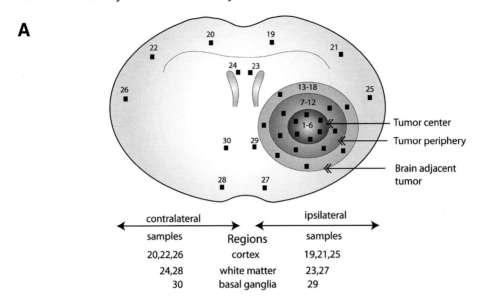

B

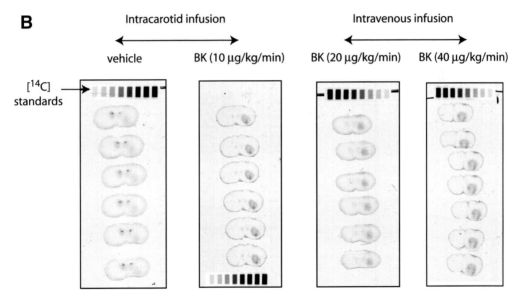

Fig. 4. Sampling of various regions of the coronal brain section through the tumor area from a digitized quantitative autoradiographic (QAR) image for measurement of regional blood–brain barrier (BBB) and blood–tumor barrier (BTB) permeabilities. (**A**) Several points are selected in the tumor center *(1–6)*, tumor periphery *(7–12)*, the brain adjacent to the tumor *(13–18)*, the ipsilateral *(19,21,25)* and contralateral *(20,22,26)* cortices, the ipsilateral *(23,27)* and contralateral *(24,28)* white matter, and the ipsilateral *(29)* and contralateral *(30)* basal ganglia for calculation of regional mean Ki values. (**B**) Representative QAR images corresponding to brain sections of rats receiving intracarotid and intravenous infusion of bradykinin (BK), co-exposed with [^{14}C] standards. These QAR films show that IC or intravenous infusion of low doses of BK significantly increases BTB permeability, but not BBB permeability. However when infused IV, a fourfold higher dose of BK is necessary to achieve a comparable increase in BTB permeability. This is confirmed by the calculated regional Ki permeability values (*see* **Table 1**).

Table 1

Effects of Intracarotid and Intravenous Infusion of Bradykinin on BBB and BTB Permeabilitites Measured by QAR in RG2 Glioma–Bearing Rats

Brain regions	Mean Ki values (µL/g/min) ± standard deviation			
	vehicle IC (n = 4)	Bradykinin IC, 10 µg/kg/min (n = 4)	Bradykinin IV, 20 µg/kg/min (n = 6)	Bradykinin IV, 40 µg/kg/min (n = 6)
Tumor center	12 ± 6	32 ± 8	17.3 ± 5.7	37.4 ± 5.7
Tumor periphery	3.0 ± 1.3	8.5 ± 4.4	15.9 ± 4.7	36.3 ± 4.3
Brain surrounding tumor	1.4 ± 0.7	1.6 ± 1.8	4.5 ± 1.7	2.8 ± 1.4
Cerebral cortex, ipsilateral	5 ± 4	4 ± 3	2.6 ± 2.1	2.4 ± 1.1
White matter, ipsilateral	3 ± 2	3 ± 1	2.7 ± 1.6	2.5 ± 1.9
Basal ganglion, ipsilateral	3 ± 2	2 ± 1	2.0 ± 1.1	2.2 ± 1.71
Cerebral cortex, contralateral	2 ± 1	2 ± 2	1.8 ± 1.2	1.8 ± 1.6
White matter, contralateral	2 ± 1	3 ± 1	1.7 ± 1.4	2.5 ± 1.7
Basal ganglion, contralateral	2 ± 1	2 ± 1	2.6 ± 1.3	1.8 ± 2.0

Intracarotid (IC) and intravenous (iv) infusion routes require different concentrations of bradykinin to produce comparable increases in Ki values. BBB, blood–brain barrier; BTB, blood–brain tumor barrier; QAR, quantitative autoradiography.

2. If cells are implanted at 4–5 mm lateral to the bregma in young rats, or at 3 mm lateral to the bregma in heavier rats, the site of implantation can be either the ventricle or farther away, and the tumor may grow at an undesirable location. It is critical that during the craniotomy the dura is not injured or perforated. This takes practice, and can be accomplished by using a 2-mm drill bit. With the drill switched on, approach the bone with the drill held vertical, and a mere touch to the skull should create the hole. Accidental perforation of the dura at this step frequently results in leakage and loss of implanted cells during withdrawal of the needle following injection of cells.

3. When allowed to grow for 7–9 d, the tumor size from intracranially implanted RG2, C6 and 9L cells attains a size of 7–10 mm in diameter, without appreciable tissue necrosis in the tumor core.

4. Because of its very long half-life (approx 5,730 yr), the ^{14}C radioactive waste is disposed off by a special handling process other than that used for shorter-lived radioisotopes.

Acknowledgments

The authors thank Shyam Goverdhana for technical assistance; Drs. Karen L. McKeown, Tripathi B. Rajavashisth, and Terrence M. Doherty for critically reading the manuscript, and Kolja Wawrowski for assistance with preparation of figures. This work was supported by a research grant from Neurogen Corporation (to K. Asotra) and an NINDS grant NS32103 (to K. L. Black).

References

1. Black, K. L., King, W. A., and Ikezaki, K. (1990) Selective opening of the blood-tumor barrier by intracarotid infusion of leukotriene C$_4$. *J. Neurosurg.* **72,** 912–916.
2. Inamura, T., and Black, K. L. (1994) Bradykinin selectively opens blood-tumor barrier in experimental brain tumors. *J. Cereb. Blood Flow Metab.* **14,** 862–870.
3. Inamura, T., Nomura, T., Ikezaki, K., Fukui, M., Pöllinger, G., and Black, K. L. (1994) Intracarotid histamine infusion increases blood tumor permeability in RG2 glioma. *Neurol. Res.* **16,** 125–128.
4. Nomura, T., Inamura, T., and Black, K. L. (1994) Intracarotid infusion of bradykinin selectively increases blood-tumor permeability in 9L and C6 brain tumors. *Brain Res.* **659,** 62–66.
5. Neuwelt, E. A., Barnett, P. A., Bigner, D. D., and Frenkel, E. P. (1982) Effects of adrenal cortical steroids and osmotic blood-brain barrier opening on methotrexate delivery to gliomas in the rodent: The factor of the blood-brain barrier. *Proc. Natl. Acad. Sci. USA* **79,** 4420–4423.
6. Matsukado, K., Inamura, T., Nakano, S., Fukui, M., Bartus, R. T., and Black, K. L. (1996) Enhanced tumor uptake of carboplatin and survival in glioma-bearing rats by intracarotid infusion of bradykinin analog, RMP-7. *Neurosurgery* **39,** 125–133; discussion, 133–134.
7. Matsukado, K., Nakano, S., Bartus, R., and Black, K. L. (1997) Steroids decrease uptake of carboplatin in rat gliomas: Uptake improved by intracarotid infusion of bradykinin analog, RMP-7. *J. Neurooncol.* **34,** 131–138.
8. Matsukado, K., Sugita, M., and Black, K. L. (1998) Intracarotid low dose bradykinin infusion selectively increases tumor permeability through activation of bradykinin B2 receptors in malignant gliomas. *Brain Res.* **792,** 10–15.
9. Sugita, M., and Black, K. L. (1998) Cyclic GMP-specific phosphodiesterase inhibition and intracarotid bradykinin infusion enhances permeability into brain tumors. *Cancer Res.* **58,** 914–920.
10. Nakano, S., Matsukado, K., and Black, K. L. (1996) Increased brain tumor microvessel permeability after intracarotid bradykinin infusion is mediated by nitric oxide. *Cancer Res.* **56,** 4027–4031.

11. Liu, Y., Hashizume, K., Chen, Z., Samoto, K., Ningaraj, N., Asotra, K., and Black, K. L. (2001) Correlation between bradykinin-induced blood-tumor barrier permeability and B2 receptor expression in experimental brain tumors. *Neurol. Res.* **23**, 379–387.

12. Black, K. L. (1995) Biochemical opening of the blood-brain barrier. *Adv. Drug Delivery Rev.* **15**, 37–52.

13. Asotra, K., and Black, K. L. (2000) Blood-brain barrier as portal for drug delivery. *Adv. Clin Neurosci.* **10**, 323–339.

14. Asotra, K., and Black, K. L. (2001) Blood-brain barrier research: Getting cancer-fighting drugs past the body's natural defense mechanism. *J. Neurosc. E-Medicine Technol.* **1**, 6–11, 25.

15. Hatashita, S., and Hoff, J. T (1990) Brain edema and cerebrovascular permeability during cerebral ischemia in rats. *Stroke* **21**, 582–588.

16. Belayev, L., Busto, R., Zhao, W., and Ginsberg, M. D. (1996) Quantitative evaluation of blood-brain barrier permeability following middle cerebral artery occlusion in rats. *Brain Res.* **739**, 88–96.

17. Liu, Y., Hashizume, K., Samoto, K., Sugita, M., Ningaraj, N., Asotra, K., and Black, K. L. (2001) Repeated short-term ischemia augments bradykinin-mediated opening of the blood-tumor barrier in rats with RG2 Glioma. *Neurol. Res.* **23**, 631–640.

18. Ningaraj, N. S., Yamamoto, V., Uchida, M., Asotra, K., and Black, K. L. (2000) Nitric oxide donors increase blood-brain tumor barrier permeability via K_{Ca} channels. *Soc Neurosci.* **26**, 126.8.

19. Ningaraj, N. S., Rao, M. K., Yamamoto, V., Tsimerinov, E., Asotra, K., and Black, K. L. (2001) Role of K_{Ca} and K_{ATP} channels in blood-brain tumor barrier permeability in rats. *Soc Neurosci.* **27**, 217.4.

20. Acosta, F. L., Goverdhana, S., Uchida, M., Black, K. L., and Asotra, K. (2001). Sustained increase with nitric oxide donor but transient increase with bradykinin in blood-brain tumor barrier permeability. Abstract for 41st Annual Meeting, The American Society for Cell Biology, Washington, DC, December 8–12.

21. Uchida, M., Chen, Z., Liu, Y., Acosta, F., Asotra, K., and Black, K. L. (2001) Enhancement of bradykinin-mediated blood-brain tumor barrier permeability by overexpression of bradykinin type 2 receptor on glioma cells. Congress of Neurological Surgeons, 51st Annual Meeting, San Diego, CA, Sept. 29–Oct. 4. Poster 394.

22. Ohno, K., Pettigrew, K. D., and Rapaport, S. I. (1978) Lower limits of cerebrovascular permeability to nonelectrolytes in the conscious rat. *Am. J. Physiol.* **235**, H299–H307.

III

BLOOD–BRAIN BARRIER TRANSPORT TECHNIQUES

12

A Review of Blood–Brain Barrier Transport Techniques

Quentin R. Smith

1. Introduction

The blood–brain barrier (BBB) is a unique dynamic regulatory interface situated at the border between the blood stream and the brain extracellular (or interstitial) fluid. As the "gatekeeper" to the brain, it determines the ability of drugs to gain entrance to brain extracellular fluid and reach therapeutic concentrations within the central nervous system (CNS) (*1,2*). The BBB also protects the brain from circulating neuroactive solutes, such as glutamate, glycine, norepinephrine, epinephrine, and peptide hormones, which can increase with diet, stress, injury, or disease (*3*). The BBB has an insulating function restricting brain access of many natural toxins, metals, antibodies, and biologic complexes and, in many cases, actively removing such from brain by energy-dependent efflux (*4*). The BBB has a key regulatory role in facilitating brain uptake of essential nutrients, vitamins, and hormones to sustain cerebral growth and metabolism, and in maintaining cerebral ionic and volume balance (*3,5,6*). In the absence of a BBB, the CNS would be constantly rocked by simple alterations during the acts of daily living, such as increased levels of neurotransmitters in the "fight-or-flight" response or exposure to plant toxins (e.g., ivermectin) from novel food stuffs, that would impede higher mental function. Our view of the BBB has evolved dramatically over the past 20 years from a simple restrictive interface that impedes the diffusion of polar solutes into brain, into a dynamic, highly selective, regulatory interface that expresses a wide range of active and facilitated transport systems and that responds to alterations in environment to maintain optimal cerebral homeostasis.

The BBB is formed by a series of tissue interfaces, including the brain vasculature, choroid plexus epithelium, and arachnoid membrane. The barrier at the brain vasculature was shown elegantly by Reese and Karnovsky (*7*) and Brightman and Reese (*8*) in the late 1960s to exist at the capillary endothelium (**Fig. 1**), the cells of which are joined together by multiple bands of tight junctions (*9*). These junctions seal adjacent endothelial membranes, thereby closing off the paracellular diffusion space. This, together with the absence of brain endothelial fenestra and limited endothelial vesicular transport, creates a high electrical resistance (2000–8000 $\Omega.cm^2$) interface (*10,11*) that forces most solutes to transit the brain capillaries either by dissolving and diffusing across the

From: *Methods in Molecular Medicine, vol. 89:*
The Blood–Brain Barrier: Biology and Research Protocols
Edited by: S. Nag © Humana Press Inc., Totowa, NJ

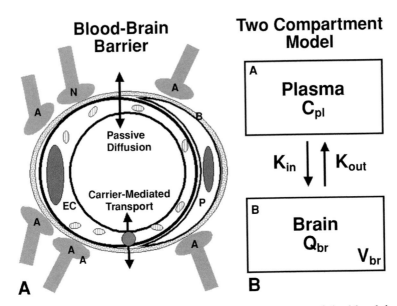

Fig. 1. **(A)** Diagram illustrating the critical cellular components of the blood–brain barrier and their relation to plasma and brain extracellular fluid. **(B)** Simplified diagram illustrating the two compartment model representing plasma and brain as well as influx and efflux. EC, endothelial cell; N, neuron; A, astrocyte; P, pericyte; B, basement membrane.

lipophilic endothelial membranes or being transported across by various energy-dependent or facilitated systems. The endothelial membranes are polarized and show selective distribution of specific transport proteins on their luminal (apical; e.g., P-glycoprotein and gamma-glutamyl transpeptidase) and abluminal (basolateral; Na^+-K^+adenosine triphosphatase [ATPase], system A) membranes that function to carry out vectorial transendothelial transport *(12)*. The capillaries also express many enzymes involved in the degradation of proteins, peptides, neurotransmitters and drugs, forming a significant metabolic barrier for uptake (the "enzymatic" BBB). In recent years, the contributory roles of astrocyte end feet, pericytes, and perivascular microglial cells on cerebrovascular function has become more evident, and the microvascular BBB concept has been expanded to include these additional cells. Solute exchange from skull and surrounding tissues into subarachnoid cerebrospinal fluid (CSF) is limited by tight junctions and polarized transport proteins at the arachnoid membrane which is part of the cerebral meninges *(13)*. The choroid plexus epithelium also plays a specialized role in secreting CSF into the brain ventricles, which acts as a diffusional sink for many polar brain metabolites and proteins *(14)*. The barrier at the choroid plexus is located not at the vascular endothelium but at the choroid plexus epithelium, which has its own blend of specialized transport carriers involved in regulation of solute exchange between blood and ventricular CSF.

In this review, the kinetics of solute transport across BBB membranes will be described and summarized, and then brief overviews of the primary methodological approaches will be presented. The chapter will provide perspective on the specific strengths and weaknesses of the various BBB transport methods presented in this section.

2. Transport Kinetics

2.1. Flux Rates of Transport In and Out of Brain

Solute transport across isolated BBB membrane vesicles or more complicated in vivo or in vitro barrier cellular systems is based on kinetic analysis of solute flux or exchange. The critical parameters in this analysis are the unidirectional flux of solute in one direction (e.g., from plasma to brain extracellular fluid or A → B), unidirectional flux in the opposite direction (e.g., from brain extracellular fluid to plasma or B → A), and net flux (e.g., the difference between the two unidirectional flux rates). Flux can be across a single cell membrane, such as in cellular uptake from a bathing medium, or transcellular across a polarized cellular system.

Very commonly in the BBB literature, flux from plasma to brain or brain extracellular fluid is referred to as "influx" (J_{in}; mass/time/weight brain) and flux from brain or brain extracellular fluid to plasma is referred to as "efflux" (J_{out}; mass/time/weight brain) (*see* **Fig. 1**) *(15)*. Flux rates for passive diffusion are directly related to concentration via a transfer coefficient (K, volume/time/weight brain), which is similar to a clearance constant, such that,

$$J_{in} = K_{in}C_{pl} \tag{1}$$

and

$$J_{out} = K_{out}C_{br} \tag{2}$$

where C_{br} = solute concentration in brain or brain extracellular fluid and C_{pl} = solute concentration in plasma. In both fluids, concentration ideally refers to the unbound or "free" concentration of solute.

For saturable transport, flux often takes the form of a Michealis-Menten equation

$$J = V_{max}C / (K_m + C) \tag{3}$$

where V_{max} = the maximal saturable transport rate (mass/time/weight brain) and K_m = the half-saturation concentration (mass/volume). When C<< K_m, flux varies linearly with concentration as $J = V_{max} C/K_m$, whereas when C > K_m, flux rate falls off from linearity and eventually reaches an asymptote (V_{max}) as C → ∞. The maximal saturable transport rate occurs at the concentration where there is no remaining "free" transporter to contribute to solute transport. **Equation 3** has readily been applied to analyses of saturable influx of solutes across the BBB into brain to obtain estimates of K_m and V_{max} *(3,16)*. It has been more difficult to provide the same kinetic detail for saturable efflux due to the problem of controlling solute concentration in brain extracellular fluid.

2.2. Kinetic Modeling of BBB Transport

In most in vivo BBB studies of solute uptake, the time course of net solute appearance in brain is analyzed following solute introduction into plasma using a two-compartment model, which incorporates influx and efflux (*see* **Fig. 1**) *(15)*,

$$dQ_{br} / dt = J_{in} - J_{out} = K_{in}C_{pl} - K_{out}C_{br} \tag{4}$$

where t = time of exposure. If significant cerebral metabolism of the compound occurs, a component of metabolic loss can be added ($K_{metab}C_{br}$) to **Eq. 4**. However, as both efflux and metabolic rate depend on C_{br}, they often are brought together in an overall loss (K_{out}) constant (i.e., $K_{out} = K_{metab} + K_{efflux}$).

$$dQ_{br}/dt = J_{in} - J_{out} = K_{in}C_{pl} - (K_{metab} + K_{efflux})C_{br} \qquad (5)$$

As the true solute concentration in brain is difficult to measure, most studies express their results in terms of the quantity (Q) of solute in brain per gram weight of brain (mass/weight brain). In this analysis, $C_{br} = Q_{br}/V_{br}$, where V_{br} = the effective brain volume of distribution (volume/weight brain). Given that Q_{br} is more readily measured in many brain transport studies than C_{br}, **Eq. 4** can be revised as

$$dQ_{br}/dt = K_{in}C_{pl} - k_{out}Q_{br} \qquad (6)$$

where k_{out} is the rate coefficient for brain efflux defined as $k_{out} = K_{out}/V_{br}$ and has units of inverse time (1/time). If transport at the BBB is symmetrical ($K_{in} = K_{out}$) and there is essentially no metabolic loss of solute from brain, then **Eq. 6** can be simplified as

$$dQ_{br}/dt = K_{in}(C_{pl} - Q_{br}/V_{br}) \qquad (7)$$

which states that the net rate of solute accumulation or loss is directly proportional to the effective concentration gradient of solute between plasma and brain, i.e., $(C_{pl} - Q_{br}/V_{br})$. The constant of proportionality is the transfer constant (K_{in}), which is determined largely by the properties of the BBB system.

If solute concentration is held constant in plasma (C_{pl} = constant) and there is no solute in brain at t = 0, then **Eq. 7** can be integrated to give,

$$Q_{br} = C_{pl}V_{br} (1 - e^{-KinT/Vbr}) \qquad (8)$$

where T = time. If solute transport is asymmetric, then Eq. 6 can be integrated with constant C_{pl} to give

$$Q_{br} = C_{pl}(K_{in}/k_{out})(1 - e^{-koutT}) \qquad (9)$$

Either **Eq. 8** or **9** can be fit by nonlinear least squares to time dependent determinations of Q_{br} for constant C_{pl} data to estimate K_{in}, k_{out}, and/or V_{br}. K_{in} and k_{out} give direct quantitative measures of the ability of a solute to move between plasma and brain as well as the distribution of the solute within the CNS.

2.3. Initial Uptake and the Patlak Equation

Because of the difficulty of making multiple Q_{br} determinations over a wide range of solute exposure times in in vivo experiments, two critical simplifications of the above analysis were introduced in the late 1970s and early 1980s. First, Bradbury (*16*) and Rapoport (*17*) described the "initial uptake" simplification which stated that if exposure time was limited so that only a small amount of solute accumulated in brain relative to plasma ($Q_{br}/V_{br} \ll C_{pl}$), backflux could be ignored ($k_{out}Q_{br} \approx 0$) and **Eqs. 6** and **7** simplified to

$$dQ_{br}/dt \approx K_{in}C_{pl} \qquad (10)$$

Integration of **Eq. 10** over the solute exposure time yields the following equations depending on whether plasma solute concentration in the experiment is constant or varies with time, such as after single dose intravenous injection.

$$Q_{br} \approx K_{in}C_{pl}T \quad \text{(if } C_{pl} = \text{constant)} \tag{11}$$

and

$$Q_{br} \approx K_{in}\int C_{pl}dt \quad \text{(if } C_{pl} = \text{not constant)} \tag{12}$$

Equations 11 and **12** can be rearranged to provide K_{in} explicitly in terms that are readily determined,

$$K_{in} \approx Q_{br}/C_{pl}T \quad \text{(if } C_{pl} = \text{constant)} \tag{13}$$

and

$$K_{in} \approx Q_{br}/\int C_{pl}dt \quad \text{(if } C_{pl} = \text{not constant)} \tag{14}$$

In the above analysis, Q_{br} is the quantity of solute that has left the vascular compartment and has entered brain parenchyma (i.e., crossed the BBB). As most analytical methods measure total brain solute content (Q_{tot}), comprising compound from both the brain parenchymal and vascular compartments, a vascular correction is generally required in order to obtain an accurate estimate Q_{br}.

$$Q_{tot} = Q_{br} + Q_{vas} \tag{15}$$

This adjustment is often made either by a) removing vascular tracer at the termination of the initial uptake period via brain vascular perfusion in situ with solute-free saline for 15–60 s *(18)* or by b) calculating Q_{vas} from the brain vascular volume (V_v) as measured with a poorly-penetrating BBB reference tracer such as [^{125}I]albumin, [^{14}C]dextran, or [^{3}H]inulin *(19,20)*. V_v is defined as the ratio of the measured content of vascular reference tracer in brain (Q_{vas}) at the end of the experiment divided by the matching plasma (C_{pl}) or blood (C_{bl}) concentration of vascular reference tracer. If the test solute under study does not enter blood cells, then one is interested primarily in a brain plasma correction (V_{pl}) and Q_{br} can be estimated as,

$$Q_{br} = Q_{tot} - V_{pl}C_{pl} \tag{16}$$

However, most solutes reside both in plasma and blood cells, and therefore it is more appropriate to make a blood correction (V_{bl}) in most instances,

$$Q_{br} = Q_{tot} - V_{bl}C_{bl} \tag{17}$$

Equations 16 and **17** adjust only for solute trapped within residual blood in the brain vasculature at the time of measurement, and does not take into account solute that may be bound to the vascular endothelium or trapped within BBB endothelial cells. This latter source can be estimated for large, polar solutes that permeate cell membranes slowly with the "capillary depletion" technique of Triguero *(21)*. With the capillary depletion method, the brain sample is homogenized in cold buffer and centrifuged in a 12–20% dextran solution that pellets the brain microvessels out of the homogenate. This method has been extremely valuable in helping to estimate peptide and protein transfer across the BBB in initial uptake experiments.

An alternate approach pioneered by Patlak and colleagues *(22,23)* is to use the solute itself as its own vascular marker, to express **Eqs. 13** and **14** in terms of Q_{tot}, and to divide each equation by the plasma solute concentration,

$$Q_{tot} \approx K_{in}C_{pl}T + Q_{vas} \quad \text{(if } C_{pl} = \text{constant)} \tag{18a}$$
$$Q_{tot}/C_{pl} \approx K_{in}T + Q_{vas}/C_{pl} \tag{18b}$$
$$Q_{tot}/C_{pl} \approx K_{in}T + V_i \tag{18c}$$

and

$$Q_{tot} \approx K_{in}\int C_{pl}dt + Q_{vas} \quad \text{(if } C_{pl} = \text{not constant)} \tag{19a}$$
$$Q_{tot}/C_{pl} \approx K_{in}[\int C_{pl}dt/C_{pl}] + Q_{vas}/C_{pl} \tag{19b}$$
$$Q_{tot}/C_{pl} \approx K_{in}[\int C_{pl}dt/C_{pl}] + V_i \tag{19c}$$

where V_i = the effective vascular space (Q_{vas}/C_{pl}) of the compound including endothelial binding and accumulation as well as intravascular distribution.

Equations 18c and **19c** are termed the "Patlak" equations and as they are in the form of a simple straight line (y = mx + b) provide a useful means to study solute uptake into brain and obtain estimates of K_{in}. As noted in the original 1983 publication *(22)*, the analysis is not dependent on a specific compartmental model. The data (Q_{tot}/C_{pl}) are plotted versus time (t) in constant C_{pl} experiments or "effective" time ($\int C_{pl}dt/C_{pl}$) in variable C_{pl} experiments, and the results are inspected for the initial linear portion where backflux is negligible. With this analysis, the slope of the linear regression curve is K_{in} and the y-intercept is Q_{vas}/C_{pl}. In concept, deviations or fall off from linearity can be readily assessed, and the method is thought to provide the most accurate assessment of vascular correction. Limitations to this approach also exist and will be discussed in greater detail in **Subheading 3.**

If experience determines that a simple vascular wash *(18)* or volume measurement using an impermeant vascular marker (e.g., [^{14}C]dextran, or [^{3}H]inulin)*(15)* provides an accurate correction for brain residual vascular tracer for a test solute, then K_{in} can be measured in single animal or assay experiments as,

$$K_{in} \approx (Q_{tot} - V_{bl}C_{bl})/C_{pl}T \quad \text{(if } C_{pl} = \text{constant)} \tag{20}$$

and

$$K_{in} \approx (Q_{tot} - V_{bl}C_{bl})/\int C_{pl}dt \quad \text{(if } C_{pl} = \text{not constant)} \tag{21}$$

Equations 20 and **21** are especially useful in pathological conditions where K_{in} and V_{bl} can vary significantly from animal to animal and within brain, so that the ability to make a single determination with simultaneous measurement of solute uptake and vascular volume is critical. In fact, they are the mainstay of many in vivo BBB permeability studies.

2.4. Blood Flow and BBB Permeability Coefficients

The transfer coefficient, K_{in}, gives an excellent index of the ease with which a solute can move from plasma into brain, but it is not a permeability coefficient. This is because, if the solute is sufficiently permeant, the rate by which the solute is presented to brain can also impact the solute concentration in blood as it transits from the arterial end to the venous end of the cerebral capillary (**Fig. 2**). At the far extreme, a compound with infinite BBB permeability would be rapidly depleted from blood in a single pass through the brain capillaries. As a result, the measured input concentration (C_{pl}) at the arterial end would not be the same as the average capillary concentration (*see* **Fig. 2**). The net result is that brain uptake (or efflux) also depends on cerebral blood or plasma flow (F), which varies within and between regions of the brain and is altered by mental activity,

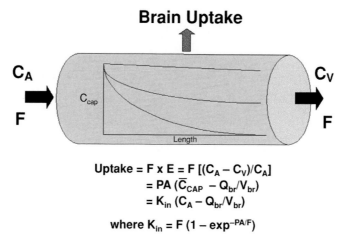

Brain Uptake

$$\text{Uptake} = F \times E = F\,[(C_A - C_V)/C_A]$$
$$= PA\,(\overline{C}_{CAP} - Q_{br}/V_{br})$$
$$= K_{in}\,(C_A - Q_{br}/V_{br})$$

where $K_{in} = F\,(1 - \exp^{-PA/F})$

Fig. 2. Diagram illustrating the relation between capillary fluid concentration, arterial input concentration (C_A), venous output concentration (C_V), flow (F), and capillary length. The difference between the capillary fluid concentration and brain fluid concentration (Q_{br}/V_{br}) is the driving force for solute diffusion across the BBB. The Crone Renkin equation relates K_{in} to F and PA.

drugs, and disease. Brain uptake is also influenced by the capillary surface area (A), which differs among brain regions and in rat cortical gray matter equals approximately 97–148 cm²/g brain *(24)*. The value in grey matter exceeds that in white matter by three- to fourfold.

Renkin *(25)* and Crone *(26)* modeled this flow dependence using the Krogh single capillary model, and derived the following theoretical relationship between flow (F), capillary permeability (P), and capillary surface area (A),

$$K_{in} = F[\,1 - \exp^{-PA/F}] \qquad (22)$$

Thus, K_{in} depends on three parameters: P, A, and F. Two limiting conditions in **Eq. 21** are worth noting. When F >> PA, $K_{in} \rightarrow$ PA, and when PA >> F, $K_{in} \rightarrow$ F. Functionally, $K_{in} \approx$ PA with less than 10% error when F > 5 x PA, and $K_{in} \approx$ F with less than 10% error when PA > 2.3 x F *(15)*. Thus, K_{in} is an acceptable estimate of BBB PA when F/PA > 5. When F/PA < 5, K_{in} is influenced by both PA an F. P is often expressed in the form of a permeability-surface area product (PA) in in vivo studies due to the difficulty of measuring regional A.

K_{in} can be converted to PA using a rearranged form of the Crone Renkin equation as,

$$PA = -F \ln\,(1 - K_{in}/F) \qquad (23)$$

F is measured commonly with radioactive iodoantipyrine, microspheres, or diazepam. Care needs to be placed in the choice of plasma or blood flow for F in **Eqs. 22** and **23**. If the compound in blood distributes only in plasma, then plasma F would be the more appropriate choice. However, if the compound distributes significantly in blood cells and blood cell exchange is rapid, then blood F may be more appropriate. Anesthetics can significantly alter F and thus care should be taken in the choice of an appropriate agent.

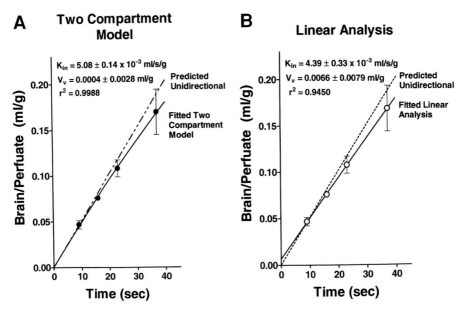

A   **Two Compartment Model**

$K_{in} = 5.08 \pm 0.14 \times 10^{-3}$ ml/s/g
$V_v = 0.0004 \pm 0.0028$ ml/g
$r^2 = 0.9988$

Predicted Unidirectional

Fitted Two Compartment Model

Y axis: Brain/Perfuate (ml/g)
X axis: Time (sec)

B   **Linear Analysis**

$K_{in} = 4.39 \pm 0.33 \times 10^{-3}$ ml/s/g
$V_v = 0.0066 \pm 0.0079$ ml/g
$r^2 = 0.9450$

Predicted Unidirectional

Fitted Linear Analysis

Y axis: Brain/Perfuate (ml/g)
X axis: Time (sec)

Fig. 3. Plot of the time course of rat brain uptake of [^{14}C]iodoacetamide during perfusion with physiologic saline fluid. Data were analyzed and fit using a two compartment model (**Eq. 8**) (**A**) and by simple linear Patlak plot (**B**) using the initial uptake assumption. The Y axis is the ratio of Q_{br} divided by C_{pf} to obtain the brain uptake "space" or brain/perfusate ratio (mL/g). The slope of the linear regression line is K_{in}. The perfusate contained [^{3}H]inulin to correct for intravascular tracer that is trapped in perfusion fluid within brain capillaries. Q_{br} was calculated as $(Q_{tot} - V_v Q_{pf})$ where V_v was obtained from the Q_{tot} / C_{pf} of [^{3}H]inulin. Data are mean ± SD for $n = 3$–4. The dotted line in both curves shows expected uptake for unidirectional influx (no backflux) using the K_{in} estimated using the two compartment model.

Finally, the single pass extraction (E) of a solute from blood as it passes from the arterial end to the venous end of a brain capillary can be calculated as,

$$E = K_{in} / F = (C_A - C_V) / C_A \qquad (24)$$

where C_A is the solute concentration at the arterial end and C_V is the concentration at the venous end of the brain capillary. In experiments that measure E directly, peripheral arterial blood is used for C_A and superior saggital sinus blood is used for C_V.

3. Example: Brain Uptake of [^{14}C]Iodoacetamide

Kinetic analysis of a typical initial brain uptake experiment is illustrated (**Fig. 3**). This data was obtained using the *in situ* rat brain perfusion technique that delivers solute to brain at a constant concentration and flow rate (*see* Chapter 13). [^{14}C]Iodoacetamide, the test tracer, is a small, moderately lipid soluble substance (octanol/water partition coefficient ≈ 1) that exhibits minimal plasma protein binding. Brain uptake of [^{14}C]iodoacetamide from protein-free physiologic saline fluid was determined at several times over 10–40 s. At the end of the experiment, total brain [^{14}C]iodoacetamide content (Q_{tot}) was measured and corrected for residual intravascular tracer using [^{3}H]inulin. Intravascular-corrected brain data (Q_{br}) were then expressed as brain/perfusate ratios (Q_{br} / C_{pl}) and plotted vs time on a graph.

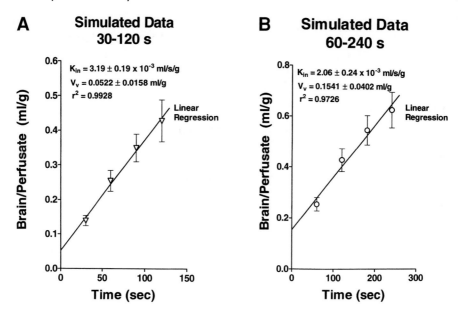

Fig. 4. **A,B,** Patlak plots of predicted data obtained using the two compartment model and the best-fit K_{in}, 5.08×10^{-3} mL/s/g and assuming ± 10% random error. Data are mean ± SD for n = 3–4. Data were analyzed by least-squares linear analysis to obtain best fit values for K_{in} and V_v.

Figure 3A illustrates the fit of the two compartment model to the data using a brain distribution volume (V_{br}) of 0.80 (i.e., brain water content) and allowing for an vascular component (V_i) in addition to that measured by [³H]inulin,

$$Q_{br} = C_{pl}V_{br} \left(1 - e^{-K_{in}T/V_{br}} \right) + V_i \qquad (25)$$

The mean brain [³H]inulin space equaled 0.0108 ± 0.0006 mL/g consistent with previous reports *(16,20)*. As intravascular tracer was already subtracted from brain, the calculated V_i obtained by fitting the model to the data was expected to approach zero.

[¹⁴C]Iodoacetamide was taken up into brain at a fairly significant rate reaching a brain-perfusion fluid Q_{br}/C_{pf} ratio of 0.15–0.20 mL/g in only 40 s. The calculated BBB K_{in} value obtained by fitting the complete two-compartment model **(Eq. 25)** to the data equaled $5.08 \pm 0.14 \times 10^{-3}$ ml/s/g. The simultaneously derived V_i was very small (0.0004 ± 0.0028 ml/g) and did not differ significantly from zero, as expected for the tracer. The best fit r^2 was 0.9988. The matching analysis **(Fig. 3B)** using the linear uptake equation **(Eq. 18c)** provided a slightly lower BBB K_{in} value, $4.39 \pm 0.33 \times 10^{-3}$ mL/s/g, and a slightly higher V_v estimate of 0.0066 ± 0.0079 ml/g ($r^2 = 0.9450$). Fall off from unidirectional uptake is shown in both plots and biases the linear analysis to slightly underestimate K_{in} and overestimate V_i. The artifactual increase in V_i makes the reduction in K_{in} greater than if a V_i of zero had been assumed. Thus, the simple linear analysis worked reasonably well, although a bias toward reduced K_{in} and increased V_i was noted.

However, the assumption that demonstration of linear uptake ensures unidirectional influx and protects one from significantly underestimating K_{in} was questioned in further modeling of the data. As shown in **Fig. 4**, predicted uptake data, obtained using the two-compartment model and the best fit K_{in} and V_i values, over two extended time ranges (30–120 s and 60–240 s) also were quite "linear" with r^2 values of 0.97–0.99 when a sampling of four uptake times were used that were evenly spaced across the graph. K_{in} decreased to approx 30% of the original value and V_i increased to 0.15 mL/g as the uptake time was extended to longer intervals. Therefore, demonstration of "linear uptake" does not guarantee an accurate K_{in} determination and appropriate care should be employed to further analyze the unidirectional influx assumption, especially in experiments that give significantly positive V_i intercepts after vascular correction. In the case of [^{14}C]iodoacetamide, with a V_{br} of 0.80 and measured Q_{br}/C_{pl} ratios of 0.15–0.20 at 40 s, one would expect that the effective concentration of tracer in brain at 40 s would be 19–25% of that in perfusion fluid, and thus backflux during those 40-s experiments may lead to an underestimation in the unidirectional uptake value by $100 \times (0.19–0.25)/2 = 9.5–12.5\%$. These values are within a reasonable range. In the extended time plots (*see* **Fig. 4B**), Q_{br}/C_{pl} ratios for many of the later time data points fell in 0.3–0.6 mL/g range, in which case backflux would be much greater. Detection of nonlinearity, in this instance, would depend on whether one or more time points was chosen in the 8–20 s range. Clearly, more evidence than simple linear uptake, as in the Patlak plot, is needed in order to document unidirectional BBB transport. This often can be obtained by supplemental measurements of V_{br} or brain efflux.

4. BBB Transport Methods

A range of experimental techniques exist to examine solute transport across the BBB in vivo or in vitro. Each has particular strengths, as well as limitations that need to be watched. Seven methods will be discussed briefly below in a presentation that is meant to complement the specific techniques outlined later in this section.

4.1. Intravenous Injection Method

The "gold standard" for all BBB work is the intravenous (iv) injection method where a solute is administered intravenously to an animal and solute concentration is determined at different times in brain, plasma, and CSF *(27–29)*. The data can be utilized to provide estimates of BBB K_{in}, k_{out}, V_{br}, F, and PA and lend themselves to model testing. The method has the advantage that the system is intact, all transporters, junctional proteins, and enzymes are present at their physiologic concentrations, the unique architecture of the blood vessels and perivascular cells is present and undisturbed, cerebral metabolic pathways are not compromised, and regulatory pathways, neuronal input, and second messenger systems have not been damaged or altered. Further, the iv injection method also allows accurate incorporation of exchange and contributions with CSF and choroid plexus pathways. In effect, in vivo is the real thing that is extremely difficult to reconstitute in vitro. It lends itself well to correlative studies with cerebral blood flow and metabolism, and can readily be used with quantitative autoradiography, magnetic resonance imaging, and positron emission tomography for analysis at the regional level. It is essentially the approach used by Ehrlich to demonstrate originally the presence of a BBB more than 100 years ago and was used again most recently to help establish the critical importance of efflux at the BBB in the studies of drug distribution in P-glycoprotein (P-gp) knockout mice *(4,30)*.

Limitations of the approach are also related to its strength. Because the in vivo BBB system is so complex, it is often difficult to dissect out roles of individual components.

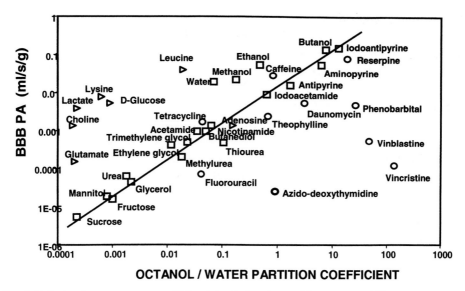

Fig. 5. Relation between measured blood–brain barrier capillary permeability and surface area (BBB PA) and the octanol/water distribution coefficient for reference tracers, drugs, and nutrients. BBB PA data are all from *in situ* brain perfusion in the absence of plasma protein binding. Each point represents a mean for $n = 3–8$ perfusions.

Control of circulating plasma levels of hormones, ions, nutrients, and proteins is limited and time consuming, which impedes studies that characterize the roles of individual transport proteins and carriers. The method often requires multiple samples, the analysis of which can be costly in time and resources. Solute metabolism in liver and other organs can complicate data acquisition and analysis, particularly when radiotracers are used. And finally, the method provides no ready means to modulate brain solute concentrations in brain extracellular fluid for studies of BBB efflux.

4.2. In Situ *Perfusion*

The *in situ* perfusion method (*see* Chapter 13), complements the iv injection method with which it holds a number of factors in common *(20,31–33)*. The perfusion method utilizes the in vivo structure of the BBB and cerebral tissues, and simply superimposes its own vascular perfusion fluid in replacement to the animal's circulating blood. The particular advantage of this method is the ready control of perfusate solute concentration that can be altered at will over a much larger range than generally tolerated in vivo. Concentration can be changed virtually instantaneously and maintained constant, which simplifies kinetic analysis. Other constituents of perfusion fluid can also be varied allowing ready characterization of saturable transport systems, plasma protein binding and the effects of regulatory modifiers, hormones, and neurotransmitters that can be presented to the brain at defined concentrations. Furthermore, because solute goes directly from perfusate to brain, without transit to other organs, artifacts that arise due tracer metabolism are minimized. The perfusion approach, originally developed for rats *(32–33)*, has been expanded for mice *(34)*, guinea pigs *(35)*, and rabbits *(36)*. The method readily lends itself for use with genetically modified animals, such as the P-gp knockout mice *(37)* and can also be used for efflux studies *(18)*. The relation between BBB PA and octanol water distribution coefficient (pH 7.4) is shown for PA data obtained via perfusion measurement for reference tracers, drugs, and nutrients in **Fig. 5**.

Drawbacks of the perfusion technique include a number of those listed for iv injection where the complexity of the system can be limiting. Whereas permeant solutes can be readily loaded into brain for efflux studies, the method does not lend itself to efflux studies of poorly penetrating solutes and provides no ready means to control brain extracellular solute concentrations for characterization studies of efflux transport systems. Furthermore, similar to iv injection, complete kinetic analysis can involve a number of animals and require significant analytical time and commitment. The time investment is considerably reduced once the kinetics of a given solute is known and the method can be simplified to the single time point approach.

4.3. Brain Efflux Index

Mechanisms of brain-to-blood efflux can also be investigated with the brain efflux index technique of Kakee et al. *(38)*, which is described in Chapter 14. This method involves direct microinjection of test solute and impermeant reference tracers into brain *(38,39)*. At various times thereafter, the ratio of test tracer to impermeant reference marker (R) is determined in brain and expressed as a brain efflux index (BEI) value, defined as

$$BEI = 100 \times (1 - [(R)_{brain} / (R)_{injectate}]. \tag{26}$$

From this data a rate coefficient for efflux (k_{out}) from brain is calculated and is converted to a transfer coefficient (i.e., clearance) for efflux, as $K_{out} = k_{out} \times V_{br}$ where V_{br} as determined from the steady-sate distribution of test solute in brain slices in vitro. Caution must be exercised with the technique as BBB damage from needle tract injections may alter BBB transport or blood flow. Similarly, the necessity in some experiments to use very high levels (e.g., 50–100 m*M*) of competitor or transport inhibitor in the injection solution may also have adverse effects. Finally, the calculated K_{out} is dependent on accurate determination of V_{br} which may change with in vitro tissue slice incubation or due to the effects of transport inhibitors.

The brain efflux index method has produced some very valuable insight on the role and characterization of a number of efflux transport systems at the BBB *(39–41)*.

4.4. Isolated Membrane Vesicles

The introduction of brain microvessel membrane vesicles for separate analysis of the transport properties of isolated luminal and abluminal membranes of the BBB by Sanchez del Piño *(12)* opened an extremely important door to verify the presence of specific transport systems at the BBB and study their polarized function under controlled conditions (*see* Chapter 15). The method allows determination of the activity of transporters from separate luminal and abluminal membranes. It has verified the selective distribution of several key active transporters to the abluminal membrane, including Na^+-K^+-ATPase, the system A and system B^{o+} sodium-dependent amino acid transporters, and the sodium-dependent glutamate transporters, EAAT1, EAAT2, and EAAT3 *(42,43)*. It has also been used to examine the distribution and function of critical BBB-facilitated exchange carriers, such as the BBB glucose transporter (GLUT-1) and the sodium-independent large neutral amino acid transporter (LAT-1) *(44,45)*. Limitations to this approach include the fact that the membranes fractions are not completely pure with regard to luminal vs abluminal membrane origin, so that corrections must be made for contamination, and the fact that the membrane protein characteristics and regulation may be altered by the isolation procedure.

4.5. Intracerebral Microdialysis

Intracerebral microdialysis (*see* Chapter 16) offers the distinct advantage of an internal portal to the extracellular fluid of the brain. Thus, although most in vivo methods lump the brain parenchyma, including extracellular and intracellular spaces, into a single unit, brain microdialysis samples selectively, using a semipermeable membrane, from the brain interstitial compartment and at steady state allows determination of the brain interstitial fluid concentration (*46,47*). The method has been used extensively to document the kinetics of solute exchange between plasma and brain interstitial fluid and offers the advantage of protein-free sample fluids for ready determination of solute concentration by LC-MS or HPLC. Limitations to the method include the damage to BBB and brain cell membranes that accompany microdialysis probe insertion in the CNS and can cause overestimation of transport rates for polar compounds (*48–50*). In addition, dialysis probe efficiency can vary under different conditions (*51*) and, because only one or two small probes are generally used per animal, the method does not lend itself to efficient regional comparison of brain uptake or efflux. However, its time resolution is excellent and a significant amount of data can be obtained from each animal experiment.

4.6. Choroid Plexus and CSF

Lastly, the transport roles of choroid plexus and CSF in solute uptake and distribution in brain remain unresolved with many important questions. The choroid plexuses contribute significantly through the CSF to brain uptake of certain ions (e.g., Na, Ca) and metals. Their quantitative role may differ between species based upon differences in brain size and diffusion distance between brain core and ventricle. Chapter 19 presents methods for characterizing choroid plexus transport in vivo and in vitro.

References

1. Pardridge, W. M. (2001) BBB genomics: creating new openings for brain-drug targeting. *Drug Discovery Today* **6**, 381–383.
2. Lee, G., Dallas, S., Hong, M., and Bendayan, R. (2001) Drug transporters in the central nervous system: brain barriers and brain parenchymal considerations. *Pharmacol. Rev.* **53**, 569–596.
3. Smith, Q. R. (2000) Transport of glutamate and other amino acids at the blood-brain barrier. *J. Nutrition* **130**, 1016S–1022S.
4. Schinkel, A. H., Smit, J. J., van Tellingen, O., et al. (1994) Disruption of the mouse mdr1a P-glycoprotein gene leads to a deficiency in the blood-brain barrier and to increased sensitivity to drugs. *Cell* **77**, 491–502.
5. Keep, R. F., Ennis, S. R., and Betz, A. L. (1998) Blood-brain barrier ion transport. In; *Introduction to the Blood-Brain Barrier.* Pardridge, W. M., ed. Cambridge University Press, Cambridge, pp. 207–213.
6. Lee, W. J., Peterson, D. R., Sukowski, E. J., and Hawkins, R. A. (1997) Glucose transport by isolated plasma membranes of the bovine blood-brain barrier. *Am. J. Physiol.* **272**, C1552–1557.

7. Reese, T. S., and Karnovsky M. J (1967) Fine structural localization of a blood-brain barrier to exogenous peroxidase. *J. Cell Biol.* **34,** 207–217.

8. Brightman, M. W., and Reese, T. S. (1969) Junctions between intimately apposed cell membranes in the vertebrate brain. *J. Cell Biol.* **40,** 648–677.

9. Huber, J. D., Egleton, R. D., and Davis, T. P. (2001) Molecular physiology and pathophysiology of tight junctions in the blood-brain barrier. *Trends Neurosci.* **24,** 719–725.

10. Smith, Q. R., and Rapoport, S. I. (1986) Cerebrovascular permeability coefficients to sodium, potassium, and chloride. *J. Neurochem.* **46,** 1732–1742.

11. Butt, A. M., and Jones, H. C. (1992) Effect of histamine and antagonists on electrical resistance across blood-brain barrier in rat brain-surface microvessels. *Brain Res.* **569,** 100–105.

12. Sanchez del Pino, M. M., Hawkins, R. A., and Peterson, D. R. (1995) Biochemical discrimination between luminal and abluminal enzyme and transport activities of the blood-brain barrier. *J. Biol. Chem.* **270,** 14907–14912.

13. Johanson, C. E. (1998) Arachnoid membrane: subarachnoid CSF and pia-glia. In: *Introduction to the Blood-Brain Barrier.* Pardridge, W. M., ed. Cambridge University Press, Cambridge, pp. 259–269.

14. Segal, M. (1998) The blood-CSF barrier and the choroid plexus. In: *Introduction to the Blood-Brain Barrier.* Pardridge, W. M., ed. Cambridge University Press, Cambridge, pp. 251–258.

15. Smith, Q. R. (1989) Quantitation of blood-brain barrier permeability. In: *Implications of the Blood-Brain Barrier and Its Manipulation.* Vol. 1. Neuwelt, E. A., ed. Plenum Press, New York, pp 85–118.

16. Bradbury, M. W. B. (1979) *The Concept of a Blood-Brain Barrier*, Wiley, Chirchester, 1979.

17. Rapoport, S. I., Ohno, K., and Pettigrew, K. D. (1979) Drug entry into the brain. *Brain Res.* **172,** 354–359.

18. Allen, D. D., and Smith, Q. R. (2001) Characterization of the blood-brain barrier choline transporter using the *in situ* rat brain perfusion technique. *J. Neurochem.* **76,** 1032–1041.

19. Smith, Q. R., Ziylan, Z., Robinson, R. J., and Rapoport, S. I. (1988) Kinetics and distribution volumes for tracers of different sizes in the brain plasma space. *Brain Res.* **462,** 1–9.

20. Takasato Y., Rapoport S. I., and Smith Q. R. (1984) An *in situ* brain perfusion technique to study cerebrovascular transport in the rat. *Am. J. Physiol.* **247,** H484–H493.

21. Triguero, D., Buciak, J., and Pardridge, W. M. (1990) Capillary depletion method for quantification of blood-brain barrier transport of circulating peptides and plasma proteins, *J. Neurochem.* **54,** 1882–1888.

22. Patlak, C. S., Blsberg, R. G., and Fenstermacher, J. D. (1983) Graphical evaluation of blood-to-brain transfer constants from multiple-time uptake data. *J. Cereb. Blood Flow Metab.* **3,** 1–7.

23. Blasberg, R. G., Fenstermacher, J. D., and Patlak, C. S. Transport of α-aminoisobutyric acid across brain capillary and cellular membranes. *J. Cereb. Blood Flow Metab.* **3,** 8–32.

24. Gross, P. M., Sposito, N. M., Pettersen, S. E., and Fenstermacher, J. D. (1986) Differences in function and structure of the capillary endothelium in gray matter, white matter, and a cricumventricular organ of rat brain. *Blood Vessels* **23,** 261–270.

25. Renkin, E. M. (1959) Transport of potassium from blood to tissue in isolated mammalian skeletal muscles. *Am. J. Physiol.* **197,** 1205–1210.

26. Crone, C. (1963) Permeability of capillaries of various organs as determined by use of the "indicator diffusion" method. *Acta Physiol. Scand.* **58,** 292–305.

27. Ohno, K., Pettigrew, K. D., and Rapoport, S. I. (1978) Lower limits of cerebrovascular permeability to nonelectrolytes in the conscious rat. *Am. J. Physiol.* **235,** H299–H307.

28. Duncan, M. W., Villacreses, N., Pearson, P. G., et al. (1991) 2-Amino-3-(methylamino)-propanoic acid (BMAA) pharmacokinetics and blood-brain barrier permeability in the rat. *J. Pharmacol. Exp. Ther.* **258,** 27–35.

29. Bickel, U., Schumacher O. P., Kang, Y. S., and Viogt, K. (1996) Poor permeability of morphine 3-glucuronide and morphine 6-glucuronide through the blood-brain barrier in the rat. *J. Pharmacol. Exp. Ther.* **278,** 107–113.

30. Schinkel, A. H., Wagenaar, E., Mol, C. A. A. M., and van Deetmer, L. (1996) P-glycoprotein in the blood-brain barrier of mice influences the brain distribution of many compounds. *J. Clin. Invest.* **97,** 2517–2524

31. Takasato, Y., Rapoport, S. I., and Smith, Q. R. (1982) A new method to determine cerebrovascular permeability in the anesthetized rat, *Soc. Neurosci. Abstr.* **8,** 850.

32. Momma, S., Aoyagi, M., Rapoport, S. I., and Smith, Q. R. (1987) Phenylalanine transport across the blood-brain barrier as studied with the *in situ* brain perfusion technique. *J. Neurochem.* **48,** 1291–1300.

33. Smith, Q. R. (1996) Brain perfusion systems for studies of drug uptake and metabolism in the central nervous system. *Pharmaceutical. Biotechnol.* **8,** 285–308.

34. Murakami, H., Takanaga, H., Matsuo, H., Ohtani, H., and Sawada, Y. (2000) Comparison of blood–brain barrier permeability in mice and rats using *in situ* brain perfusion technique. *Am. J. Physiol.* **279,** H1022–H1028.

35. Zloković, B. V., Begley, D. J., Djuričić, B. M., and Mirtovic, D. M. (1986) Measurement of solute transport across the blood-brain barrier in the perfused Guinea Pig brain: method and application to N-methyl-α-aminoisobutyric acid. *J. Neurochem.* **46,** 1444–1451.

36. Hervonen, H., and Steinwall, O. (1984) Endothelial surface sulfhydryl-groups in blood-brain barrier transport of nutrients, *Acta Physiol. Scand.* **121,** 343–351.

37. Cisternino, S., Rousselle, C., Dagenais, C., and Scherrmann J. M. (2001) Screening of multidrug resistance sensitive drugs by in situ brain perfusion in P-glycoprotein-deficient mice. *Pharm. Res.* **18,** 183–190.

38. Kakee, A., Terasaki, T., and Sugiyama, Y. (1996) Brain efflux index as a novel method of analyzing efflux transport at the blood-brain barrier. *J. Pharmacol. Exp. Ther.* **277,** 1550–1559.

39. Kakee, A., Terasaki, T., and Sugiyama, Y. (1996) Selective brain to blood efflux transport of para-aminohippuric acid across the blood-brain barrier: In vivo evidence by use of the brain efflux index method. *J. Pharmacol. Exp. Ther.* **283,** 1018–1025.

40. Kakee, A., Takanaga, H., Terasaki, T., Naito, M., Tsuruo, T., and Sugiyama, Y. (2001) Efflux of a suppressive neurotransmitter, GABA, across the blood-brain barrier. *J. Neurochem.* **79,** 110–118.

41. Takanaga, H., Tokuda, N., Ohtsuki, S., Hosoya, K., and Terasaki, T. (2002) ATA2 is predominantly expressed as system A at the blood-brain barrier and acts as brain-to-blood efflux transport of proline. *Mol. Pharmacol.* **61,** 1289–1296.

42. Sánchez del Pino, M. M., Hawkins, R. A., and Peterson, D. R. (1995) Neutral amino acid transport characterization of isolated luminal and abluminal membranes of the blood-brain barrier. *J. Biol. Chem.* **270,** 14913–14918.

43. O'Kane, R. L., Martínez-López, I., DeJoseph, M. R., Viña, J. R., and Hawkins, R. A. (1999) Na+-dependent glutamate transporters (EAAT1, EAAT2, and EAAT3) of the blood-brain barrier. *J. Biol. Chem.* **274,** 31891–31895.

44. Sánchez del Pino, M. M., Hawkins, R. A., and Peterson, D. R. (1992) Neutral amino acid transport by the blood-brain barrier. *J. Biol. Chem.* **267,** 25951–25957.

45. Simpson, I. A., Vannucci, S. J., DeJoseph, M. R., and Hawkins, R. A. (2001) Glucose transporter asymmetries in the bovine blood-brain barrier. *J. Biol. Chem.* **276,** 12725–12729.

46. Wang, Y., Wong, S. L., and Sawchuk, R. J. (1993) Microdialysis calibration using retrodialysis and zero-net flux: application to a study of the distribution of zidovudine to rabbit cerebrospinal fluid and thalamus. *Pharmacol. Res.* **10,** 1411–1419.

47. de Lange, E. C. M., Danhof, M., de Boer, A. G., and Breimer, D. D. (1997) Methodological considerations of intracerebral microdialysis in pharmacokinetic studies of drug transport across the blood-brain barrier. *Brain Res. Rev.* **25,** 27–49.

48. Morgan, M. E., Singhal, D., and Anderson, B. D. (1996) Quantitative assessment of blood-brain barrier damage during microdialysis. *J. Pharmacol. Exp. Ther.* **277,** 1167–1176.

49. Major, O., Shdanova, T., Duffek, L., and Nagy, Z. (1990) Continuous monitoring of blood-brain barrier opening to Cr^{51}-EDTA by microdialysis following probe injury. *Acta Neurochir.* **51,** 46–48.

50. Westergren, I., Nystrom, B., Hamberger, A., and Johansson, B. (1995) Intracerebral dialysis and the blood-brain barrier. *J. Neurochem.* **64,** 229–234.

51. Sun, H., Bungay, P. M., and Elmquist, W. F. (2001) Effect of capillary efflux transport inhibition on the determination of probe recovery during in vivo microdialysis in the brain. *J. Pharmacol. Exp. Ther.* **297,** 991–1000.

13

In Situ Brain Perfusion Technique

Quentin R. Smith and David D. Allen

1. Introduction

Many techniques have been developed to study transport across the blood–brain barrier (BBB). One that offers particular advantage is the *in situ* brain perfusion technique. The primary objective of the *in situ* brain perfusion technique is to take over the circulation to the brain via direct infusion of artificial blood, plasma, or saline into the heart or major vessels leading to the brain, such as the carotid arteries *(1–4)*. The solute of interest is mixed into the perfusion medium at a known concentration and delivered to the brain via the perfusion fluid for a defined interval. At preselected times, the perfusion is stopped and the amount of solute in brain is determined. From these measurements, the kinetics of brain uptake are analyzed and appropriate transport or permeability constants calculated *(5)*.

There are a number of advantages of this approach for studies of BBB transport. First, the kinetics of brain uptake via perfusion are quite simple because the compound of interest is delivered to the brain at a set, known concentration that does not vary with time. Second, because of the investigator's control of perfusion uptake time, transport can be measured over a broad range of intervals (anywhere from 5 s to >30 min), allowing accurate determination of permeability over a 10^5 range *(2)*. Third, because solute is not delivered first to peripheral metabolizing organs (e.g., liver, lung or kidney), the investigator can generally be confident that the compound is presented to brain in the intact form and thus minimizes chances for error due to brain uptake of permeable metabolites, especially ^3H-H_2O and ^{14}C-CO_2 *(6,7)*. Fourth, because the concentration of solute as well as potential competitors, blockers, ions (e.g., Na^+), and plasma proteins can be readily manipulated in the perfusion medium, the method is ideal for in vivo characterization of the kinetics of saturable transport across the BBB and the effects of plasma protein binding *(4,8–10)*. And finally, the method can be used to study solute efflux from brain after preloading by switching the perfusate to solute-free saline *(3,10)*.

A number of perfusion systems have been published over the past 20 yr for study of drug transport into brain. This chapter will present the most widely used method—the *in situ* rat brain perfusion technique (or "Takasato" method).

From: *Methods in Molecular Medicine, vol. 89:*
The Blood–Brain Barrier: Biology and Research Protocols
Edited by: S. Nag © Humana Press Inc., Totowa, NJ

2. Materials

2.1. Surgical Preparation

1. Sodium pentobarbital, or a suitable animal anesthetic (Sigma Chemical, St. Louis, MO).
2. Surgical shaver (e.g., electric clippers) (Stoelting, Wood Dale, IL or Fisher Scientific, Pittsburgh, PA).
3. Heating blanket or light—to maintain rat body temperature during surgery (Stoelting or Harvard Apparatus, Holliston, MA).
4. Indicating Controller (Yellow Springs Instruments, Yellow Springs, OH).
5. Temperature probe (Harvard Apparatus, Stoelting or Yellow Springs Instruments).

2.2. Surgery

1. Rodent surgical board (Harvard Apparatus or Stoelting).
2. Binocular dissecting microscope (up to power 40x) (Carl Zeiss Surgical, Thornwood, NY).
3. Fiber optic light illuminator—for dissecting microscope (Carl Zeiss Surgical).
4. Surgical instruments (e.g., microdissecting tweezers and forceps, retractors, microvascular clamps and applicator, spring scissors, microdissecting scissors, operating scissors, retractors, hemostatic forceps, and scalpel blades and handles) (Roboz Surgical Instruments, Rockville, MD; Daigger, Vernon Hills, IL).
5. Surgical silk (4-0, 6-0) (Ethicon, Johnson & Johnson Health Care Systems, Piscataway, NJ).
6. Polyethylene tubing (PE- 60 or PE-50)—for catheters (Becton Dickinson, Sparks, MD).
7. Plastic syringes (1 mL and 5 mL) (Becton Dickinson).
8. Syringe needles—20-gage for PE-60 tubing and 23-gage for PE-50 (Becton Dickinson).
9. Plastic ultra three-way stopcocks (Medex, Hilliard, OH) or four-way stopcock (Hamilton, Reno, NV).
10. Heparin (Sigma, St. Louis, MO).
11. Bacteriostatic 0.9% sodium chloride solution (Abbott Laboratories, North Chicago, IL).
12. Malis Bipolar coagulator (Codman, Raynham, MA).
13. Cotton-tipped applicators (Allegiance, MacGraw Park, IL).
14. Absorbent paper (Fisher Scientific).

2.3. Perfusion Fluid

1. Perfusion buffer or fluid: If physiologic saline fluid is used, excellent results are obtained with a perfusion buffer of the following composition: 128 mM NaCl, 24 mM NaHCO$_3$, 4.2 mM KCl, 2.4 mM NaH$_2$PO$_4$, 1.5 mM CaCl$_2$, and 0.9 mM MgCl$_2$ (Sigma or Fisher Scientific).
2. D-Glucose: added to perfusion fluid on day or experiment (Sigma).
3. Bovine serum albumin or 70 k dextran if oncotic agent is used; added to perfusion fluid on day or experiment (Sigma).
4. Red cells: collected fresh on the day of the experiment in physiologic blood perfusions.
5. 95% oxygen/5% carbon dioxide gas.
6. Glass infusion syringe; 5–50 mL, depending on required infusion rate for infusion pump (Becton Dickinson).
7. Circulating water bath: capable on holding temperature of perfusion fluid and syringes at 37–39° C (Lauda, Daigger or Fisher Scientific).

2.4. Perfusion Procedure

1. Constant or variable rate infusion pump (Harvard Apparatus, Holliston, MA).
2. Physiologic pressure transducer (Gould, Valleyview, OH).
3. Amplifier and chart recorder for pressure transducer (Gould).

4. Microcentrifuge: for centrifuging blood samples (Fisher Scientific).
5. Analytical balance: for weighing samples (Sartorius, Daigger).
6. Kimwipes (Kimberly-Clark, Roswell, GA).
7. Ice bucket with ice (Fisher Scientific).
8. Glass petri dish (12–20 cm): to dissect brain on (Fisher Scientific).
9. Filter paper (Whatman, Fisher Scientific).
10. Vials (plastic or glass): for sample collection and holding (Fisher Scientific).
11. Stop watch or timer (Fisher Scientific or Daigger).

3. Methods

The specific steps required to carry out a brain perfusion experiment are described from the point of anesthesia and surgical preparation of the animal through sample collection and data analysis

3.1. Surgical Preparation

1. Anesthetize an adult rat (body weight 220–380 g) with sodium pentobarbital (40–50 mg/kg, 50–60 mg/mL, intraperitoneally [i.p.]) or other suitable anesthetic agent.
2. Once surgical anesthesia is obtained, the neck of the animal is shaved using electric clippers and the area is cleaned in preparation for surgery.
3. Attach the animal to the surgical board taking care not to inhibit respiration or blood flow to the head, limbs or thoracic cavity.
4. The animal's body temperature is maintained at 37°C during surgery and perfusion through the use of a heating pad or lamp that is linked to a feedback device (Indicating Controller) with a temperature probe placed in the rectum.

3.2. Surgery

1. A surgical incision is made on the ventral surface of the neck and the subcutaneous tissue and muscles are moved laterally to expose either the right or left carotid artery system (**Fig. 1**, *see* **Note 1**). Choice or right or left carotid artery depends on preference of the surgeon.
2. The common carotid artery is encircled and ligated with surgical silk (4-0) and then a cannula (15 cm of PE-60 containing 100 IU/mL heparin in 0.9% NaCl) is placed in the common carotid artery (*see* **Fig. 1**) distal to the ligature for delivery of perfusion fluid to brain (*see* **Notes 2** and **3**).
3. The corresponding external carotid artery is also ligated with surgical silk (6-0) to reduce the volume of perfusion fluid that goes to extracerebral tissues.
4. At the end of the surgical procedure, the wound site is cleaned and body rectal temperature is checked to ensure that it is approx 37°C. Rat rectal temperature is maintained following surgery through the use of a heating pad or heat lamp connected to a feedback device and a rectal probe.

3.3. Perfusion Fluid

1. The most frequently used perfusion fluid is a bicarbonate-buffered physiologic saline *(8)* (*see* **Subheading 2.3.** and **Note 4**).
2. Five to 10 m*M* D-glucose is usually added to the perfusate on the day of the experiment to maintain cerebral glucose metabolism and adenosine triphosphate (ATP) production.
3. Plasma oncotic agents and oxygen-carrying agents are also added as required (*see* **Note 5**).
4. Care should be taken to filter the perfusion medium (0.2 μm) and heat it to 37°C.
5. Filtered erythrocytes (if used) are added to a hematocrit of 10–40% *(2,3)*(*see* **Note 6**).

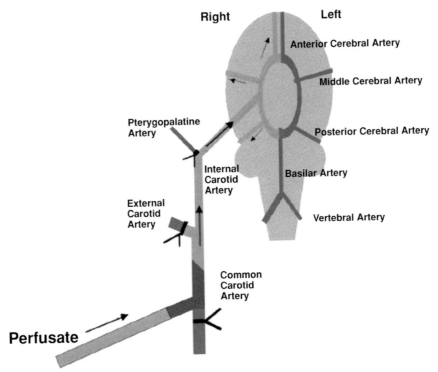

Fig. 1. Diagram of the modified Takasato method with catheter placement in the common carotid artery and ligation of the external carotid artery. The pterygopalatine artery can be either ligated or left open (*6,10,12*). The heart is stopped just before the start of perfusion to minimize mixing.

6. The solution is bubbled gently with humidified 95% O_2/5% CO_2 to pH 7.4 (1–3 min) to attain steady state gas levels within the solution.
7. The test compound of interest is added to the perfusate at known concentration.
8. The solution is drawn up within a heated (37°C) glass syringe and maintained at 37°C throughout the experiment through the use of a circulating water bath and copper syringe jackets (*see* **Note 7**).
9. The perfusion syringe is placed within the perfusion pump and the pump is set for an infusion rate of 5–20 mL/min (*see* **Note 8**).
10. The perfusion catheter is attached to the perfusion syringe through a 20-gage needle and a three-way stopcock. The three-way stopcock can be used to attach a pressure transducer in line to monitor perfusion pressure during perfusion (*2*). In addition, a four-way valve can be put in line with a dual syringe infusion pump if "pre-perfusion" or "post-perfusion" with a different fluid is to be used as part of the uptake measurement.

3.4. Perfusion Procedure

1. The thoracic cavity is opened and the cardiac ventricles severed to stop blood flow from the systemic circulation.
2. Then, perfusion fluid is infused up the common carotid artery at a rate (5–20 mL/min) sufficient to perfuse the ipsilateral cerebral hemispheres at a normal pressure (approx 80–120 mm Hg) and the time of perfusion is monitored with a stopwatch. **Figure 2** illustrates a typical set up for a perfusion experiment.

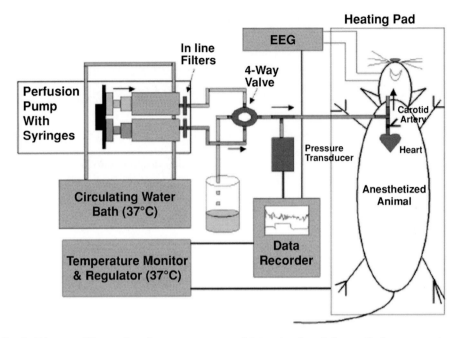

Fig. 2. Diagram illustrating the arrangement of the animal and the perfusion pump, temperature pad and regulator, and circulating water bath. Perfusion pressure can be measured with an in line pressure transducer linked to a recorder (Gould). The electroencephalograph can also be monitored with the same recorder used for the perfusion pressure.

3. Perfusion fluid that has passed through the brain vasculature bed and exited to the venous system accumulates in the thoracic cavity.
4. A prewash of the brain vasculature with drug or test-solute free fluid may be performed via perfusion for 10–45 s to clear the cerebral vessels of red cells and plasma constituents before onset with fluid containing test solute.
5. Because only one carotid artery is perfused, flow in the ipsilateral cerebral hemisphere exceeds that in the contralateral cerebral hemisphere by five- to tenfold. Bilateral perfusion can be achieved by cannulating and infusing fluid into both common carotid arteries (right and left).
6. Because the heart is stopped, there is no contribution to flow from the systemic circulation and the brain flow rate can be set at will for studies of the flow-dependent transport or metabolism.
7. At the end of the perfusion period, the pump is turned off and the animal is decapitated.
8. The brain is removed from the skull and placed on a filter paper (Whatman no. 1) moistened with 0.9% NaCl and sitting on a Petri dish cooled on ice (within an ice bucket).
9. Surface blood vessels are removed from the brain using cotton swabs.
10. The ipsilateral hemisphere is dissected into regions using surgical scalpel blades. Tissue sections are placed in sample tared vials and reweighed to obtain tissue weight.
11. Corresponding samples of perfusion fluid are also collected, placed in vials, and analyzed to obtain sample weight.
12. The total concentration of solute of interest in brain is then determined by scintillation counting, LC-MS, HPLC, GC or other suitable analytical means. When metabolism is a factor, it is often necessary to perform chromatography to carefully measure intact compound. Passage of solute across the blood-brain barrier can be demonstrated by separating

the cerebral microvessels from brain parenchyma using the capillary depletion technique *(11,12)*.

13. At the end of perfusion, the amount of compound trapped in perfusion fluid in the brain blood vessels (Q_{vas}) is either removed from the brain by a short washout (10–15 s) with perfusion fluid that does not contain the compound of interest or it is measured indirectly using a vascular marker, such as [^{3}H]-sucrose or [^{14}C]-inulin which is included in the perfusion fluid *(2,8)*.

3.5. Calculations

1. The BBB transfer coefficient for unidirectional uptake of solute into brain from perfusion fluid (K_{in}) is calculated as,

$$K_{in} = (Q_{tot} - V_v Q_{pf}) / C_{pf} T \tag{1}$$

where C_{pf} = the concentration of solute in the perfusion fluid (mass drug/volume fluid), Q_{tot} = total amount of solute in brain (mass/weight brain), V_v = brain vascular volume (mL/g), and T= perfusion time.

2. The perfusion uptake time is calculated as the total time the solute was infused into the carotid artery minus the delay time it takes the perfusion fluid to reach the brain once the perfusion pump is started *(2)* (*see* **Note 9**).

3. The total quantity of solute in brain (Q_{tot}) is the sum of what has left the perfusate and crossed into brain (Q_{br}) and what remains in perfusate in the brain blood vessels (Q_{vas}) *(2,5)* (*see* **Notes 10** and **11**),

$$Q_{tot} = Q_{br} + Q_{vas} \tag{2}$$

4. K_{in} can be converted into a cerebrovascular permeability-surface area product (PA) if the flow (F) of perfusion fluid in brain is known (*see* **Note 12**). PA is calculated from K_{in} using the Crone Renkin equation as *(2,3,5)*:

$$PA = -F \ln (1 - K_{in} / F) \tag{3}$$

5. F can be measured in brain perfusion experiments with radioactive diazepam or iodoantipyrine *(2,3)*. [^{3}H]- or [^{14}C]Diazepam are most commonly used for such measurements and have the advantage that uptake is linear for 60 s *(2,3,6)*. However, uptake with diazepam can become flow limited at high flow rates in the presence of plasma protein, as diazepam shows significant plasma protein binding.

6. The rate of drug uptake into brain (J_{in}) is given by the following equation,

$$J_{in} = K_{in} C_{pf} \tag{4}$$

7. If the interval over which brain uptake is unidirectional is unknown, this can be checked by measuring uptake over several different time periods and then plotting the data as Q_{tot}/C_{pf} vs T and looking for the initial, linear portion of uptake curve *(8,10–12)*. K_{in} can be obtained by fitting the following equation to the linear portion of the uptake curve using least-squares regression:

$$Q_{tot} / C_{pf} = K_{in} T + Q_{vas} / C_{pf} \tag{5}$$

8. The brain perfusion technique can also be used to measure the rate of drug removal from brain by preloading the drug or solute of interest into the brain by vascular perfusion and then switching the perfusion fluid to drug-free fluid *(10)*.

3.6. Applications

1. **Figure 3** illustrates the application of the *in situ* rat brain perfusion method to measure the rates of brain uptake and vascular intercepts of L-[^{14}C]alanine and [^{14}C]thiourea.

2. Data in **Fig. 3** are plotted as Q_{br}/C_{pf} values to obtain effective "brain uptake spaces" with units of mL/g. Measured brain L-[^{14}C]alanine and [^{14}C]thiourea contents were corrected for residual intravascular radioactivity (i.e., tracer sitting within the brain blood vessels) using [^{3}H]inulin (1–2 μCi/mL) which was included in the perfusion fluid along with either [^{14}C]thiourea or L-[^{14}C]alanine (0.25–0.50 μCi/mL). The mean brain [^{3}H]inulin space equaled 0.0126 ± 0.0005 ml/g in the [^{14}C]thiourea and L-[^{14}C]alanine experiments and was comparable to previous reports (2). [^{3}H]Inulin has a very low BBB K_{in} and is frequently used to measure brain vascular volume because it essentially does not cross the BBB in short experiments.

3. Brain uptake data were analyzed by linear regression to obtain slope and y-intercept values for the initial "linear" portion of the brain uptake graphs. Over the initial "linear" portion of the uptake plots, the slope is equivalent to the K_{in} and the y-intercept is equivalent to the residual vascular space (Q_{vas}/C_{pf}), separate from that measured by [^{3}H]inulin.

4. The brain uptake of L-[^{14}C]alanine showed evidence of significant brain vascular accumulation (**Fig. 3A**), as noted by a positive y-intercept (Q_{vas}/C_{pf}). Rat brain was perfused with saline containing L-[^{14}C]alanine at tracer concentration in the absence of added unlabeled L-alanine or other amino acid competitors. Under these conditions, brain L-[^{14}C]alanine uptake slightly exceeded that of [^{14}C]thiourea, reaching brain/perfusion Q_{br}/C_{pf} ratio values of approx 0.03 mL/g with a 60 s perfusion exposure. Brain L-[^{14}C]alanine uptake was linear and showed a best-fit regression K_{in} of $3.9 \pm 0.3 \times 10^{-4}$ mL/s/g which decreased by 67% to $1.3 \pm 0.3 \times 10^{-4}$ mL/s/g upon addition of 10-mM unlabeled L-alanine to the perfusion fluid ($p < 0.05$). Such concentration-dependent reduction in K_{in} is suggestive of brain L-[^{14}C]alanine uptake by a saturable transport carrier. However, unlike [^{14}C]thiourea, the calculated y-intercept Q_{vas}/C_{pf} value for L-[^{14}C]alanine (0.0082 ± 0.0012 mL/g), significantly exceeded zero, suggesting rapid and selective accumulation of L-[^{14}C]alanine in the brain endothelium. This putative brain vascular component was concentration-dependent, as addition of 10-mM unlabeled L-alanine to the perfusate reduced the L-[^{14}C]alanine y-intercept to essentially zero (0.0006 ± 0.0016 mL/g) (*see* **Fig. 3A**).

5. [^{14}C]Thiourea, as a fairly polar nonelectrolyte, crossed the BBB slowly reaching a brain-perfusion fluid Q_{br}/C_{pf} ratio of only 0.02 mL/g in 60 s of perfusion (**Fig. 3B**). The calculated BBB K_{in} value obtained by least squares linear regression equaled $2.9 \pm 0.3 \times 10^{-4}$ mL/s/g. The calculated y-axis intercept value of Q_{vas}/C_{pf} using **Eq. 5** equaled 0.0016 ± 0.0011 mL/g and did not differ significantly from zero ($p > 0.05$), suggesting that for [^{14}C]thiourea there is no additional binding or association with the vascular endothelium above that measured by [^{3}H]inulin. Thus, for [^{14}C]thiourea, [^{3}H]inulin appears to serve as a suitable marker of the vascular intercept space ($Q_{vas}/C_{pf} - Vv \approx 0$).

6. Given that the measured cerebral perfusion fluid flow rate (F = 0.1 mL/s/g) in the experiment greatly exceeded K_{in} for both [^{14}C]thiourea and L-[^{14}C]alanine, no correction is required for flow-limited uptake and K_{in} essentially equals PA for both tracers (*see* **Eq. 3**). In general, flow limitations only become important when $K_{in}/F \geq 0.2$ (*see* **Note 12**).

7. The results demonstrate the use of the *in situ* brain perfusion technique to quantitate and characterize solute transport across the BBB. This same method can be used in genetically modified rats and mice to study the effects of specific gene modifications on BBB function, such as knockout of the P-glycoprotein drug efflux transporter.

4. Notes

1. All surgery is performed using clean surgical technique. Deep anesthesia is required for surgery on the carotid circulation. Care should be taken to maintain anesthesia throughout the experiment and to insure that air bubbles and particulate matter is not introduced into

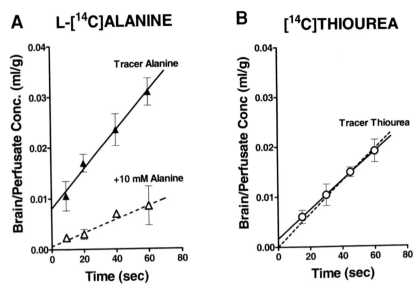

Fig. 3. Plot of the time course of brain uptake of L-[14C]alanine (**left**) and [14C]thiourea (**right**) during perfusion with physiologic saline fluid. The Y axis is the ratio of Q_{br} divided by C_{pf} to obtain the brain uptake "space" or brain/perfusate ratio (mL/g). The slope of the linear regression line is K_{in}, which is reduced for L-[14C]alanine during perfusion with fluid containing 10-mM unlabeled L-alanine due to saturation of a transport carrier. The perfusate in both experiments contained [3H]inulin to correct for intravascular tracer that is trapped in perfusion fluid within brain capillaries. Q_{br} for L-[14C]alanine and [14C]thiourea was cal-culated as $(Q_{tot} - V_v Q_{pf})$ where V_v was obtained from the Q_{tot}/C_{pf} of [3H]inulin. Data are mean ± SD for $n = 3$–4.

the blood leading to the brain. The saline in the perfusion catheter should contain 100 IU/mL heparin sodium to minimize the formation of blood clots.

2. If only one of the two carotid arteries is ligated and cannulated, adequate blood flow to brain can be maintained prior to perfusion in normal animals via cross over of blood from the contralateral carotid artery at the Circle of Willis, which joins the vessels that feed the brain (*2,3*). This prevents hypoxic brain damage in the intervening minutes as the perfusion cannula is connected to the pump and the perfusion is begun.

3. Some groups have found that it possible to perfuse via the left cardiac ventricle or aorta (*13–15*). Such systems provide equal perfusion to both hemispheres of the brain, but require higher infusion rates.

4. A number of perfusion fluids have been developed for use with the *in situ* brain perfusion technique. The optimum fluid depends upon the experimental goal. The BBB will accept considerable variation in perfusate composition for short intervals (<5 min) as long as the pH (7.0–7.5) and osmolarity (280–320 mOsm/kg) are maintained within normal limits.

5. Serum proteins are not required in the perfusate for oncotic pressure, as the brain capillaries have a low permeability to both large and small solutes. However, serum proteins often serve critical roles with regard to drug binding and transport, and may be added for that purpose, as well as to maintain oncotic balance in the "nonbarrier" regions or tissues. Albumin is frequently used for this purpose (2.7–4%). High molecular weight, nonprotein polymers, including dextran, can also be used.

6. For long perfusions (>60 s), an oxygen-carrying agent should be added to the perfusate to sustain normal cerebral metabolic and electrical activity. Most use washed erythrocytes to

a hematocrit of 20–40% or whole rat blood. With good oxygenation, cerebral oxygen metabolism and EEG can be maintained for more than 1 h.

7. In-line filters are valuable for removal of particulate matter and clots from the perfusion fluid *(2)*. But, care should be taken because they also can bind drugs (e.g., diazepam) and lead to lower concentration of test drug in the infusion solution coming out of the filter.

8. The recommended infusion rates for the modified Takasato method (cannulated common carotid artery—ligated external carotid artery) with saline perfusion fluid is 5–20 mL/min. In general, saline has a lower viscosity than blood and requires a higher infusion rate to obtain the same pressure. Pressures in excess of 200 mm Hg should be avoided because they may damage the BBB.

9. Often 1–5 s are required after start of the perfusion pump for the perfusion fluid to reach the brain. This delay time should be subtracted from the total perfusion time to obtain the net perfusion time (i.e., the time that perfusion fluid was actually passing through the brain). The delay time will differ depending on the size and length of the perfusion catheter and the infusion rate. The delay time can be measured by creating a cranial window in the skull of an animal immediately before perfusion and then determining how long it takes for the blood to clear from the cerebrovascular surface vessels after start of the perfusion pump.

10. Investigators should conduct all perfusion experiments so that more than half of the drug in brain has crossed the BBB ($Q_{tot}/C_{pf} \geq 2 \times Q_{vas}/C_{pf}$). Otherwise, small errors in intravascular drug distribution can appreciably affect the brain permeability calculations *(3,5)*. Further, it may be appropriate in some instances to confirm that the test solute does not modify BBB permeability. The latter has been shown to occur for some vasoactive peptides and drugs.

11. Some solutes bind significantly to the luminal surface of the cerebral blood vessels and can lead to overestimation of brain uptake if an appropriate vascular correction is not made. Drugs that exhibit this behavior tend to be either cationic (e.g., choline) *(10,16)* or highly hydrophobic. The loosely bound vascular drug can be removed by a brief wash (10–15 s) with drug-free saline *(10)*. Bovine serum albumin (2.7 g/dL) can be added to the perfusion wash solution to enhance the efficiency of vascular drug removal.

12. If the drug or test solute binds minimally to plasma proteins ($f \approx 1.0$) and the transfer constant (K_{in}) is small relative to the flow rate (F)(i.e., $K_{in} < 0.2$ F), then K_{in} essentially equals the PA product (within <10% error) and is essentially flow independent *(5)*. On the other hand, if the K_{in} is large relative to flow ($K_{in} \geq 0.2$ F), then uptake will depend both on flow and permeability and flow measurements will be required in order to obtain accurate estimates of BBB PA.

References

1. Takasato, Y., Rapoport, S. I., and Smith, Q. R. (1982) A new method to determine cerebrovascular permeability in the anesthetized rat. *Soc. Neurosci. Abstr.* **8,** 850.
2. Takasato Y., Rapoport S. I., and Smith Q. R. (1984) An *in situ* brain perfusion technique to study cerebrovascular transport in the rat. *Am. J. Physiol.* **247,** H484–H493.
3. Smith, Q. R. (1996) Brain perfusion systems for studies of drug uptake and metabolism in the central nervous system. *Pharm. Biotechnol.* **8,** 285–307.
4. Smith, Q. R., and Nagura, H. (2001) Fatty acid uptake and incorporation in brain. *J. Mol. Neurosci.* **16,** 167–172.
5. Smith, Q.R. (1989) Quantitation of blood-brain barrier permeability. In: *Implications of the Blood-Brain Barrier and Its Manipulation.* Vol. 1. Neuwelt, E. A., ed. Plenum Press, New York, pp 85–118.
6. Hokari, M., Wu, H. Q., Schwarcz, R., and Smith, Q. R. (1996) Facilitated brain uptake of 4-chlorokynureine and conversion to 7-chlorokynurenic acid. *NeuroReport* **8,** 15–18.

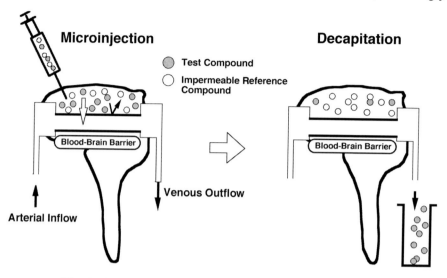

Fig. 1. Schematic diagram of the brain efflux index method.

1.1. Principle

The BEI value is defined by the following equation, the percentage of the effluxed amount being compared with the injected dose.

$$\text{BEI} = \text{Amount of drug effluxed across the BBB/Injected dose} \qquad (1)$$

Due to the difficulty in measuring the effluxed amount, the amount associated with the brain is measured. The injected dose is corrected by the recovery of a simultaneously injected reference compound of which elimination from the brain is negligible. Then, the BEI value is obtained by the following equation:

$$\text{BEI} (\%) = [1 - (X_{b, drug} / X_{b, ref}) / (X_{inj, drug} / X_{inj, ref})] \times 100 \qquad (2)$$

where X_b and X_{inj} represent the amount associated with the brain and injectate, and drug and ref represent the test and reference compounds respectively. The elimination rate constant of the test compound from the cerebrum can be obtained from the time-profile of the remaining fraction [100-BEI (%)] values versus time (**Figs. 2, 3**) (*see* **Notes 1, 2**). In the case of compounds with extensive permeability through the BBB, the efflux from the cerebrum to the systemic circulation is considered to be cerebral blood flow rate–limited because compounds excreted into the lumen of brain capillaries are cleared from the cerebrum via the blood flow (*see* **Fig. 1**).

This chapter will describe details of the BEI method. Readers are also referred to another review of this method *(21)*.

2. Materials

1. Reference compounds : [³H]inulin or [³H]dextran is used for ¹⁴C-labeled test compounds and [¹⁴C]carboxylinulin for ³H-labeled test compounds.
2. 5ul-microsyringe (Hamilton, Reno, NV) fitted with a needle (i.d. 100 um, o.d. 350 um, 27G).
3. Stereotaxic frame (Narishige, Tokyo, Japan).
4. Electric drill.

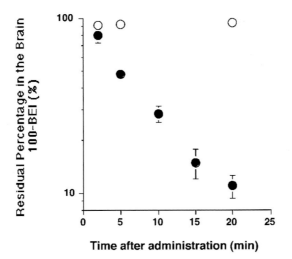

Fig. 2. Time profiles of 100-BEI(%) of L-glucose and 3-OMG after intracerebral microinjection in rats. [^{3}H]3-OMG and [^{3}H]L-glucose was microinjected into the cerebrum. At a designated time, rats were decapitated, and the radioactivity associated with the ipsilateral cerebrum was measured. The recovery of [^{3}H]3-OMG and [^{3}H]L-glucose in the ipsilateral cerebrum was corrected by that of [^{14}C]carboxylinulin used as reference compound. Closed and open symbols represent 3-OMG and L-glucose respectively. The elimination rate constant of 3-OMG was 0.129 ± 0.014 min^{-1}. No significant radioactivity of [^{3}H]3-OMG was detected in the contralateral cerebrum, cerebellum and CSF. Each points represent mean $\pm$ S.E. ($n = 4$). Cited from Kakee et al. *(9)*.

5. Decapitating instruments.
6. Anesthesia: ketamine (Sankyo, Tokyo, Japan) and xylazine (Aldrich Chemicals, Milwaukee, WI).
7. Animals: Adult male SPF Sprague-Dawley rats (Charles River, Yokohama, Japan), weighing 200 to 250 g were used. The animals had free access to food and water.
8. Buffer: 122 m*M* NaCl, 25 m*M* NaHCO$_3$, 10 m*M* D-glucose, 3 m*M* KCl, 1.4 m*M* CaCl$_2$, 1.2 m*M* MgSO$_4$, 0.4 m*M* K$_2$HPO$_4$, and 10 m*M* HEPES (pH 7.4).

3. Methods

3.1. Experimental Protocol

1. Rats were anesthetized with an intramuscular injection of 25 mg/rat ketamine and 1.22 mg/rat xylazine and placed in a stereotaxic frame.
2. After removing part of the scalp, a midline incision was made to expose the bregma.
3. A small hole was drilled at 0.2-mm anterior and 5.5-mm lateral to the bregma to allow entry of an injection needle (*see* **Note 3**).
4. The test and reference compounds ([^{14}C]carboxylinulin 0.5 nCi/rat) were dissolved in the buffer.
5. The needle was placed at a depth of 4.5 mm (Par2 region), and the solution injected (0.5 µL/rat).
6. Cerebrospinal fluid (CSF) was sampled from the cisterna magna just before decapitation and the left and right cerebrum and cerebellum were removed.
7. After measuring the wet weight of each excised cerebrum or cerebellum, they were dissolved in 2.5 mL of 2 *N* NaOH at 55°C for 3 h.
8. The radioactivity in the left and right cerebrum, cerebellum, and CSF was measured.

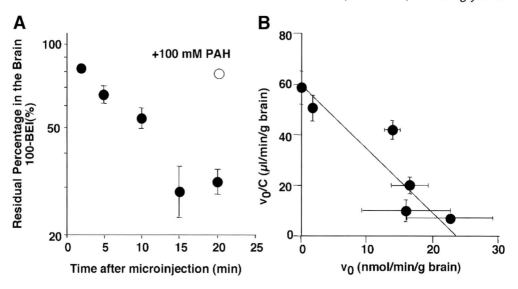

Fig. 3. Time profiles of 100-BEI(%) of PAH after intracerebral microinjection in rats (**A**) and the concentration dependence (**B**). (**A**), [³H]PAH was microinjected into the cerebrum. At designated time, rats were decapitated, and the radioactivity associated with the ipsilateral cerebrum was measured. The recovery of [³H]PAH in the ipsilateral cerebrum was corrected by that of [¹⁴C]carboxylinulin used as reference compound. Closed and open symbols represent 100-BEI(%) in the presence and absence of unlabeled PAH (100 m*M* in the injectate). The elimination rate constant of PAH was 0.0587 ± 0.0065 min^{-1}. No significant radioactivity of [³H]3-PAH was detected in the contralateral cerebrum, cerebellum and CSF. Each points represent mean $\pm$ S.E. ($n = 4$). (**B**), The concentration dependence was shown as Eadie-Hofstee plot. The 100-BEI(%) was determined at 2 and 20 min after microinjection. Unlabeled PAH was simulataneously determined, and the effective concentration was obtained by the concentration in the injectate divided by dilution factor (30.3). The K_m and V_{max} values were determined to be 396 ± 73 μ*M* and 23.4 ± 2.7 nmol/min/g brain, respectively. Solid line and each points represent fitted line and mean $\pm$ S.E. ($n = 6$). Cited from Kakee et al. (**11**).

3.2. Injection Site/Injection Volume

The recovery of [¹⁴C]carboxylinulin in the ipsilateral cerebrum is an important criterion for the reliability of the data. Detection of radioactivity of test and reference compounds associated with CSF specimens suggests that the test compound diffuses to the CSF and subsequently undergoes elimination by CSF turnover and/or efflux through the choroid plexus. It is necessary to find the injection point, which gives the maximum and minimum recovery in the cerebrum and CSF, respectively. The recovery of [¹⁴C]carboxylinulin in the ipsilateral and contralateral cerebrum and CSF was compared at the different injection sites (Par2, CA2, Ent, HiF, and Fr1) as shown (**Table 1**) (*9*). The total recovery of [¹⁴C]carboxylinulin 20 min after microinjection to the Fr1 site is relatively lower (approx 50%) than that at other sites. The recovery of [¹⁴C]carboxylinulin in the CSF was 34%, 57%, and 12% at 20 min after microinjection into the regions referred to as CA2, Ent, and HiF, whereas 2 to approx 3% of the injected dose was recovered in the CSF following microinjection into the Par2 region. Therefore, the Par2 region has been used as the site of injection in the subsequent studies. Heterogeneity of the transport sys-

Table 1

Effect of Injection Site on the Recovered Percentage of [^{14}C]Carboxylinulin in the Ipsilateral, Contralateral Cerebrum, Cerebellum, and CSF

Injection	Time (min)	Ipsilateral (%)	Contralateral (%)	Recovery Cerebellum (%)	CSF (%)	Total (%)
Par2	2	66.5 ± 10.1	0.6 ± 0.3	0.6 ± 0.4	2.8 ± 2.1	86.2 ± 8.3
	5	71.5 ± 12.0	0.1 ± 0	0.2 ± 0.1	0.7 ± 0.2	72.5 ± 11.0
	20	69.9 ± 7.9	0.2 ± 0	0.2 ± 0.0	1.9 ± 1.3	72.1 ± 8.0
CA2	2	66.7 ± 5.4	3 ± 1.4	2.3 ± 0.9	5.7 ± 4.7	77.8 ± 6.3
	5	55.5 ± 19.4	2.5 ± 1.8	1.3 ± 0.1	9.0 ± 7.7	68.3 ± 12.9
	20	47.9 ± 3.8	1.8 ± 0	2.2 ± 0.2	33.5 ± 4.6	85.5 ± 6.1
Ent	2	63.4 ± 3.0	9.1 ± 2.9	3.6 ± 0.2	2.0 ± 0.8	78.1 ± 5.9
	5	55.5 ± 14.7	5.2 ± 2.2	5.7 ± 2.7	29.0 ± 23.3	95.5 ± 13.2
	10	48.9 ± 11.8	4.4 ± 3.2	2.7 ± 1.4	46.6 ± 2	102.6 ± 12.3
	20	53.5 ± 18.5	2.4 ± 0.9	3.7 ± 1.7	57.1 ± 26.1	116.8 ± 13.3
HiF	2	77.4 ± 0.6	2.4 ± 1.6	1.4 ± 0.8	0.3 ± 0.1	81.4 ± 1.9
	5	38.5 ± 12.4	4.4 ± 1	6.5 ± 2.4	45.7 ± 15.7	95.1 ± 9.7
	20	64.7 ± 9.5	2.3 ± 1.6	1.0 ± 0.6	11.6 ± 7.9	79.7 ± 12.6
Frl	2	32.1 ± 1.3	4.4 ± 3.5	3.3 ± 3.1	1.7 ± 1.4	41.5 ± 8.9
	5	43.1 ± 4.2	6 ± 2.7	6.7 ± 3.4	6.6 ± 5.2	62.3 ± 10.7
	20	40.2 ± 9.0	0.5 ± 0.4	0.6 ± 0.5	12.0 ± 11.7	53.3 ± 8.7

Intracerebral microinjection of 0.2 µL injectate containing 0.01 µCi [^{14}C]carboxylinulin was injected for 1 s in the rat left cerebrum with a microsyringe with a needle of inner diameter 100 µm. Rats were decapitated at a designated time after microinjection. Each value represents the mean ± S.E. ($n = 3$ to approx 5). Par2, parietal cortex area; HiF, hippocampal fissure; Ent, entorhinal cortex; CA2, fild CA2 of Ammon's horn; Frl, frontal cortex area l; CSF, cerebrospinal fluid. From **ref. 9**.

tems along the BBB has not been reported. Further studies are required to confirm the uniformity and/or heterogeneity of the expression of efflux transporters along the BBB.

The effect of the injected volume was also examined over the range 0.2 to approx 2 µL/rat *(9)*. A larger injected volume tends to result in a lower recovery of [^{14}C]carboxylinulin in the ipsilateral cerebrum *(9)*. Approximately 80% was recovered when the volume of the injectate was 0.2 µL/rat. However, the recovery was 50% when the volume ranged from 0.5 to 2 µl/rat (**Fig. 1**).

3.3. Transport Systems at the BBB Characterized by the BEI Method

3.3.1. The Efflux Transport of Small Molecules Across the BBB

Stereoselective elimination was observed in 3-OMG and L-glucose (*see* **Fig. 2**) *(9)*. 3-OMG rapidly disappeared from the cerebrum (elimination half-life, 5 min) whereas the remaining percentage of L-glucose was constant, approx 100%, throughout the experiment (*see* **Fig. 2**) *(9)*. A significant reduction after preadministration of unlabeled 3-OMG and phloridizin was observed for the elimination of 3-OMG from the cerebrum *(9)*. Stereoselective elimination was also observed for L- and D-aspartate *(16)*. L-aspartate disappeared from the cerebrum with a rate constant of 0.207 min^{-1} after microinjection whereas no significant elimination of D-aspartate was observed in the cerebrum after microinjection *(16)*.

Using the BEI method, a transport system for PAH has been investigated *(11)*. PAH disappeared from the cerebrum after microinjection with a rate constant of 0.0587 ± 0.0065 min^{-1} (*see* **Fig. 3A**) *(11)*. The elimination was saturable, and Eadie-Hofstee plot indicates that the elimination of PAH from the cerebrum is accounted for by one-saturable component (*see* **Fig. 3B**) *(11)*. The K_m and V_{max} values were determined to be 396 ± 73 µM and 23.4 ± 2.7 nmol/min/g brain, respectively *(11)*. This K_m value was obtained using concentration in the injectate divided by the dilution factor (*see* **Note 4**). Briefly, because drugs microinjected into the cerebrum diffuse in the brain and undergo the uptake by brain parenchymal cells, the effective concentration becomes lower than their concentrations in the injectate. The effective concentration in the brain is roughly estimated to be the concentration divided by the dilution factor (30 to approx 40) (*see* **Note 4**).

The efflux transport of amphipathic organic anions such as TCA, estrone sulfate dehydroepiandrosterone sulfate and E217βG across the BBB has been investigated by the BEI method *(10,12–14)*. TCA, estrone sulfate, dehydroepiandrosterone sulfate, and E217βG disappeared from the cerebrum with a rate constant of 0.0233 ± 0.0025, 0.0663 ± 0.0077, 0.0268 and 0.037 ± 0.001 min^{-1}, respectively, after microinjection *(10,12–14)*. As a typical example, time profile of the elimination of E217βG from the cerebrum is shown (**Fig. 4**). A significant reduction was observed in the presence of excess unlabeled compound, except E217βG due to its low solubility in the buffer, and simultaneously administered probenecid, suggesting an involvement of transporter *(10,12–14)*. The effect of PAH on their elimination was minimal or none, suggesting that the transport system for amphipathic organic anions is different from that for PAH *(10,12–14)*.

A multispecific organic anion transporter, organic anion transporting polypeptide 2 (Oatp2; gene symbol, *Slc21a5*), has been considered to be involved in their elimination across the BBB. Oatp2 consists of 661 amino acids with 12 putative transmembrane domains, and is expressed on both the luminal and abluminal membrane of rat brain

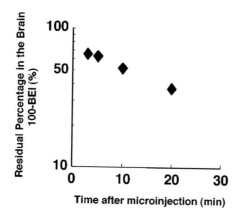

Fig. 4. Time profiles of 100-BEI(%) of E217βG after intracerebral microinjection in rats. [³H]E217βG was microinjected into the cerebrum. At designated time, rats were decapitated, and the radioactivity of [³H]E217βG and [¹⁴C]carboxylinulin associated with the ipsilateral cerebrum was measured. The recovery of [³H]E217βG in the ipsilateral cerebrum was corrected by that of [¹⁴C]carboxylinulin used as reference compound. The elimination rate constant of E217βG was 0.037 ± 0.001 min⁻¹. Each points represent mean $\pm$ S.E. ($n = 4$). Cited from Sugiyama et al. *(14)*.

capillary endothelial cells *(22)*. The substrates of Oatp2 include amphipathic organic anions such as bile acids and steroid conjugates as well as cardiac glycosides such as digoxin and ouabain, and type II organic cations *(23–26)*. The contribution of rOatp2 to the total elimination of E217βG after intracerebral microinjection was examined using BEI method and an inhibitor selective for Oatp2, digoxin *(14)*. Digoxin, a cardiotonic glycoside, is a high-affinity substrate of Oatp2 *(23)*. Simultaneous injection of digoxin reduced the elimination of E217βG to approx 60% of the control value, indicating that the contribution of Oatp2 accounts for approx 40% of the total elimination of E217βG from the cerebrum (**Fig. 5**) *(14)*. Significant inhibition by PAH at a concentration sufficient to saturate the elimination of PAH from the cerebrum was also observed (approx 20% inhibition) (*see* **Fig. 5**), suggesting that the transporter for PAH plays some role in the elimination of E217βG *(14)*. Because simultaneously administered probenecid and TCA inhibited the elimination completely (*see* **Fig. 5**), another unidentified transporter is also involved in the elimination and makes a relatively similar contribution (approx 40%) *(14)*. OATP-A (*SLC21A3*) is human isoform corresponding to rat Oatp2 and is also expressed at human brain capillary endothelial cells *(27)*. But the localization of OATP-A at the human brain capillary endothelial cells is unclear.

The presence of efflux mechanisms for small peptides across the BBB has been suggested by investigations using the BEI method *(8,10)*. BQ-123 is a cyclo-octapeptide, which antagonizes the endothelin receptor. Although the elimination of BQ-123 from the cerebrum is saturable and inhibited by TCA, mutual inhibition between BQ-123 and TCA suggests that they are eliminated from the cerebrum by different mechanisms *(10)*.

3.3.2. The Efflux Transport of Macromolecules Across the BBB

In addition to small molecules described in a previous section, the BEI technique was used to characterize the elimination of macromolecules such as IgG and transferrin *(28,29)*. The elimination of monoclonal antibody against rat transferrin receptor (OX-26) and mouse IgG$_{2a}$ from the brain was examined *(28)*. Both antibodies disappeared

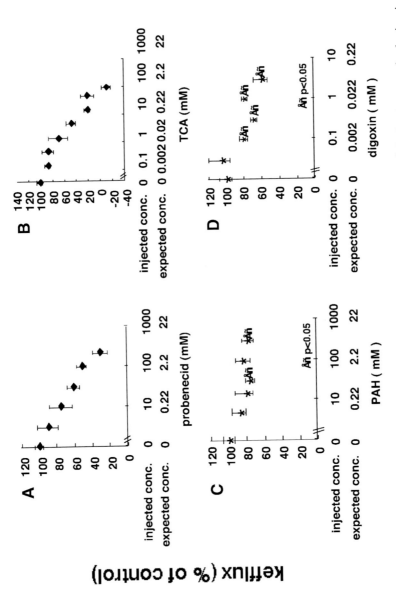

Fig. 5. Effect of probenecid, taurocholate, digoxin, and *p*-aminohippurate on the elimination of E217βG after intracerebral microinjection. Probenecid (**A**), taurocholate (**B**), *p*-aminohippurate (**C**), and digoxin (**D**) were simultaneously microinjected into the cerebrum, and their effect on the elimination of E217βG from the cerebrum after microinjection was examined at 20 min after microinjection. Each value of expected concentration is estimated by the concentration in the injectate divided by the dilution factor. Results are given as a ratio with respect to the elimination rate constant determined in the absence of unlabeled inhibitors. Each points represent mean ± S.E. ($n = 3$). *Significant difference from the control ($p < 0.05$). Cited from Sugiyama et al. (*14*).

from the cerebrum with similar elimination half-lives (48 min), faster than dextran used as reference compound. There was minimal reduction of dextran during 90 min after microinjection. This elimination of mouse IgG_{2a} from the cerebrum was saturable and the elimination rate constant was completely reduced to that of dextran by unlabeled mouse IgG_{2a} (5 mg/mL in the injectate). The elimination was completely inhibited by OX-26, mouse IgG_3 and human Fc fragment, but not by mouse F(ab')2, suggesting that Fc region is important for the elimination of the antibody from the cerebrum. Although the molecular weight of transferrin and dextran is similar (Mt 76,000 for tranferrin and Mt 70,000 for dextran), apo- and holo-transferrin (apo-transferrin+iron complex) disappeared from the cerebrum 20- and sixfold more rapidly than dextran *(29)*. Their elimination half-lives were 49 ± 4 and 170 ± 15 min, respectively. The elimination of apo-transferrin was saturable, and was completely abolished at the highest concentration examined (2400 µg/mL in the injectate). Transferrin receptor is a candidate mechanism for this elimination of apo-transferrin from the brain. These studies indicate usefulness of the BEI method for characterization of the elimination pathway of macromolecules.

4. Notes

1. Distribution volume: The efflux transport activity should be compared in terms of the efflux clearance, an intrinsic kinetic parameter describing the ability of efflux transport. Since compounds in the interstitial space undergo elimination across the BBB, the efflux clearance (CL_{eff}) is obtained by the following equation.

$$CL_{eff} = \frac{InjectedDose}{\int_0^\infty C_{br,i}dt} \qquad (3)$$

Where injected dose and $C_{br,i}$ represent the amount injected into the cerebrum and the concentration of test compound in the brain interstitial space. Amount associated with the brain (X_{br}), not the concentration in the brain interstitial space, is measured in BEI method. The elimination rate constant is obtained as follows.

$$k_{el} = \frac{Injected\ Dose}{\int_0^\infty X_{br}\ dt} = \frac{CL_{eff}}{V_{br}} \qquad (4)$$

Where V_{br} represent the fraction of $\int_0^\infty X_{br}\ dt / \int_0^\infty C_{br,i}\ dt$, and, so-called, distribution volume of test compound in the brain. Therefore, the elimination rate constant includes two parameters, the efflux clearance and the distribution volume (V_{br}) of test compound. According to **Eq. 4**, compounds with larger distribution volume exhibits slower elimination rate constant even although the efflux clearance is the same, and the BEI method is limited by its ability to detect a significant reduction in the amount associated with cerebrum after microinjection for compounds with large distribution volume in the brain and sticky compounds. Nipecotate increased the elimination rate constant of γ-aminobutyrate 1.7-fold compared with that in the absence of nipecotate *(17)*. Because nipecotate is a potent inhibitor for neuronal and glial GABA reuptake, the effect of nipecotate can be ascribed to the inhibition of the uptake by neuron and glial cells, and to the reduction of distribution volume of γ-aminobutyrate in the brain.

   The distribution volume can be estimated in a separate experiment such as an uptake study using brain slices, assuming that the distribution of test compound is rapid enough compared with the efflux across the BBB. The efflux clearance is obtained by multiplying the elimination rate constant by the distribution volume. Kakee et al. *(9)* compared the

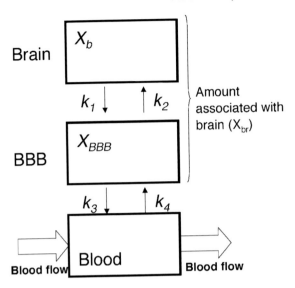

Fig. 6. Schematic diagram of mathematical model for the brain efflux index method.

efflux clearance of 3-OMG with its uptake clearance obtained by the brain uptake index method. The efflux clearance of 3-OMG was estimated to be 102 μL/min/g brain (k_{el} 0.129 ± 0.014 min^{-1}, distribution volume in the brain at steady-state 788 ± 8 μL/g brain) which is comparable with the uptake clearance of 92.6 μL/min/g brain *(9)*. This observation is consistent with the previous report by Oldendorf (1982) *(30)*.

2. Rate limiting process: As described previously, the elimination from the cerebrum after intracerebral microinjection includes the uptake through the abluminal membrane from the interstitial space and subsequent excretion into the blood. Under certain conditions where either process may be rate-limiting for overall elimination, saturation and inhibition of the transport through the opposite membrane hardly affects the net elimination time-profile. The apparent elimination rate constant from the cerebrum (k_{el}) is described by the following equation, assuming that re-uptake from the blood is negligible, i.e., $k_4 = 0$. The meanings of the parameters are described in **Fig. 6**.

$$k_{el} = \frac{k_1 \bullet k_3}{k_1 + k_2 + k_3} \tag{5}$$

If k_3 (the excretion process) is much larger than $k_1 + k_2$, the equation can be approximated to k_1. Under this condition, the uptake through the abluminal membrane is the rate-limiting step for overall elimination from the cerebrum. The results obtained by the BEI method reflect mainly the transport across the abluminal membrane. Inhibition of the luminal transport and disruption of the gene involved in the excretion hardly affect the elimination time-profile after intracerebral injection. Vice versa, if k_1 is much larger than $k_2 + k_3$, excretion is the rate-limiting process.

The procedures to obtain **Eq. 5** are as follows:
The differential equation of the amount of test compound in the brain (X_b) and brain capillary endothelial cells (X_{BBB}) is described in **Eq. 6**.

$$\frac{dX_b}{dt} = -k_1 X_b + k_2 X_{BBB}$$

$$\frac{dX_{BBB}}{dt} = k_1 X_b - (k_2 + k_3) X_{BBB} \tag{6}$$

where k represents the rate constant for the membrane transport across the luminal and abluminal membrane of the brain capillary endothelial cells (*see* **Fig. 6**). Laplace-transformed differential equations (Eq. 6) are described by the following equation:

$$(s + k_1)\tilde{X}_b - k_2 X_{BBB} = Injected\ Dose$$
$$-k_l \tilde{X}_b + (k_2 + k_3)\tilde{X}_{BBB} = 0$$

(7)

The sum of $X_b + X_{BBB}$ is described by the following Laplace transformed equation.

$$\tilde{X}_b + \tilde{X}_{BBB} = \frac{Injected\ Dose(s + k_1 + k_2 + k_3)}{(s + \alpha)(s + \beta)}$$

(8)

$$\alpha + \beta = k_1 + k_2 + k_3, \quad \alpha \cdot \beta = k_1 \bullet k_3$$

area under curve of the amount associated with the brain and BBB and the elimination rate constant (k_{el}) are obtained from **Eq. 8**

$$AUC(X_b + X_{BBB}) = \lim_{s \to 0}(\tilde{X}_b + \tilde{X}_{BBB}) = Injected\ Dose(k_1 + k_2 + k_3)/\alpha\beta$$

$$k_{el} = Injected\ Dose/AUC(X_b + X_{BBB}) = \frac{k_1 \bullet k_3}{k_1 + k_2 + k_3}$$

(9)

3. Gene knockout mouse such as Mdr1a, Mdr1a/Mdr1b, Mrp1, and Mrp1/Mdr1a/Mdr1b have recently been established *(31–34)*. Therefore, BEI technique for mice is useful to investigate the effect of gene knockout on the elimination of test compounds from the cerebrum. Banks et al. *(8)* have previously performed intracerebral microinjection in mice to characterize the efflux transport of RC-160 (a somatostatin analog) and Tyr-MIF-1 [Tyr-Pro-Leu-Gly-NH(2)] from the cerebrum, and they demonstrated carrier-mediated efflux of RC-160 from the cerebrum after microinjection. Their experimental conditions were as follows:
   Injection site: 1.0-mm lateral, 1.0-mm posterior to the bregma, and 1.5-mm deep.
   Injection volume: 0.1 μL/mice lactated Ringer's solution with 1% bovine serum albumin.
   Animals: ICR mice (from Charles River Laboratories, Wilmington, MA) weighing approx 25 g, anesthetized with intraperitoneal urethane (4g/kg).
4. The kinetic parameters (K_m and K_i values) obtained by the BEI method are apparent parameters because the net elimination from the cerebrum, which includes two processes, uptake from the cerebrum side and subsequent excretion from inside the cells into the blood (*see* **Fig. 6**), are measured in BEI method. In addition, as compounds diffuse in the cerebrum after microinjection, the effective concentration in the cerebrum becomes lower than that in the injection solution. Kakee et al. *(9)* estimated this to be approx 30- to 40-fold by examining the spread of trypan blue in the cerebrum after microinjection (39.9 ± 2.2 and 42.4 ± 5.7 at 2 and 20 min for 0.2 μL of microinjection, and 30.3 ± 4.4 and 46.2 ± 1.1 at 2 and 20 min for 0.5 μL of microinjection). Therefore, the effective concentration in the cerebrum is approx 30- to 40-fold smaller in the cerebrum than in the injection solution, assuming the dilution is the same as that of trypan blue. The uptake and adsorption by brain parenchymal cells are also factors reducing the effective concentration of substrate and inhibitors in the cerebrum. Introduction of dilution factor enables us to estimate the effective concentration of test compounds and inhibitors in the brain. But this is rough estimation, we have to carefully compare the K_m values obtained in vivo using BEI method with those obtained in vitro using isolated cells and/or complementary deoxyribonucleic acid transfectants. In the case of test compounds and inhibitors with low solubility in the buffer and/or low affinity to the efflux systems, a saturation or inhibitory effect may not be observed due to their insufficient concentration in the cerebrum. In order to overcome

    c. They are then loaded into two columns consisting of inverted plastic bottles measuring 5.5 cm in diameter by 12 cm in length (the neck is 3.5 cm in diameter).

    d. The bottom of each column is covered with 45-μm nylon mesh, which is temporarily sealed with Parafilm during the loading procedure. The beads are now added so that they fill about one-half of the column, and buffer is added immediately to prevent them from drying.

    e. Next the Parafilm seals are removed, and the above filtrate containing isolated cerebral capillaries from 10 cow brains is evenly and continuously added to the columns at a rate sufficient to maintain a continuous flow.

    f. After the capillaries have been layered upon both columns, each is rinsed with 500 mL of isolation buffer. This step may require gentle stirring of the beads with a glass rod to maintain flow. During this procedure, most of the capillaries adhere to the glass beads, and the others are retained by the nylon screen at the bottom of the columns.

    g. Red blood cells and small debris pass through the nylon filter and are removed from the capillary preparation. This filtration step is done entirely in a cold room (4°C), and takes approx 30 min.

    h. To recover the capillaries, the contents of the columns are poured into a beaker, and each column is rinsed (into the beaker) with isolation buffer. This rinsing procedure should include a careful washing of the filter, since capillaries have been retained there.

    i. Additional buffer is now added to the beads, and the mixture is stirred with a glass rod to free the capillaries.

    j. The freed microvessels are decanted and saved, and this washing procedure is repeated, until the decanted fluid is no longer cloudy.

    k. Finally, residual capillaries adhering to the glass beads are removed by using the apparatus illustrated in **Fig. 1** *(13)*. In this procedure, the beads roll through a convoluted plastic tube filled with buffer to loosen the adherent capillaries. At a juncture in the tube (Y-shaped), the beads are separated from the released capillaries, which are collected in a beaker.

    l. The procedure is initiated by filling the flask with buffer, closing the valve, and pouring isolation buffer into the funnel at the top of the apparatus until the coiled tube is also filled.

    m. After loading the system with buffer, air bubbles are removed by vigorously drawing them out with a syringe and long (2-inch) 12-gage needle, that is positioned through the neck of the funnel into the lumen of the tubing.

    n. Next the beads are spooned into the funnel, and the rate at which they flow through the system is regulated by opening and adjusting the valve. It is also important to make sure that buffer is always present in the funnel, to prevent air from entering the system and to ensure an even flow of material. If the flow rate is appropriate, the beads follow a pathway to the flask, and buffer containing the released capillaries will pass through the valve and enter the beaker (*see* **Note 3**). Fluid in the flask should be clear. If it is cloudy, this implies that separation of capillaries and beads was incomplete, and the procedure should be repeated.

    o. The suspension in the beaker is now poured through a 210-μm nylon screen to filter any residual glass beads.

12. The filtrate containing the isolated brain capillaries is next divided equally into 250 ml wide-mouth plastic bottles and centrifuged at 2000*g* for 5 min (4°C).

13. The resulting pellets are divided into two pre-weighed small (29 × 102 mm) centrifuge tubes (50 mL), filled with the isolation buffer, vortexed, and centrifuged again at 2000*g* for 5 min (4°C) in a Beckman JA-17 rotor.

14. Step 13 is repeated, and the pellets are resuspended in TSE buffer.

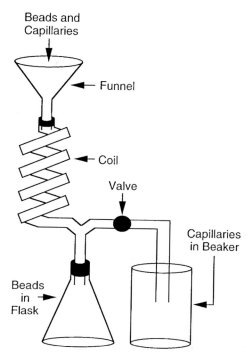

Fig. 1. This equipment is used to separate brain capillary fragments from the glass beads to which they have adhered during the isolation process. The beads with adherent capillaries are loaded into the funnel at the top of the apparatus, and the capillaries are dislodged as the beads flow through the coil. After leaving the coil, the beads fall into the flask, and the separated capillaries continue into the beaker *(13)*.

15. A final centrifugation is performed at 2000g for 5 min, after which the supernatants are decanted, and the tubes are weighed to determine the amount of tissue recovered.

16. The tubes are partially filled with freezing buffer, vortexed, labeled, and the microvessels are frozen at –80°C. The yield from 10 brains is approximately 10 g of isolated capillaries. Thus each tube will contain approx 5g of capillaries in 30–40 mL of solution.

3.2. Isolation of BBB Plasma Membrane Vesicles

To prepare plasma membranes from brain capillary endothelial cells, isolated microvessels are homogenized, and the released membrane fragments are separated in a discontinuous Ficoll gradient *(9–11)*. During fractionation, specific markers are used to identify and distinguish the luminal and abluminal membrane domains. The isolated membranes seal spontaneously to form vesicles, which may be used to measure transport processes. Each of the steps is described in **Fig. 2**. It is first necessary to remove contaminating tissue that adheres to the isolated capillaries.

1. Thus, approx 20 g of frozen capillaries are thawed at room temperature and centrifuged at 2000g for 5 min (4°C), after which the pellet is resuspended in cold TSEM buffer and re-centrifuged.

2. To remove astrocytic end-feet and pericytes, the capillary basement membrane is digested with collagenase *(11)*. For this purpose, the pellet is suspended in isolation buffer (1 g/10 mL) containing 180 units of collagenase type 1A, and the mixture is incubated

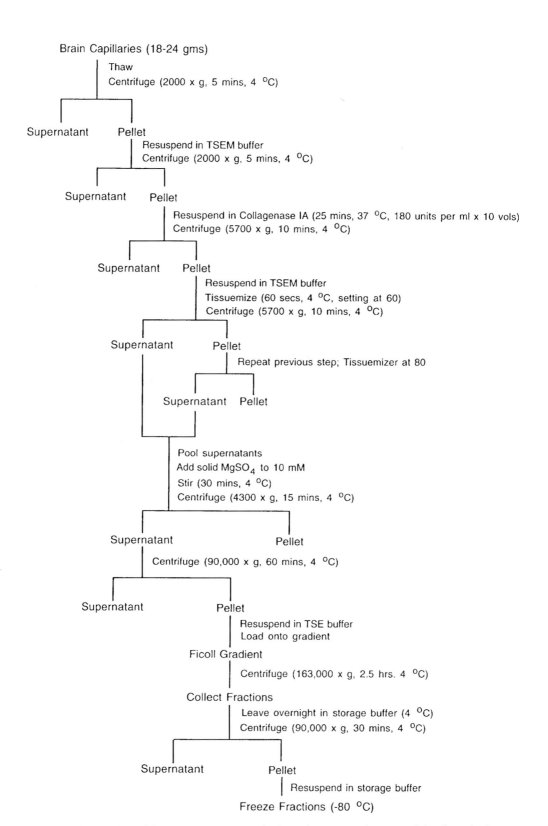

Fig. 2. Illustration of the steps necessary to isolate plasma membrane vesicles from brain capillaries. The figure is reprinted with the permission of Cambridge University Press (*9*).

in a shaking water bath for 25 min at 37°C. To recover the cleaned microvessels, the solution is centrifuged (Sorvall RC-5B centrifuge/SS-34 rotor) at 5700g for 10 min at 4°C (6–8 40-mL tubes), and each pellet is resuspended in 15 mL ice-cold TSEM buffer with protease inhibitors.

3. All subsequent steps are conducted at 4°C, unless otherwise indicated.
4. The next step is to break open the endothelial cells and release the plasma membranes. For this, aliquots of the processed microvessels are disrupted using a Polytron Tissuemizer, at a setting of 60 for 1 min.
5. The resulting suspensions are centrifuged at 5700g for 10 min, and the supernatants containing the freed membranes are pooled.
6. To retrieve membranes not extracted during the homogenization procedure, the pellets are resuspended in cold TSEM buffer with protease inhibitors (10 mL/pellet) and rehomogenized for 1 min at a setting of 80.
7. The mixture is again centrifuged at 5700g for 10 min, and the supernatants are combined with the previously freed membranes.
8. To remove additional debris, solid $MgSO_4$ is added to the pooled supernatant at a concentration of 10 mM. The solution is stirred for 30 min, centrifuged at 4300g for 15 min, and the pellet is discarded.
9. To harvest the membranes, the supernatant is placed in polycarbonate ultracentrifuge tubes (Nalgene, size 25 × 89 mm, 26.3 mL) and centrifuged at 90,000g for 1 h in a Beckman L-60 Ultracentrifuge (70Ti rotor).
10. The pelleted membranes are resuspended in 20–25 mL of TSE buffer, and disassociated by passage through a 20-gage needle, followed by a 26- or 27-gage needle. The membranes are now ready to be separated by fractionation.
11. Luminal and abluminal plasma membrane vesicles are separated in a discontinuous Ficoll (Sigma) gradient. To process the amount of tissue generated by the above procedure, six gradients are prepared (*see* **Note 4**).
    a. Each gradient consists of four concentrations of Ficoll in TSE buffer (20%, 15%, 10%, 5%), prepared by carefully layering the following volumes in each of six Beckman Quick-Seal centrifuge tubes (25 × 89 mm): 7 mL of 20%, 9 mL of 15%, 8 mL of 10%, and 10 mL of 5% (*see* **Note 5**).
    b. The dissociated membranes are distributed equally to the top of each gradient, after which the tubes are filled with TSE buffer, sealed, and centrifuged at 163,000g (70Ti rotor) for 2.5 h (*see* **Note 6**).
    c. Following separation of the membranes within the gradient, the top of each tube is opened with a razor blade, and the interfaces are collected in descending order. This is accomplished by positioning the tip of a 15-gage stainless steel needle at each interface, and drawing the suspension (4–9 mL) into a 10-mL syringe (*see* **Note 7**).
    d. Each fraction is labeled as follows: F1 is the 0/5% interface, F2 is the 5/10% interface, F3 is the 10/15% interface, F4 is the 15/20% interface, and F5 is the pellet (very small) plus the remaining gradient (i.e., pooled media between the interfaces).
    e. These fractions are diluted with storage buffer, chilled on ice overnight, and finally centrifuged in polycarbonate ultracentrifuge tubes (Nalgene, size 38 × 102 mm, 70 mL) at 90,000g (4°C) for 30 min in the ultracentrifuge (Beckman type 35 rotor).
    f. The pellets are resuspended in 1.5 mL of storage buffer and stored at –80°C.
    g. Twenty grams of isolated capillaries yield approx 10 mg of separated total protein.
12. After each batch of isolated capillaries (20 g) has been fractionated, the membranes from the same fractions for all of the batches are pooled, assayed for protein concentration (**14**), divided into aliquots of 900 µg, and stored at –80°C (*see* **Note 8**).

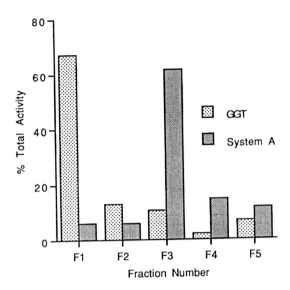

Fig. 3. The markers for luminal and abluminal plasma membranes derived from brain capillary endothelial cells are plotted as a function of fraction number, following separation in a discontinuous Ficoll gradient. GGT activity, which identifies the luminal membrane, is enriched in fraction number F1. System A amino acid transport activity is a marker for the abluminal membrane, and is observed primarily in fraction F3. This figure is reprinted with permission from Cambridge University Press *(9)*.

13. Luminal and abluminal plasma membranes are identified by specific markers. Gamma-glutamyl transpeptidase (GGT) serves as a marker for the luminal membrane (biochemical assay), and the system-A amino acid transporter is indicative of the abluminal plasmalemma *(11)*. The latter is measured by quantifying the initial rate of sodium-dependent ^{14}C-labeled *N*-(methylamino)-isobutyric acid (MeAIB) uptake *(15)* (*see* **Note 9**). Within the Ficoll gradient (**Fig. 3**), these markers localize primarily to fractions F1 and F3, respectively *(9,11,16)*.

3.3. Transport Studies with BBB Membrane Vesicles

As indicated above, fractions F1 and F3 are enriched in markers for the luminal and abluminal plasma membranes, respectively. The membranes from these fractions form sealed vesicles with characteristics that make them useful to measure transport activities. Thus, they orient primarily right-side-out *(10)*, are reasonably free of contamination by subcellular membranes *(10,11)*, and behave as osmometers, remaining relatively impermeable to the passive diffusion of sucrose *(11)*. We have used the vesicles successfully in transport studies, and in assigning the distribution of various membrane-associated enzymes *(9–13,15–19)*.

3.3.1. Measuring Transport

Transport is measured by a rapid filtration technique *(10,20)*.

1. First, frozen membranes are thawed, diluted to 0.75 mg/mL in an intravesicular medium (usually storage buffer), and equilibrated for 2–24 h on ice (4°C) (*see* **Note 10**).

2. Next, the membranes are centrifuged at $37,500g$ for 30 min (4°C), after which the pellet is resuspended in the intravesicular medium at a membrane protein concentration of 2.5–5.0 mg/mL.

3. To begin the transport process, 10 μL of uptake solution is added to 10 μL of the membrane preparation and incubated at 37°C. The basic uptake solution is storage buffer, plus radiolabeled substrate and additional co-factors that may be necessary. This medium is made hyperosmotic to the intravesicular solution, to counteract osmotic effects associated with uptake that would expand the vesicles and contribute to an over-estimation of transport rate *(15,21)*. Thus, 50–100 mM (final concentration) NaCl, KCl, or choline chloride is added to the incubation medium.

4. Transport is terminated by adding 1 mL of ice-cold stopping-solution, and rapidly filtering the mixture through a 0.45 μm Gelman Metricel filter. This is followed immediately by four 1-mL washes with cold stopping solution, after which radioactive material associated with the retained membrane vesicles is measured.

3.3.2. Determining Initial Rate

The initial rate of transport can be determined by measuring uptake of a radiolabeled substrate as a function of incubation time.

1. When uptake is plotted vs time, an equilibrium is eventually reached (**Fig. 4**), representing conditions in which net influx no longer occurs.

2. The uptake for a given substrate may exceed this equilibrium value if its transport is driven by another solute (i.e., secondary active co-transport). This is illustrated in **Fig. 4**, which compares uptake of MeAIB by abluminal membrane vesicles, in the presence and absence of an inwardly directed sodium gradient. MeAIB is co-transported with sodium by this membrane, and the initial rate of uptake is greater in the presence of sodium. Furthermore, sodium causes MeAIB uptake to "overshoot" its equilibrium value, until the sodium gradient itself is dissipated. The overshoot is not observed without sodium.

3. To quantify the initial rates of transport, a double exponential fit is used when an overshoot is present, and a single exponential curve is used without one.

3.3.3. Distinguishing Binding and Transport

Binding of substrate can be distinguished from its transport into vesicles by measuring uptake, under equilibrium conditions, as a function of 1/osmolarity of the incubation medium *(10)*.

1. As the medium osmolarity is increased from 300 to 600 mOsmol/L, vesicular volume will decrease, and uptake measured at equilibrium will be proportionately reduced.

2. By plotting uptake at equilibrium as a function of 1/osmolarity and fitting the points with a straight line (**Fig. 5**), the y-intercept represents only binding at infinite osmolarity because intravesicular space would have been eliminated. Thus, comparing the value for binding measured at the y-intercept, to that for total uptake measured at equilibrium under normal incubation conditions, the percentage of uptake due to surface binding may be estimated.

3. Treating the membranes with pronase (5 μg/mL) can also be used to remove bound material (*see* **Fig. 5**) (*see* **Note 11**).

3.3.4. Measuring Kinetic Constants

Kinetic constants of transport may be measured by plotting the velocity of uptake (initial rates) as a function of substrate concentration, using non-linear regression analysis.

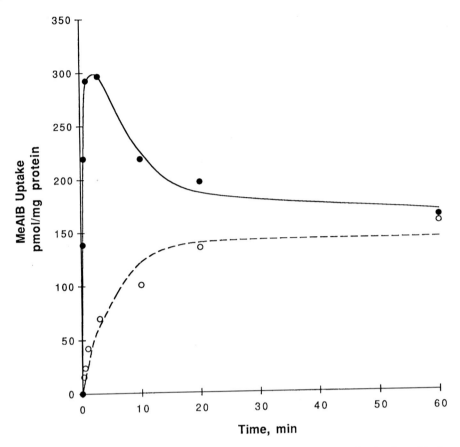

Fig. 4. Uptake of radiolabeled MeAIB by abluminal membrane vesicles is plotted in the presence and absence of an inwardly directed sodium gradient, as a function of incubation time. The presence of sodium increases the initial rate of transport and causes MeAIB uptake to overshoot the equilibrium value, indicative of a sodium co-transport mechanism. This figure is reprinted with permission from Cambridge University Press *(9)*.

1. Transport of L-phenylalanine *(10)* is shown (**Fig. 6**), along with the raw data, as well as calculated saturable and nonsaturable components (dotted lines).
2. The apparent affinity (Km), maximum velocity (Vmax), and diffusion constant (Kd) may be determined by using the Michaelis-Menten equation:

$$V = [(Vmax \times S)/(Km + S)] + (Kd \times S)$$

In this equation, S is the substrate concentration, and the term Kd x S represents the nonsaturable component.

3. To save radiolabel when doing kinetic measurements of carrier-mediated transport, uptake of a tracer amount of labeled substrate may be quantified as a function of unlabeled material. The latter competes for transport, and the rate of uptake for radiolabeled material is reduced as the concentration of unlabeled substrate is increased in the incubation medium. This is illustrated in **Fig. 7**, which compares sodium dependent and independent glucose transport into abluminal membrane vesicles derived from the BBB *(18)*.

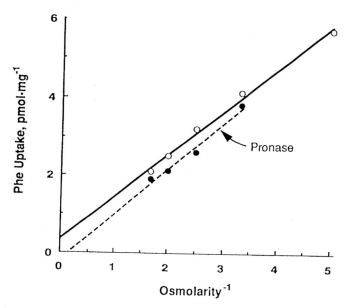

Fig. 5. Uptake of radiolabeled phenylalanine by luminal membrane vesicles is plotted as a function of the reciprocal of incubation medium osmolarity, under equilibrium conditions. The y-intercept represents surface binding, since intravesicular volume would be eliminated at infinite osmolarity. The results indicate that binding accounts for a small percentage of uptake measured under normal conditions, at an osmolarity of 300 mOsmol/L. Repeating the measurements in the presence of pronase (5 µg/mL) eliminates the binding component. This figure is reprinted with permission from The American Society for Biochemistry and Molecular Biology (10).

4. A modified Michaelis-Menten equation (22) is used to fit these data:

$$V_{tracer} = [Vmax \times (T)]/[Km + (T) + (S)]$$

If the fitted curve is not inhibited to a value of 0, indicative of a nonsaturable component, the value representing uninhibited transport must be subtracted from each point prior to determining the kinetic constants, using the above equation.

3.3.5. Determining the Distribution of Transport Activities

After the transport properties of a given substrate have been characterized for both luminal and abluminal membranes, the distribution of functional transporters between these two membrane domains may be determined by calculating an f-value (11).

1. To make this calculation, one must know the percentage of luminal and abluminal membrane in fractions F1 and F3, as well as the total units of transport activity there.
2. The following equation is used:

$$f = 1/1 + R, \text{ where } R = (UiLj - UjLi)/(UjAi - UiAj)$$

where U is total units of activity, L and A are the percentages of luminal and abluminal membrane, and i and j represent the two fractions.
3. L and A may be calculated by measuring GGT and system-A marker activities throughout the gradient, and expressing total activity (specific activity × mg of protein) for fraction F1 and F3 as a percentage of total activity for the entire gradient.

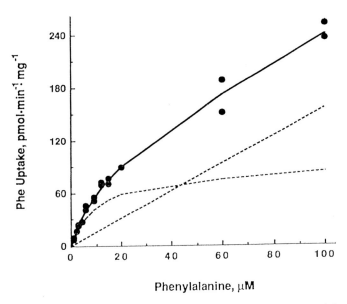

Fig. 6. Uptake of radiolabeled phenylalanine by luminal membrane vesicles is plotted as a function of substrate concentration. The dotted lines represent the saturable and non-saturable components of transport, as determined by the Michaelis-Menten equation. The data indicate that the apparent affinity for phenylalanine transport is 11.8 μM. This figure is reprinted with permission from The American Society for Biochemistry and Molecular Biology (*10*).

4. An *f*-value of 1.0 means that measured transport activity is present only in the luminal membrane. A value of 0 implies only abluminal activity, and numbers in between describe the relative distribution of functional transporters in both membrane domains. This analysis is derived from a mathematical model described in detail elsewhere (*11*), and is pertinent to transporters in the plasma membrane.

We have found this technique to be useful in describing specific transport properties for each plasma membrane domain of the blood-brain barrier (*9*). Substrates studied have included amino acids (*10,11,15–17*), glucose (*18*), the peptide glutathione (*12*), and electrolytes (*19*).

4. Notes

1. Peeling the meninges away from the brain can be very difficult and time-consuming, especially if blood has clotted. It helps to begin peeling immediately after the brains have been immersed in cold buffer (4°C), which serves both to preserve the tissue and reduce clotting of blood.
2. When passing capillaries through the 70-μm mesh, stir them with a rapid side-to-side motion to prevent material from settling on and clogging the mesh. For this purpose, we use a glass rod that has been formed into the shape of a miniature hockey stick.
3. The speed at which buffer flows through the spiral apparatus is very important. If it runs too fast, glass beads will back up into the Y-fork in the tube. Then it must be stopped completely until the beads have settled. If it flows too slowly, capillaries will fall into the flask, with the glass beads.

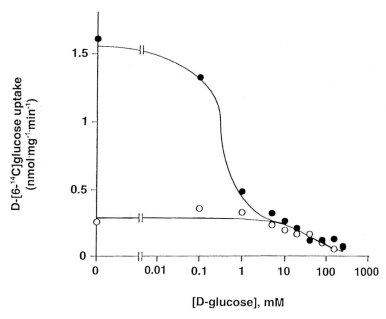

Fig. 7. Uptake of a tracer amount of radiolabeled glucose by abluminal membrane vesicles is plotted as a function of unlabeled substrate, in the presence and absence of an inwardly directed sodium gradient. With this protocol, uptake of radiolabeled material is reduced in the presence of competing unlabeled substrate, and a modified Michaelis-Menten equation may be used to analyze kinetics. The data indicate that both sodium-dependent and independent glucose transporters are present in the abluminal membrane of the blood–brain barrier. This figure is reprinted with permission from The American Physiological Society (*18*).

4. To save time, the Ficoll gradients should be prepared in the cold room during collagen treatment of the capillaries and the subsequent centrifugations, because this procedure requires great care and is time-consuming.

5. Layers of the Ficoll gradient are applied using a 10-mL syringe and a 10-cm-long 15-gage needle, bent at 30° approx 2 cm above the tip. Between layers, the needle and syringe should be rinsed with 1 mL of the Ficoll solution to be added next. Each layer is added by placing the tip of the needle against the inside of the tube wall just at the surface of the previous layer.

6. The membrane suspension is added on top of the 5% Ficoll layer, using a 10-mL syringe and a 6.5-cm-long 18-gage needle, bent at 30° 2 cm above the tip. The end of the needle is placed against the inside wall of the tube just above the surface of the gradient, and membrane is ejected slowly and smoothly, avoiding mixing into the Ficoll solution. The top 0% layer is prepared by adding TSE buffer to the tissue, drawing the tip of the needle upward as the neck of the tube is filled. During this procedure, the membrane solution will mix with the TSE buffer. All air bubbles must be removed before sealing the tube, or it will collapse during centrifugation.

7. Following centrifugation, a straight 10-cm long 15-gage needle is used to withdraw each fraction. When removing tissue accumulated at the interface between two layers of the gradient, it is imperative not to contaminate it with adjacent material from the layers above and below.

8. The corresponding fractions from all isolations are eventually pooled and assayed for protein concentration, using the Bradford technique *(14)*. Each pooled fraction is then divided into 900-μg aliquots in 1.5 mL siliconized microcentrifuge tubes, and stored at –80°C. The rationale for using 900-μg aliquots stems from the fact that transport studies usually require about 900 μg of tissue.

9. Caution should be exercised when using radiolabeled materials. In general, a lab coat and rubber gloves should be warn, as well as a badge to detect potential exposure to radioactivity. Depending on the isotope and nature of its radioactive emission, lead shielding may be required. Care should be taken to protect against spillage, or dispersion into the air. All radioactive waste must be properly disposed of, in accordance with existing regulations.

10. When an aliquot of frozen tissue is thawed for use in an experiment, the membranes should be allowed to reseal in ice-cold storage buffer for at least 2 h.

11. If it becomes necessary to measure specific binding sites or enzyme activities on the inner surface of the membrane vesicles, detergents may be used to open them. We have found that Triton X-100 (0.1%, v/v) and deoxycholate (0.06%, w/v) are useful to permeablize the luminal and abluminal membranes, respectively *(10)*.

Acknowledgments

This work was supported by grants from the National Institutes of Health (NS31017), and the American Heart Association.

References

1. Betz, A. L. (1986) Transport of ions across the blood-brain barrier. *Fed. Proc.* **45**, 2050–2054.

2. Betz, A. L., and Goldstein, G. W. (1986) Specialized properties and solute transport in brain capillaries. *Ann. Rev. Physiol.* **48**, 241–250.

3. Betz, A. L., Firth, J. A., and Goldstein, G. W. (1980) Polarity of the blood-brain barrier: Distribution of enzymes between the luminal and antiluminal membranes of brain capillary endothelial cells. *Brain Res.* **192**, 17–28.

4. Reese, T. S., and Karnovsky, M. J. (1967) Fine structural localization of a blood-brain barrier to exogenous peroxidase. *J. Cell Biol.* **34**, 207–217.

5. Brightman, M. W., and Reese, T. W. (1969) Junctions between intimately apposed cell membranes in the vertebrate brain. *J. Cell Biol.* **40**, 648–677.

6. Pardridge, W. M. (1983) Brain metabolism: a perspective from the blood-brain barrier. *Physiol. Rev.* **63**, 1481–1535.

7. Betz, A. L., and Goldstein, G. W. (1978) Polarity of the blood-brain barrier: Neutral amino acid transport into isolated brain capillaries. *Science* **202**, 225–226.

8. Pardridge, W. M., Eisenberg, J. and Yamada, T. (1985) Rapid sequestration and degradation of somatostatin analogues by isolated brain microvessels. *J. Neurochem.* **44**, 1178–1184.

9. Peterson, D. R. and Hawkins, R. A. (1998) Isolation and behavior of plasma membrane vesicles made from cerebral capillary endothelial cells. In *Introduction to the Blood-Brain Barrier*. Pardridge, W. M., ed. Cambridge University Press, London, pp. 62–70.

10. Sanchez del Pino, M. M., Hawkins, R. A., and Peterson, D. R. (1992) Neutral amino acid transport by the blood-brain barrier: Membrane vesicle studies. *J. Biol. Chem.* **267**, 25951–25957.

11. Sanchez del Pino, M. M., Hawkins, R. A., and Peterson, D. R. (1995) Biochemical discrimination between luminal and abluminal enzyme and transport activities of the blood-brain barrier. *J. Biol. Chem.* **270**, 14907–14912.

12. Peterson, D. R., Rambow, J., Sukowski, E. J., and Zikos, D. (1999) Glutathione transport by the blood-brain barrier. *FASEB J.* **13**, A709.

13. Sanchez del Pino, M. M. (1994) Neutral amino acid transport by the blood-brain barrier using isolated membrane vesicles. Ph.D. Thesis. Finch University of Health Sciences/The Chicago Medical School, N. Chicago, IL.
14. Bradford, M. M. (1976) A rapid and sensitive method for the quantitation of microgram quantities of protein utilizing the principle of protein-dye binding. *Anal. Biochem.* **72,** 248–254.
15. Sanchez del Pino, M. M., Peterson, D. R., and Hawkins, R. A. (1995) Neutral amino acid transport characterization of isolated luminal and abluminal membranes of the blood-brain barrier. *J. Biol. Chem.* **270,** 14913–14918.
16. Lee, W-J., Hawkins, R. A., Peterson, D. R., and Vina, J. (1996) Role of oxoproline in the regulation of neutral amino acid transport across the blood-brain barrier. *J. Biol. Chem.* **271,** 19129–19133.
17. Lee, W-J., Hawkins, R. A., Vina, J. R. and Peterson, D. R. (1998) Glutamine transport by the blood-brain barrier: a possible mechanism for nitrogen removal. *Am. J. Physiol.* **274,** C1101–C1107.
18. Lee, W-J., Peterson, D. R., Sukowski, E. J., and Hawkins, R. A. (1997) Glucose transport by isolated plasma membranes of the blood-brain barrier. *Am. J. Physiol.* **272,** C1552–C1557.
19. Peterson, D. R., Rambow, J., Sukowski, E. J., and Zikos, D. (2000) Mechanisms for sodium transport by the blood-brain barrier. *FASEB J.* **14,** LB74.
20. Skopicki, H. A., Fisher, K., Zikos, D., Flouret, G., and Peterson, D. R. (1989) Low-affinity transport of pyroglutamyl-histidine in renal brush-border membrane vesicles. *Am. J. Physiol.* **257,** C971–C975.
21. Hopfer, U. (1989) Tracer studies with isolated membrane vesicles. *Methods Enzymol.* **172,** 313–321.
22. Malo, C., and Berteloot, A. (1991) Analysis of kinetic data in transport studies: new insights from kinetic studies of Na^+-D-glucose cotransport in human intestinal brush-border membrane vesicles using a fast sampling, rapid filtration apparatus. *J. Membr. Biol.* **122,** 127–141.

16

Drug Transport Studies Using Quantitative Microdialysis

Haiqing Dai and William F. Elmquist

1. Introduction

Cellular barriers in the central nervous system (CNS) present a formidable challenge in the delivery of drugs to the brain. These barriers include the blood–brain barrier (BBB) and the blood–cerebrospinal fluid barrier (BCSFB). Of these, given surface area and diffusional distance considerations, the BBB may be the more important barrier to drug transport to the bulk of the brain parenchyma. With the development of increasing numbers of new compounds to treat CNS diseases, quantitative methods to examine the transport of drugs in the CNS are necessary. Several methods in the past have been used, e.g., whole brain homogenates, quantitative autoradiography, *in situ* perfusion, noninvasive imaging techniques such as positron emission tomography (PET), and in vivo microdialysis. Although all these techniques have distinct advantages and disadvantages, this chapter will focus on the use of in vivo microdialysis in the rat brain to examine the mechanisms of drug transport through the barriers of the CNS.

The theory of the microdialysis technique is quite simple. A semi-permeable dialysis hollow fiber membrane (the microdialysis "probe") is placed in a tissue of interest (in this case the CNS), and a physiological perfusate is pumped through the probe at low flow rates. Compounds in the extracellular fluid of the tissue then diffuse into the perfusate (now termed dialysate), and the concentration of the compound is measured by some analytical technique, frequently high-performance liquid chromatography (HPLC). In essence, the microdialysis probe is acting like an artificial blood capillary, where substances can diffuse into, and out of, the probe (**Fig. 1**). Consequently, the tissue concentration of solutes can be measured and, conversely, the solute can be delivered to the local tissue site *(1)*.

There are several advantages to the use of the microdialysis technique. In the quantitative analysis of the transport of compounds across the BBB, there is a need to study the kinetics of the transport process. Microdialysis sampling allows the measurement of the extracellular concentration of a compound over time in specific areas of the CNS, like the cerebrospinal fluid (CSF) in the lateral ventricle or the extracellular fluid (ECF, i.e., the interstitial fluid) in the frontal cortex. Having continuous measurements over time allows a kinetic analysis within the same animal, thus avoiding the interanimal variability that confounds the data from single-time point brain homogenate studies that use several animals to construct a single CNS concentration-time profile *(1)*. Moreover,

From: *Methods in Molecular Medicine, vol. 89:*
The Blood–Brain Barrier: Biology and Research Protocols
Edited by: S. Nag © Humana Press Inc., Totowa, NJ

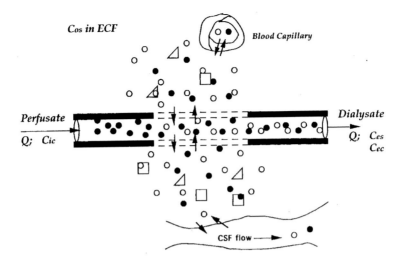

Microenvironment of Microdialysis Probe
Microdialysis
Retrodialysis

Fig. 1. The principle of microdialysis sampling. The microenvironment within and surrounding the microdialysis probe in vivo (not drawn to scale). The solid and dashed line segments schematically represent the nonpermeable probe wall and semipermeable membrane, respectively. Open and closed circles represent the molecules of the solute of interest and a retrodialysis calibrator, respectively. Squares and triangles represent proteins that may bind the solute and/or calibrator, but are not recovered by the dialysis process. Arrows indicate the direction of mass transport. Q is the perfusate flow, Cic is the concentration of calibrator in, Cos is the concentration of the solute on the outside of the probe in the tissue, Ces and Cec are the concentrations of the solute of interest and the calibrator, respectively, in the outflow dialysate. (Adapted from **ref. *1*.)

the microdialysis sampling technique measures the free extracellular concentration of the compound, which can allow the quantitative determination of the transport processes across the barrier. Other important considerations include the fact that the microdialysis technique can be performed in awake, freely moving animals and crossover studies in the same animal can be performed. Furthermore, the ability to simultaneously measure the concentration of compound in the blood and the brain extracellular fluid yields data that make the technique particularly attractive to determine the kinetics of influx and efflux from the brain *(2)*.

One important aspect of the use of microdialysis in BBB studies is the accurate determination of the recovery of the microdialysis probe. The recovery is defined as the ratio of the concentration measured in the dialysate to the actual concentration in the extracellular fluid at the site of measurement. For in vivo microdialysis sampling to be truly useful in the estimation of BBB permeability, an in vivo recovery determination is necessary *(1,3,4)*. This aspect of the technique separates this use of microdialysis sampling from previous uses in the CNS to measure the relative change in neurotransmitter concentrations, which is still an important application *(5)*.

Over the past decade, quantitative microdialysis has been used to examine the brain penetration of drugs in numerous studies *(6–9)*. Several differences in the method can be seen in each study, such as different probe design and manufacture, different perfusion rates, different probe placement techniques, and different means of determining probe recovery. In many cases, each difference in the general application of the microdialysis method can be attributed to the special needs of the experiment, the solute to be measured, and the experimental goal. In this chapter we will describe the method as used in our hands to determine drug distribution to the CNS, but there may be other special considerations that will necessarily be applied to specific research questions and solutes.

2. Materials

2.1 Microdialysis Probe Materials

There are several commercially available probes from various manufacturers. Materials listed below are those currently used in the described method, however, other probes can be used that are self-made or purchased.

2.1.1. Intracerebral Probes

1. CMA-12 microdialysis probe, 3-mm membrane length for the cortex, 1-mm membrane length for the lateral ventricle (part no. 012-8309563, CMA/Microdialysis, North Chelmsford, MA).
2. CMA-12 guide cannula (part no. 012-8309025, CMA/Microdialysis).
3. Teflon inlet and outlet microdialysis tubing (part no. 030-8409501, CMA/Microdialysis).
4. Tubing adapter (part no. 030-3409500, CMA/Microdialysis).
5. Anchor screws (part no. 7431021, CMA/Microdialysis).

2.1.2. Intravascular Probes

1. Hollow dialysis fiber, Hospal AN69 HF polyacrylonitirile/sodium methyl sulfonate copolymer (Multiflow 60 hemodialyzer, Gambro Renal Care Products, Lakewood, CO). These fibers are stored dry at 4°C.
2. Inlet fused silica tubing, OD-165 μm, ID-98 μm (Polymicro Technologies, Phoenix, AZ).
3. Outlet fused silica tubing, OD-210 μm, ID-145 μm (Polymicro Technologies).
4. PE-10 and PE-50 tubing, (Intramedic polyethylene tubing, Clay Adams Brand, Becton-Dickinson, Sparks, MD).

2.2. Surgical Materials

1. Stereotaxic apparatus, Kopf 900 standard small animal stereotaxis (part no. 900, David Kopf Instruments, Tujunga, CA).
2. Rat adapter for small animal stereotaxis (part no. 1220, David Kopf Instruments).
2. Stereotaxic adapter for probe holder (part no. 130-8309005, CMA/Microdialysis, North Chelmsford, MA).
3. Dremel rotary drill with flexible shaft, model 732 part no. 58650, with carbide dental drill bit (HP-2), 1.00 mm in diameter, part no. 51455-4 (Stoelting, Wood Dale, IL).
4. Standard scapel (Harvard Apparatus, Holliston, MA).
5. Vascular scissors (Roboz Surgical Instruments, Harvard apparatus no. 60-3921).
6. Dental cement (part no. 51456, Stoelting).
7. Nylon bolts with nuts, size no. 2-56 × 1 inch (part no. MN-256-16B, Small Parts, Miami Lakes, FL).

8. Stainless steel spring tethers (Harvard Apparatus, Holliston, MA).
9. Betadine solution, and 70% ethanol.
10. Oster small animal clipper, size 40 blade.
11. Pentobarbital sodium (Nembutal, Abbott Labs., North Chicago, IL).
12. Buprenorphine HCl (Buprenex Injectable, Reckitt and Colman Pharmaceuticals, Richmond, VA).
13. Penicillin G procaine suspension (Wycillin, Wyeth-Ayerst, Philadelphia, PA).
14. PE-10 and PE-50 polyethylene tubing (Intramedic polyethylene tubing, Clay Adams brand, Becton-Dickinson).

2.3. Experimental Materials

1. Artificial cerebrospinal fluid for perfusion: 119.5 mM NaCl, 4.75 mM KCl, 1.27 mM $CaCl_2$, 1.19 mM KH_2PO_4, 1.19 mM $MgSO_4$, and 1.6 mM Na_2HPO_4, pH 7.4 (*see* **Note 1**).
2. Microprocessor-controlled syringe pump, Harvard 22 (Harvard Apparatus).
3. Precision Exmire glass microsyringe, 2.5 mL, 7.24 mm diameter, (part no. 100-8309021, CMA/Microdialysis, North Chelmsford, MA).
4. Square plexiglass cage, $30 \times 30 \times 30$ cm, with cover (constructed in house) (*see* **Note 2**).

2.4. Materials for On-Line Analysis of Microdialysates (see Note 3)

1. Multiport two-position injection valve, with electric actuator model E-36 (Valco Instruments C, Houston, TX).
2. Digital valve sequence programmer, model DVSP 2 (Valco Instruments C., Houston, TX).
3. Solvent delivery pump for the HPLC analysis, Shimadzu LC-10AD (Shimadzu Scientific Instruments, Columbia, MD) (*see* **Note 4**).
4. Detector for HPLC analysis, depending on the analyte one can use UV-VIS, fluorescence, electrochemical, or mass spectrometry.
5. Data integrator for response from detector, Shimadzu CR501 (Shimadzu Scientific Instruments).

3. Methods

The methods described here outline 1) the stereotaxic placement of the intracerebral guide cannula for the microdialysis probe, 2) the placement of microdialysis probe in guide cannula and vascular access for blood sampling and drug dosing, 3) the placement of the intravascular microdialysis probe (alternative to vascular blood draw for sampling), 4) the construction of the intravascular microdialysis probe, and 5) general comments on experimental design and data analysis. A specific application of the technique, the role of p-glycoprotein on the distribution of quinidine in the rat brain, is also presented.

3.1. Placement of the Intracerebral Guide Cannula

1. Wistar male rats weighing 280–320 g are used.
2. Administer 50 mg/kg pentobarbital anesthesia by an i.p. injection. Wait for 20–30 min and check the eye blink and paw squeeze reflex to determine the depth of anesthesia.
3. Use Oster clippers (size 40 blade) to clip hair on rat head.
4. Place the rat on a Kopf 900 Standard stereotaxic frame placing ear bars and assuring the "skull flat" position before inserting the incisor bar. Move the incisor bar to keep head level.
5. Scrub scalp with betadine solution and then 70% ethanol.
6. Make a 2–3 cm incision to expose the skull.

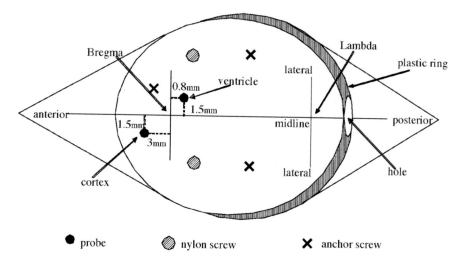

● probe   ◉ nylon screw   ✕ anchor screw

Fig. 2. Schematic diagram of the rat skull with positioning of: 1) the bore holes for the intracerebral guide cannula (●), 2) the anchor screws (✕), and 3) the nylon screws to attach to the tethers (◉).

7. Separate the soft tissues using the blunt edge of the scalpel. The incision should accommodate the plastic ring (circa 2 cm diameter) used to retain the dental cement that will stabilize the probe guide cannula.

8. To assure "skull flat" position, place guide cannula in probe holder in the stereotaxic apparatus, measure depth to the bregma (front suture intersection) and the lambda (rear suture intersection) to get each within 0.5 mm of one another. If not, adjust incisor bar to move rat head up or down *(10)*.

9. To place the probe in the left frontal cortex, put the guide cannula tip at the bregma, record the posterior-anterior, medial-lateral, and depth value for the bregma using the vernier scales on the stereotaxic apparatus. Then raise the tip from the skull, move using anterior-posterior screw adjustment to +3 mm (anterior), move medial-lateral screw adjustment to +1.5 mm (lateral) and mark the tip with ink. Carefully lower the guide cannula to just touch the skull. The guide cannula must not be moved in the probe holder, or coordinates will be inaccurate (*see* **Fig. 2**).

10. Once the mark is placed on the skull, remove the stereotaxic holder and drill a hole through the skull at the mark.

11. Drill using carbide drill bit until just through the skull without disturbing the meninges.

12. Then use a 23-gage beveled needle to cut through the meninges.

13. Drill three guide holes in skull which do not penetrate the skull for placement of anchor screws in the skull (*see* **Fig. 2**).

14. Place the stereotaxic probe holder back to the original depth. We use routinely 1mm depth for the distance from skull surface to the brain surface. So a depth of -1 mm into the cortex would be the same as -2 mm from the original coordinates.

15. For lateral ventricle, the coordinates are 0.8 mm posterior, 1.5 mm lateral from the bregma and 3 mm from the brain surface (*see* **Fig. 2**).

16. Mix dental cement and carefully place some cement to surround probe cannula shaft.

17. Allow the cement to dry for about 10 min and then remove the stereotaxic holder from the probe guide cannula being careful not to move the guide cannula.

18. Then place a 2-cm plastic ring with a small predrilled hole at the posterior end for the tubings from vascular access. Then fill the ring with dental cement. Cover the anchor screws by placing a cotton swab into the hole from the rear, and place another cotton swab on the inside of the plastic ring at the site of the hole to maintain their patency when the dental cement is placed within the ring (*see* **Fig. 2**).

19. Before the cement solidifies, place two 1-inch nylon screws in the cement on each side of the cement ring to hold the tethers (see **Fig. 2**).

20. Leave animal in the stereotaxic device for another 20 min until cement has hardened (it will become more transparent).

21. Place one suture stitch at the posterior side of the ring so that the skin is closed snugly around the ring.

22. Administer 22,000 U/kg of procaine penicillin G I.M. in the flank.

23. Remove the animal from the stereotaxic frame and allow it to regain consciousness on a heating pad (37°C).

24. The animal is then placed in a rat cage (single housing) to recover for a minimum of 3 d before probe insertion and vascular cannulation.

3.2. Placement of Microdialysis Probe in Guide Cannula and Vascular Access

1. After 3 d of recovery after guide cannula placement, re-anesthetize the rat using 50 mg/kg sodium pentobarbital by an i.p. injection.

2. Shave hair in the left groin area and in the scapular region using the Oster 40 clipper. Apply povidone-iodine solution and then 70% ethanol to the shaved area.

3. Make approximately a 0.5 cm incision in the scapular region for later passage of the vascular cannula or intravenous microdialysis probe tubing.

4. Then make a 1.5 cm incision along the inguinal crease.

5. Use blunt dissection to separate the femoral artery, vein, and nerve bundle taking care not to injure the nerve.

6. Using 4-0 silk 9 (*see* **Note 5**), place suture material under the vessel of interest (artery or vein), ligate the distal portion and make a small cut with vascular scissors. Insert a PE-10 tip secured in PE-50 tubing that is filled with 20 U/mL heparin in saline for injection. Advance the PE-10 tip into the vessel for approx 4 cm (*see* **Note 6**).

7. Using a hollow stainless steel trochar, insert the cannula subcutaneously in the scapular region and feed cannula through the hole previously drilled in the dental cement ring. Close the inguinal incision using 4-0 silk, suturing the muscle first, then closing the skin with an additional suture line. Use dental cement to secure cannula to the dental cement crown. Place a stitch of 4-0 silk to secure cannula to the skin at the point of cannula emergence.

8. Prepare the brain microdialysis probe using the manufacturer's recommendations (CMA-12, 3 mm probe). This includes soaking the probe in 70% ethanol for 5 min to remove the glycerin, flushing the probe with the perfusate to be used, and checking the probe integrity using a stereomicroscope.

9. Upon removing the guide cannula stylet, very slowly place probe into the guide cannula until a snug fit is achieved (*see* **Note 7**).

10. Attach the spring tethers (*see* **Note 8**) to the nylon screws and feed the probe inlet and outlet Teflon tubing through one tether and the vascular access cannula (vein and artery) through the other tether.

11. Treat the animal again with IM procaine penicillin G (22,000 U/kg) and 0.05 mg/kg buprenorphine SC for analgesia.

12. Place rat in the plexiglass cage and allow it to recover for at least 16 h after probe insertion (*see* **Note 7**) before beginning the experimental protocol (i.e., administration of the drug substance under study).

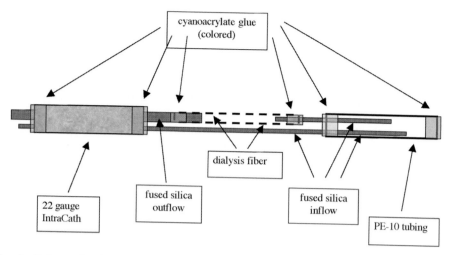

Fig. 3. Schematic drawing of the intravascular microdialysis probe. To aid in a clear representation of the different probe components, the various parts of the probe are not drawn to exact scale.

3.3. Placement of Intravascular Microdialysis Probe (Alternative to Vascular Blood Draw for Sampling)

The simultaneous sampling of both blood and brain by microdialysis is particularly attractive to obtain concentration-time profiles that will allow a complete characterization of the time course for distribution of a drug to, and from, the brain. Therefore, it is sometimes useful to use intravascular microdialysis sampling in conjunction with intracerebral microdialysis sampling *(11,12)*. The following describes the placement of the blood probe into the inferior vena cava or the dorsal aorta.

1. Locate the artery and vein in the left inguinal region of the rat and cannulate these vessels as described in **Subheading 3.2., steps 5** and **6**.
2. Using the self-made intravascular microdialysis probe the same as a PE-10/PE-50 cannula, insert the probe tip (PE-10, *see* **Subheading 3.4**) into the vessel of interest and advance approximately 4 cm (*see* **Note 9**).
3. Using a hollow stainless steel trochar, route the probe inlet and outlet fused silica tubings (both in protective PE-50 sheath) subcutaneously to the scapular region to emerge and then route through the hole in the rear of the dental cement crown on the skull of the animal.
4. Secure the probe to the flank muscle with one stitch of 4-0 silk before closing the incision.
5. If an intravascular probe is placed and the experimental protocol calls for intravenous infusion, then the artery or vein will be cannulated for direct vascular access for dosing.

3.4. Construction of the Intravascular Probe

There are commercially available blood probes e.g., CMA/Microdialysis, Bioanalytical Systems, however, we and others have developed probes that are made in-house for blood microdialysis sampling *(13,14)*. The schematic diagram of our current probe can be seen in **Fig. 3**.

1. PE-10 tubing is used for the probe tip, and plugged with cyanoacrylate glue 24 h before probe construction (*see* **Note 10**).

2. Inlet and outlet fused silica tubings of approx 90 cm lengths are placed in PE-50 cannulas. This length can be modified depending on use, e.g., online vs. offline collection, in vivo vs in vitro, etc.

3. Optional: 2 cm segments of intracath tubing (22 gage) are cut with a sharp scalpel so that the ends are straight. This tubing can be placed on the inlet and outlet fused silica distal to the active probe membrane. It adds some stability to the probe, especially when placing the probe in vessels or spaces that offer some resistance to advancing the probe.

4. Place Intracath tubing on the fused silica, inflow fused silica will extend 15 mm from the Intracath, and the outflow fused silica (larger internal bore) will extend approx 3 mm cm from Intracath.

5. Hollow dialysis fibers are cut with a sharp (fresh) scalpel on a plastic surface to avoid crimping and so that a straight cut can be made. The total length of the membrane is 0.9 cm.

6. Cut the previously prepared PE10 tips into segments 0.8 cm in length with a scalpel so the ends are straight.

7. Using the stylet from a 19-gage Intracath, carefully expand the open end of the PE-10 tip to allow the fused silica inlet tubing to fit for approximately a 0.2 cm distance. This is the distance that the inlet tubing will be inserted into the PE-10 tip of the probe.

8. Check the ends of the fused silica inlet and outlet tubings, to make sure that the ends are cut straight and that there are no burrs or sharp jagged edges.

9. Carefully insert the two inlet fused silica tubing into the PE-10 tip (approx 0.2 cm). Be careful that the tubing does not touch the glue plug. The short "inlet" tubing, i.e., the one that leaves the PE-10 tip to enter the dialysis membrane, has an initial working length of approx 4–5 cm which will be trimmed later. The final length of this inlet fused silica will be approx 0.3 cm long.

10. Probes can be made in batches of at least three probes at a time.

11. The next step is to place the dialysis hollow fiber membrane on the inlet and outlet fused silica tubing. Wear powder free gloves when handling the dialysis membrane. Do not use forceps to avoid damage to the membrane.

12. First place the hollow fiber on the outlet tubing (larger bore), carefully advancing the membrane fiber to approx 0.2 cm on the fused silica tubing.

13. Then place the other end of the hollow fiber dialysis membrane onto the inflow fused silica tubing (coming from the PE-10 tip), advancing the hollow fiber approx 0.2 cm, leaving an effective length for dialysis of approx 0.4 to 0.5 cm. These lengths can be modified depending on the analyte and the desired in vivo recovery.

14. Now use the colored cyanoacrylate glue at the various points of connection. These include the distal and caudal ends of the dialysis fiber to the outlet and inlet fused silica, respectively, and the two "inlet" fused silica tubings to the PE-10 tip, and each end of the 22 gauge Intracath stabilizer, that contains the long inlet and outlet tubings. Using a stereomicroscope (Nikon SMZ-2T), check the probe to examine exactly where the glue has been applied (*see* **Note 11**).

15. If glue placement passes inspection, tape the probes to a box or a counter edge with the constructed ends protruding level into the air to cure the glue for at least 24 h without the glue flowing up or down from the spot of application.

16. Once glue is completely dry, probe tips can be protected using a PE-190 sheath, and stored dry until use (*see* **Fig. 3**).

3.5. Experimental Design and Data Analysis

The design of experimental protocols will depend on the goal of the experiment. General considerations are addressed in the following sections emphasizing dosing protocols, dialysate analysis, and data analysis. No specific step-by-step detail is given in this section due to the variability in protocol that will come from the differences in the chemical and biological properties of the compound and the purpose of the experiment. However, an overall guide in preparing and executing a quantitative microdialysis experiment in the CNS will be illustrated by an example of using quantitative micro-dialysis in the transport studies of quinidine in rats (*see* **Subheading 3.5.4.**).

3.5.1. Dosing Protocols

If the equilibrium distribution ratio (tissue partitioning or distribution) of a drug or test compound between the blood and the brain is the focus, then frequently a steady-state experiment is most appropriate. This would involve using a dosing proto-col where the drug is infused to steady state; where the time to reach the steady-state concentration in blood and tissue would be dependent on the elimination half-life of the drug (*15*). In this type of design, the free concentration of the drug in the blood and in the extracellular fluid of the cortex or cerebrospinal fluid of the lateral ventricle can be compared at steady state and the true partitioning that is due to processes directly at the BBB or the BCSFB can be obtained. In some cases the use of a loading dose and an infusion, or an exponentially changing infusion rate, has been used to achieve a pseudo-steady state condition faster than would be achieved giving a constant rate infusion (*16*). The sampling of the extracellular fluid concentration does not represent the total concentration in the tissue because there may be selective partitioning into the intracellular space (*2*).

Many studies have reported the use of a single intravenous injection of the test sub-stance, followed by frequent sampling in the blood and brain. To obtain the equilibrium distribution ratio from this type of design would require a complete and accurate char-acterization of the areas under the curve (AUCs) for both the blood and brain. This can be difficult if the concentrations in the blood are more rapidly changing with respect to time than those in the brain. Moreover, if the concentration in the blood or tissue is changing rapidly, the fact that microdialysis sampling is an integral collection, not a discrete sample at a time point, can lead to problems in the estimation of the average concentration of the sample over the interval for collection. Therefore, if a single injec-tion dosing protocol is employed and the concentration at the site of the probe is chang-ing rapidly (either due to elimination or a distribution process), it is necessary to ensure that the assay is sensitive enough to tolerate low collection volumes that are collected over short collection periods. It is also necessary to be aware that the determination of the relative recovery of the probe may be compromised by transient conditions, depend-ing on the diffusion of the solute through the tissue (*4*).

3.5.2. Collection and Chemical Analysis of the Microdialysate

The collection of microdialysis samples can be done "on-line," i.e., collecting the dialysate flow directly into an injection loop for HPLC analysis, or it can be done "off-line," where the microdialysate is collected in a small sample vial or tube for further analysis either immediately or after storage. Depending on the analyte, the experimental goals, and the analytical sensitivity, one or the other of these collection methods will be better suited to apply to the experiment. There are excellent reviews discussing the various aspects of sample collection and analytical processing (*17,18*). In our experi-ments to determine the CNS distributional kinetics of a variety of solutes, we have routine used the "on-line" collection system, where the dialysate is collected at 20–30 min into the HPLC loop (50–200 µL, Valco), and the injection is made automatically using the Valco multiport valve with electric actuation (Valco E-36) and an automatic sequence programmer (Valco DVSP-2). Solutes are then separated by reverse-phase HPLC and detected by either UV or fluorescence detectors.

The perfusate flow rates also are an important experimental consideration. Typical flow rates in our hands are 0.2–1.0 µL/min. While increasing the flow rate will increase the sample volume over a given time, doing so will also decrease the relative recovery of the probe, and may lead to changes in net water flux across the probe membrane (the desirable condition is no net water flux into or out of the probe perfusate). Flow rates should be as low as possible, while still allowing for adequate sample volume, and short enough intervals to have resolution and precision in experiments that have tissue concentrations that are rapidly changing.

3.5.3. Data Analysis

The analysis of the concentration-time data obtained from either on-line or off line microdialysis experiments is generally straightforward, however there are a few caveats that must be kept in mind. One important aspect of the analysis of concentration-time profiles is that data collected from a microdialysis experiment is collected over individual time intervals, and as such represent the time-average concentration over that interval (integral sampling). This is different than traditional single-time point sampling, where the concentration measured is a reflection of the concentration in the tissue at that discrete point (discrete sampling). This difference can have important ramifications on the handling of the data to determine important parameters such as the AUCs and the lag time for an absorption or distribution process to occur *(19,20)*.

Determining the relative recovery of the microdialysis probe is critical for use of quantitative microdialysis to study drug distribution to the CNS. As mentioned previously, the recovery is defined as the ratio of drug concentration measured in the microdialysate to the drug concentration in the extracellular fluid at the sampling site. This ratio is best measured under steady-state conditions *(4)* and it is critical to measure the recovery at the site of interest, because it has been frequently shown that recovery determinations in vitro overestimate the true recovery in vivo *(3)*. Several methods exist to measure the in vivo recovery of the probe at the sampling site; two methods we have employed include the zero-net flux method and the use of a calibrator to measure the loss of a structurally similar analog, retrodialysis *(21)*. Once the tissue concentration-time profile has been corrected for the in vivo recovery of the probe, then a pharmacokinetic analysis of the distributional kinetics of the compound of interest can be performed. The pharmacokinetic analysis can include something as simple, but important, as a tissue-to-plasma steady state concentration ratio (the distributional equilibrium constant or tissue partition coefficient) or it can include a detailed analysis of the kinetics and mechanism of distribution using standard compartmental models or physiologically based pharmacokinetic models *(22–25)*.

3.5.4. An Example of the Use of Quantitative Microdialysis in Drug Transport Studies

To illustrate the technique, the following describes an in vivo rat microdialysis study of quinidine distribution in the CNS. This experiment examines the effect of a specific p-glycoprotein inhibitor *(26)*, LY335979, on the distribution of quinidine in the CNS. It is well known that p-glycoprotein, a drug efflux transporter at the BBB, can influence the CNS distribution of several drugs *(27)*. In this study, six rats received quinidine alone (control phase) and quinidine plus LY335979 (treatment phase) in a balanced crossover fashion, which means the first phase is control phase and the second phase is treatment phase in one group of rats ($n = 3$) and vice versa in the other group ($n = 3$).

1. The microdialysis probe in an instrumented rat (*see* **Subheadings 3.1.–3.3.**) is perfused with artificial cerebrospinal fluid, containing 50 ng/mL of hydroquinine as the retrodialysis calibrator, at a flow rate of 0.4 µL/min using a syringe pump.
2. Prior to collection online for HPLC analysis, the dialysate outflow from the microdialysis is collected in small vials and carefully weighted to ensure the integrity of the microdialysis probe and that there was no net water flux either into or out of the probe (*see* **Note 9**).
3. The dialysate is collected every 15 minutes and the concentrations of calibrator are measured by on-line HPLC analysis to determine the loss of the calibrator (for future probe recovery calculation, *see* **Subheading 3.5.4., step 9**).
4. While the microdialysis probe is being perfused, rats receive quinidine with or without LY335979 by intravenous infusion. For the control phase, 25 mg/kg of quinidine is administered by intravenous infusion through femoral vein over a 150-min dosing interval.
5. The dialysates are collected every 15 min from the brain probe and measured for the concentrations of quinidine and calibrator by the on-line HPLC assay.
6. 0.25 mL of blood was sampled from the vascular access every 30 min during and after quinidine infusion. Plasma was collected and stored at –20°C until analysis by off-line HPLC.
7. For the treatment phase, rats received an intravenous bolus (10 mg/kg) and then intravenous infusion of the p-glycoprotein inhibitor, LY335979-3HCl (1.25 mg/kg/h) for 16–18 h before quinidine dosing (the same dose of quinidine by intravenous infusion over the same time period as described above).
8. The infusion of LY335979 is maintained until the brain quinidine peak is no longer detected by HPLC assay (approx 10–12 h after quinidine dose). Similarly, the dialysate and blood samples are collected for on-line or off-line HPLC assay. The interval between these two phases (control and treatment phases) is approx 48 h to ensure adequate washout between the treatment arms.
9. The loss of the calibrator is calculated according to the following equation:

$$\text{Loss} = (C_{in} - C_{out}) / C_{in} \qquad (1)$$

where C_{in} and C_{out} are the concentrations of calibrator (hydroquinine) in the artificial CSF and dialysate, respectively.
10. The concentration of quinidine in the brain ECF is calculated using the following equation:

$$C_{ecf} = C_{dialysate} / \text{Recovery} \qquad (2)$$

where C_{ecf} and $C_{dialysate}$ are the concentrations of quinidine in brain ECF and collected dialysate, respectively. Recovery is the gain of the quinidine from the extracellular fluid to the probe dialysate and this gain has been shown to be equal to the loss of the hydroquinine from the probe in both in vitro and in vivo experiments.
11. The brain and plasma quinidine concentration-time curves were obtained and are shown in **Fig. 4**. The AUC from 0 to the last concentration time point ($AUC_{brain,}$ and $AUC_{plasma,}$) for both brain and plasma were calculated using linear trapezoidal method *(15)*.
12. The effect of LY335979 on the distributional kinetics of quinidine across the BBB was quantified as a brain distribution enhancement parameter (DE) using the following equation:

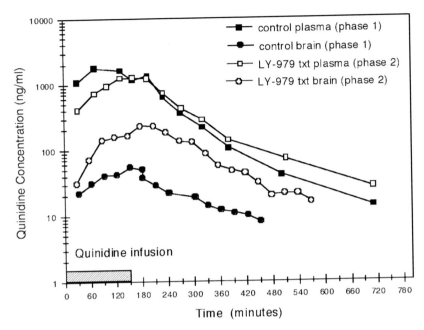

Fig. 4. Concentration-time profiles of quinidine in the plasma (squares) and cortical extra-cellular fluid (circles) after a short intravenous infusion of quinidine in the rat. The solid sym-bols indicate the control phase (without P-glycoprotein inhibition) and the open symbols indicate the treatment phase (P-glycoprotein inhibition with LY335979).

$$DE = \frac{(AUC_{brain,treated} / AUC_{plasma,treated})}{(AUC_{brain,control,} / AUC_{plasma,control})},\quad (3)$$

where $AUC_{brain,control}$ and $AUC_{brain,treated}$ are AUC of quinidine in brain ECF for control phase and treated phase, respectively. $AUC_{plasma,control}$ and $AUC_{plasma,treated}$ are AUC of quinidine in plasma for control phase and treated phase, respectively.

4. Notes

1. Perfusion media used in microdialysis experiments can vary widely in their composition and pH. The ideal perfusate would have the pH, ionic strength, composition, and osmotic pressure the same as that of the external medium, i.e., the extracellular fluid at the sampling site. Several "recipes" for different perfusate have been presented by Benveniste and Huttemeier (28).
2. The cage size and geometry will depend on the purpose of the experiment. For some experiments where activity and/or behavior is being monitored at the same time as drug concentration, larger cages, or computerized activity monitors may be used. For simple drug CNS partitioning experiments, we have found this size adequate to give the rat enough room to move freely without causing problems with the tethers (*see* **Note 8**).
3. Several options exist for the analysis of the microdialysates. The analysis can be done "off-line" where the dialysate is collected in microvials for later processing, e.g., by HPLC, radioimmunoassay, enzyme-linked immunosorbent assay, and liquid scintillation counting. Frequently, the microdialysates can be collected directly into an injection loop for HPLC analysis. This method is useful in obtaining real time data; for instance, to examine whether or not the concentration at the sampling site has reached a steady-state value.

4. For many of the HPLC analyses of microdialysis samples, sensitivity becomes an important issue. Therefore narrow bore or micro bore chromatography is frequently used. The smaller diameter columns require a lower flow rate that is not pulsatile, and the solvent delivery pumps used allow very low flow rates.

5. The choice of surgical silk as the suture material is sometimes questionable. For long-term placements, it may be better to use monofilament suture material to avoid infection at the incision site.

6. Advancing the vascular cannula 4 cm in a 280–300 g rat should result in the cannula tip being placed in the vena cava or the dorsal aorta approximately at the level of the diaphragm. It is useful to have these cannula in a large vessel to avoid problems with placement that lead to difficulty in drawing blood. The tips of these cannula should be cut off perpendicular to the longitudinal axis to further reduce the chance that the cannula tip will be drawn up against the vessel wall, again making blood sampling difficult.

7. Placing the microdialysis probe into the guide cannula slowly can help reduce the damage caused by pushing the probe through the dense capillary bed in the parencyhma of the frontal cortex. The dimensions of the probe (typically 500 µm in diameter) are such that placing the probe in the brain parenchyma will cause damage to the capillaries that comprise the BBB. It has been suggested that the damage caused by the probe insertion is long lasting *(29)* and in one study, it was shown that the chemical selectivity in the rate of transfer across the barrier between urea and sucrose was lost, and the suggestion was that this was indicative of the loss of BBB integrity *(30)*. Other reports indicate that the barrier will reanneal after the insertion trauma *(31)*. Using the procedures outlined in this chapter, we have examined the chemical selectivity of tritiated water and ^{14}C-mannitol regarding the permeability of the BBB following probe implantation in the frontal cortex. The simultaneous plasma and cortical extracellular fluid concentration-time profiles from a representative experiment are shown in **Fig. 5**. As would be expected for the distribution of water, a rapid and complete equilibrium was achieved between plasma and cortical extracellular fluid. However, throughout the 6-h experiment the cortical mannitol concentrations remaining significantly lower than the plasma levels.

8. The spring tethers will be directly attached to the nylon screws that protrude from the dental cement cap. We use two tethers to allow for more room to run vascular cannula and the probe tubing, and very importantly, the two tethers give a mild rotational torque, which prevents the animal from tangling the lines.

9. Although the vascular microdialysis probe can be placed in either the artery or the vein, in recent experiments we have routinely placed the probe in the artery (dorsal aorta) for two reasons. The first is that the arterial concentration of drug is the actual concentration that is perfusing the tissue (brain), and the second is that the intravascular hydrostatic pressure in the artery is higher than in the vein, thus diminishing the pressure gradient across the probe membrane and reducing the possibility of water flux out of the probe at higher perfusate flow rates. This is particularly valuable when using the retrodialysis technique to determine probe relative recovery.

10. It is important that the glue be completely dry before fused silica tubings are placed in the tip. It is helpful to add color to the glue with a colored marker by mixing glue on colored marker spot on glass. This helps to decide how much glue to use, and where the glue is in the probe.

11. It is critical at this stage not to use too much glue to avoid the glue reaching an area that would block flow, or have the glue run up the hollow fiber membrane and block the active surface area for analyte diffusion across the membrane. Using the colored glue with a fine stick applicator can avoid these complications. Placement of the glue is the most critical step in the entire probe construction procedure that will lead to probe failure if not done properly.

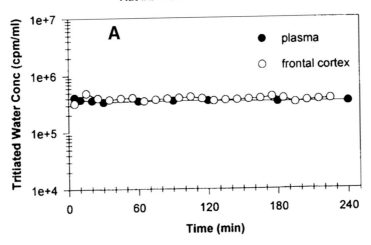

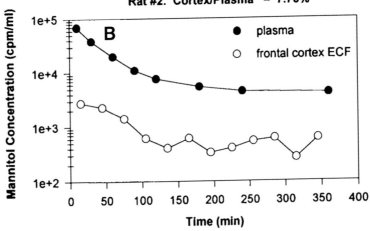

Fig. 5. Concentration-time profiles for plasma and cortical ECF concentration of tritated water (**A**) and ^{14}C-labeled mannitol (**B**).

Acknowledgments

The authors thank Hua Yang, Qin Wang, and Haiying Sun; graduate students who helped develop and establish the intracerebral microdialysis techniques described herein. We also acknowledge support from Pfizer, Eli Lilly, Novartis, and the NIH grants CA75466 and CA71012 that made these studies possible.

References

1. Elmquist, W. F., and Sawchuk, R. J. (1997) Application of microdialysis in pharmacokinetic studies. *Pharm. Res.* **14**, 267–288.
2. Wang, Y., and Welty, D. F. (1996) The simultaneous estimation of the influx and efflux blood-brain barrier permeabilities of gabapentin using a microdialysis-pharmacokinetic approach. *Pharm. Res.* **13**, 398–403

3. Stenken, J. A., Lunte, C. E., Southard, M. Z., and Stahle, L. (1997) Factors that influence microdialysis recovery. Comparison of experimental and theoretical microdialysis recoveries in rat liver. *J. Pharm. Sci.* **86,** 958–966.

4. Bungay, P. M., Dedrick, R. L., Fox, E., and Balis, F. M. (2001) Probe calibration in transient microdialysis in vivo. *Pharm. Res.* **18,** 361–366.

5. Kehr, J., Yoshitake, T., Wang, F. H., et al. (2001) Microdialysis in freely moving mice: Determination of acetylcholine, serotonin and noradrenaline release in galanin transgenic mice. *J. Neurosci. Methods* **109,** 71–80.

6. Scism, J. L., Powers, K. M., Artru, A. A., Lewis, L., and Shen, D. D. (2000) Probenecid-inhibitable efflux transport of valproic acid in the brain parenchymal cells of rabbits: A microdialysis study. *Brain Res.* **884,** 77–86.

7. Yang, Z., Brundage, R. C., Barbihaiya, R. H., and Sawchuk, R. J. (1997) Microdialysis studies of the distribution of stavudine into the central nervous system in the freely-moving rat. *Pharm. Res.* **14,** 865–872.

8. Bouw, M. R., Xie, R., Tunblad, K., and Hammarlund-Udenaes, M. (2001) Blood-brain barrier transport and brain distribution of morphine-6-glucuronide in relation to the antinociceptive effect in rats—pharmacokinetic/pharmacodynamic modelling. *Br J Pharmacol.* **134,** 1796–1804.

9. Yang, H., Wang, Q., and Elmquist, W. F. (1996) Fluconazole distribution to the brain: A crossover study in freely-moving rats using in vivo microdialysis. *Pharm. Res.* **13,** 1570–1575.

10. Paxinos, G. and Watson.(eds.) (1998) *The Rat Brain in Stereotaxic Coordinates*. 4th ed. Academic Press, San Diego, CA.

11. Kau, Y. C., Wong, K. M., Shyr, M. H., Lee, Y. H., and Tsai, T. H. (2001) Simultaneous determination of unbound ropivacaine in rat blood and brain using microdialysis. *J. Chromatogr. B Biomed. Sci. Appl.* **760,** 107–112.

12. Evrard, P. A., Ragusi, C., Boschi, G., Verbeeck, R. K., and Scherrmann, J. M. (1998) Simultaneous microdialysis in brain and blood of the mouse: Extracellular and intracellular brain colchicine disposition. *Brain Res.* **786,** 122–127.

13. Evrard, P. A., Deridder, G., and Verbeeck, R. K. (1996) Intravenous microdialysis in the mouse and the rat: Development and pharmacokinetic application of a new probe. *Pharm. Res.* **13,** 12–17.

14. Yang, H., Wang, Q., and Elmquist, W. F. (1997) The design and validation of a novel intravenous microdialysis probe: application to fluconazole pharmacokinetics in the freely-moving rat model. *Pharm. Res.* **14,** 1455–1460.

15. Gibaldi, M. and Perrier, D. (1982) *Pharmacokinetics*. 2nd ed. Marcel Dekker, New York.

16. Hammarlund-Udenaes, M., Paalzow, L. K., and de Lange, E. C. (1997) Drug equilibration across the blood-brain barrier—pharmacokinetic considerations based on the microdialysis method. *Pharm. Res.* **14,** 128–134.

17. Davies, M. I., Cooper, J. D., Desmond, S. S., Lunte, C. E., and Lunte, S. M. (2000) Analytical considerations for microdialysis sampling. *Adv. Drug Deliv. Rev.* **45,** 169–188.

18. Lunte, S. M., and Lunte, C. E. (1996) Microdialysis sampling for pharmacological studies: HPLC and CE analysis. *Adv. Chromatogr.* **36,** 383–432.

19. Stahle, L. (1993) Zero and first moment area estimation from microdialysis data. *Eur. J. Clin. Pharmacol.* **45,** 477–481.

20. Stahle, L. (1992) Pharmacokinetic estimations from microdialysis data. *Eur. J. Clin. Pharmacol.* **43,** 289–294.

21. Wang, Y., Wong, S. L., and Sawchuk, R. J. (1993) Microdialysis calibration using retrodialysis and zero-net flux: application to a study of the distribution of zidovudine to rabbit cerebrospinal fluid and thalamus. *Pharm. Res.* **10,** 1411–1419.

22. Dukic, S., Heurtaux, T., Kaltenbach, M. L., et al. (1999) Pharmacokinetics of methotrexate in the extracellular fluid of brain C6-glioma after intravenous infusion in rats. *Pharm. Res.* **16,** 1219–1225.

23. Tsai, T. H., Lee, C. H., and Yeh, P. H. (2001) Effect of P-glycoprotein modulators on the pharmacokinetics of camptothecin using microdialysis. *Br. J. Pharmacol.* **134,** 1245–1252.

24. Ma, J., Pulfer, S., Li, S., Chu, J., Reed, K., and Gallo, J. M. (2001) Pharmacodynamic-mediated reduction of temozolomide tumor concentrations by the angiogenesis inhibitor TNP-470. *Cancer Res.* **61,** 5491–5498.

25. Zamboni, W. C., Houghton, P. J., Hulstein, J. L., et al. (1999) Relationship between tumor extracellular fluid exposure to topotecan and tumor response in human neuroblastoma xenograft and cell lines. *Cancer Chemother. Pharmacol.* **43,** 269–276.

26. Dantzig, A. H., Shepard, R. L., Law, K. L., et al. (1999) Selectivity of the multidrug resistance modulator, LY335979, for P-glycoprotein and effect on cytochrome P-450 activities. *J Pharmacol Exp Ther.* **290,** 854–862.

27. Schinkel, A. H., Wagenaar, E., Mol, C. A., and van Deemter, L. (1996) P-glycoprotein in the blood-brain barrier of mice influences the brain penetration and pharmacological activity of many drugs. *J Clin Invest.* **97,** 2517–2524.

28. Benveniste, H., and Huttemeier, P. C. (1990) Microdialysis—theory and application. *Prog. Neurobiol.* **35,** 195–215.

29. Groothuis, D. R., Ward, S., Schlageter, K. E., et al. (1998) Changes in blood-brain barrier permeability associated with insertion of brain cannulas and microdialysis probes. *Brain Res.* **803,** 218–230.

30. Morgan, M. E., Singhal, D., and Anderson, B. D. (1996) Quantitative assessment of blood-brain barrier damage during microdialysis. *J. Pharmacol. Exp. Ther.* **277,** 1167–1176.

31. de Lange, E. C., Danhof, M., de Boer, A. G., and Breimer, D. D. (1997) Methodological considerations of intracerebral microdialysis in pharmacokinetic studies on drug transport across the blood-brain barrier. *Brain Res. Rev.* **25,** 27–49.

17

Single-Pass Dual-Label Indicator Method

Blood-to-Brain Transport of Glucose and Short-Chain Monocarboxylic Acids

Michelle A. Puchowicz, Kui Xu, and Joseph C. LaManna

1. Introduction

The brain requires an adequate and constant supply of glucose or alternate energy substrates to support its metabolic demands. Because of the special conditions imposed by the presence of the blood–brain barrier (BBB), specific transport mechanisms are required for the influx of water-soluble substrates. Alterations in energy substrate transport or availability has profound consequences that may result in inadequate energy supply and possible cell death. To study these transport mechanisms, quantitative methods of substrate influx and blood flow have been developed. These methods are based on a model of unidirectional tracer influx at the endothelial cell boundary from the blood during capillary transit (*1*).

1.1. Transport-Mediated Substrate Uptake

The dual-label single-pass method (*2*), a variation of the brain uptake index (BUI) method, was first used to investigate carrier mediated BBB transport of short-chain monocarboxylates such as pyruvate and lactate (*3*). Pellegrino et al. (*4*), reviewed and critiqued this method in detail with respect to glucose transport. They compared the results obtained by this in vivo method with other methods, such as the BUI, and discussed the physiologic interpretation of the data (*see* **Note 1**). The advantage of the dual-label single-pass method is that it simultaneously measures both substrate influx and blood flow, which enables the study of glucose, lactate, and other substrates transported in the brain as well as the blood flow in different pathophysiologic states.

Carrier-mediated facilitative diffusion of glucose was first reported by Crone (*5*) and further confirmed by Dick et al. (*6*) in their studies on glucose transport in isolated microvessels. This transport is characterized by measuring the transfer of a unidirectional tracer influx of a blood–brain substrate across the BBB from which kinetic transport constants are derived (*see* **Subheading 3.8.4.**). The derivation of these constants is based on Michaelis-Menten kinetics (**Fig. 1**), but may differ between substrates. Individual considerations are needed for those substrates that have unique characteristics different from glucose. For example, the transport properties of glucose (*7*) and lac-

From: *Methods in Molecular Medicine, vol. 89:*
The Blood–Brain Barrier: Biology and Research Protocols
Edited by: S. Nag © Humana Press Inc., Totowa, NJ

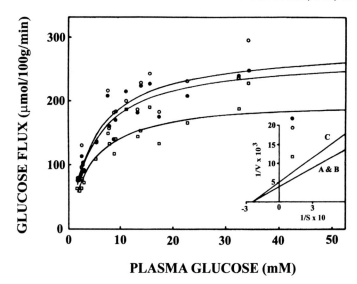

PLASMA GLUCOSE (mM)

Fig 1. Influx of glucose from blood to brain (J) as a function of arterial plasma glucose content for frontal cortex (●), cerebellum (O), and hippocampus (□). The lines represent computer-generated nonlinear least-square fits to the Michaelis-Menten equation:, $J = \frac{T_{max} \times G}{K_t + G}$, where T_{max} is maximal transport and K_t is the glucose concentration at half-maximal transport. T_{max} for the cerebral cortex and cerebellum were not different but T_{max} for the hippocampus was significantly less than for all other regions at $p < 0.05$. K_t values for all regions ranged between 4.4 and 5.1 mM and were not significantly different from each other *(7)*. The inset showing lines of regression of double reciprocal plots is another way of presenting the data where the x-intercept is equal to $-1/K_t$ and the y-intercept is equal to $1/T_{max}$. (*A*, frontal cortex; *B*, cerebellum; *C*, hippocampus.) Reprinted from **ref. 7**, with permission from Elsevier Science.

tate *(8)* differ in the diffusional component and substrate stereospecificity (e.g., both glucose and lactate have D- and L-isomers but the glucose transporter is specific for D-glucose, whereas the transporter for lactate is nonspecific).

1.2. Dual-Label Single-Pass Indicator Fractionation Method

This method was developed to characterize the transport efficiency and transport capacity of a substrate across the BBB and is based on the following assumptions:

1. There is efficient mixing of the injected bolus of the tracer with the parent compound (endogenous substrate being measured) and blood before influx into the brain so that the prepared tracer bolus does not interfere with blood flow *(8)*.
2. Loss of the radiolabeled tracer from the blood-vascular compartment does not occur during the time between the bolus injection and decapitation, e.g., tracers should not be taken up by erythrocytes or leak out of the vascular system *(1)*.
3. Brain–blood capillary volume remains constant and the time between bolus injection and decapitation occurs within mean circulation time, so that recycling of tracer does not occur *(9)*.
4. Metabolic steady state tracer transport is unidirectional and the net transport of the parent compound is constant along the length of the capillary *(1)*.

2. Materials

2.1. Tracers

n-[1-³H]butanol (1.0 mCi/mL; American Radiolabeled Chemicals (ARC), St. Louis, MO) with one of the following, D- or L-[^{14}C]glucose, D- or L-[^{14}C]lactate (NEN, Boston MA) or β-[1-^{14}C]hydroxybutyrate (ARC).

2.2. Anesthetics

1. 4% chloral hydrate (Fisher Scientific, Pittsburgh, PA).
2. 2% Lidocaine hydrochloride (Abbot Laboratories, N. Chicago, IL), Marcaine hydrochloride (Astra Pharmaceutical Products, Westborough, MA).
3. Gas mixture containing 2–5% halothane (Halocarbon Laboratories, River Edge, NJ) and 70% N_2O in O_2.

2.3. Surgical Supplies

1. Polyethylene tubing, PE 50, 0.023-inch i.d., 0.038-inch o.d (Becton Dickinson, Sparks, MD).
2. Silastic catheter, 0.025-inch i.d., 0.047-inch o.d (Fisher Scientific).
3. Heparinized micro-hematocrit capillary tubes (Fisher Scientific).

2.4. Equipment and Radioassay Supplies

1. Rectal thermister probe (Yellow Springs Instrument, Yellow Springs OH).
2. Syringe-pump withdrawal system: pump is plugged into a timer, syringe and tubing.
3. Blood gas analyzer (ABL5, Radiometer Copenhagen, Denmark).
4. Guillotine.
5. 1600 TR Liquid Scintillation Analyzer (Packard Instrument, Meriden, CT).
6. Scintillation vials (20 ml) with cap (Fisher Scientific).
7. Soluene 350 (Packard Instrument).
8. Scintillation cocktail (Ultima Gold, Packard Instrument).

3. Methods

3.1. Animal Preparation

1. All procedures performed on animals are approved by the Case Western Reserve University Institutional Animal Care and Use Committee.
2. Male Wistar rats are anesthetized with either chloral hydrate (10cc/ kg, i.p) or halothane gas mixture for the duration of the experiment (*see* **Note 2**).
3. The surgical procedure for the placement of catheters is previously described *(10)*. Cannulae are placed in (a) a femoral artery using polyethylene tubing (PE-50) for the purpose of monitoring systemic arterial blood pressure and obtaining blood samples, and (b) an external jugular vein as far as the right atrium using a silastic catheter for bolus administration of tracers.
4. Incision sites are infiltrated with a local anesthetic such as 2% lidocaine HCl or for longer acting anesthesia marcaine HCl is used and the incision sites are sutured closed.
5. Body temperature is maintained near 37°C by a rectal thermistor probe connected to an infrared-feedback controlled lamp.
6. The following physiological parameters are measured: blood gases (PCO_2 and PO_2), pH, hematocrit, plasma substrate concentrations (mM) for glucose, lactate, and ketones.

7. Regional blood-to-brain substrate transport and brain blood flow are measured simultaneously approx 60 min after the induction of anesthesia or until the mean arterial blood pressure is at least 85 mm Hg (near baseline, 100 mm Hg).

3.2. Preparation of Tracers and Substrates

The dual-label, single-pass, right atrial bolus injection is performed using two radiolabeled tracers, [^{3}H] and [^{14}C]. Depending on the commercial availability of the tracers, either labeling can be used as the blood flow indicator or the substrate (*see* **Note 3**). For example, the blood flow radiolabeled indicator tracer, n-[^{3}H]butanol is prepared with [^{14}C]glucose, [^{14}C]lactate, or β-[^{14}C]hydroxybutyrate. Other studies have used 1-[methyl-^{3}H]-4 phenyl-1,2,3,6-tetra-hydropyrdine (MPTP) as the blood flow indicator for uptake studies measuring lactate, using D- or L-[^{14}C]lactate (*8*). The combination of n-[^{14}C]butanol with D- or L-[^{3}H]glucose has also been used for simultaneously measuring blood flow and glucose uptake (*7*). Either MPTP or n-butanol are equally efficient for use in studies that are designed where the time between the bolus administration and decapitation is within the washout time of the blood flow indicator (*see* **Note 4**). [^{14}C]Iodoantipyrine is another blood flow indicator commonly used in combination with autoradiography techniques for measuring blood flow (*11*) and can also be used in the dual-label method.

The purpose for using substrate analogues such as L-glucose or D-lactate is to determine the stereospecificity of the transporter as they are not metabolized to intermediates or end products in mammals. With regards to ketone body kinetics, there are two constraints in determining ketone body kinetics in brain. First, the monocarboxylate transporter (MCT) is apparently not stereospecific, as the glucose transporter (GLUT) is for D-glucose and not L-glucose. This makes it difficult to account for nonspecific transport. Also, both β-hydroxybutyric acid isomers, R- and S-, are metabolized in the liver to end products (*12*), whereas the metabolism of the S-isomer is yet to be determined in the brain.

1. Special precautions are observed when handling radioactive materials (*see* **Note 5**).
2. The tracer bolus is prepared by first pipetting into a 10 × 75 mm test tube either, D- or L-[^{14}C]glucose, D- or L-[^{14}C]lactate, or β-[1-^{14}C]hydroxybutyrate, then adding the blood flow indicator, n-[1-^{3}H]butanol.
3. If the substrate is packaged in ethanol, then the bolus is first prepared by evaporating the substrate tracer to dryness and then adding the n-[1-^{3}H]butanol (loss of n-butanol may occur as a result evaporation, because of its volatility).
4. A final volume of 150 mL is achieved by adding normal saline and unlabeled substrate to match that of the blood (*see* **Note 6**). For example, the final bolus volume (150 μL) contains a mixture of 50 μL (50 μCi) of D- or L-[1-^{14}C]glucose and 10 μL (10 μCi) of n-[1-^{3}H]butanol (1.0 mCi/mL) and an amount of nonradioactive D-glucose to match the concentration of glucose in the blood.
5. Load the bolus into a 1-mL syringe with approx 0.100 mL of air. Be sure that the air is at the top of the syringe so that it is captured in the catheter tubing following the administration of the tracer bolus.
6. The air volume can be determined by first measuring the volume in the atrial catheter before surgical placement. The purpose is to ensure that all of the tracer bolus is administered to the animal and that none is left in the tubing.
7. The amount of radioactivity used is dependent on both the sensitivity of the β-scintillation analyzer and the amount taken up by the tissue. We recommend using 10–50 μCi for each

of the tracers, the radiolabeled substrate or blood flow indicator, per rat (200–600 g). The μCi ratio (^{3}H:^{14}C) in tissue should be approx 1:1–5 (*see* **Note 7**).

3.3. Substrate Transport Measurements

The measurement of regional blood flow and the transport of glucose, lactate, or ketones are determined by giving doses of the unlabeled substrate, intravenously or intraperitoneally, such that the blood concentrations range from low to high levels (the high level should exceed maximal transport, *see* **Notes 6**, **8**). For example, the group receiving the dose (either by infusion or bolus) of the natural form of the substrate, e.g., D-glucose, L-lactate, or R-β-hydroxybutyrate, would be given a bolus of the radiolabeled tracer, either D-[^{14}C]glucose, L-[^{14}C]lactate, or R-[^{14}C]-β-hydroxybutyrate. The diffusional component and stereospecificity for a substrate can be determined by giving the unnatural form of the radiolabeled tracer, D-[^{14}C]lactate or L-[^{14}C]glucose to another group of animals.

3.4. Experimental Protocol

1. The arterial cannula is connected to the tubing, which is connected to a syringe fitted to a pump that is calibrated to withdraw blood at a rate of 1.60 mL/min. The tubing connected to the cannula should be long enough to contain the volume of blood collected in 10–12 s, approx 0.020 mL.
2. The pump is plugged into a clock-timer set to automatically stop the withdrawal after 10–12 s. Choosing 10 or 12 s depends on the mean circulation time (from heart to tissue to heart).
3. For brain, we have found that between 10 and 12 s is the optimal time for the substrate to be taken up by the tissue.
4. The pump is turned on 3 s before the tracers are given. During this time, verify that blood is collecting in the tubing before continuing with the experiment. If there is no flow or a delay in the collection of the blood, then stop the pump (before giving the tracer bolus), disconnect the line from the pump, and flush the arterial line with saline. Also, check whether the arterial line is twisted by gently rotating the arterial line and observe if there is a "pulsing" in the line near the incision. This is a good indicator that there is a good flow and that you can proceed with the experiment.
5. Inject the bolus containing the tracers into the right atrium.
6. The pump is set to shut off by the timer 10–13 s after the bolus is administered and the rat is simultaneously decapitated.
7. The femoral line is first disconnected from the rat and then flushed with 10 mL of water into a pre-weighed 20 mL screw-cap vial. The total contents (blood and tap water) are re-weighed and the weight recorded.

3.5. Sample Preparation and Analysis of Radioactivity

1. Immediately following decapitation, venous blood oozing from the foramen magnum is collected in heparinized tubes and an aliquot of the plasma obtained is used to estimate the radioactive content of the cerebral intravascular compartment. This accounts and corrects for the residual vascular radioactive content *(2)*.
2. The brains are then rapidly removed and dissected and the regional bilateral tissue samples (cortex, parietal, hippocampus, cerebellum, striatum, and brainstem) are collected into pre-weighed vials and a final weight is recorded.
3. The brain tissue is then solubilized by adding 0.600 mL of Soluene 350 to each of the vials. The vials are tightly closed and shaken overnight or until the tissue is completely dissolved, (approx 8 h).

3.8.3. Regional Permeability-Surface Area and Substrate Clearance Rates

Regional substrate clearance rates can be estimated from the regional permeability-surface area product (PS). PS is calculated from the regional plasma flow, F_{pl} **(Eq. 4)** and the extraction fraction, E, **(Eq. 5)**, *(5,15)*:

$$PS = -F_{pl} \times ln(1\text{-}E)$$

(6)

From the PS of the natural substrate (e.g., D-glucose) and unnatural substrate (e.g., L-glucose) clearance can be estimated by summing the two values (*see* **Note 10**).

3.8.4. Estimation of Substrate Transport Constants

To characterize the stereospecificity for the physiologic enantiomer, estimations of the substrate transport constants, extraction fraction (E_f) and diffusion (K_d), are calculated *(8)*.

The estimation of E_f involves the determination of the diffusional component of the blood-to-brain substrate influx. This requires the measurement of influx using a non-physiologic enantiomer (e.g., L-glucose, D-lactate). E_f is calculated as the ratios of the dual-labels in the tissue to the withdrawn arterial blood (as described in **Eq. 5**). This implies that E_f is independent of the blood substrate concentrations. However, when lactate is the substrate, E_f appears to have an indirect dependence. The effects of substrate concentrations on E_f are presumed to be a consequence of altered metabolic state that results in changes in blood flow and thus the amount of radioactive labeled substrate crossing the BBB (*see* **Notes 6,8,10**).

The regional substrate diffusion constant, K_d, is calculated by regression analysis from the linear slope of the substrate influx plotted as a function of plasma substrate concentration. From this K_t, E_{max}, and T_{max} are determined *(7,8)*. First, K_t is defined as the substrate plasma level at half-maximal transport, E_{max} , and is estimated from the exponential relationship expressed as a function of extraction fraction (minus the component of E_f for the unnatural enantiomer, e.g., $E_{f\,L\text{-}\underline{lactate}} - E_{f\,D\text{-}lactate}$):

$$E_{[substrate]pl} = E_{max\,e}{}^{-K_{t[substrate]pl}}$$

The theoretical maximum extraction fraction, E_{max} , is calculated by letting the [substrate]$_{pl}$ equal zero (this is when E_{max} is equal to $E_{[substrate]pl}$). From this, the maximal stereospecific transport capacity (T_{max}) is estimated by nonlinear regression (*see* **Fig 1.**):

$$J = \frac{T_{max} \times [substrate]_{pl}}{K_t + [substrate]_{pl}} + (K_d \times [substrate]_{pl})$$

(8)

The regional maximal transport for glucose (T_{max}) is reported to be in the range of 2.0–3.0 µmol/g/min and 4.4 –5.0 mM for the K_t and for the lactate the T_{max} ranges from 23–40 µmol/g/min and 6.0–14.0 mM for the K_t *(7,8)*. To correct for the diffusional component, the additive component on the right side of **Eq. 8** ($K_d \times$ [substrate]$_{pl}$) is then subtracted from the influx (J), as calculated in **Eq. 3**.

3.9. Applications

Information concerning substrate availability and metabolism in the brain is fundamental in establishing appropriate treatments during critical events such as ischemic/hypoxic insult. The dual-label single-pass method enables one to simultaneously measure regional blood flow and substrate influx across the BBB in normal and pathophysiological conditions in small animals, such as rat. This method is useful for obtaining a quantitative estimate of the capacity of test substances such as glucose *(7,14)*, lactate *(8)*, or leucine *(14)*, to cross the BBB. It allows correction for the vascular space, and if an inactive analog is available, calculation of specific transport rates. We have used this method extensively in studies of substrate influx across the BBB in rat. Substrate influx has important or significant implications in a number of pathophysiological conditions such as, acute and chronic states of hyperglycemia *(4,7,21)*, hypoxic *(9)*, and ischemic insults *(10)*, and epilepsy. We have also modified the dual-label single-pass method to measure plasma, sucrose, or albumin space *(10,22–24)*. In addition to metabolic energy substrate transport, the method also lends itself to studies of drug *(16)* and hormone BBB influx, as well as experiments using animal models to test clinically relevant molecules. In particular, the influx rate of glucose in the brain can be limiting in cases where there are genetic defects in the endothelial GLUT1 transporter *(25)*. Accurate data on blood flow and metabolism in humans can only be obtained through expensive and not readily available techniques, such as PET and MR based approaches. Thus, it is valuable to obtain information from animal studies using the dual-label single-pass method prior to evaluating human disease states. Furthermore, with the addition of autoradiographic protocols, this method can be adapted to provide micro-regional and topographic analyses, depending on the availability of suitable tracers.

4. Notes

1. An important distinction between the two methods is the difference in the mixing and dilution of the bolus of radiolabeled tracers in its passage through the brain circulation. Because of the differences between the two methods the results from studies using the BUI method may be underestimated *(1,2,4)*. This may be explained by the following: the BUI method requires the use of an intracarotid bolus site, resulting in a *"bolus wave front"* upon reaching the brain, leading to an apparent underestimation of maximal transport, as reported for glucose *(14)* and lactate *(8)*. It is thought that the intracarotid bolus technique interferes with capillary bed perfusion, resulting in changes in blood flow. The intra-atrial injection method is designed to eliminate or minimize the effect of the tracer bolus on blood flow. This is achieved by i) preparing the bolus to match the concentration of the substrate in the blood, and ii) intra-atrial administration of the tracer bolus, rather than in the carotid artery. The bolus then mixes with the circulating blood and as a result the time the bolus wavefront reaches the brain, the effects of the bolus on blood flow are minimized. Studies using this technique report data similar to studies using the BUI method.

2. When using alternative anesthetics such as chloral hydrate (4%, 10 cc/kg, i.p), wait at least 60 min after induction and avoid additional doses of anesthetic when possible, as this delays the experiment. A single dose is effective for approx 1.5–2 h. Large doses of anesthesia have an effect on regional blood flow and substrate uptake.

3. There are a variety of commercially available tracers for use. Any ^{3}H/^{14}C tracer is acceptable for use in the single-pass dual-label indicator method; cost and availability are the major considerations. We have used a variety of tracer mixtures: For the blood flow indicator, n-[1-^{14}C]butanol, either D- or L-[^{3}H]glucose *(7)* or D- or L-[^{3}H]lactate can be used. For n-[1-^{3}H]butanol, D- or L-[^{14}C]glucose, D- or L-[^{14}C]lactate, or β-[1- ^{14}C] -hydroxybutyric acid. Other studies have used 1-[methyl-^{3}H]-4 phenyl-1,2,3,6-tetrahydropyrdine (MPTP) as the blood flow indicator with D- or L-[^{14}C]lactate *(8)*. Because n-butanol is volatile, only

purchase the compound in a normal saline storage solvent and not a solvent that would require evaporation before administration.

4. Both MPTP and butanol are almost completely extracted by all regions of the brain on the first pass. They are equally efficient for studies using a single-pass extraction technique for measuring blood flow. The main difference between the two blood flow indicators is that butanol rapidly washes out of the brain, whereas MPTP is retained by the brain for a longer period of time. MPTP may be more appropriate in studies where the bolus-to-decapitation time needs to be extended. In the case of butanol, such studies should be designed with a bolus-to-decapitation time of approx 0.21 min (or less), whereas 0.33 min is maximal for MPTP but could be longer (*16*).

5. The handling and use of radioactive material during this procedure conform to the guidelines set forth by the Nuclear Regulatory Commission:
   a. Personal protective equipment (gloves, lab coat, goggles) is worn at all times during the experiment.
   b. To minimize contamination, areas where radioactive materials are placed during the experiment are designated and covered with bench pads and ^{14}C and ^{3}H doses are taken from 0.50-μCi aliquots.
   c. The NRC only allows 20 μCi of ^{14}C and ^{3}H sewer disposal per day and this limits the number of procedures that can be performed daily.
   d. The carcasses are put into radiation bags and tagged with the date, activity and authorized user name, then taken to our institution's radioactive animal disposal area. Tubing, catheters, and pipet tips coming into contact with the radioactive solutions are disposed of as are dry SHARPS waste in the laboratory's SHARPS waste container designated for ^{14}C and ^{3}H waste. The bench pads and other non-SHARP waste are disposed of as dry waste in our laboratory's ^{14}C and ^{3}H dry waste container.
   e. All areas and equipment are washed clean with No Count Decontaminant Surface Cleaner (Fisher) and a wipe test for ^{14}C and ^{3}H is performed before and after experimental procedures to make sure the area and equipment are kept uncontaminated.

6. Unlabeled substrates can be administered as a bolus or infusion, either iv or i.p. However, in some cases perturbations of the physiologic state can occur, leading to changes in PaO_2 and $PaCO_2$, resulting in changes in blood flow (*2,8*). We also recommend that the concentration of the tracer bolus be matched to the plasma concentration of glucose (or substrate of interest) to avoid changes in blood flow and erroneous extraction fraction values.

7. Measuring a sample containing two radionuclides requires additional considerations than measuring one radionuclide. When determining the amount of tracer (μCi) administered: i) the dpm value in the arterial blood and tissue samples should be within the sensitivity and efficiency of the liquid β-scintillation analyzer (1,000–10,000 dpm is the approximate range we recommend), and ii) the prepared tracer bolus should contain a $^{3}H^{14}C$ ratio of approx 1:1-5 μCi, when the substrate is ^{14}C labeled or 1-5:1 when the substrate is ^{3}H labeled. The reason for this is to account for the substrate uptake by the brain. The resulting ratio in tissue should be 1:1 dpm. These considerations are based on the theory of dual label dpm measurements (see the Packard manual). Briefly, each of the beta-emitting radionuclides contributes to one spectrum from zero to beta-maximum. The composite spectrum is the sum of the ^{3}H and ^{14}C spectra. The ^{14}C has higher energy than ^{3}H but the spectra are indistinguishable from each other, making it difficult to estimate the contribution of each radionuclide.

8. In the case of lactate, there is no apparent stereospecificity or saturation kinetics. The influx kinetics for D- and L-lactate, measured up to a 12 m*M* plasma concentration, indicate a linear correlation that is consistent with the lack of stereospecificity and saturation kinetics (*8*). However, in the case of glucose there is a competition for the natural enantiomer,

D-glucose. Such that the influx of the nonphysiological radioactive labeled enantiomer (L-glucose) exhibited Michaelis-Menten saturation kinetics, with increasing blood concentrations of D-glucose. This is indicative of a stereospecific-saturable transporter for D-glucose.

9. It has been reported that there are regional differences in brain vascular volumes in awake vs choral hydrate-anesthetized rats, suggesting that blood-to-brain substrate transport may also depend on local tissue perfusion. Thus, V_{pl} may differ when physiologic conditions are altered and we recommend that it might be necessary to measure V_{pl} in the current conditions when using V_{pl} corrections.

10. LaManna et al. *(8)* report a cortical clearance of 0.09 mL/g/min for lactate using the dual-label fractionation bolus method. These values are similar to previously reported values using the BUI method in rat *(17)* and human *(18)*, but 35% lower compared to a study using an autoradiographic method in rat *(19)*. These variations may be ascribed to differences in the PS products of the natural substrate from the unnatural enantiomer. For example, the PS product for L-lactate has been reported to be twice that for D-lactate in all regions of the brain *(8)*. Also, estimations of PS products can vary if the metabolic state is acutely altered. It has been previously suggested that the PS may vary in rats infused with lactate for the purpose of achieving nonphysiologic concentrations (for the purpose of varying the blood concentrations). Such conditions may alter physiologic state, resulting in changes in $PaCO_2$ and PaO_2 *(8)*.

Acknowledgments

The authors thank Sami I. Harik, MD, for his helpful comments and Max Neal for helping with the manuscript preparation.

References

1. Gjedde, A. (1983) Modulation of substrate transport to the brain. *Acta Neurol. Scand.* **67**, 3–25.
2. Sage, J. I., Van Uitert, R. L., and Duffy, T. E. (1981) Simultaneous measurement of cerebral blood flow and unidirectional movement of substances across the blood-brain barrier: theory, method, and application to leucine. *J. Neurochem.* **36**, 1731–1738.
3. Oldendorf, W. H. (1973) Carrier-mediated blood-brain barrier transport of short-chain monocarboxylic organic acids. *Am. J. Physiol.* **224**, 1450–1453.
4. Pelligrino, D. A., LaManna, J. C., Duckrow, R. B., Bryan, R. M. Jr., and Harik, S. I. (1992) Hyperglycemia and blood-brain barrier glucose transport. *J. Cereb. Blood Flow Metab.* **12**, 887–899.
5. Crone, C. (1965) Facilitated transfer of glucose from blood into brain tissue. *J. Physiol.* **181**, 103–113.
6. Dick, A. P., Harik, S. I., Klip, A., and Walker, D. M. (1984) Identification and characterization of the glucose transporter of the blood-brain barrier by cytochalasin B binding and immunological reactivity. *Proc. Natl. Acad. Sci. U.S.A* **81**, 7233–7237.
7. LaManna, J. C., and Harik, S. I. (1985) Regional comparisons of brain glucose influx. *Brain Res.* **326**, 299–305.
8. LaManna, J. C., Harrington, J. F., Vendel, L. M.,Abi-Saleh, K., Lust, W. D., and Harik, S. I. (1993) Regional blood-brain lactate influx. *Brain Res.* **614**, 164–170.
9. Shockley, R. P., and LaManna, J. C. (1988) Determination of rat cerebral cortical blood volume changes by capillary mean transit time analysis during hypoxia, hypercapnia and hyperventilation. *Brain Res.* **454**, 170–178.

10. Crumrine, R. C., and LaManna, J. C. (1991) Regional cerebral metabolites, blood flow, plasma volume, and mean transit time in total cerebral ischemia in the rat. *J. Cereb. Blood Flow Metab.* **11,** 272–282.

11. Sakurada, O., Kennedy, C., Jehle, J., Brown, J. D., Carbin, G. L., and Sokoloff, L. (1978) Measurement of local cerebral blood flow with iodo [14C] antipyrine. *Am. J. Physiol.* **234,** H59–66.

12. Lincoln, B. C., Des Rosiers, C., and Brunengraber, H. (1987) Metabolism of *S*-3-hydroxybutyrate in the perfused rat liver. *Arch. Biochem. Biophys.* **259,** 149–156

13. Lowry, O. H., and Passonneau, J. V. (1972) *A Flexible System of Enzymatic Analysis.* Academic Press, New York.

14. LaManna, J. C. and Harik, S. I. (1986) Regional studies of blood-brain barrier transport of glucose and leucine in awake and anesthetized rats. *J. Cereb. Blood Flow Metab.* **6,** 717–723

15. Crone, C. (1977) Transport of solutes and water across the blood-brain barrier [proceedings]. *J. Physiol.* **266,** 34P–35P.

16. Riachi, N. J., LaManna, J. C., and Harik, S. I. (1989) Entry of 1-methyl-4-phenyl-1,2,3, 6-tetra-hydropyridine into the rat brain. *J. Pharmacol. Exp. Ther.* **249,** 744–748.

17. Pardridge, W. M., Connor, J. D., and Crawford, I. L. (1975) Permeability changes in the blood-brain barrier: Causes and consequences. *CRC Crit. Rev. Toxicol.* **3,** 159–199.

18. Knudsen, G. M., Paulson, O. B., and Hertz, M. M. (1991) Kinetic analysis of the human blood-brain barrier transport of lactate and its influence by hypercapnia. *J. Cereb. Blood Flow Metab.* **11,** 581–586.

19. Lear, J. L., and Kasliwal, R. K. (1991) Autoradiographic measurement of cerebral lactate transport rate constants in normal and activated conditions. *J. Cereb. Blood Flow Metab.* **11,** 576–580.

20. Gjedde, A., and Crone, C. (1975) Induction processes in blood brain transfer of ketone bodies during starvation. *Am. J. Physiol.* **229,** 1165–1169.

21. Harik, S. I., and LaManna, J. C. (1988) Vascular perfusion and blood-brain glucose transport in acute and chronic hyperglycemia. *J. Neurochem.* **51,** 1924–1929.

22. LaManna, J. C., McCracken, K. A., and Strohl, K. P. (1989) Changes in regional cerebral blood flow and sucrose space after 3–4 weeks of hypobaric hypoxia (0.5 ATM). *Adv. Exp. Med. Biol.* **248,** 471–477.

23. LaManna, J. C., Kikano, G. E., and Harik, S. I. (1989) Brain blood flow and sucrose space in acute and chronic hyperglycemia. In: *Neurotransmission and Cerebrovascular Function I.* Elsevier, Amsterdam.

24. Kikano, G. E., LaManna, J. C., and Harik, S. I. (1989) Brain perfusion in acute and chronic hyperglycemia in rats. *Stroke* **20,** 1027–1031.

25. De Vivo, D. C., Trifiletti, R. R., Jacobson, R. I., Ronen, G. M., Behmand, R. A., and Harik, S. I. (1991) Defective glucose transport across the blood-brain barrier as a cause of persistent hypoglycorrhachia, seizures, and developmental delay. *N. Engl. J. Med.* **325,** 703–709.

18

Protein Transport in Cerebral Endothelium

In Vitro Transcytosis of Transferrin

Laurence Fenart and Roméo Cecchelli

1. Introduction

Brain capillary endothelial cells forming the blood–brain barrier (BBB) are sealed by complex tight junctions and possess few pinocytotic vesicles. These characteristics, added to a metabolic barrier, restrict the passage of most small polar molecules and macromolecules from cerebrovascular circulation to the brain. Many endothelial functions that include the diffusion or transport barrier of brain microvessels have been defined by studies in whole animals and in isolated capillaries in vitro. The ability to grow central nervous system microvascular endothelial cells in culture has opened the door to many new experimental approaches for studying the transendothelial transport of substances across the in vitro BBB. However, several lines of evidence suggest that cultured brain endothelial cells rapidly lose the characteristics of a differentiated BBB in vitro (1). Furthermore, Risau and Wolburg (2) suggested that long-term cultures of brain endothelial cells may not provide a good model system for the BBB in vitro. Nevertheless, we have described subcultures up to the 50th generation of bovine brain capillary endothelial cells (BBCECs) that maintain both endothelial and some of the BBB markers (tight junctions, low rate of pinocytosis, and monoamine oxidase activity but a loss of γ-glutamyl transpeptidase activity) (3).

Debault and Cancilla (4) demonstrated the importance of the coculture of glial cells with brain capillary endothelial cells in the reinduction of the barrier properties in endothelial cells. To reconstruct some of the complexity that exists in vivo, we have developed an in vitro model consisting of a coculture of BBCECs and glial cells (**Fig. 1**) (5), to assess drug transport across the BBB. This model, which closely resembles the conditions in vivo and dramatically reduces the paracellular transport pathway (6–8), allows the study of the cellular and molecular mechanism involved in the intracellular transport of proteins such as transferrin (Tf) (9–13).

2. Materials

2.1 Labeling of Transferrin

1. Bovine holoTf (diferric transferrin) and apoTf (non-iron-loaded tranferrin) (Sigma, Steinheim, Germany).

From: *Methods in Molecular Medicine, vol. 89:*
The Blood–Brain Barrier: Biology and Research Protocols
Edited by: S. Nag © Humana Press Inc., Totowa, NJ

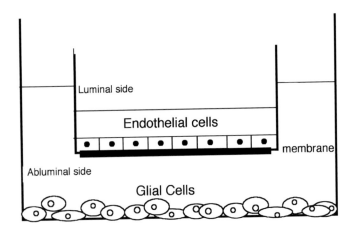

Fig. 1. In vitro model of the blood–brain barrier. Brain capillary endothelial cells are cultured in the upper compartment on a collagen-coated filter and glial cells in the lower compartment on a plastic Petri dish (*5–13*).

2. Iodogen (1,3,4,6-tetrachloro-3a,6a-diphenylglycoluril) (Perbio Science, Bezons, France) precoated tube: Dissolve 1mg of Iodogen in 1 mL dichloromethane and pipet 20 μL into polypropylene tubes. Rotate the tubes in a 37°C water bath until the solvent has evaporated, leaving a film of Iodogen in the bottom. These prepared tubes can be stored dessicated for several weeks before use.
3. Phosphate-buffered saline (PBS): 150 mM NaCl, 2.7 mM KCl, 1.3 mM KH_2PO_4, 1 mM $Na_2HPO_4.7H_2O$, pH 7.4.
4. [^{125}I]Na carrier free (ICN, Costa Mesa, CA) (*see* **Note 1**).
5. Sephadex G25 column (PD 10, Amersham, Orsay, France).
6. Fluorescein 5-isothiocyanate, Isomer I (FITC, Sigma, Steinheim, Germany) is first dissolved in 20 μL DMSO and then 980 μl of PBS are added to obtain a 125 μM FITC solution.
7. Bicarbonate buffer 0.1 M pH 9.5: Mix a $NaHCO_3$ solution at 0.1 M with a Na_2CO_3 at 0.1 M to obtain a final solution at pH 9.5.

2.2. Binding Experiments

1. Supplemented Hanks balanced salt solution (HBSS, GIBCO-BRL, Cergy Pontoise, France) with $NaHCO_3$ (2 g/L), 0.5% saponin (wt/vol), 0.1% bovine serum albumin, cultured tested (BSA, ICN, Costa Mesa, CA), 1 mM PMSF, and 1 μg/mL leupeptin. The pH is set at 6.8 with diluted HCl.
2. Supplemented Dulbecco's modified Eagle's medium (DMEM, GIBCO-BRL, Cergy Pontoise, France) with 25 mM sodium acetate (pH 5.4), 1 mM phenylmethyl sulfonyl fluoride (PMSF, Acros Organics, Noisy le grand, France), and 1 μg/mL leupeptin (Acros Organics, Noisy le grand, France).
3. Medium A: HBSS containing 5 mM HEPES (pH 7.2), 0.1% BSA, 1 mM PMSF, and 1 μg/mL leupeptin.
4. Gamma Counter (Cobra, Packard).
5. Enzfitter nonlinear regression data analysis program (Elsevier BIOSOFT, Cambridge, UK).

2.3. Endocytosis Experiments

1. Ringer HEPES: 150 mM NaCl, 5.2 mM KCl, 2.2 mM CaCl$_2$, 0.2 mM MgCl$_2$, 6 mM NaHCO$_3$, 2.8 mM glucose, 5 mM HEPES, pH 7.4.
2. Fixatives: 4% Paraformaldehyde (Sigma, Steinheim, Germany) in PBS or 4% paraformaldehyde and 0.1% glutaraldehyde (Sigma) in PBS.
3. Mowiol mountant (Hoescht, Frankfurt, Germany) containing 0.1% P-phenylenediamine (Sigma) anti-quenching agent.
4. Fluorescence microscope (DMRB; Leica Mikroskopie ans Systeme GmbH, Wtzlar, Germany).
5. Monoclonal anti-rat Tf receptor antibody (clone MRC OX-26, Serotec, Oxford, UK), 1/100 dilution in PBS.
6. PBS/FCS: PBS containing 10% of fetal calf serum (FCS, Hyclone laboratories, Logan, UT).
7. Phosphate buffer 0.2 M Stock solution: Prepare separately a solution of 0.2M Na$_2$HPO$_4$ and a solution of 0.2 M NaH$_2$PO$_4$. Before use, the two solutions are mixed in a 4:1 ratio to obtain a final solution at pH 7.4.
8. Glycine (Sigma, Steinheim, Germany) 50 mM is prepared in 0.1 M phosphate buffer.
9. 7.5% Gelatin (Merck, Darmstadt, Germany) in 0.1 M phosphate buffer. After stirring for 10 min at room temperature the solution is warmed to 60°C for 2 h. When all the gelatin dissolves the solution is cooled to 37°C, and 200 µL of a 10% azide (Sigma) solution is added. The homogenous 10% gelatin solution is poured into 5 mL vials and stored at 4°C.
10. Cryoprotectant: 2.3 M of sucrose (Sigma) in 0.1 M phosphate buffer. Stir until the sucrose is completely dissolved and aliquot in 1 mL vials.
11. Formvar solution: Formvar powder (Merck, Darmstadt, Germany) is kept under vacuum or at 60°C in the presence of silicagel. Formvar, 1.2 g, is put into a volumetric flask and 100 mL of chloroform is added while stirring. We make a film using a clean microscope slide introduced upright in a glass column filled with the formvar solution. A stopcock allows draining of the formvar over the slide, leaving a thin film of formvar on the slide. The draining time determines the thickness of the film, and is usually 12–15 s.
12. Polyclonal rabbit anti-FITC IgG antibody (Zymed, Montrouge, France), 1/50 dilution in PBS/FCS.
13. Gold-labeled (10 nm) goat anti-rabbit antibody (EMS, Fort Washington).
14. Methyl cellulose-uranyl acetate, pH4: 20 mL of the methyl cellulose solution is replaced by 20 mL of 4% uranyl acetate and gently mixed.
    a. Methyl cellulose: For a final volume of 200 mL, 196 mL of distilled water is heated to a temperature of 90°C and 4g of methyl cellulose (25 centipoise; Sigma) is added while stirring. The solution is rapidly cooled on ice while stirring, until the solution has reached a temperature of 10°C.
    b. 4% aqueous uranyl-acetate (Merck, Darmstadt, Germany) solution: The pH is set to 4.0 with 25% ammonium-hydroxide. The latter is added drop by drop, to prevent formation of insoluble precipitates.
15. Electron microscope (420; Philips, Eindhoven, The Netherlands).

2.4. Transendothelial Transport Experiments

1. RPMI-1640 (GIBCO BRL, Cergy Pontoise, France).
2. Trichloroacetic acid (TCA) (Sigma) 20% in water.
3. AgNO$_3$ (Sigma) 5% in water.

4. ^{14}C-sucrose (Amersham, Orsay, France) is used as a tracer (80 µM). The stock solution is diluted in Ringer HEPES to obtain an activity of 0.1 µCi/2 mL. All safety precautions have to be followed (*see* **Note 1**). Protection from beta emissions does not require lead shielding and use of a dosimeter.

5. ^{59}Fe (Amersham) (*see* **Note 1**).

6. Tris-bicarbonate buffer 0.1 M pH7,6 : Mix a bicarbonate solution 0.1 M with a Tris (acid) solution 0.1 M to obtain a final solution at pH 7.6.

7. Chelex 100 (Sigma).

2.5. Characterization of Tf After Transcytosis

1. 4–20% Sodium dodecyl sulfate (SDS)-polyacrylamide gel (Lifescience Biorad Laboratories, CA).

2. Autoradiography film (Kodak X-OMAT AR Film).

3. Methods

The methods described below outline the different steps for studying transcytosis of proteins through EC monolayers. This chapter will focus only on the transport of Tf. This includes 1) the preparation of labeled Tf, 2) the binding experiment, 3) the detection of internalized Tf by fluorescence or electron microscopy, and 4) the transcellular and efflux experiments. All these experiments can be applied to other macromolecules such as low-density lipoprotein *(10)*, lactoferrin *(12)*, and cyclophylin B *(13)*.

3.1. Labeling of Transferrin

3.1.1. Preparation of Radiolabeled Transferrin

1. Bovine holoTf or apoTf (non-iron–loaded transferrin) is iodinated using Iodogen iodination procedure. Tf (500 µg) dissolved in PBS is placed in an iodogen-precoated tube. 0.3 mCi Na ^{125}I carrier free (*see* **Note 2**) is dispensed in the tube in a fume cupboard prepared for radioactive work and incubated for 30 min at 4°C with gentle stirring.

2. Free iodine is removed on a Sephadex G25 column, followed by an extensive dialysis against PBS.

3. Generally, the specific activity range is 0.35–0.5 µCi/µg of protein (*see* **Note 3**). Protein dosage is carried out by Peterson's method *(14)*.

4. The iodinated proteins are stored at 4°C for as long as 1 wk without significant radiolysis (*see* **Note 4**).

3.1.2. Preparation of FITC-Labeled HoloTf

FITC labeling of holoTf is performed as follows:

1. HoloTf (25 µM) is dissolved in 0.1 M bicarbonate buffer, pH 9.5.

2. A fivefold excess of FITC (125 µM) is added dropwise with gentle stirring.

3. After 6 h at room temperature in the dark, uncoupled FITC is removed using a Sephadex G25 column equilibrated with PBS, and the labeled protein is dialyzed against PBS at 4°C.

4. Then, the amount of FITC is determined by measurement of fluorescence and Tf amount by protein dosage *(14)*. The ratio of fluorescein to protein is estimated by measuring the absorbances at 496 nm and 280 nm *(15, see* **Note 5**).

5. The FITC-Tf solution can be stored for 1 mo at 4°C.

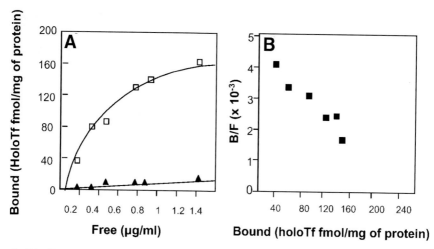

Fig. 2. Binding (**A**) and Scatchard plot (**B**) of bovine holoTf binding to bovine brain capillary endothelial cells (BBCECs). BBCECs were preincubated 1 h at 25°C in Hanks' balanced salt solution in absence (▲) or presence (☐) of 0.5% saponin. Then increased concentration of ^{125}I-labeled holoTf were added on the luminal side of cells, and binding experiments were performed at 4°C for 2 h. Each point represents the mean of triplicate inserts, which were corrected for nonspecific binding in the presence of 100-fold excess unlabeled holoTf. B/F, bound to free ratio. The curves are representative of three series of independent experiments.

3.2. Binding Experiments

Binding studies are performed according to the method of Raub and Newton (*16*) with some modifications.

1. BBCEC monolayers are preincubated for 1 h at 25°C in supplemented HBSS to permeabilize cellular membranes and gain access to all Tf receptors with minimal degradation (*see* **Note 6**).
2. The BBCEC monolayers are washed in supplemented DMEM (2x10 min) to remove iron from endogenous Tf, and twice in Medium A to remove endogenous apoTf.
3. Following the safety precautions (*see* **Note 1**) binding experiments are carried out for 2 h at 4°C in Medium A with ^{125}I-holoTf concentration ranging from 300 to 1400 ng/mL (3.75–17.5 n*M*). Similar experiments are carried out with a 100-fold excess of unlabeled holoTf to determine the nonspecific binding of holoTf to endothelial cell membranes.
4. The BBCECs are washed carefully and cell associated radioactivity is determined by removing the membrane of the culture insert and counting it in a gamma counter.
5. Results are expressed in milligrams of ^{125}I-holoTf bound per milligram of cell proteins (*14*).
6. The amount of bound Tf is calculated using the equation in the Enzfitter nonlinear regression data analysis program.

$$F(x) = x.(B_{max}) / (Kd + x)$$

where $F(x)$ represents the amount of bound Tf, x is free Tf, B_{max} is the total Tf receptor concentration, and Kd is the concentration at 50% receptor saturation. B_{max} and Kd are calculated using Scatchard transformation with the same data analysis program (**Fig. 2**).

^{125}I-holoTf binds to the ECs with a saturation kinetic (**Fig. 2A**). The saturability of the binding indicates that the interaction of Tf with a specific receptor has occurred. In transformed Scatchard plots (**Fig. 2B**), the data suggested evidence for a single binding site with a dissociation constant of 11.3 ± 2.1 nM and a total capacity of approx 35,000 receptors per cell. These results suggest that differentiated BBCECs (after 12 d in coculture with glial cells) exhibit the specific receptor for holoTf.

3.3. Endocytosis Experiments

After specific holoTf binding to its receptor expressed on the EC membrane, the process of transcytosis requires ligand internalization. The latter can be studied by fluorescence or/and electron microscopy. Preliminary studies have to be done to determine that holoTf and its receptor interaction occurs at the BBCEC level and that it is specific.

3.3.1. Detection of Endocytosis by Fluorescence Microscopy

3.3.1.1. FITC LABELED HOLOTF INTERNALIZATION

In preliminary experiments, we tested the endocytotic activity of BBCECs with holoTf conjugated with FITC exposed to the luminal surface of the cells.

1. BBCEC monolayers are washed twice with Ringer HEPES at 37°C.
2. Then, BBCECs are incubated in the dark with FITC labeled holoTf (200 µg/mL) for 45 min at 37°C in prewarmed Ringer HEPES with 0.1% BSA.
3. Incubation is terminated by three washes in ice-cold Ringer-HEPES buffer and immediately fixed with 4% paraformaldehyde for 20 min at 4°C (*see* **Note 7**).
4. After final washes, the filters and their attached monolayers are mounted on glass microscopic slides using Mowiol mountant containing 0.1% p-phenylenediamine, an antiquenching agent.
5. Specimens are visualized using a fluorescence microscope.
6. A significant accumulation of FITC-labeled holoTf is observed within the BBCECs (**Fig 3**; *see* **Note 8**).

3.3.1.2. SPECIFICITY OF THE ENDOCYTOTIC PROCESS

To verify that holoTf accumulation in the BBCECs is specific, experiments using mouse anti-rat Tf receptor monoclonal antibody (Mab OX-26) are performed (*see* **Note 9**).

1. BBCEC monolayers are incubated with Mab OX-26 solution for 30 min at 37°C.
2. Then, 70 µg/mL of FITC-labeled holoTf is added to the medium, and the cells are incubated for an additional 45 min at 37°C in the dark.
3. This results in a total inhibition of FITC-labeled holoTf uptake.
4. As a negative control, the same experiment is carried out using a monoclonal antibody nonspecific for the Tf-receptor. This antibody did not block FITC-labeled holoTf uptake.

3.3.2. Detection of Endocytosis by Electron Microscopy

Endocytosis can be studied by electron microscopy to obtain precise localization of holoTf. It is important to use labeling markers compatible with the particle size. Therefore, the classical immunolabeling method using gold conjugated-particles was not used and holoTf was localized using an anti-FITC IgG antibody on ultrathin cryosections.

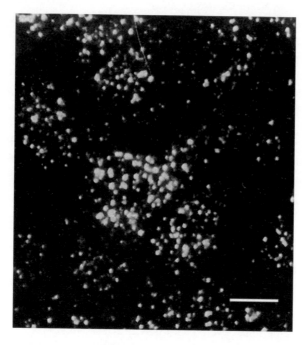

Fig. 3. Fluorescein isothiocyanate-labeled holoTf endocytosis in bovine brain capillary endothelial cells (45 min at 37°C). Bar = 15 μm.

1. BBCECs are incubated in the dark, first with FITC-labeled holoTf (200 μg/mL) for 45 min at 37°C in prewarmed Ringer HEPES.
2. BBCECs are then fixed with paraformaldehyde 4% and 0.1% glutaraldehyde in PBS for 1 h at 4°C.
3. BBCEC monolayers are washed with PBS and then incubated for 90 min in PBS/FCS at 4°C.
4. After washing, the BBCECs are collected by scraping in PBS, and centrifuged.
5. The cell pellet is washed in 50 m*M* glycine in PBS to quench aldehydes. Three washes are necessary to remove fixative.
6. Then BBCECs are embedded in 7.5% gelatin at 37°C *(17)*, (*see* **Note 10**) for less than 5 min. BBCECs are centrifuged to remove most of the gelatin. If too much gelatin is left in the tube, the final concentration of cells will be too low. The gelatin-cell suspension is solidified on ice (*see* **Note 11**).
7. The BBCEC pellet is gently removed from the tube, transfered to a microscope slide covered with a piece of parafilm, and cut into suitable blocks.
8. Gelatin-embedded specimens are infused with 2.3 *M* sucrose by rotating the small vials for at least 2 h at 4°C *(18)*, (*see* **Note 12**).
9. The blocks mounted on the holders are immediately frozen in liquid nitrogen.
10. Ultrathin sections are cut at –110°C using a UltraCryo Microtome and sections are placed on Formvar carbon coated grids.
11. The immunolabeling procedure is performed by floating the grids on successive drops of PBS/FCS followed by polyclonal rabbit anti-FITC IgG antibody for 1 h at room temperature, followed by gold-labeled (10 nm) goat anti-rabbit antibody (*see* **Note 13**).
12. Finally the immunolabeled cryosections are contrasted and embedded in methylcellulose-uranyl acetate for electron microscopy.

These results clearly demonstrate that, in contrast to what happens in most cells, FITC-holoTf is not present in the lysosomal compartment in the differentiated brain capillary endothelial cells but is present in multivesicular endosomal structures.

In most organs, the existence of receptor-mediated processes that bypass the lysosome compartment seems to be a feature of continuous endothelia. Indeed, as continuous endothelia, BBCECs are linked to each other by junctional complexes that constitute a major barrier to the bidirectional exchange of macromolecules. Specific receptors on the cell surface play a significant role in the transendothelial transport of plasma molecules to tissues. To further characterize this process, transendothelial transport experiments using ^{125}I-holoTf are carried out.

3.4. Transendothelial Transport Experiments

3.4.1 Apical to Basolateral Transport of Tf Across the BBCEC Monolayers

1. Cells (BBCEC monolayers and glial cells) are placed in RPMI-1640 medium at 37°C for 4 h to deplete the cells of endogenous Tf.
2. One insert covered with BBCECs is set into a six-well dish with 2 mL of Ringer HEPES with 0.1 % BSA added to each well. Safety precautions are followed (*see* **Note 1**). Bovine ^{125}I-holoTf is added in the presence or absence of 100-fold excess of unlabeled protein to the upper side of the filter covered with BBCECs (*see* **Note 14**).
3. At various times, the insert is transferred to another well to avoid possible reendocytosis of Tf by the abluminal side of the BBCECs.
4. At the end of the experiment, intact Tf is assessed using TCA precipitation of lower mediums:
   a. 600 µL of the medium added to 300 µL of TCA solution and 50 µL of BSA (40 mg/mL) is spun down at 1100*g* for 15 min. The supernatant fraction (S1) is removed for treatment by AgNO$_3$ precipitation.
   b. The pellet is washed once with 600 µL H$_2$O and 300 µL TCA solution and centrifuged at 1100*g* for 15 min.
   c. The pellet is counted in a gamma counter. This pellet corresponds to the intact Tf fraction.
5. Protein degradation is assessed with AgNO$_3$ precipitation:
   a. 600 µL of S1 added to 300 µL of AgNO$_3$ solution is centrifuged at 1100*g* for 15 min.
   b. The pellet corresponds to the ^{125}I free fraction and the supernatant reveals the Tf degradation.
6. All results are expressed in Tf equivalent flux (ng/cm^2), which represents TCA-precipitable radioactivity recovered in the lower compartments (**Fig. 4**).
7. The transport of labeled holoTf from the luminal to the abluminal compartment is reduced severely by an excess of unlabeled Tf, suggesting that Tf transport from the apical to the basal side of the cells was specific (**Fig. 4A**).
8. To assess the influence of iron on the transport of Tf across the BBCEC monolayers, the same experiments are done with exposure of ^{125}I-apoTf on the luminal surface of BBCECs. This experiment demonstrates that in contrast to holoTf, no specific accumulation of ^{125}I apoTf reached the abluminal compartment (**Fig. 4B**). ApoTf has a low affinity for the Tf receptor, and the absence of demonstrated transport of apoTf across the BBCEC monolayers in contrast to holoTf provides additional evidence for a receptor-mediated transcytosis of holoTf across cerebral BBCECs from the luminal to the abluminal side.

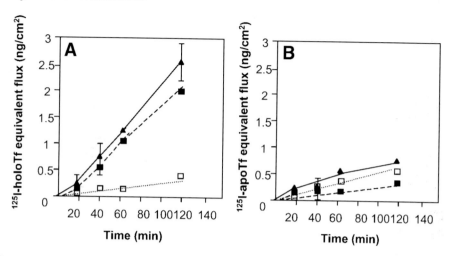

Fig. 4. Apical-to-basolateral transport of ^{125}I-holoTf (**A**) and ^{125}I-apoTf (**B**) across the bovine brain capillary endothelial cell monolayers grown on a porous filter. ^{125}I-Tf (1400 ng/mL) was added to the luminal side of cells. All values (mean of triplicate inserts ± SE [bars]; $n = 3$) represent radioactivity that was TCA precipitable. Total Tf flux (▲) was corrected for nonspecific Tf flux (□) giving the specific Tf flux (■). The curves are representative of five series of independent experiments.

3.4.2 Effect of Temperature on the Tf Transport

1. The transport experiment described above is then carried out at 4°C with holoTf incubated with ^{14}C-sucrose, as a marker of paracellular transport.
2. The effect of temperature on the transport of both molecules from the apical to the basal compartment is compared (**Fig. 5**).
3. A decrease in the incubation temperature from 37°C to 4°C slightly affects the passage of sucrose (**Fig. 5A**), whereas a dramatic decrease in holoTf transport through the monolayer is observed (**Fig. 5B**).
4. These results suggest that Tf reaches the abluminal side of the cell via an energy-dependent, receptor-mediated transcytotic process.

3.4.3. Polarized Efflux of Tf

A pulse-chase experiment measuring the asymmetric efflux of Tf is carried out to quantify possible Tf recycling to the luminal side of the cells subsequent to endocytosis. Experiments are performed as described by Raub and Newton (*16*).

1. BBCECs are allowed to accumulate radiotracers from the apical surface for 1 h at 37°C after the addition of 1400 ng/mL of ^{125}I-labeled holoTf.
2. BBCECs are washed four times with Ringer HEPES containing 0.1% BSA at 4°C for 5 min each to remove nonspecific binding of tracers on cells; the cells are returned to a fresh medium at either 37°C or 4°C for an additional 30 min.
3. At the end of the experiment, the media bathing the luminal and abluminal sides of the insert are TCA precipitated and counted in a gamma counter, as are the cells on filters. The results are expressed as ^{125}I-holoTf equivalent efflux (in ng) (**Fig. 6**).

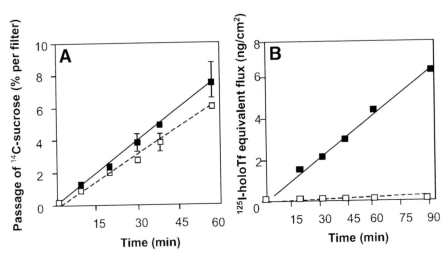

Fig. 5. Effect of a diminution of temperature from 37° (■) to 4°C (□) on transport of sucrose (**A**) and holoTf (**B**) across bovine brain capillary endothelial cells. Results expressed as percentage of sucrose recovered in lower compartments. For Tf, they are expressed as an equivalent Tf flux (ng/cm^2), and values are means of triplicate inserts. The curves are representative of two series of independent experiments.

4. Ten percent of the endocytosed ^{125}I-holoTf at 37°C is recycled to the luminal side, whereas 75% is transcytosed to the abluminal compartment.
5. In the control condition (4°C), the major part of the Tf accumulating in the BBCECs remained in the cells, thereby demonstrating that the efflux of holoTf is energy dependent and that the majority of endocytosed holoTf is transferred by a transendothelial receptor-mediated pathway.

3.4.4. Transendothelial Iron Transport Studies Using ^{59}Fe-Tf and ^{125}I-^{59}Fe-Double-Labeled Tf

All our experiments clearly demonstrate that Tf, a 80-kDa glycoprotein that binds iron in the blood, is able to go through the BBCEC monolayer by a receptor-mediated process. The importance of iron for central nervous system function is now well documented *(19)*, therefore, it is of interest to determine whether the transcytosis of Tf through the BBB delivers iron to brain tissue. Additional experiments are carried out to examine iron transport across the brain capillary EC monolayers.

3.4.4.1. PREPARATION OF RADIOLABELED DIFERRIC TF

1. For ^{59}Fe labeling, 1.5 μg of iron/mg of Tf is used. Follow safety precautions (*see* **Note 1**). ^{59}Fe (18 nmol) is mixed with a solution containing 30 μg of nitriloacetic acid and 19 μg of NaOH for 5 min at room temperature. The pH of the solution is adjusted to 8.2 with NaOH.
2. ApoTf (37 nmol) dissolved in Tris-bicarbonate buffer is added to the solution, which is then incubated for 2 h at room temperature.
3. Unbound ^{59}Fe is removed by adding 50 μL of Chelex 100 preequilibrated with Tris-bicarbonate buffer and the mixture is stirred gently for 30 min.
4. This method gives an average specific activity of 5 μCi/mg of protein and a yield of Tf saturation with iron equal to 99%.

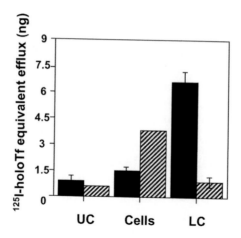

Fig. 6. Asymmetric efflux of ^{125}I-Tf from bovine brain capillary endothelial cell monolayers grown on a porous filter. Cells were allowed to accumulate ^{125}I-holoTf (1400 ng/mL) from their luminal side for 1 h at 37°C. Then they were carefully washed and put in fresh medium at either 37°C (black bars) or 4°C (hatched bars) for 30 min. Amount of cell-associated radioactivity and TCA precipitated radioactivity in upper (UC) and lower (LC) compartment were measured. All values are means of triplicate inserts ± SE (bars; $n = 3$) and the curves are representative of three series of independent experiments.

5. For dual labeling of Tf, protein is first labeled with ^{59}Fe and then iodinated, giving a specific activity of 0.49 µCi/µg of protein for ^{125}I and 4 µCi/mg of protein for [59]Fe.

3.4.4.2. TRANSENDOTHELIAL IRON TRANSPORT EXPERIMENTS

These experiments are carried out as described for ^{125}I-Tf (*see* **Subheading 3.4.1.**). ^{59}Fe-Tf (0.26 µ*M*) is added to the upper compartment and the integrity of the monolayer is evaluated. No increase in the permeability of sucrose is detected. A specific transendothelial transport occurs over 90 min, suggesting that the holoTf receptor could take part in the transport. To examine whether iron is transcytosed through BBCECs with holoTf, double-labeling experiments are carried out.

The transport experiments are performed with I-^{59}Fe-double-labeled Tf at a concentration of 21 µg/mL. The ^{125}I-holoTf equivalent flux represents TCA precipitable radioactivity. ^{59}Fe-Tf equivalent flux is calculated from ^{59}Fe flux, because Tf binds 2 mol of iron. The ^{59}Fe-holoTf equivalent flux (0.088 pmol.cm$_2$.h$_{-1}$), which is equal to one-half the ^{59}Fe flux (0.176 pmol.cm$_2$.h$_{-1}$), is within the same range of the observed ^{125}I-holoTf equivalent flux (0.093 pmol.cm$_2$.h$_{-1}$). These results suggest that iron bound to Tf crossed the BBCEC monolayers, because for every 2 mol of iron that crosses the BBCECs, 1 mol of Tf is found in the lower compartment.

3.5. Characterization of Tf After Transcytosis

To address the question whether holoTf undergoes degradation within BBCECs during transcytosis, electrophoretic characterization is carried out.

1. After 2 h at 37°C, ^{125}I-labeled holoTf degradation in the upper and the lower compartments is determined by TCA precipitation, followed by AgNO$_3$ precipitation.

2. The apical and basolateral compartment solutions are collected and analyzed with 4–20% SDS-polyacrylamide gel electrophoresis.
3. After electrophoresis, the gel is dried and exposed for autoradiography for 2 h at –80°C.
4. Results suggested that in all instances, approx $3 \pm 1\%$ of ^{125}I-labeled holoTf degradation is recovered in the upper and lower compartments.
5. The data supports the premise that holoTf is transported across BBCECs without any degradation.

The development in our laboratory of a cell culture system consisting of coculture of BBCECs on one side of a porous filter and glial cells plated on the plastic dish that mimics the properties of the BBB provides an additional approach to define BBB events in iron Tf handling. With the use of this model, we provide evidence that after binding to BBCECs, there is a specific mechanism for the transport of iron-loaded Tf across the endothelial monolayer from apical to the abluminal surface. This mechanism might best be explained by the process of receptor-mediated transcytosis. By this pathway, bypassing the lysosomal compartment, Tf can deliver iron, which is essential for normal growth and function of the brain cells.

4. Notes

1. Safety precautions are required when using radioactive compounds. Protection from gamma-emitting isotopes is achieved by lead shielding. Radioactive material should be used by authorized persons and only in authorized areas. Care should be taken to prevent ingestion or contact with skin or clothing. Protective clothing, such as laboratory overalls, safety glasses, and gloves should be worn whenever radioactive materials are handled. The operator has to wear a dosimeter to measure radiation dose to the body. Ampules containing volatile radioactive compounds such as ^{125}I should be opened only in a well-ventilated fume cabinet. Work should be carried out on a surface covered with absorbent material. Working areas should be monitored regularly. Any spills of radioactive material should be cleaned immediately and all contaminated materials should be decontaminated or disposed of as radioactive waste via an authorized route. Contaminated surfaces should be washed with a suitable detergent to remove traces of radioactivity.
2. Na ^{125}I is an inexpensive radioisotope that can be obtained with very high specific activity (carrier free, about 2000 Ci/matom) and used to radiolabel both soluble protein and cells by various simple procedures. Labeling occurs by electrophilic addition of cationic iodine (I^+) to tyrosine residues and to a lesser extent to histidine and tryptophan. The monoiodinated derivative is the major product under the conditions normally used.
3. This incorporation of iodine is usually a little lower than with chloramine T but the reaction time can be extended to improve this. Following our procedure, the incorporation of iodine is always around 90%.
4. Fresh solution are advisable because Na ^{125}I is oxidized to molecular iodine upon storage especially if the vial is opened repeatedly. Storage at 4°C increases the rate of decomposition. Iodinated proteins tend to lose activity upon storage faster dictated by the half-life of the isotope. This is attributable to several factors: 1) radiation destruction (the products of radioactive decay of ^{125}I can cause damage to the protein), 2) loss of iodine (iodinated proteins tend to break down to yield free iodine and unlabeled proteins), 3) protein deterioration (some proteins tend to aggregate upon iodination and during subsequent storage).

5. The ratio of the dye to protein is calculated as follows:

$$R = \frac{\text{FITC concentration}}{\text{Tf concentration}} = \frac{6.5.\ 10^{-3} \times DO_{496nm}}{0.74 \times (DO_{280nm} - 0.367 \times DO_{496nm})}$$

   To give the best results, this ratio has to be, in the case of FITC and Tf, near 2.7.

6. Tf binding experiments are first performed with ^{125}I-holoTf on intact BBCECs at 4°C. Binding of ^{125}I-holoTf is not significantly different from background, as already demonstrated by Raub and Newton *(16)*. That is the reason why experiments are carried out by treating the cells with 0.5% (wt/vol) saponin to permeabilize the cellular membranes, thus permitting access of the ^{125}I-holoTf to Tf receptors on the surface and within intracellular pools.

7. The choice of fixative is crucial in maintaining good ultrastructure and immunoreactivity. Glutaraldehyde and paraformaldehyde can be used alone or combined. Glutaraldehyde has two reactive aldehyde groups, which primarily react with lysins. It irreversibly crosslinks amino acids within or between proteins. Paraformaldehyde contains only one reactive aldehyde group. It crosslinks by the formation of methylene bridges which are not stable particularly at concentrations below 5%. Its binding is reversible and therefore this fixative is a relatively weak fixative. When deciding on the fixative of choice for an unknown antigen/ antibody combination, different fixatives should be first tested to achieve optimal results.

8. The FITC-labeled Tf repartition throughout the cytoplasm observed in **Fig. 3** clearly demonstrates that Tf is not directed to the lysosomal compartment. Indeed, when a molecule accumulates in the lysosomal compartment, a characteristic-intense staining around the nucleus is noted. This observation is characteristic of the molecule degradation pathway.

9. Before using rat OX26 antibody in our experiment, the capacity of this antibody to cross-react with bovine Tf receptors is verified by immunofluorescence labeling of bovine lymphocytes.

10. Cultured cells need support to handle them prior to sectioning. Support can be achieved by embedding the specimens generally in 10% gelatin *(17)*. Due to its large molecular weight, it cannot enter the cells and will therefore fill the extracellular spaces with a "cytoplasma-like" gel. Embedding our BBCECs in a 10% gelatin solution makes it difficult to cut good quality sections while these problems are not observed with a 7.5% gelatin solution.

11. Generally, at the time that gelatin embedding was introduced, the gelatin was fixed. In the current protocol, this step is omitted because fixatives crosslink gelatin to plasma membrane antigens, thus reducing the labeling efficiency.

12. A major breakthrough in cryo-ultramicrotomy came with the introduction of sucrose as a cryoprotectant. Initially, sucrose at a concentration of 1 *M* was used but Geuze and Slot *(18)* increased the concentration to 2.3 *M*. It has been shown that no crystals are formed in concentration >1.6 *M* but only vitreous ice that facilitates sectioning.

13. If sections dry during the immunoincubation, high background levels or other contamination is observed.

14. Before studying the transcellular transport of a molecule, the integrity of the BBCEC monolayer has to be verified. Using ^{14}C-sucrose as a test substance for a possible paracellular pathway, we determined that in the presence of 100-fold excess of holoTf, no leakiness in barrier function occurs. In each conditions, endothelial permeability coefficient (Pe) is determined *(11)* (Pe = $0.85 \pm 0.05 \times 10_{-3}$ cm/min and $0.80 \pm 0.09 \times 10_{-3}$ cm/min for control and in the presence of 140 µg/mL of holoTf, respectively), demonstrating that holoTf does not have a toxic effect on the BBB.

(CSF-side) to blood vessels in intact CP. This approach remains, however, limited to fluorescent compounds. This chapter will focus on two different methods, namely the *in situ* isolated perfused CP and the in vitro choroidal epithelium, that allow the study of true transepithelial transfer mechanisms. The protocols to perform the transport experiments and the calculations will be described in detail.

1.1. The Isolated Perfused Choroid Plexus

This model provides a good approximation of the in vivo BCSFB, while allowing simultaneous access to both blood and CSF faces of the tissue. Because this requires perfusion of choroidal vasculature, a large animal model is necessary, and a sheep is chosen for scale and because it is relatively cheap and easy to handle compared to other large animal models. The model allows estimates of rapid unidirectional uptake of a range of radiolabeled molecules from blood to CP using the single-pass technique for 10–60 s *(3)*, steady-state blood to CP and blood to CSF transport (over 1–4 h), transport from CSF face to blood (1–4 h), and measurement of CSF secretion *(4–6)*.

1.2. The In Vitro Choroidal Epithelium

In vitro models reconstituting the polarized and impermeable choroidal epithelium have been established in the last decade using various animal species (*see* **Note 1**). When cultured on a microporous permeable membrane in cell culture inserts, choroidal epithelial cells form a tight monolayer and maintain their specific barrier properties, thus mimicking the in vivo BCSFB and delimiting two liquid compartments. The bicameral device allows independent access to both sides of the cell monolayer and provides a simple versatile tool to investigate transfer across the epithelial cells in both blood to CSF and CSF to blood directions **(Fig. 1)**. Passive diffusion, facilitated or vectorial transport mechanisms can be approached using this type of in vitro model of the BCSFB. Other applications include the possibility of studying the metabolism of molecules en passage, and the polarity of excretion of the resulting metabolites.

We use the in vitro choroidal epithelium in two ways: 1) to estimate the global transepithelial permeability of compounds which result from passive diffusion and /or facilitative transport and/or active transport and/or metabolism, 2) to obtain evidence for active transport mechanisms, by investigating the ability of the choroidal monolayer to create an imbalance between both compartments containing the same concentration of the compound of interest and to accumulate this molecule against a concentration gradient.

2. Materials

2.1. Isolated Perfused Choroid Plexus

1. Thiopentone sodium (make fresh for each procedure 50 mg/mL).
2. Heparin, 25,000 IU/mL.
3. Cleaver or bone saw/drill.
4. Ringer solution containing: 120 mM NaCl, 5.4 mM KCl, 2.35 mM CaCl$_2$, 1.13 mM MgCl$_2$, 26.2 mM NaHCO$_3$, 5 mM glucose, and 4% bovine serum albumin, Fraction V (BSA). Ca^{2+}, glucose, and BSA are added on the day of the experiment, otherwise Ringer is stable for 1–2 wk.
5. 0.5% Evans Blue labeled albumin (EBA): Prepared in a ratio of 1:7 Evans blue to BSA and dissolved in distilled water. This is placed in dialysis tubing and dialysed for 24 h in distilled water.

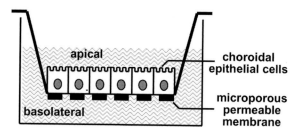

Fig. 1. The in vitro choroidal epithelium. Primary cultured epithelial cells are grown in a bicameral device on a microporous membrane. The cells reproduce a polarity as in vivo, with an apical brushborder membrane in contact with the fluid in the upper chamber, while the basolateral membrane lies on the laminin-coated microporous membrane through which it is in free communication with the fluid in the lower compartment. Thus, in this configuration, the cerebrospinal fluid and stromal blood compartments correspond respectively to the upper and lower chambers.

6. Artificial CSF containing: 123.5 mM NaCl, 2.9 mM KCl, 2.5 mM CaCl$_2$, 1.8 mM MgCl2, 0.25 mM Na$_2$HPO$_4$, and 26 mM NaHCO$_3$. Glucose, 5 mM, is added on the day of use.
7. Gas, 95% O$_2$/5% CO$_2$.
8. Silicone tubings having internal diameters of 1 mm and 0.5 mm.
9. PVC tubing having a 0.5 mm outside diameter.
10. Three peristaltic pumps.
11. Water bath with circulator and heating pad/jacket attachments.

2.2. In Vitro Choroid Plexus

1. Orbital shaker in a 37°C incubator.
2. Paracellular marker: radiolabeled mannitol, sucrose, or inulin.
3. Ringer-Hepes buffer (RH): 150 mM NaCl, 5.2 mM KCl, 2.2 mM CaCl$_2$, 0.2 mM MgCl$_2$, 6 mM NaHCO$_3$, 2.8 mM glucose, and 5 mM Hepes, pH 7.4. Sterilize by 0.22 μm filtration. This solution can be kept at 4°C for 1–3 wk.
4. Serum-free culture medium: DMEM/Ham's F12 (1:1).

3. Methods

3.1. Isolated Perfused Choroid Plexus

3.1.1. Anesthesia and Surgery

1. Anesthetize sheep with thiopentone sodium (20 mg/Kg iv) via a forelimb vein, followed by heparin (20,000 IU), cannulation of one common carotid artery, and exsanguination.
2. Remove the head and quickly and carefully open the back of the skull without damaging the dura using a cleaver at a shallow angle, or bone drill/saw followed by rongeurs.
3. Remove the dura, cut the tentorium on either side, and section the cerebellum and hindbrain in an axial plane with a scalpel. The cut should pass cleanly through the center of the cerebellum. Avoid pulling the tissue at this point or the venous outflow will be damaged.
4. Carefully lift the brain starting at the remaining cerebellum, and reflect forward 0.5–1 cm to visualize the internal carotid arteries, which can be seen running from the vessels of the circle of Willis at the base of the brain, toward the base of the skull. Cleanly cut the arteries close to the base of the skull and avoid any unnecessary pulling as the plexuses attached to the circle of Willis via the internal choroidal arteries can be easily damaged at this stage.
5. Cut through the pituitary stalk, loosen the brain from the remaining dura and remove from the cranial cavity.

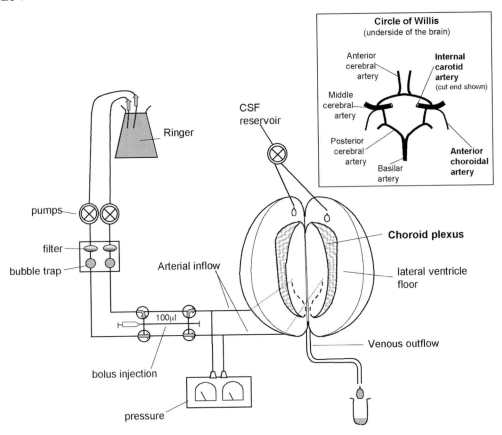

Fig. 2. Perfusion circuit for the isolated perfused choroid plexus. The lateral ventricle choroid plexuses remain in situ and the roofs of the ventricles reflected to gain access to the apical surface. Compounds for transport assessment can be included either in the Ringer reservoir (blood to cerebrospinal fluid [CSF] transfer) or the CSF reservoir (CSF to blood transfer) or in the 100-µL bolus injection site (single-pass experiment). A diagram of the blood vessels of the Circle of Willis at the base of the brain is shown in the insert.

3.1.2. Perfusion of the Choroid Plexuses

1. The cut ends of both internal carotid arteries can be seen at the base of the brain. Cannulate the arteries (0.5 mm outside diameter PVC cannulae tips in silicone tubing) and start perfusion with Ringer. Ligate the circle of Willis both proximal and distal to the cannulae to direct flow into the choroidal arteries, which arise from the circle of Willis close to the internal carotid arteries (**Fig. 2**).
2. Turn the brain over and locate the corpus callosum by blunt dissection. Make a horizontal slice through each cerebral hemisphere starting at the corpus callosum at the level of the roof of the lateral ventricles, taking care to locate and avoid touching the lateral ventral CPs. The roof of each ventricle can then be reflected revealing intact perfused CPs lying on the floor of the ventricles.
3. Superfuse the exposed tissues with artificial CSF to keep them moist and allow them to float slightly as they would in vivo.
4. Carefully remove the remaining cerebellum from around the great vein of Galen, which carries the venous drainage from both CPs through the corpus callosum to the hindbrain and cerebellum. Cannulate the vein using a 19G venous cannula to receive out flow.

5. Perfuse the vascular system with the mammalian Ringer solution (*see* **Note 2**), which allows complete control of the vascular environment of the CPs. The Ringer contains 0.5% EBA to measure CSF secretion rate. All solutions are gassed with 95% O_2/5% CO_2 at 37°C to a pH of 7.4.

6. Ringer and CSF are maintained in a water bath at 37°C and passed via peristaltic pumps (one to perfuse each CP; one for CSF) into a heat exchanger to maintain temperature. The Ringer additionally passes through a filter and bubble trap before entering the choroidal arteries. All tubing is low gas permeable silicone.

7. Tissue temperature is maintained by a water jacket around the brain, heated tray under the entire perfusion circuit, heat lamp above the brain, and monitored with a thermister probe between the two CPs.

8. Perfusion pressure is maintained at 40–60 mm Hg above that of the perfusion circuit giving a flow rate of approx 1mL/min^{-1}, and monitored by pressure transducers connected between the bubble trap and choroidal artery cannulae (*see* **Fig. 2**).

9. The preparation is viable for 3–6 h. The first indication of loss of viability is a fall in CSF secretion rate, followed by a rise in perfusion pressure and a fall in venous outflow rate.

3.1.3. CSF Secretion Rate

This is used to check the continued viability of the preparation throughout the experiment, and to test effects of potential modulators of CSF secretion (*6*). The concentration of EBA is determined every 10–15 min in arterial inflow and venous outflow from CP, using a visible light spectrophotometer at 625 nm. Because the dye cannot leave the vascular circulation, increase in venous concentration represents CSF secretion and is calculated using the formula:

$$CSF\ secretion\ (\mu L.min^{-1}.g^{-1}) = Fv[(Ab_V/Ab_A)-1]$$

Where Fv = venous flow rate ($\mu L.min^{-1}.g^{-1}$), Ab_V and Ab_A = venous and arterial absorbance, respectively (*3,7*).

3.1.4. Global Transfer

Transfer by diffusion, facilitated diffusion, or active transport in either direction across the CP (blood to CSF, or CSF to blood) can be studied using steady-state techniques, as can uptake by the CP (blood to CP).

3.1.4.1. BLOOD TO CHOROID PLEXUS

1. Include test and reference molecules labeled with different radioisotopes in the bulk Ringer (typically 40 μCi ^{3}H, 20 μCi ^{14}C per 100 mL). The reference molecule remains in the vascular and/or extracellular space, giving an index of CP "leak" and paracellular transport (*see* **Note 3**).

2. Perfuse the tissue for at least 60 min (up to 4 h is possible) until steady state is achieved, i.e., no change in the concentrations of test and reference in the venous outflow.

3. Steady-state extraction for test and reference are calculated separately from:

$$Extraction\ \% = 100\ .\ (Fa\ A^* - Fv\ V^*)\ /\ Fa\ A^*$$

where Fv = venous flow ($\mu L.min^{-1}.g^{-1}$); A* and V* are radiolabel activity in arterial and venous perfusates respectively; Fa = arterial flow. Fa is greater than Fv because CSF secretory activity removes arterial fluid. Fa is therefore calculated from $Fv(Ab_V/Ab_A)$.

4. Extractions of the test and reference molecules can then be compared, and the net extraction for the test marker calculated from the difference between them, Enet.

3.1.4.2. BLOOD TO NEWLY FORMED CSF

1. Begin perfusion as above, and in addition completely cover the CP and brain in a light silicone oil.
2. As new CSF forms, it collects in droplets under the oil and may be aspirated with a fine pipet. Collect small volumes of CSF in this way.
3. Beware of contamination by any small leaks of Ringer. These may be accounted for by the presence of EBA in the sample, or by including a high molecular weight-labeled reference (e.g., ^{3}H-dextran 70,000) that should normally remain in the vascular space *(6)*.

3.1.4.3. CSF TO BLOOD

1. Place radiolabeled test and reference molecules (typically 80 µCi ^{3}H, 40 µCi ^{14}C per 100 mL) in artificial CSF and superfuse over the CPs.
2. Appearance in Ringer confirms passage across the CP, although care must be taken to exclude metabolism of the tracers.
3. The appearance of test and reference molecules in Ringer are expressed as a percentage of that present in an equivalent volume of CSF, termed $R_{test}\%$, $R_{ref}\%$ and the two compared to compensate for paracellular leaks and diffusion between choroidal cells.

3.1.5. Kinetics of Rapid Uptake by Choroid Plexus from Blood

The single circulation paired tracer dilution technique has been used to study the characteristics of movement of rapidly transported molecules across a variety of tissue barriers including brain, placenta, salivary gland *(8,9)*. The technique modified for this preparation measures rapid uptake of a radiolabeled test compound compared to a reference molecule that is restricted to the extracellular space (interstitial and/or vascular spaces) (*see* **Note 3**). Uptake of the test compound, over and above uptake of the reference, is taken to indicate cellular uptake. For transport studies there are several advantages in using this system, which allows several test substances being studied on the same preparation (20–30 single-circulation tests are possible in 3–5 h). In addition, this technique can separate unidirectional uptake from backflux and complete Michaelis-Menten type kinetic constants can be derived from each preparation *(10)*.

1. Inject a 100 µL bolus of perfusate containing the test and reference compounds (typically ^{3}H-labeled, 4 µCi and ^{14}C-labeled, 2 µCi) in a calibrated sidearm of the perfusion circuit that can be filled independently and switched into either CP via a system of taps (**Fig. 2**). This system prevents build up of pressure from simply injecting into the circuit.
2. Allow approx 20 s to clear the dead-space of the tubing before taking venous samples (1 drop samples every 2–3 s up to 20 drops). It is necessary to carry out a preliminary run with concentrated EBA in the bolus to visualise transit through the CP and tubing and estimate the time needed to clear the dead-space and complete the collection. The final sample of each run is longer (3–10 min) to collect all test and reference washed through.
3. The recovered test and reference markers in each drop are expressed as a percentage of that in the bolus. Recovery typically ranges from less than 1% in the first drops, to a peak of 5% mid-way through the collection (**Fig. 3**). The lower recovery of test molecule (in this example, amino acid) relative to the reference (mannitol) indicates retention by the choroidal cells.
4. An uptake measure **U%** for each drop is calculated from:

U% for each drop = 100. (% reference recovered – % test recovered) / (% reference recovered).

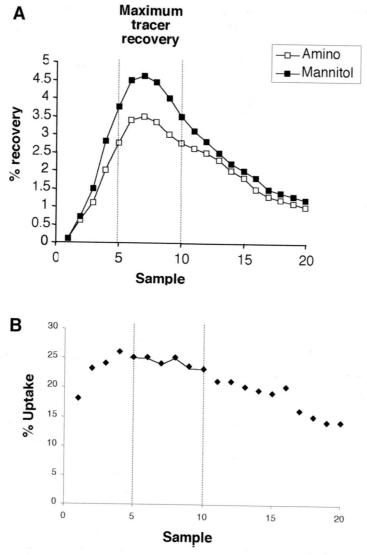

Fig. 3. Example of data from one single-pass experiment. **(A)** Recovery of labeled test (^{14}C-amino acid) and reference (^{3}H-mannitol) compounds in sequential venous outflow drops, 20 s after bolus injection. **(B)** Uptake (U%) is calculated for each drop. U% during peak recovery is averaged to calculate Umax%.

5. An index of the rapid unidirectional maximal cellular uptake, commonly termed **Umax**, is calculated from the average U% for samples close to peak reference recovery, since counting efficiency is best for these sample (*see* **Fig. 3**).

6. For carrier-mediated transport systems, kinetic constants **Vmax** (maximal transport) **Km** (half saturation constant), **Kd** (diffusion constant) may be estimated by repeated measurements of Umax with increasing concentrations of competing unlabeled test molecule in the Ringer and bolus. The **Flux** at each concentration is first calculated using the formula:

$$\text{Flux (nmol.min}^{-1}.\text{g}^{-1}) = -F \ln(1-\text{Umax}) \, S$$

where F = perfusate flow (μL.min^{-1}.g^{-1}), S = total test concentration, mM (labeled plus unlabeled molecule) *(10)*. A plot of Flux against S is described by the Michaelis-Menten equation,

$$Flux = (Vmax\ S/Km + S) + Kd\ S$$

3.2. Transport Across the In Vitro Choroidal Epithelium

3.2.1. Selection of the Culture Insert

A broad range of inserts with different types of membrane and of variable porosity are available from several manufacturers (*see* **Note 4**). As some compounds may be adsorbed and retained by the coated microporous membrane, the selection of the cell culture chamber insert is a crucial step. It is therefore necessary to run a preliminary experiment for each compound of interest, in order to evaluate its rate of transfer on different types of filters, coated with the appropriate basal lamina components, in the absence of any choroidal epithelial cells, using the protocol described in **Subheading 3.2.2.** The filter yielding the largest permeability x surface area product coefficient (PS$_f$) will be preferred for further permeability studies performed on the in vitro BCSFB.

3.2.2. Global Transfer

Prepare the solute(s) solution in RH (*see* **Note 5**) and pre-warm all reagents at 37°C.

3.2.2.1. Apical to Basolateral Transfer

1. Prepare transfer plates by filling wells of multiwell plates with warm RH. The volume will depend on the diameter of the insert, and should be adjusted so that volumes in both the donor and acceptor chambers are leveled (*see* **Note 6**).
2. Rinse both compartments of the culture inserts with RH, and place empty inserts into the transfer multiwell plate.
3. Add the solute(s) of interest in the inner chamber compartment and place at 37°C at constant stirring on a orbital rotator (200 rpm, *see* **Note 7**).
4. At regular intervals thereafter (*see* **Note 8**), transfer the inserts to another well.
5. Perform simultaneously the same experiment on matrix-coated filters without cells, transfer these inserts at shorter intervals because of the relatively high rate of flux.
6. Sample aliquots of the acceptor solution for each time point and of the donor solution at the end of the experiment and analyze them by an adapted technique (liquid scintillation counting, chromatography, spectrophotometry, or fluorescence detection) to determine the concentrations of the compound(s) in both compartments (*see* **Note 9**).

3.2.2.2. Basolateral to Apical Transfer

1. Distribute RH containing the compound(s) of interest in wells of a multiwell plate.
2. Rinse both compartments of the inserts with RH, and place empty inserts into the transfer plate.
3. Add RH in the upper compartment and place at 37°C on the orbital rotator.
4. At regular intervals thereafter, remove an aliquot (the volume will depend on the insert size, but should be as large as possible without risk of damaging the cell monolayer) from the apical chamber and replace it with an equal volume of fresh RH.
5. Process each sampled aliquot from the acceptor chamber as well as an aliquot taken from the donor chamber at the end of the experiment, to determine the solute(s) concentrations.
6. Filters without cells are run simultaneously, as described above.

7. For both types of studies, it is advisable to run a control filter with cells, in the absence of any solute, as various molecules secreted by the epithelial cells may interfere with the analytic procedure.

3.2.2.3. CALCULATION OF FLUX

1. The flux of material across the monolayer is estimated as the amount cleared from the donor fluid *(11)*. The volume clearance is given by the following equation:

$$\text{Volume cleared} = C_a V_a / C_d$$

where C_a is the concentration in the acceptor solution at the time of sampling, V_a is the volume of the acceptor solution, and C_d is the concentration in the donor solution. The latter is corrected for each sampling period by adjusting its value for the amount of molecule cleared during the previous time point. This correction is essentially insignificant for small polar molecules but may become important for highly lipophilic compounds, or for measurements of compound flux across filters without cells.

2. For apical to basolateral flux measurement, as the filter is transferred to fresh medium at each time point, the concentration in the acceptor fluid is therefore zero at the beginning of each sampling period.

3. For basolateral to apical flux experiments, the acceptor solution is usually sampled and renewed only partly, and from the second to the last time point, C_a is corrected to account for the amount of compound remaining from the previous sampling period.

4. For hydrophilic to mildly lipophilic compounds, backflux is negligible in our experimental conditions and the clearance volume increases in a linear manner with time (*see* **Note 10**). The rate of clearance, equal to the slope of a plot of the cumulative volume over time, is determined by least squares regression analysis. C_d, can be assumed constant over each sampling period. With a backflux considered as negligible, the rate of clearance becomes equal to the permeability x surface area product (PS in $\mu L.min^{-1}.filter^{-1}$).

5. For more lipophilic compounds such as naphthol or caffeine, backflux can be taken into account and the underestimated amount of material cleared during each sampling period is corrected by using half of the concentration value reached in the acceptor compartment at the end of that period as the concentration available for backflux transfer from the acceptor to the donor compartment.

6. As for electrical resistances in series, the reciprocals of the PS products of the serially arranged layers composing the cell monolayer-laminin-filter system are additive *(11)* and verify the following equation

$$1/PSt = 1/PSf + 1/Pse$$

where PSt and PSf are the PS products determined for filters with and without epithelial cells respectively, and PSe is the permeability x surface area product of the epithelial monolayer. The permeability coefficient of the epithelial cells, Pe ($cm.min^{-1}$) is obtained by dividing the calculated PSe value by the surface area of the filter.

3.2.2.4. GENERAL CONSIDERATIONS

1. The reproducible tightness of each cell preparation should be ascertained by measuring the paracellular permeability of radiolabeled markers such as sucrose, mannitol, or inulin on some filters. We try to include one of these markers on every single filter, if analytical conditions are compatible. Thereby not only the quality of each monolayer is controlled, but the possible adverse effects exerted by the transferred compounds or by the additives required for their solubility is monitored. The paracellular marker selected must have approximately the same molecular weight as the molecules being tested.

2. It is important to compare the amount initially added to the donor compartment with the sum of the amounts remaining in that chamber and those transferred to the acceptor chamber during the experiment. A lack of concordance may give evidence of either adsorption to the insert and well plastic, cellular accumulation, or even intracellular metabolism of the solute. In the former event, a similar lack of mass balance should be observed on filters without cells. Cellular accumulation can be directly tested by measuring the compound concentration in the harvested cells. If metabolism is suspected, analytical conditions should be developed to detect and identify the metabolites.

3. When evidence for a transporter-mediated process is obtained, the Michaelis-Menten kinetic parameters can be determined by measuring the Pe for various concentrations of the solute. The Pe data are then fitted to the following equation:

$$P_e = V_{max} (K_m + C) + K_{diff}$$

where C is the concentration, K_m is the affinity constant, V_{max} is the maximum velocity of the transporter, and K_{diff} is the transporter-independent diffusional constant. Care should be taken when applying high concentrations of the solutes, as these may alter the barrier integrity and increase the paracellular pathway. Sucrose permeability should be estimated simultaneously, or on separate filters, to assess that point.

3.2.3. Active Transport

1. Rinse filters on both sides with the transfer medium.
2. Apply a precise volume of this medium (either RH or serum free medium, *see* **Note 11**), containing the compound of interest in both chambers, and incubate at 37°C.
3. Collect the medium from both chambers at the end of the incubation and determine the solute concentration.
4. The active component of the clearance (in microliters) of a given compound can be calculated using the following equation:

$$Cl = [(C_a - C_d) \times V_a \times V_d] / [C_a \times V_a + C_d \times V_d]$$

where C_a and C_d are the concentrations in the acceptor and donor solutions respectively at the time of sampling, and V_a and V_d are the volumes of the acceptor and donor solutions, respectively.

The kinetic parameters can be determined by increasing the compound concentration in the incubation medium, and measuring the clearance within the linearity time frame of the transport process.

4. Notes

1. Detailed protocols to isolate and culture choroidal epithelial cells have been recently published by several groups (*12–17*), and will not be reviewed in this chapter. Briefly, choroid plexuses are submitted to a mild enzymatic digestion step, which yields a suspension of choroidal epithelial cells. Other cell types are removed by differential attachment, or alternatively through the use of fibroblast growth inhibitors. The epithelial cells are seeded on laminin-coated inserts, as this basal lamina component was shown to favor attachment of choroidal epithelial cells (over that of stromal fibroblasts) and then cultured for a few days after confluence is reached. In the bicameral device, choroidal cells exhibit a high degree of morphological and functional differentiation and polarization, most likely because of the bilateral access of cells to the media components. It is crucial however, when reproducing one of these culture protocols, to evaluate the characteristics and properties of the monolayer obtained in vitro. Besides assessing the purity and the choroidal features of the

cells, it is particularly important to consider other criteria such as the maintenance of specific transport proteins and a low paracellular pathway (demonstrated by a low sucrose or mannitol permeability), which is a requisite to investigate transport mechanisms in the in vitro model of the BCSFB.

2. Perfusion fluids and CSF secretion: An alternative blood perfusate can be prepared from filtered homologous sheep blood diluted in a ratio of 1:1 with a solution containing 50% Dextran 40 in saline, 50% arterial perfusate (NaCl 4.8 g/L, KCl 0.7 g/L, 4.7 mL/L of a 1 M CaCl$_2$ solution, NaHCO$_3$ 4.4 g/L). The advantages are oxygen carriage capacity and viscosity are closer to that present in vivo. The disadvantages are potential metabolism or uptake of the molecules of interest by plasma enzymes and red blood cells before they reach the CP (**10**). If blood perfusate is possible, red cell haematocrit can be used as an alternative to EBA for estimates of CSF secretion and arterial flow rate. Blue dextran is often used in perfusion studies, but is toxic when perfused through the CP and should not be used in place of EBA.

3. Selection of a reference marker: Modifications when selecting the test and reference molecules allows estimation of extraction of various substances from plasma, as well as providing estimates of the extravascular compartment size for diffusible substances, and the identification of carrier-mediated transport systems. The protocol and modifications used must however, satisfy some basic criteria: the flow rate must remain constant during collection of venous outflow; the same transit times apply for the test and reference molecules through the vascular bed; there is no recirculation of the reference molecule; as many characteristics of the test and reference compounds must be matched as possible, e.g., molecular weight, diffusion characteristics, since hindrance to passage of test compound through intercellular clefts for example, could give rise to apparent "negative" uptakes if the reference molecule has a less impeded passage. Any separation of the test and reference molecules due to different diffusional or flow properties are evident in the rapid uptake single circulation technique where peak recovery of the two tracers does not coincide (the Taylor effect). Examples of appropriate reference molecules that can be radiolabelled and have limited (or no) choroidal transport, include mannitol (180 Da), Polyethylene glycol (available at several molecular weights, e.g., 4000 Da), and dextran (70,000 Da).

4. The microporous membrane of the insert and the extracellular matrix components (laminin, +/– collagen) applied for attachment of choroidal epithelial cells, may lead to the nonspecific absorption and the retention of some compounds. Different inserts can be tested among the following types, listed in **Table 1**.

   Inserts are also available from Millipore and Nunc. However, in our experience, choroidal epithelial cells from newborn rats do not attach to laminin-coated inserts from Millipore, and we have not tested inserts from Nunc. Laminin-coated inserts from Costar and Falcon yield similar results in terms of cell attachment, cell differentiation, and transport properties. Using precoated Transwell-COL membranes does not improve any of those criteria, even when treated with laminin. Transparent membranes have the advantage of allowing the assessment of monolayer formation using phase contrast microscopy.

   The 0.4 μ porosity is well suited to study transfer of solutes as such pores will allow free diffusion of chemicals or even proteins across the filter. In theory, restricted diffusion will not occur if the pore diameter is greater than 20-fold the effective diameter of the solute (**18**). As an example, for a medium-sized protein such as albumin (MW 70,000 Da), the effective diameter is around 6 nm.

5. It may be necessary to use additives to prepare some poorly water soluble molecules. The absence of effect of these compounds on the cell monolayer should however be ascertained. In the case of DMSO or ethanol, we observe no deleterious effect on the barrier, when the concentration is kept below 0.5%.

Table 1
Membrane Characteristics of Commercially Available Cell Culture Inserts

Manufacturer		Membrane	Pore size (μm)
Costar	Transwell-Clear	Transparent polyester	0.4, 3.0
	Transwell-PC	Translucent polycarbonate	0.4, 3.0, 5.0, 8.0, 12.0
	Transwell-COL	Transparent collagen-coated Teflon	0.4, 3.0
Falcon		Polyethylene terephtalate	
	PET - LD	transparent low pore density	0.4, 1.0, 3.0, 8.0
	PET - HD	translucent high pore density	0.4, 1.0, 3.0, 8.0

6. Inserts are available in different sizes as indicated in **Table 2**. The volumes of transfer medium may be varied in both compartments, depending on the experimental requirements (*see* **Table 2**). For example, it may be necessary to keep the volume as small as possible in the acceptor chamber for sensitivity reasons, or inversely, use a maximal volume in the donor chamber to maintain a constant concentration in the case of a high capacity transport process. The volume in the opposite chamber should then be adjusted consequently, to prevent the formation of a hydrostatic pressure gradient. Because the in vitro choroidal epithelium forms a hydrodynamic barrier, the fluid balance is not critical when measuring solute transfer on filters with cells. However, it becomes critical for PSf determination on filters without cells, or for permeability studies on cell monolayers on which the barrier properties have been experimentally altered.
   Typical volumes that can be used for various inserts are given in **Table 2**.
7. Stirring of the solutions in both chambers should be performed in order to reduce the unstirred layer of fluid lying adjacent to the cell monolayer and the microporous membrane of the insert, and acting as a barrier per se. Although stirring does not greatly influence the flux of polar molecules, it has a major impact on the flux of highly permeable molecules. We observed a 40% increase in the epithelial permeability coefficient for naphthol when the orbital stirring applied to the inserts was increased from 50 to 200 rpm. At this higher speed, no adverse effect is observed on the cell monolayer.
8. Several criteria have to be considered when selecting the duration of sampling intervals, such as a) the detection limit of the assay, b) the maintenance of a constant donor concentration, c) and the need to minimize the solute backflux from the acceptor chamber.
9. A large panel of drug metabolizing enzymes are highly active in the choroidal tissue (*19*), and solutes can be metabolized en passage through the epithelial cells. The influence of this specific property on the transepithelial permeability of compounds can be approached in the in vitro choroidal epithelium, which reproduces the metabolic capacity of the in vivo tissue (*17*). If metabolism of a solute is probable or suspected, a chromatographic detection method is to be preferred over liquid scintillation counting or fluorescence and UV absorbance measurements, as the parent compound identity can be ascertained and metabolites can be distinguished.
10. An increase over time in the calculated clearance volume may result from different factors. It could be explained by a progressive toxic effect of the solute on the epithelial cells, leading to an increase in the paracellular permeability of the monolayer. If metabolism of the solute occurs, the increase in the clearance could also result from the depletion of an intracellular cofactor required for the metabolic reaction, or from a progressive substrate-mediated inhibition of the metabolizing enzyme.

Table 2
Recommended Volumes of Fluid for Commercially Available Bicameral Systems

	Insert diameter	Multiwell plate	Growth area (cm²)	Volume in insert (µL)	volume in well (µL)
Costar	6.5 mm	24 well	0.33	0.15–0.25	0.6–0.8
	12 mm	12 well	1.0	0.4–0.8	1.0–1.5
	24 mm	6 well	4.7	1.5–2.0	2.5–3.0
Falcon	6.5 mm	24 well	0.3	0.2–0.35	0.7–0.9
	12 mm	12 well	0.9	0.4–1.0	1.4–2.3
	24 mm	6 well	4.2	1.5–2.5	2.7–3.2

11. In case of prolonged incubation periods, the RH can be replaced by serum free medium and the experiment has to be performed in an CO_2 incubator. However, the time of collection should be kept to the minimal to limit the backflux diffusion that will arise from the creation of a gradient. A time course should be performed to assess the linearity of the transport process.

Acknowledgments

The authors would like to thank Dr. Malcolm Segal for expert advice, and Dr. J.-F. Ghersi-Egea for fruitful discussion and advice. The work was supported by the Welcome Trust and the BBSRC to J.E. Preston, and PRISME (INSERM), ARSEP, Ligue SEP and ANRS to N. Strazielle.

References

1. Davson, H., and Segal, M. B., eds. (1996) *Physiology of the CSF and Blood-Brain Barriers.* CRC Press, Boca Raton, Fla.
2. Strazielle, N., and Ghersi-Egea, J. F. (2000) Choroid plexus in the central nervous system: biology and physiopathology. *J. Neuropathol. Exp. Neurol.* **59(7)**, 561–574.
3. Preston, J. E., Segal, M. B., Walley, G. J., and Zlokovic, B. V. (1989) Neutral amino acid uptake by the isolated perfused sheep choroid plexus. *J. Physiol. (Lond)* **408**, 31–43.
4. Preston, J. E., and Segal, M. B. (1990) the steady-state amino acid fluxes across the perfused choroid plexus of the sheep. *Brain Res.* **525**, 275–279.
5. Preston, J. E. (2001) Ageing choroid plexus-cerebrospinal fluid system. *Micr. Res. Tech.* **52**, 31–37.
6. Thomas, S. A., Preston, J. E., Wilson, M. R., Farrell, C. L., and Segal M. B. (2001) Leptin transport at the blood-cerebrospinal fluid barrier using the perfused sheep choroid plexus model. *Brain Res.* **895**, 283–290.
7. Pollay, M., Stevens, A., Estrada, E., and Kaplan, R. (1972) Extracorporeal perfusion of the choroid plexus. *J. Appl. Physiol.* **32**, 612–617.
8. Yudilevich, D. L., and Mann, G. E. (1982) Unidirectional uptake of substrates at the blood-tissue interface of secretory epithelia: stomach, salivary gland and pancreas. *Fed. Proc.* **41**, 3045–3053.
9. Bustemante, J. C., Mann, G. E., and Yudilevich, D. L. (1981) Specificity of neutral amino acid uptake at the basolateral side of the epithelium in the cat submandibular gland in situ. *J. Physiol. (Lond)* **313**, 65–79.

10. Segal, M. B., Preston, J. E., and Zlokovic, B. V. (1990) Comparison between blood and saline perfusion on the uptake of amino acids by choroid plexus of the sheep. *Endocrinol. Exp.* **24,** 29–36.

11. Siflinger-Birnboim, A., Del Vecchio, P. J., Cooper, J. A., Blumenstock, F. A., Shepard, J. M. and Malik, A. B. (1987) Molecular sieving characteristics of the cultured endothelial monolayer. *J. Cell Physiol.* **132,** 111–117.

12. Southwell, B. R., Duan, W., Alcorn, D., Brack, C., Richardson, S. J., Kohrle, J., and Schreiber, G. (1993) Thyroxine transport to the brain: role of protein synthesis by the choroid plexus. *Endocrinology* **133,** 2116–2126.

13. Ramanathan, V. K., Hui, A. C., Brett, C. M., and Giacomini, K. M. (1996) Primary cell culture of the rabbit choroid plexus: An experimental system to investigate membrane transport. *Pharm. Res.* **13,** 952–926.

14. Gath, U., Hakvoort, A., Wegener, J., Decker, S., and Galla, H. J. (1997) Porcine choroid plexus cells in culture: expression of polarized phenotype, maintenance of barrier properties and apical secretion of CSF-components. *Eur. J. Cell. Biol.* **74,** 68–78.

15. Villalobos, A. R., Parmelee, J. T., and Pritchard, J. B. (1997) Functional characterization of choroid plexus epithelial cells in primary culture. *J. Pharmacol. Exp. Ther.* **282,** 1109–1116.

16. Zheng, W., Zhao, Q., and Graziano, J. H. (1998) Primary culture of choroidal epithelial cells: characterization of an in vitro model of blood-CSF barrier. *In Vitro Cell. Dev. Biol. Anim.* **34,** 40–45.

17. Strazielle, N., and Ghersi-Egea, J. F. (1999) Demonstration of a coupled metabolism-efflux process at the choroid plexus as a mechanism of brain protection toward xenobiotics. *J. Neurosci.* **19,** 6275–6289.

18. Pappenheimer, J. R., Renkin, E. M., and Borrero, L. M. (1951) Filtration, diffusion and molecular sieving through peripheral capillary membranes. A contribution to the pore theory of capillary permeability. *Am. J. Physiol.* **167,** 13–46.

19. Ghersi-Egea, J. F., and Strazielle, N. (2001) Brain drug delivery, drug metabolism, and multidrug resistance at the choroid plexus. *Microsc. Res. Tech.* **52,** 83–88.

IV

IN VITRO TECHNIQUES

Table 1

Characteristics and Applications of Immortalized Cell Line Models of the Blood–Brain Barrier

Cell line	Species (transfection)	Characterization	Reported applications
MBEC-4	Mouse (1)	AP, γGT, AcLDL upt., polarized *mdr1b* Pgp expression but not *mdr1a*	Modulation by glial factors, choline and GSH transporters, drug efflux mechanisms (Pgp, MRPs), uptake and vectorial transport studies, membrane vesicle studies
S5C4	Mouse (2)	vWF, BSI, UEA-1, PECAM-1	Role of gap junctions in maturation of BECs
TM-BBB4	Mouse (1)	vWF, AP, γGT, Pgp, GLUT-1, AcLDL upt. $P_{sucrose}$ ~140 × 10^{-6} cm/s	Functional expression of GLUT-1, Pgp, OATP2
RBE4	Rat (2)	vWF, AP, γGT, BSI, Pgp, MRP1, L-system AA transport, Ns transporters, GLUT1, 5-HT transporter, carnitine transporter, CYP450s, UGT, SOD, receptors for ATP/UTP, BK, ANP, NA, Tf, secretion of NO, ET-1, PGE2 & TB2; TEER~10–150 Ωcm², $P_{sucrose}$=3–200 × 10^{-6} cm/s	Modulation by glial factors, mechanistic investigation of carrier-mediated transport of numerous endogenous and exogenous agents, structure-affinity relationships for AA and Ns carriers, identification of Pgp substrates, mechanisms of toxin actions and oxidative damage, receptor-mediated modulations of second messengers in BECs, TfR-mediated drug delivery strategies, study of antioxidative defense mechanisms in BECs, study of metabolic enzymes, genetically modified RBE4 cells as vector for CNS gene therapy
CR-3	Rat (1)	vWF, BSI, ACE, AcLDL upt., weak expression of γGT and Pgp, poor monolayer formation, lack of tight junctions	Regulation of BBB properties
GP8	Rat (1)	vWF, BSI, γGT, AP, AcLDL upt., GLUT-1, TfR, PECAM-1, secretion of PGE2	Modulation by glial factors, study of cell adhesion molecules on BECs, mechanisms of T-cell migration, regulation of MHC class I and II, ICAM-1 and VCAM-1 expression by cytokines, role of signal transduction pathways
GPNT	Rat (1)	As GP8 plus high levels of Pgp, no expression of MRP1, $P_{fluorescein}$ = 120 × 10^{-6} cm/s	Polarized drug efflux by Pgp, drug-induced Pgp activation, regulation of Pgp activity via second messengers
RBEC1	Rat (1)	vWF, AP, γGT, AcLDL upt. TfR, weak expression of Pgp	MCT1-mediated transport functions, MRP1-mediated drug efflux

RCE-T1	Rat (3)	vWF, γGT, ACE, tumorigenic	Cellular mechanisms involved in transition from normal to neoplastic endothelium
t-BBEC-117	Bovine (1)	AP, Pgp, GLUT-1, AcLDL upt., $P_{L-glucose}$ = 1×10^{-4} cm/s	Modulation by glial factors, modulation of transendothelial permeability
SV-BEC	Bovine (1)	vWF, AP, γGT, BSI, AcLDL upt., TfR, ACE, TEER~40 Ωcm²	Functional expression of adrenergic receptors (β1, β2), GSH transporters, secretion of ET-1
BBEC-SV	Bovine (1)	vWF, γGT, AcLDL upt., CAIV	Cytokine-induced expression of adhesion molecules (ECAM and VCAM), LPS-induced cell injuries
PBMEC/C1-2	Procine (1)	vWF, γGT, AP, BSI, AcLDL upt., GLUT-1 Apo A1, TEER~250–300 Ωcm²	Transendothelial permeability studies
SV-HCEC	Human (1)	VWF, γGT, AP, AcLDL upt., TfR, TEER~40–150 Ωcm², $P_{sucrose}$ = 5×10^{-6} cm/s	Modulation by glial factors of AP but not Pgp and GLUT-1, study of cell surface antigens, GSH transport mechanisms, role of surface antigens on inflammatory cytokines
t-HBMEC	Human (1)	vWF, γGT, LDL upt., TEER~60 Ωcm²	Mechanisms of cytotoxicity to BECs by bacteria
HBEC-5I	Human (1)	Expr. of ICAM-1, VCAM-1, thrombospondin (but no E-selectin and CD36)	Regulation of cell surface receptors and their role in cytoadhesion of parasitized red blood cells
BB19	Human (4)	vWF, confluent monolayers	Cytokine-regulated expression of VCAM, ICAM-1, E-selectin, CD36;
BB19/MDR	Human (4)	cytoadherence of parasites as BB19 plus high levels of Pgp (non-polarised expression)	Pgp efflux and modulation
ECV304/C6	Human, non-brain endothelial	Endothelial and epithelial characteristics, co-culture with C6 glioma: TEER→>200 Ωcm², $P_{sucrose}$~5 × 10^{-6} cm/s, Pgp, TfR, GLUT-1	Induction of BEC characteristics (AP, γGT, Pgp, TfR, selectin, CD36; AA transport and polarized Ns transport, cytokine modulation of paracellular permeability and ICAM-1 expression
MDCK	Dog, kidney epithelial	Epithelial characteristics, TEER~200–400 Ωcm²; $P_{mannitol}$~4 × 10^{-7} cm/s	Higher throughput permeability screen, MDR1 transfected variant (MDCK- MDR1) used to identify Pgp substrates and inhibitors

315

Vectors used for transfections: (1) SV40 large T antigen, (2) Adenovirus E1A gene, (3) Rous sarcoma virus, (4) Human papilloma E6E7 gene. Abbreviations: AA, Amino acid; Ns, nucleoside; GSH, glutathione; 5-HT, serotonin; ANP, atrio-natriuretic peptide; SOD, superoxide dismutase; TB2, thromboxane B2; MHC, major histocompatibility complex; CAIV, carbonic anhydrase IV; ApoA1, Apolipoprotein A1; PECAM, ICAM, ECAM, and VCAM are cell adhesion molecules. (*See* **Subheading 3.3.; 31.**)

The three bovine BEC lines: SV-BEC *(42)*, BBEC-SV *(43)* and t-BBEC-117 *(44)* and one porcine BEC line: PBMEC/C1-2 *(45)*, all were generated by SV40 T antigen transfection. All systems retain a number of basic BEC characteristics and have been used to study changes in protein expression following induction by astrocytes and neuroinflammatory responses. Recent reports on the PBMEC/C1-2 model show that the junctions are sufficiently tight (TEER: 250–300 Ωcm^2) for the study of drug permeation *(46)*.

The generation of murine cell lines has been most popular in Japan and there are now reports on three immortalized BEC mouse lines: MBEC-4 *(47)*, S5C4 *(48)*, and TM-BBB4 *(49)* with the MBEC-4 cell system being used for mechanistic studies on BBB drug transport and efflux mechanisms.

To date, at least six different immortalized cell lines have been generated by transfection of primary rat BECs: RCE-T1 *(50)*, RBE4 *(51)*, CR-3 *(52)*, GP8 *(53)*, GPNT *(54)*, and RBEC1 *(55)*. The RBE4 and GP8/GPNT cell lines are by far the most widely used cell line models of the BBB, and have been found useful in the study of a broad array of topics ranging from mechanistic transport studies to receptor-mediated modulation and inflammatory responses.

A human in vitro system would be an invaluable asset for BBB studies, and several attempts have been made to generate immortalized human BECs suitable for examination of the physiology, pharmacology, and pathology of the human BBB in vitro and as a screening tool for CNS penetration. Although immortalization has proved to be much more difficult for BECs of other species, there are reports on four human cell line models and one retransfected cell line. The system, which is best characterized in terms of a BBB phenotype is the SV-HCEC cell line *(56)*. The model has been used to study cell adhesion properties *(57)* and glutathione transport *(58)* but does not appear to grow reproducibly and junctions are not sufficiently tight to be used in permeability studies (TEER: 40 Ωcm^2). Three other human cell lines HBEC-51 *(59)*, t-HBMEC *(60)*, and BB19 *(61)*, have been used mainly for parasitological studies. As the BB19 cell line does not express Pgp, it has undergone further transduction with the MDR1 gene to produce a model suitable for investigation of drug-Pgp interactions *(62)*.

It is generally difficult to make BEC cell lines switch from the exponential growth phase after cell seeding to a more static phase of cell differentiation after the cells have reached confluence. Therefore, immortalized cell lines are less useful for studies requiring a tight in vitro barrier, but they have proved particularly useful for mechanistic and biochemical studies requiring large amounts of biological material, e.g., to establish structure-affinity relationships for BBB-specific carrier systems *(63)*. As continuous cell lines may gradually deviate from the normal BEC phenotype, both the basic characteristics and the particular feature of interest should be monitored through successive passages and the results validated with primary cells or in vivo data in order to assure the relevance of the information obtained.

One of the greatest limitations of currently available immortalized BEC lines is their insufficient tightness, rendering these systems unsuitable for use as BBB permeability screens. Therefore, some groups have turned to other cell lines which, although of non-brain origin, either express sufficient brain endothelial features for functional and permeation studies such as ECV304/C6 *(64)*, or prove on validation to be useful predictors of passive CNS penetrability of compounds such as the MDCK *(25)*.

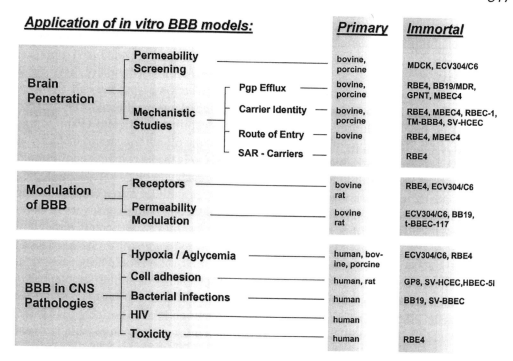

Fig. 4. Cladistic diagram of the applications of different in vitro models of the blood–brain barrier. See text for abbreviations of the immortalized cell line models. SAR, structure-affinity relationships.

3.4. Choroid Plexus Epithelial Cells

In vitro cell model systems also exist for the blood-cerebrospinal fluid barrier (BCSFB), which is located at the choroid plexus epithelium of the brain ventricles. In contrast to the BBB, which is made up of endothelial cells, choroid plexus (CP) epithelial cells form the basis of this barrier and thus are used in culture as a model system for the BCSFB. Both primary rat (*65,66*) and primary porcine (*67*) CP epithelial cells have been successfully cultured to examine the specific transport properties of the BCSFB. Recently, a CP epithelial cell line, TR-CSFB3, has been introduced (*68*).

4. Selected Applications of In Vitro Models

In vitro models of the BBB have proven invaluable for a number of applications covering both basic and applied research interests. Specific fields of interest include i) BBB permeability, ii) BBB-specific transport processes, iii) receptor-mediated modulation of BBB functions, iv) pharmacology and toxicology of the BBB, v) role of the BBB in the pathophysiology of CNS diseases, and vi) the use of genetically engineered BEC for gene therapy. A cladistic diagram to help guide selection of an appropriate in vitro system suitable for particular applications and research interests is given (**Fig. 4**).

One of the most popular applications of in vitro BBB models is the prediction of CNS penetration. There are a large number of permeability assays but all are variants of the

the system in vivo, so precluding simple extrapolation back to the more complex situation in vivo. As the level of expression of many BBB features (e.g., tight junctions, receptors, transporters, enzymes) is not as high in vitro as it is in vivo the findings are essentially of relative (qualitative) value. In order to be of absolute relevance, more quantitative knowledge concerning the consequences of the differences between in vitro and in vivo models is needed to facilitate better extrapolation to the situation at the BBB in the intact organisms (both animal and human).

Future research into in vitro BBB models may focus more on culture conditions for BECs that better mimic the in vivo situation such as use of multiple cell co-cultures involving neurons and pericytes and the use of advanced bioanalytical methods such as matrix-assisted laser desorption ionization time-of-flight mass spectrometry and molecular biological techniques such as gene arrays to identify the induction factor(s) needed to make the in vitro models more in vivo-like. Thus new knowledge and technologies will make it possible to develop refined in vitro models for studying not only the basic molecular and biochemical functions of the normal BBB, but also the specific involvement of the BBB in human neurological disorders.

References

1. Joo, F. (1992) The cerebral microvessels in culture, an update. *J. Neurochem.* **58,** 1–17.
2. Pardridge, W. M. (1998) Isolated brain capillaries: an in vitro model of blood-brain barrier research. In *Introduction to the Blood-Brain Barrier: Methodology, Biology and Pathology.* Pardridge, W. M., ed. University Press, Cambridge, pp. 49–61.
3. Brendel, K., Meezan, E., and Carlson, E. C. (1974) Isolated brain microvessels: a purified, metabolically active preparation from bovine cerebral cortex. *Science* **185,** 953–955.
4. Panula, P., Joo, F., and Rechardt, L. (1978) Evidence for the presence of viable endothelial cells in culture derived from dissociated rat brain. *Experimentia* **34,** 95–97.
5. Kramer, S. D., Abbott, N. J., and Begley, D. J. (2001) Biological models to study blood-brain barrier permeation. In *Pharmacokinetic Optimization in Drug Research: Biological, Physicochemical and Computational Strategies.* Testa, B., van de Waterbeemd, H., Folkers, G., and Guy, R., eds. Wiley-VCH, Weinheim, pp. 127–153.
6. Garberg, P. (1998) In vitro models of the blood-brain barrier. *ATLA* **28,** 821–847.
7. DeBault, L. E., and Cancilla, P. A. (1980) Gamma-glutamyl transpeptidase in isolated brain endothelial cells: induction by glial cells in vitro. *Science* **207,** 653–655.
8. Cecchelli, R., Dehouck, B., Descamps, L., et al. (1999) In vitro model for evaluating drug transport across the blood-brain barrier. *Adv. Drug Del. Rev.* **36,** 165–178.
9. De Boer, A. G., and Sutanto, W. (eds.) (1997) *Drug Transport Across the Blood-Brain Barrier.* Harwood, Amsterdam.
10. Engelbertz, C., Korte, D., Nitz, T., et al. (2000) The development of the in vitro models of the blood-brain and blood-CSF barriers. In *The Blood-Brain Barrier and Drug Delivery to the CNS.* Begley, D. J., Bradbury, W. M., and Kreuter, J., eds. Marcel Dekker, New York, pp. 33–63.
11. Miller, D. S., Nobmann, S. N., Gutmann, H., Toeroek, M., Drewe, J., and Fricker, G. (2000) Xenobiotic transport across isolated brain microvessels studied by confocal microscopy. *Mol. Pharm.* **58,** 1357–1367.
12. Audus, K. L., Ng, L., Wang, W., and Borchardt, R. T. (1996) Brain microvessel endothelial cell culture systems. In *Models for Assessing Drug Absorption and Metabolism.* Borchardt, R. T., Smith, P. L., and Wilson, G., eds. Plenum, New York, pp. 239–258.

13. Fricker, G. (2002) Drug transport across the blood-brain barrier. In *Pharmacokinetic Challenges in Drug Discovery*. Pelkonen, O., Baumann, A., and Reichel, A., eds. Springer, Berlin, pp. 13–154.

14. Li, Y. J., Boado, R. J., and Pardridge, W. M. (2001) Blood-brain barrier genomics. *J. Cereb. Blood Flow Metab.* **21,** 61–68.

15. Sipos, I., Domotor, E., Abbott, N. J., and Adam Vizi, V. (2000) The pharmacology of nucleotide receptors on primary rat brain endothelial cells grown on a biological extracellular matrix: effects on intracellular calcium concentration. *Br. J. Pharmacol.* **131,** 1195–1203.

16. Abbott, N. J., Hughes, C. C. W., Revest, P. A., and Greenwood, J. (1992) Development and characterisation of a rat brain capillary endothelial culture: towards an in vitro blood-brain barrier. *J. Cell Sci.* **103,** 23–37.

17. Liebner, S., Kniesel, U., Kalbacher, H., and Wolburg, H. (2000) Correlation of tight junction morphology with the expression of tight junction proteins in blood-brain barrier endothelial cells. *Eur. J. Cell Biol.* **79,** 707–717.

18. Abbott, N. J., Revest, P. A., Greenwood, J., et al. (1997) Preparation of primary rat brain endothelial cell culture. In *Drug Transport Across the Blood-Brain Barrier.* de Boer, A.G., Sutanto, W., eds. Harwood, Amsterdam, pp. 5–16.

19. Bowman, P. D., Ennis, S. R., Rarey K. E., Betz, A. L., and Goldstein, G. W. (1983) Brain microvessel endothelial cells in culture: a model for study of blood-brain barrier permeability. *Ann. Neurol.* **14,** 396–402.

20. Audus, K. L., and Borchardt, R. T. (1986) Characterisation of an in vitro blood-brain barrier model system for studying drug transport and metabolism. *Pharm. Res.* **3,** 81–87.

21. Miller, D. W., Audus, K. L., and Borchardt, R. T. (1992) Application of cultured endothelial cells of the brain microvasculature in the study of the blood-brain barrier. *J. Tiss. Cult. Meth.* **14,** 217–224.

22. Raub, T. J., Kuenzel, S. L., and Sawada, G. A. (1992) Permeability of bovine brain endothelial microvessel endothelial cells in vitro. *Exp. Cell Res.* **1999,** 330–340.

23. Priya Eddy, E., Maleef, B. E., Hart, T. K., and Smith, P. L. (1997) In vitro models to predict blood-brain barrier permeability. *Adv. Drug Del. Rev.* **23,** 185–198.

24. Glynn, S. L., and Yazdanian, M. (1998) In vitro blood-brain barrier permeability of nevirapine compared to other HIV antiretroviral agents. *J. Pharm. Sci.* **87,** 306–310.

25. Polli, J. W., Humphreys, J. E., Wring, S. A., et al. (2000) A comparison of Madin-Darby canine kidney cells and bovine brain endothelial cells as a blood-brain barrier screen in early drug discovery. In *Progress in the Reduction, Refinement and Replacement of Animal Experimentation*. Balls, M., van Zeller, A. M., and Halder, E. M., eds. Elsevier, New York, pp. 271–289.

26. Zhang, Y., Han, H., Elmquist, W. F., and Miller, D. W. (2000) Expression of various multidrug resistance-associated protein (MRP) homologues in brain microvessel endothelial cells. *Brain Res.* **876,** 148–153.

27. Batrakova, E. V., Li, S., Vinogradov, S. V., Alakhov, V. Y., Miller, D. W., and Kabanov, A. V. (2001) Mechanism of pluronic effect on Ppg-efflux system in the blood-brain barrier: contribution of energy depletion and membrane fluidization. *J. Pharm. Exp. Ther.* **299,** 483–493.

28. Rochat, B., Baumann, P., and Audus, K. L. (1999) Transport mechanisms for the antidepressent citalopram in brain microvessel endothelium. *Brain Res.* **831,** 229–236.

29. Letrent, S. P., Polli, J. W., Humphreys, J. E., Pollack, G. M., Brouwer, K. R., and Brouwer, K. L. R. (1999) P-glycoprotein-mediated transport of morphine in brain capillary endothelial cells. *Biochem. Pharmacol.* **58,** 951–957.

64. Hurst, R. D., and Fritz, I. B. (1996) Properties of an immortalized vascular endothelial/ glioma cell co-culture model of the blood-brain barrier. *J. Cell Physiol.* **167,** 81–88.

65. Villalobos, A. R., Parmelee, J. T., and Pritchard, J. B. (1997) Functional characterization of choroid plexus epithelial cells in primary culture. *J. Pharm. Exp. Ther.* **282,** 1109–1116.

66. Strazielle, N., and Ghersi-Egea, J. F. (1999) Demonstration of a coupled metabolism-efflux process at the choroid plexus as a mechanism of brain protection toward xenobiotics. *J. Neurosci.* **19,** 6275–6289.

67. Hakvoort, A., Haselbach, M., and Galla, H. J. (1998) Active transport of porcine choroid plexus cells in culture. *Brain Res.* **795,** 247–256.

68. Kitazawa, T., Hosoya, K., Watanabe, M., et al. (2001) Characterization of the amino acid transport of new immortalized choroid plexus epithelial cell lines: a novel in vitro system for investigating transport functions at the blood-cerebrospinal fluid barrier. *Pharm. Res.* **18,** 16–22.

69. Audus, K. L., Rose, J. M., Wang, W., and Borchardt, R. T. (1998) Brain microvessel endothelial cell culture systems. In *Introduction to the Blood-Brain Barrier: Methodology, Biology and Pathology.* Pardridge, W. M., ed. University Press, Cambridge, pp. 86–93.

70. Adson, A., Burton, P. S., Raub, T. S., Barsuhn, C. L., Audus, K. L., and Ho, N. F. (1995) Passive diffusion of weak organic electrolytes across Caco-2 cell monolayers: uncoupling the contributions of hydrodynamic, transcellular and paracellular barrier. *J. Pharm. Sci.* **84,** 1197–1204.

71. Johnson, M. D., and Anderson, B. D. (1999) In vitro models of the blood-brain barrier to polar permeants: comparison of transmonolayer flux measurements and cell uptake kinetics using cultured cerebral capillary endothelial cells. *J. Pharm. Sci.* **88,** 620–625.

72. De Vries, H. E., Kuiper, J., de Boer, A. G., van Berkel, T. J. C., and Breiber, D. D. (1997) The blood-brain barrier in neuroinflammatory diseases. *Pharm. Rev.* **49,** 143–155.

73. Abbott, N. J. (1998) Role of intracellular calcium in regulation of brain endothelial permeability. In *Introduction to the Blood-Brain Barrier: Methodology, Biology and Pathology.* Pardridge, W.M., ed. University Press, Cambridge, pp. 345–353.

74. Hipkiss, A. R., Preston, J. E., Himsworth, D. T., et al. (1998) Pluripotent protective effects of carnosine, a naturally occurring dipeptide. *Ann. NY Acad. Sci.* **854,** 37–53.

75. Abbott, N. J. (2000) Inflammatory mediators and modulation of blood-brain barrier permeability. *Cell. Mol. Neurobiol.* **20,** 131–147.

76. Rubin, L. L., and Staddon, J. M. (1999) The cell biology of the blood-brain barrier. *Annu. Rev. Neurosci.* **22,** 11–28.

77. O'Kane, R. L., Martinez-Lopez, I., De Joseph, M. R., Vina, J. R., and Hawkins, R. A. (1999) Na-dependent glutamate transporters (EAAT1, EAAT2 and EAAT3) at the blood-brain barrier. *J. Biol. Chem.* **274,** 31,891–31,895.

78. Stannes, K. A., Neumaier, J. F., Sexton, T. J., et al. (1999) A new model of the blood-brain barrier: co-culture of neuronal, endothelial and glial cells under dynamic conditions. *Neuroreport* **10,** 3725–3731.

79. Duport, S., Robert, F., Muller, D., Grau, G., Parisi, L., and Stoppini, L. (1998) An in vitro blood-brain barrier model: co-cultures between endothelial cells and organotypic brain slice cultures. *Proc. Natl. Acad. Sci. USA* **95,** 1840–1845.

21

Isolation and Characterization of Human Brain Endothelial Cells

Katerina Dorovini-Zis, Rukmini Prameya, and Hanh Huynh

1. Introduction

The ability to isolate and culture endothelial cells derived from brain microvessels has led to a considerable increase in our understanding of the biology of these cells over the last two decades. Most of these in vitro systems are derived from various animal sources and provide a valuable tool for investigating the function of cerebral endothelial cells and their reaction to injury under controlled conditions. Recent advances in vascular biology, however, indicate that endothelial cells from different species and vascular beds differ in their morphology, function, permeability, and immunological properties (1). The need to investigate the function of the human cerebral endothelium and its role in the pathogenesis of diseases unique to the human central nervous system led to the development, over the last decade, of methods for the isolation and cultivation of human brain-derived endothelial cells (2–7).

A relevant in vitro model of the human blood–brain barrier (BBB) should consist of cells that can be reproducibly isolated and grown in culture and that retain both their endothelial phenotype and important morphological, functional, and barrier properties of the cerebral endothelium in vivo, namely, the presence of high resistance interendothelial tight junctions and paucity of cytoplasmic vesicles. The cells should maintain these characteristics during the cultivation period and should form confluent, contact-inhibiting monolayers of high transendothelial electrical resistance that restrict the paracellular passage of macromolecules.

Material for the isolation of human brain microvessel endothelial cells (HBMEC) is usually obtained from temporal lobectomy specimens, which, because of their small size, often necessitate transformation and immortalization or serial passaging of the isolated cells (8). In general, HBMEC lose the BBB and endothelial characteristics after several passages, and co-culture with astrocytes appears to partly restore some of the BBB properties in these cultures. Immortalized cell lines retain certain morphological and physiological characteristics of the BBB and exhibit accelerated proliferation rates without proliferative senescence unlike primary HBMEC cultures. Alternatively, HBMEC can be harvested from normal brains at autopsy following a reasonably short post-mortem interval. The isolation procedure generates a large number of viable

From: *Methods in Molecular Medicine, vol. 89:*
The Blood–Brain Barrier: Biology and Research Protocols
Edited by: S. Nag © Humana Press Inc., Totowa, NJ

HBMEC thus decreasing the need for serial passaging and generation of cell lines. Difficulty in obtaining human autopsy material is a limiting factor in this approach.

This chapter describes the method for the isolation, culture, and characterization of human cerebral endothelial cells. Despite limitations inherent in an artificial in vitro system, primary HBMEC cultures constitute an invaluable tool for the investigation of physiological and pathological responses at the human BBB.

2. Materials

2.1. Isolation of HBMEC

1. Sterile petri dishes.
2. Instruments: Scalpels and blades, Forceps.
3. Sterile pipettes: 1, 5, and 10 mL.
4. Sterile beakers.
5. Sterile 500-mL bottles.
6. Polycarbonate 250-mL bottles.
7. Polycarbonate 50-mL tubes (Nalgene Centrifuge Ware).
8. Polystyrene 50-mL centrifuge tubes.
9. Medium 199 (M199), 1X and 10X (Invitrogen, Burlington, Ontario).
10. Horse plasma-derived serum (Hyclone Laboratories, Logan, UT).
11. 25 mM HEPES (Sigma, St. Louis, MO).
12. 10 mM Sodium bicarbonate.
13. Dispase (Roche Diagnostics, Laval, Quebec).
14. Dextran (70,000 mW) (Sigma).
15. Collagenase/Dispase (Roche).
16. Percoll and Percoll density beads (Pharmacia, Piscataway, NJ).
17. Trypan blue.
18. Hemacytometer.

2.2. Culture of Endothelial Cells

1. Plastic multiwell plates (4, 24, or 96 well plates) (Corning Plastics, Corning, NY).
2. Fibronectin (Sigma), 100 µg/mL.
3. Antibiotics (penicillin, streptomycin and amphotericin B) (Invitrogen).
4. 10% horse plasma-derived serum (Cocalico, Reamstown, PA).
5. Endothelial cell growth supplement (Sigma), 20 µg/mL.
6. Heparin (Sigma), 100 µg/mL.
7. Glutamine (Sigma), 292 µg/mL.
8. CO$_2$ incubator.

2.3. Cryopreservation

1. 20% Dimethylsulphoxide in M199 with 10% horse serum kept cold on wet ice.
2. Cryovials (1-mL capacity).

2.4. Characterization of Human Brain Endothelial Cells

2.4.1. Immunoperoxidase Staining for FVIIIR:Ag and Ulex Europeaus I Lectin

1. Phosphate-buffered saline (PBS).
2. 0.05 M, Tris-HCl, pH 7.6, with and without Tween-20.
3. Endogenous peroxidase blocking solution: 1:4 3% H$_2$O$_2$ prepared from a dilution of 30% H$_2$O$_2$/100% Methanol.

 4. Fixative: 1:1 (v:v) acetone: 100% ethanol, refrigerated.
 5. Rabbit antiserum to FVIIIR:Ag (Dako), 1:100 dilution.
 6. Goat antirabbit IgG (Jackson Immunoresearch Laboratories), 1:400 dilution.
 7. UEA-1 lectin (Vector, Mississauga, Ontario), 1:400 dilution.
 8. Rabbit antiserum to UEA-1 lectin (Dako), 1:100 dilution.
 9. 3,3′-diaminobenzidine (DAB) (Sigma): 0.5 mg/mL Tris-HCl buffer without Tween 20.
10. Hematoxylin and base.

2.4.2. Uptake of Acetylated Low Density Lipoprotein (DiI-Ac-LDL)

1. DiI-Ac-LDL (Biomedical Technologies Inc., Stoughton, MA).
2. 3% formaldehyde in PBS.
3. 90% glycerol in PBS.
4. Fluorescence microscope with standard rhodamine excitation:emission filters.

2.4.3. Detection of Alkaline Phosphatase Activity

1. Fixative: 1 mL 37% formaldehyde, 97.5 mL distilled water, 0.5 g sodium barbital, 1.0 g anhydrous $CaCl_2$, pH 7.0.
2. Incubation buffer: 0.10 g sodium barbital, 0.196 g anhydrous $CaCl_2$, 0.12 g β-glycerophosphate (Sigma), 20 mL distilled water. Just before use, adjust to pH 9.0–9.2.
3. 1.5% lead nitrate solution: 0.225 g lead nitrate in 15 mL distilled water.
4. 2% ammonium sulfide solution (freshly made and kept in a fume hood): Make a 1/10 dilution of stock solution: 0.5 mL of 21.1% ammonium sulfide in 4.5 mL distilled water.
5. Hematoxylin and eosin for counterstaining.

2.4.4. Transmission Electron Microscopy (TEM) of HBMEC Cultures

1. Karnovsky's fixative: 2.5% glutaraldehyde and 2% paraformaldehyde in 0.2 M sodium cacodylate buffer, pH 7.35, refrigerated (*see* **Note 1**).
2. 1% Osmium tetroxide (OsO_4) (Marivac Ltd., NS) in 0.2 M sodium cacodylate buffer.
3. Sodium acetate buffer: 15 mL 0.2 N sodium acetate, 5 mL 0.2 N acetic acid and 20 mL dd H_2O. Keep refrigerated.
4. Uranyl magnesium acetate (Ted Pella Inc., Tustin, CA): 1 g uranyl Mg acetate, 25 mL 0.2 N sodium acetate, 75 mL dd H_2O. Adjust pH to 5.0 using glacial acetic acid. Keep refrigerated.
5. Graded series of methanol (30%, 50%, 70%, 95%, 100%).
6. Epon-Araldite (Fullam Inc., Latham, NY/Polysciences Inc., Warrington, PA).
7. Uranyl acetate (Fisher Scientific, Fair Lawn, NJ).
8. Lead citrate (Ladd, Research Ind., Burlington, VT).

2.4.5. Scanning Electron Microscopy (SEM) of HBMEC Cultures

1. Hank's balanced salt solution.
2. Sodium cacodylate buffer 0.05 M, pH 7.2.
3. Fixative: 2.5% glutaraldehyde in 0.05 M sodium cacodylate buffer pH 7.2.
4. 1% Osmium tetroxide (OsO_4) in 0.05 M cacodylate buffer.
5. 1% tannic acid in 0.05 M cacodylate buffer.
6. Graded series of ethanol up to 100% (10%, 30%, 50%, 70%, 100%).
7. 5% Uranyl acetate in 70% ethanol.
8. Liquid CO_2.

2.4.6. Measurement of Transendothelial Electrical Resistance

1. Endohm electrical resistance measuring apparatus (Volt-ohm meter) (World Precision Instruments, Sarasota, FL).
2. Sterile forceps.
3. Cellagen discs (ICN, Cleveland, OH).
4. Endothelial cell growth medium at 37°C (M199 supplemented with 10% horse plasma-derived serum (PDS) (*see* **Note 2**), 20 µg/mL endothelial cell growth supplement, 100 µg/mL heparin (*see* **Note 3**), 292 µg/mL glutamine, 25 mM HEPES, 10 mM sodium bicarbonate and antibiotics (100 µg/mL penicillin, 100 µg/mL streptomycin and 2.5 µg/mL amphotericin B).

2.4.7. Monolayer Permeability to Macromolecules

1. Horseradish peroxidase (Sigma) diluted at 1mg/mL in M199.
2. 3,3'-diaminobenzidine: 0.5 mg/mL in 50 mM Tris-HCl buffer (pH 7.6) with 0.01% H_2O_2.
3. Materials for TEM (*see* **Subheading 2.4.4.**).

3. Methods
3.1. Isolation of HBMEC

1. The cerebral cortex obtained from temporal lobectomy specimens or from normal brains at autopsy is transferred to the laboratory in cold M199 containing antibiotics (*see* **Note 4**).
2. After carefully removing the leptomeninges and their blood vessels, the brain tissue is chopped into 1–2-mm pieces in a sterile glass petri dish using two scalpels.
3. Place each 50 g of minced cortex in a 500-mL glass bottle containing 125 mL of Dispase (0.5%) and incubate for 3 h at 37°C in a shaking water bath (100–120 rpm) (*see* **Note 5**).
4. Centrifuge the contents of each bottle in 50-mL polystyrene tubes at 1000g for 10 min.
5. After centrifugation, discard the supernatants and suspend the pink pellets in 15% Dextran in M199 in 250 mL polycarbonate bottles, each containing 125 mL Dextran. Separation of microvessels from other tissue elements is accomplished by centrifugation at 5800g for 10 min. Wash the pellets containing the microvessels in M199 and collect by centrifugation.
6. In order to remove the basement membranes and pericytes, incubate the microvessel fragments in M199 + 5% horse serum containing 1 mg/mL collagenase/dispase in a shaking water bath (100 rpm) for 14–16 h at 37°C.
7. Centrifuge at 800g for 6 min, wash microvessels in M199 + 5% horse serum, pellet at 800g for 6 min and resuspend in 5% horse serum.
8. Layer 2 mL of the suspension over 35-mL Percoll gradients containing 50% Percoll in M199 *(11)* (*see* **Note 6**) in 50-mL polycarbonate tubes and centrifuge at 1000g for 10 min in order to separate endothelial cells from pericytes, red blood cells, and cellular debris.
9. Aspirate the band corresponding to a density of 1.052–1.055 g/mL containing the endothelial cells, with a syringe and 16-gage needle, suspend in M199 + 10% horse serum and collect by centrifugation at 1000g for 8 min. Repeat the wash with M199 + 10% horse serum.
10. Suspend the final pellet in M199 + 10% horse serum. Determine cell numbers by counting in a hemocytometer and cell viability by trypan blue dye exclusion. The yield from 2–4 g of surgical material is approx 4–7 × 10^5 cells and from autopsy brain (400–600 g) is 2 × 10^6 to 1 × 10^7 cells.

3.2. Cryopreservation

1. Dimethylsulfoxide (20% in 10% horse serum) is added slowly, dropwise, to the suspension of freshly isolated HBMEC (≈1 × 10^6 cells/mL) kept on wet ice, to a final concentration of 10%, and mixed gently.

2. The cells are frozen in 1-mL cryovials using controlled rate freezing vessels (–1°C/min) and stored in liquid nitrogen for periods up to 8 yr.
3. Required cells are removed from the liquid nitrogen storage tank and thawed rapidly in a 37°C water bath. One mL of cell suspension is gently transferred to a 50-mL centrifuge tube and 9 mL of M199 containing 10% horse serum are added dropwise.
4. The suspension is then centrifuged at 1000*g* for 8 min. After one more wash, the pellet is suspended in growth medium containing 10% horse serum, endothelial cell growth supplement, heparin, glutamine, and antibiotics and the cells are plated on fibronectin-coated wells as described below.

3.3. Establishment and Maintenance of Primary HBMEC Cultures

1. The freshly isolated small clumps of endothelial cells are seeded onto plastic wells coated with fibronectin (*see* **Note 7**) at a density of 50,000 cells/cm².
2. The endothelial cells are cultured in nutrient medium (*see* **Subheading 2.2.**) and maintained in a 5% CO_2/95% air incubator at 37°C.
3. Culture medium is changed every 2–3 d.
4. Four to six hours after plating, endothelial cells start migrating from the initial small clumps to form small islands and then start dividing. Confluent contact-inhibiting monolayers are formed 7–10 d after plating (**Fig. 1A**).

3.4. Characterization of Human Brain Endothelial Cells

3.4.1. Immunoperoxidase Staining for FVIIIR:Ag and Ulex Europeaus I Lectin

1. Wash cultures with warm M199, two times, 5 min each and then once with PBS for 5 min.
2. Fix in acetone: 100% EtOH (1:1) for 7 min at 4°C.
3. Air dry.
4. Wash with Tris-Tween-20 buffer three times, 3 min each.
5. Block endogenous peroxidase for 30 min in 3% H_2O_2: 100% MeOH (1:4).
6. Wash with Tris-Tween-20 buffer three times, 5 min each.
7. Incubate with 1:100 dilution of rabbit α-human FVIII antibody for 60 min at room temperature.
8. For the demonstration of UEA-1 lectin, incubate the cells with 1:400 dilution of UEA-1 lectin for 2 h at room temperature, wash with Tris-Tween-20 buffer three times, 3 min each, and incubate with rabbit α-human UEA-1 antibody at 1:100 dilution for 60 min at room temperature.
9. Wash with Tris-Tween-20 buffer three times, 5 min each.
10. Incubate with HRP-conjugated goat α-rabbit IgG at 1:400 dilution for 90 min at room temperature.
11. Wash with Tris-Tween-20 buffer two times, 5 min each.
12. Wash once with Tris buffer without Tween-20.
13. Incubate in DAB for 20 min, covered, at room temperature (*see* **Note 8**).
14. Wash with H_2O three times, 5 min each.
15. Stain with hematoxylin for 60 s. Wash. Leave in base for 60 s. Wash. (*See* **Fig. 1B**.)
16. Cover with crystal mount (*see* **Note 9**).

3.4.2. Uptake of Acetylated Low Density Lipoprotein (DiI-Ac-LDL)

1. Aseptically dilute the DiI-Ac-LDL to 10 μg/mL in M199.
2. Replace culture medium with the DiI-Ac-LDL solution and incubate for 4 h in a 37°C incubator (*9*).
3. Remove DiI-Ac-LDL from the media and wash cells five times, 3 min each with M199.

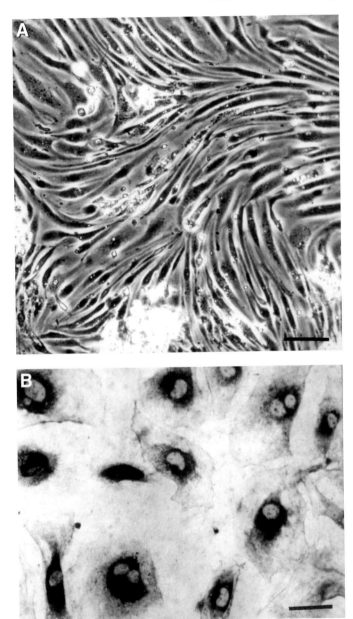

Fig. 1. (**A**) Human brain microvessel endothelial cells in primary culture form confluent, contact inhibiting monolayers composed of elongated cells. Refractile remnants of the initial clumps are still present 6 days after plating. (**B**) Immunoperoxidase staining for FVIIIR:Ag demonstrates intense, granular, predominantly perinuclear staining of all cells in primary culture. Bars: A = 14 µm, B = 2.3 µm.

4. Fix in 3% paraformaldehyde in PBS for 20 min at room temperature.
5. Wash for 5 s in dd H$_2$O.
6. Drain excess water onto Kim-wipe.
7. Cut the walls of the well using a hot scalpel blade.

8. Place a drop of 90% glycerol in PBS over the well containing the cultured cells and place a glass coverslip on top.
9. Visualize using standard rhodamine excitation filters.

3.4.3. Detection of Alkaline Phosphatase Activity

1. Wash wells three times with PBS at room temperature.
2. Fix cells (*see* **Subheading 2.4.3.**) for 1.5 h at 4°C.
3. Rinse cells two times 3 min each with cold distilled water.
4. Incubate cells in buffer for 2 h at 37°C.
5. Wash quickly three times with distilled water.
6. Add 1.5% lead nitrate for 5 min.
7. Rinse with two changes of distilled water.
8. Add 2% ammonium sulfide for 3.5 min.
9. Rinse three times, 5 min each with distilled water over a period of at least 20 min.
10. Counterstain with hematoxylin and eosin.

3.4.4. TEM of HBMEC Cultures

1. Wash cultures with warm (37°C) serum free M199 three times, 5 min each.
2. Fix in Karnowsky's fixative for 1 h at 4°C.
3. Wash with cold 0.2 M sodium cacodylate buffer, pH 7.4, three times, 10 min each.
4. Post-fix in 1% OsO_4 in 0.2 M sodium cacodylate buffer for 1 h at 4°C.
5. Wash with cold 0.2 M sodium cacodylate buffer three times, 10 min each.
6. Wash with cold acetate buffer for 30 min (three changes).
7. Block stain with uranyl Mg acetate overnight at 4°C.
8. Wash with cold acetate buffer for 30 min (three changes).
9. Dehydrate in graded series of methanol (*see* **Subheading 2.4.4.**).
10. Embed in Epon-Araldite (*see* **Note 10**) (**Fig. 2**).

3.4.5. SEM of HBMEC Cultures

1. Wash confluent cultures grown on plastic wells with warm (37°C) Hank's balanced salt solution three times, 5 min each (**10**).
2. Fix cells in 2.5% glutaraldehyde in 0.05 M sodium cacodylate buffer pH 7.2 for 1 h at 4°C.
3. Wash with 0.05 M sodium cacodylate buffer for 30 min (three changes).
4. Post-fix in 1% OsO_4 in 0.05 M sodium cacodylate buffer for 1 h at 4°C.
5. Wash with 0.05 M sodium cacodylate buffer for 30 min (three changes).
6. Stain with 1% tannic acid in 0.05 M sodium cacodylate buffer for 1 h.
7. Wash with 0.05 M sodium cacodylate buffer for 30 min (three changes).
8. Dehydrate in graded series of ethanol (10%, 30%, 50%, 70%) for 10 min each.
9. Stain with 5% uranyl acetate in 70% ethanol overnight.
10. Wash with 70% ethanol three to five times (10 min each) until ethanol is clear.
11. Dehydrate in 90% ethanol for 12 min, 100% ethanol for 30 min (3×10 min changes).
12. Critical point dry with liquid CO_2. Place in dessicator.
13. Coat dry specimen with gold (**Fig. 3**).

3.4.6. Measurement of Transendothelial Electrical Resistance

1. Cultivate HBMEC to confluence on permeable substrate (Cellagen discs) in a double chamber.
2. Sterilize the ohm-meter chamber for 30 min with 70% ethanol. Dry.

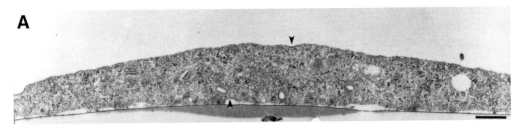

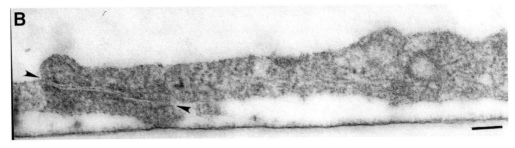

Fig. 2. Transmission electron micrographs of HBMEC in primary culture. (**A**) Endothelial cells are elongated with long, thin processes, dense cytoplasm and no apparent cytoplasmic vesicles. Clathrin-coated pits are present along the apical and basal cell surface (arrows). (**B**) Adjacent endothelial cells are bound together by tight junctions (arrows). A discontinuous basal lamina-like material is present along the basal cell surface. Bars: A = 0.57 μm, B = 0.1 μm.

3. Add 1 mL warm (37°C) M199 to chamber. Replace top of chamber and read resistance as zero (background).
4. Transfer the cellagen disc from the double chamber culture system to the Volt-ohm chamber. Secure lid. Read resistance.
5. Calculate the transendothelial electrical resistance by subtracting the resistance of the cellagen disc without endothelial cells (*see* **Note 11**).
6. Remove lid and return HBMEC-coated cellagen disc to tissue culture plates.

3.4.7. Monolayer Permeability to Macromolecules

1. Cultures are grown to confluence on 4 or 24 well plates.
2. Gently wash monolayers with warm (37°C) serum-free M199 three times, 5 min each.
3. Fix, without prior washing, in Karnovsky's fixative (*see* **Subheading 2.4.4.**) for 1 h at 4°C.
4. Wash with 0.2 M sodium cacodylate buffer for 30 min (three changes).
5. Incubate cultures with 3,3'-diaminobenzidine for 1 hr at 4°C.
6. Wash with 0.2 M sodium cacodylate buffer for 30 min (three changes).
7. Post-fix in 1% OsO_4 in 0.2 M sodium cacodylate buffer for 30 min.
8. Wash with 0.2 M sodium cacodylate buffer for 30 min (three changes).
9. Wash with cold acetate buffer for 30 min (three changes).
10. Block stain in uranyl Mg acetate at 4°C overnight.
11. Wash with cold acetate buffer for 30 min (three changes).
12. Dehydrate in graded series of methanol (*see* **Subheading 2.4.4.**).
13. Embed in Epon-Araldite.
14. Examine thin sections without heavy metal staining (**Fig. 4**).

3.5. Applications

Primary cultures of HBMEC form highly organized, contact-inhibiting monolayers that express human endothelial cell markers and maintain important morphological and

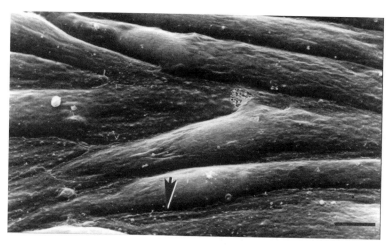

Fig. 3. Scanning electron microscopy of endothelial cells 6 d after plating. Closely packed cells display marginal folds in areas of contact (arrow). Bar = 10 μm.

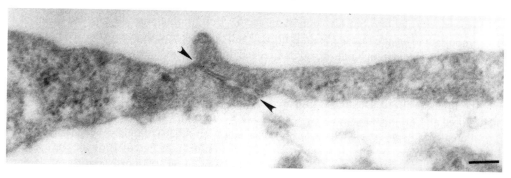

Fig. 4. Confluent monolayers were incubated with HRP for 5 min and then processed for electron microscopy. Tight junctions along intercellular contacts prevent the passage of HRP (arrows). The basal cell surface contains no tracer. Bar = 0.14 μm.

functional properties of brain endothelium in vivo including the presence of tight junctions and paucity of cytoplasmic vesicles. Confluent monolayers have high electrical resistance and exclude the paracellular passage of macromolecules. This in vitro system can be used to investigate the expression, function, and regulation of surface-associated and intracellular molecules, gene expression, the barrier and transport functions of cerebral endothelial cells, the structure and function of tight junctions, and the immunological properties of the human cerebral endothelium.

4. Notes

1. Prepare Karnovsky's fixative by first dissolving 2 g paraformaldehyde in 50 mL dd H_2O heated in water bath to 65–70°C. Add NaOH drop by drop until the solution becomes crystal clear. Cover and cool in refrigerator. Add 3.5 mL of 70% glutaraldehyde stock solution, 46.5 mL 0.2 M sodium cacodylate buffer and 0.0120 g $CaCl_2$. Adjust pH to 7.15. Filter. The fixative should be made fresh prior to use.
2. HBMEC in primary culture form compact sheets of elongated, closely associated cells that exhibit density-dependent inhibition of growth after reaching confluence. Occasionally,

13. Thornton, S. C., Mueller, S. N., and Levine, E. M. (1983) Human endothelial cells: use of heparin in cloning and long-term serial cultivation. *Science* **222,** 623–625.
14. Bowman, P. D., Betz, A. L., Ar, D., Wolinsky, J. S., Penney, J. B., Shivers, R. R., and Goldstein, G. (1981) Primary culture of capillary endothelium from rat brain. *In Vitro* **17,** 353–362.

22

Human Brain Microvessel Endothelial Cell and Leukocyte Interactions

Jacqueline Shukaliak-Quandt, Donald Wong, and Katerina Dorovini-Zis

1. Introduction

The recruitment of leukocytes from the blood into secondary lymphoid and peripheral organs is a key process in both leukocyte homeostasis and the initiation and maintenance of immune responses (1). Within the past decade, several important advances have been made in identifying factors involved both in normal leukocyte homing and recruitment to sites of inflammation. These studies have identified adhesion molecules as instrumental in tethering leukocytes to endothelial surfaces and potentiating subsequent adhesion and migration. In addition, soluble chemoattractant cytokines (known as chemokines) have been identified that can direct the site and often the composition of the inflammatory infiltrate, as well as proteinases, which dismantle the barrier that leukocytes are crossing (2). In several of these studies, in vivo models have been used to block the activity of specific molecules in order to determine the global outcome and the involvement of these molecules in secondary lymphoid organ trafficking or disease pathogenesis. Such models, however, become of limited use when trying to focus on relative contributions to specific events such as direct cell–cell interactions occurring at the onset of inflammatory cell extravasation. Because of the complicated nature and multitude of cells and factors present in in vivo systems, in vitro models have become a powerful and more simplistic way of analyzing events by allowing for better control of the environment. For this purpose, in vitro models employing endothelial monolayers have been instrumental in trying to characterize factors which influence vascular permeability and the sequence of events during leukocyte-endothelial adhesion and transmigration. By culturing organ-specific endothelial cells on extracellular matrices in either one-dimensional (planar) or two-dimensional (double chamber) systems, one can create a model that mimics in vivo leukocyte-endothelial interactions across a blood–tissue interface. To ensure the relevance of such models it is important that the cultured cells maintain most, if not all, of the in vivo characteristics of the endothelial barrier under study.

In vitro models have been particularly useful in studying interactions between leukocytes and endothelial cells forming the blood–brain barrier (BBB) (3–5). Increased per-

From: *Methods in Molecular Medicine, vol. 89:*
The Blood–Brain Barrier: Biology and Research Protocols
Edited by: S. Nag © Humana Press Inc., Totowa, NJ

2.7. PMN Migration Across HBMEC

1. Confluent HBMEC monolayers grown on collagen discs.
2. M199. Warm to 37°C.
3. M199 with 10% HS. Filter sterilize.
4. Karnovsky's fixative: 2.5% glutaraldehyde and 2% paraformaldehyde in 0.2 M sodium cacodylate buffer (*see* Chapter 21 for details on preparation).
5. 1% OsO_4. Make 2 mL for each collagen disk.

3. Methods

This chapter will describe 1) how to establish confluent primary HBMEC cultures, 2) how to isolate T-cells and their subsets, 3) the T-cell migration assay, 4) PMN isolation, 5) how to assess PMN adhesion to HBMEC, and 6) how to assess PMN migration across HBMEC.

3.1. Establishment of Confluent Primary HBMEC Cultures on Cellagen Disks

1. Fill the appropriate wells of a 24-well plate with enough ECM to cover the well and the bottom of the cellagen membrane once inserted (approx 350 µL) (*see* **Note 1**).
2. In a laminar flow hood, open the individually packaged sterile inserts and with sterile forceps place the inserts into the wells of the 24-well plate.
3. Pre-wet the membrane by adding approx 200 µL of ECM to the tissue culture insert (*see* **Note 2**).
4. Return plate to incubator for at least 30 min to allow for stabilization/wetting of the collagen membranes.
5. During this time, cryopreserved endothelial cells can be thawed and enumerated for plating. Resuspend cells to a sufficient plating concentration for the required surface area (at least 50,000 cells/cm^2) in a volume of 200 µL/cellagen disc.
6. Gently aspirate the 200 µL of ECM from the tissue culture insert using a glass pipet, taking extra care not to touch or damage the membrane. Immediately add the HBMEC, and using the sterile forceps gently swirl the insert to allow for even dispersal of the cells over the membrane. Return the insert to the 24-well plate.
7. Return the cells to the incubator. Change ECM every other day (*see* Chapter 21) until the monolayers become semi-confluent (*see* **Note 3**). Transendothelial resistance can then be measured as described in Chapter 21.

3.2. Isolation of T-Cells and T-Cell Subsets

1. Draw 6.5 mL of venous blood into EDTA tubes.
2. Layer blood carefully over an equal volume of Histopaque 1077 in 10 mL-polycarbonate tubes and centrifuge at 400g for 30 min at room temperature.
3. Aspirate and discard the plasma layer.
4. Aspirate the interface into 10 mL-polycarbonate centrifuge tubes containing approx 5 mL PBS. Centrifuge at 250g for 10 min.
5. Decant supernatants. Resuspend cells and pool in approx 5 mL of PBS. Centrifuge at 250g for 10 min.
6. Decant supernatants and wash cells once more in PBS. Centrifuge at 250g for 10 min.
7. Suspend white cell pellet in 2% FCS in PBS.
8. Count in hemocytometer and determine viability by the trypan blue exclusion test.
9. Pass the lymphocyte suspension through human T-cell recovery columns to negatively select for CD8 and CD4 T-cell subsets according to the manufacturer's recommendations.

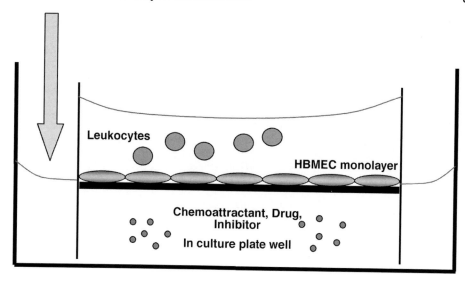

Fig. 1. Human brain microvessel endothelial cells (HBMEC) monolayers grown on collagen membrane inserts form an in vitro model of the blood–brain barrier. HBMEC are grown to confluence on a permeable collagen membrane devoid of pores. Leukocyte adhesion and transendothelial migration can be studied in the presence of inhibitors or chemoattractants that are added to the upper or lower chamber representing the intravascular and central nervous system compartments, respectively.

3.3. T-Cell Adhesion Assay

1. Isolate or thaw previously cryopreserved T-lymphocytes or T-lymphocyte subsets, washing twice. Resuspend in appropriate volume of TCM to 1×10^6 cells/mL.
2. Wash the upper and lower chambers of the chemotaxis system gently with 10% HS in M199 to remove any cell debris (*see* **Note 4**).
3. Replace media in the lower chamber with 350 µL of 10% HS in M199 at 37°C (with or without the addition of potential chemoattractants).
4. Aspirate media from the upper chamber carefully to avoid disturbing the HBMEC monolayer.
5. Add immediately 100 µL of T-cell suspension containing $1–2 \times 10^6$ cells/mL to the upper chamber (**Fig. 1**). Use the forceps to swirl the insert to allow for even dispersal of the T-cells across the monolayer. After an interval of 2 min add the lymphocytes to the inserts (*see* **Note 5**).
6. Return the cells to the incubator for 1 h. T-cells and HBMEC monolayer can be viewed and/or photographed periodically throughout this time, or before fixation (**Fig. 2A**).
7. Prepare wash buffers and fixatives.
8. Wash gently the non-adherent cells four times at the four lateral walls (three times with 37°C M199 and once with 37°C PBS).
9. Fix top and bottom of collagen membrane in fixative by quickly aspirating the PBS in the upper and lower chambers and replacing with cold fixative.
10. Fix for 7 min at room temperature. Remove the fixative from both chambers and air dry.
11. Wash upper and lower chambers with Tris buffer with Tween-20, twice for 3 min each.
12. Block endogenous peroxidase with 100 µL of 1:4 3% H_2O_2 : 100% MeOH for 30 min at room temperature (*see* **Note 6**).

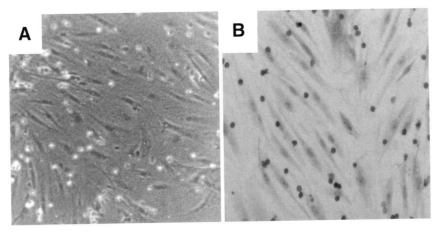

Fig. 2. (**A**) Phase contrast microscopy allows monitoring of lymphocyte adherence to human brain microvessel endothelial cells (HBMEC) after 1 h. The darker, spindly cells represent the HBMEC, whereas the small, refractile cells represent adherent lymphocytes. (**B**) LCA positive adherent T-cells are readily enumerated against a backdrop of blue HBMEC nuclei in a confluent monolayer following washing, fixation, and immunohistochemistry.

13. Wash with Tris containing Tween-20, three times for 5 min each.
14. Incubate with HRP-linked-mouse anti-human leukocyte common antigen antibody for 45 min with agitation on the shaker (1 min) every 15 min (*see* **Note 7**).
15. Wash cells three times for 5 min each with Tris (without Tween-20).
16. Incubate with DAB for 20 min at 4°C, covered.
17. Wash three times for 5 min each with ddH$_2$O.
18. Counterstain with hematoxylin for 20 s (*see* **Note 8**).
19. Wash three times with ddH$_2$O. Aspirate H$_2$O. Add base for 20 s. Wash.
20. Cut the collagen membranes from the plastic walls using a scalpel and transfer onto labeled glass slides.
21. Dehydrate with five dips successively in 95% EtOH, 100% EtOH, and a quick dip in xylene before adding one drop of Entellan non-aqueous mounting medium. Avoid bubbles, coverslip, and allow to dry (*see* **Note 9**).
22. Count the number of LCA positive T-cells adherent to the HBMEC monolayer in one central and four to eight peripheral fields using a 1 mm^2 ocular grid and a 20× objective (**Fig. 2B**).

3.4. T-Cell Migration Assay

1. Follow **steps 1–5** of the adhesion assay.
2. Return cells to the incubator for a 3–4 h incubation period (*see* **Note 10**).
3. Wash cells gently as in **Subheading 3.3., step 8**.
4. Fix in Karnovsky's fixative for 1 h at 4°C.
5. Wash with 0.2 *M* sodium cacodylate buffer three times for 10 min each.
6. Fix in 1% OsO$_4$ for 1 h at 4°C, sealed with parafilm.
7. Wash in 0.2 *M* sodium cacodylate buffer three times for 10 min each.
8. Wash in Na acetate buffer three times for 10 min each.
9. Block stain with uranyl *M*g acetate overnight at 4°C.
10. Wash in Na acetate buffer three times for 10 min each at 4°C.

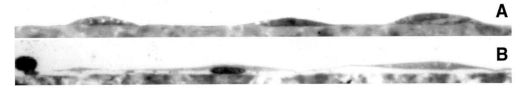

Fig. 3. **(A)** 1-μm thick cross sections through human brain microvessel endothelial cells (HBMEC) monolayer grown to confluence on collagen membrane are stained with toluidine blue and analyzed under a light microscope. The nuclei of three HBMEC are readily visible. Their processes become thin as they reach toward and contact other cells in the monolayer. **(B)** Two intensely stained lymphocytes are visible against the pale cytoplasm of HBMEC. One lymphocyte is in the process of migrating by inserting a cytoplasmic process between adjacent HBMEC, whereas another has already migrated and is seen sandwiched between the HBMEC monolayer and the dense collagen membrane.

11. Dehydrate in graded series of 30%–90% Methanol solutions at 4°C as follows:
    a. 30% MeOH twice for 3 min each.
    b. 50% MeOH twice for 3 min each.
    c. 70% MeOH twice for 3 min each.
    d. 90% MeOH twice for 6 min each.
    e. 100% MeOH three times for 10 min each at room temperature.
12. Add Epon-Araldite (EA) mixtures as follows:
    a. 60:40 EA:MeOH overnight.
    b. 80:20 EA:MeOH overnight.
    c. 100% EA overnight.
13. Drain the EA, remove and cut the collagen membranes into strips (1–1.5 mm).
14. Align cell side up in plastic embedding molds. Add fresh EA to molds, and polymerize at 60°C in a vacuum oven for 3 d.
15. Cut 100 serial semithin sections having a thickness of 1 μm at 40-μm intervals of the block.
16. Stain with toluidine blue for 15–30 s, rinse with water, and dry.
17. Calculate the total number of adherent/migrated cells per mm of membrane (**Fig. 3**).
18. Cut and stain the thin sections with uranyl acetate and lead citrate for transmission electron microscopy examination (**Fig. 4**).

3.5. PMN Isolation

1. Draw venous blood into a EDTA vacutainer.
2. Allow the blood and the Lymphocyte-poly to reach room temperature.
3. Layer the blood carefully onto the same volume of Lympholyte-poly in a 15-mL tube. Do not disrupt the Lympholyte-poly beneath or mix the blood with the Lympholyte-poly.
4. Centrifuge at 550*g* for 40 min at room temperature.
5. Aspirate and discard the distinct mononuclear band on top of the PMN band. Collect the PMN band, without getting near the RBC pellet, from each tube into individual graduated 15-mL tubes. Note the volume collected.
6. Add equal volume of 0.45% NaCl (0.45 g NaCl in 100 mL ddH$_2$O) and double the volume of 0.9% NaCl (0.9 g NaCl in 100 mL ddH$_2$O) to the PMNs. Resuspend and spin at 400*g* for 10 min.
7. Aspirate supernatant. Pool pellets. Repeat the wash with 8 mL 0.9% NaCl.
8. Suspend the PMN pellet in 1 mL 10% HS in M199, count in a hemocytometer and determine viability with the trypan blue exclusion test.

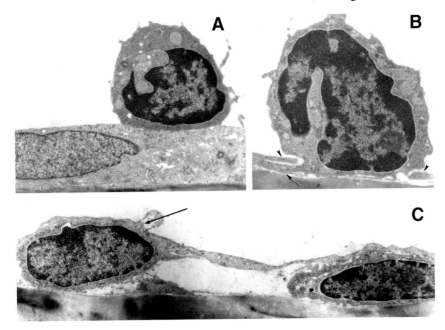

Fig. 4. Cross-sections of the human brain microvessel endothelial cells (HBMEC) monolayer on collagen membrane are viewed under an electron microscope. (**A**) An adherent lymphocyte with several visible processes is attached to the HBMEC monolayer. (**B**) Following firm adhesion, this lymphocyte begins its transendothelial migration by inserting a long cytoplasmic process (arrow) between the processes of two HBMEC (arrowheads) that have separated but are in close contact with the migrating T-cell. (**C**) Two lymphocytes have migrated across the in vitro BBB. The cell to the left has just migrated and a small cytoplasmic projection appears caught between two endothelial cell processes (arrow).

9. 100 µL of PMN at 6×10^6 cells/mL are required for each well of a 24-well plate (i.e., 6×10^5 cells per well) (*see* **Note 11**).

3.6. PMN Adhesion to HBMEC

1. Grow HBMEC to confluence on fibronectin coated 24-well plates. Confluence of the monolayers can be assessed visually with an inverted phase contrast light microscope.
2. Wash EC with M199 (37°C) three times for 5 min each. Makeup adhesion blocking antibodies in 10% horse serum in M199 at 37°C.
3. Isolate PMNs using lympholyte-poly. Count cells using a hemacytometer. Dilute to 6×10^6 cells/mL in 10% HS in M199. If adhesion blocking antibodies are to be used, incubate EC with blocking antibody at 37°C for 30 min and wash once with M199 at 37°C.
4. Place 100 µL of PMN suspension over EC in each well. Work as quickly as possible. Shake the plate gently to evenly distribute the cells. Incubate for up to 30 min at 37°C.
5. Wash with warm M199 at each of the four corners of each well.
6. Fix with 1 : 1 acetone : ethanol (100%) for 7 min at 4°C. Air dry (approx 10 min).
7. Wash with Tris Tween-20 buffer two times for 3 min each.
8. Block endogenous peroxidase for 30 min at room temperature. Wash with Tris, three times for 5 min each.
9. Incubate with anti-LCA antibody (180 µL per well) in Tris Tween-20 buffer for 90 min at room temperature with slight agitation on an orbital shaker table for 1 min every 15 min.

10. Wash cells three times for 5 min each with Tris without Tween-20.
11. Incubate with DAB for 20 min at 4°C, covered with foil.
12. Wash three times for 5 min each with ddH$_2$O.
13. Counterstain with hemotoxylin.
14. Wash three times with ddH$_2$O.
15. Aspirate H$_2$O. Leave in base for 20 s.
16. Wash with ddH$_2$O. Air dry. Count the number of adherent LCA positive PMNs in four peripheral fields and one central field. Express as number of cells per mm^2.

3.7. PMN Migration Across HBMEC

1. Grow HBMEC to confluence on collagen membranes.
2. Isolate PMNs and resuspend at 6×10^6 cells/mL for each cellagen disc.
3. Wash EC with M199 at 37°C three times for 5 min each.
4. Add 200 μL PMNs to each cellagen disc. Incubate for 2–3 h at 37°C.
5. Wash cells four times for 5 min each with warm M199.
6. Fix in Karnovsky's fixative for 1 h at 4°C. Seal with parafilm (*see* **Note 12**).
7. Wash in 0.2 *M* sodium cacodylate buffer three times for 10 min each.
8. Post-fix in 1% OsO$_4$ in 0.2 *M* sodium cacodylate buffer for 1 h at 4°C, sealed with parafilm.
9. Wash in 0.2 *M* sodium cacodylate buffer three times for 10 min each, then in Na acetate buffer three times for 10 min each at 4°C.
10. Block stain with uranyl *M*g acetate overnight at 4°C.
11. Wash in Na acetate buffer three times for 10 min each at 4°C.
12. Dehydrate in graded series of methanol and embed in Epon-Araldite as given in **Subheading 3.4., steps 11** and **12**.

3.8. Applications

Cultivation of HBMEC on a permeable substrate in a double chamber chemotaxis system provides a useful and reliable in vitro system for studying the interactions between peripheral blood leukocytes and cerebral endothelium under closely controlled conditions. This system allows for the investigation of the effects of various chemoattractants and inflammatory mediators present in the upper (luminal side) or lower (abluminal side) chambers on leukocyte adhesion and transendothelial migration, monitoring of the junctional permeability during transmigration, and the step-by-step dissection of the migration process. Localization of membrane-associated molecules on endothelial cells and/or leukocytes by light and electron microscopy can be performed and further contribute to our understanding of the cell trafficking events at the human BBB.

4. Notes

1. Tissue culture inserts or transwell systems typically consist of a plastic insert with a membranous or porous bottom that fits inside regular culture plates. A wide variety of inserts are available commercially that vary in two main membrane parameters: pore size and coating. Several employ uncoated membranes with or without pores ranging from very small up to 10 μm in size. Depending on the goal of the assay, some are more appropriate than others. For a pure chemotaxis assay and to assess the attractant properties of a molecule, one might choose an uncoated membrane with a 3- or 10-μm pore size to allow the passage of cells from one chamber to the other. Transwell inserts with pore sizes less than 1 μm are most applicable to cell co-culture experiments where cell-to-cell contact is undesirable when searching for soluble mediators. The coating or consistency of the membrane

can also be important in the design of experiments. In cell migration assays, components of the basement membrane may be required for adhesive and migratory events. In addition, when endothelial cells are to be cultured on the membrane, basement membrane components are often required for reasonable growth. Commercially available inserts come with coatings ranging from fibronectin, collagen, or laminin to combinations of these to form a complex extracellular matrix on which cells are cultured. The inserts described here consist of a dense, solubilized type I collagen membrane. The lack of pores restricts the passage of cells through the membrane, but still allows for the diffusion of soluble mediators (**Fig. 1**). HBMEC grown on these collagen membranes display similar characteristics as those cultured on fibronectin in plastic wells. Importantly, these cells form a confluent, contact-inhibiting monolayer that displays high transendothelial electrical resistance. In migration assays leukocytes traverse the HBMEC monolayers and are then sandwiched between the collagen layer and the EC, which allows for T-cell enumeration in cross section. It is also important to note that while this procedure describes the use of inserts to fit a 24-well plate system that is compatible with the use of an ohmmeter, transwell systems are available in both larger and smaller formats that can fit in a 96-well plate. The latter requires less cells and may be more suitable if Cr^{51} labeling of T-cells and subsequent lysis or other quantitative techniques are used.

2. Prewetting of the membrane is essential for an even distribution of the seeded clumps of HBMEC across the collagen membrane. Clumps of cells will stick to the membrane if added to a dry surface, and will not spread evenly over the entire surface.

3. Cells should be monitored daily under a phase contrast microscope. As described in Chapter 21, cells will migrate out from the initial clump, proliferate, and form a confluent monolayer. Once the cultures become subconfluent, daily measurements of the electrical resistance across these monolayers can be performed.

4. Gentle washing of the HBMEC monolayer is important to remove cellular debris that could otherwise interfere with the adhesion assay. Utmost care should be taken not to disturb the monolayer, as gradients established in subsequent steps could be greatly compromised should areas of the monolayer be disrupted. This step is particularly important if inflammatory cytokines or other compounds have been added to alter the permeability/activation status of the monolayers, as these compounds could have effects on the added leukocytes.

5. While the addition of the leukocytes to the monolayer can be done quickly, the steps required to later remove non-adherent cells and fix the monolayer require much more time. Two minutes is an estimate of the time later required to wash the monolayer at the four lateral walls and add fixative, but will depend on the investigator and should be determined by trial runs. If performing experiments on 20 transwells for a 1 h adhesion assay, on the first few inserts it may matter less to be off by a minute or two, but by the time one reaches the 20th insert, leukocytes here may have been adherent for up to an extra hour, thus substantially affecting the timing of the assay.

6. Endogenous peroxidase activity should be blocked during immunocytochemistry if an HRP-linked secondary antibody is to be used for immunostaining. The standard 3% solution of H_2O_2 in methanol is very effective. Exercise caution and wear gloves when handling 30% H_2O_2, which is caustic and causes burns.

7. Antibodies to surface markers on leukocytes can be general or specific, depending on the assay. An antibody to CD45, a pan-leukocyte marker, will stain all leukocytes but could be substituted for markers to stain T-cells alone (CD3), CD4 or CD8 T-cell subsets, B cells, or monocytes/macrophages. These can be purchased unlabeled, labeled, or linked to an enzyme such as HRP. Use of the former will require the additional step of a secondary antibody, such

as goat anti-mouse antibody, which would have to be labeled or linked to an enzyme. This decision should be made after consideration of both time and cost. Labeled antibodies require 30–45 min incubation, and after washing one can proceed directly to the reagent/development step. The use of a primary and secondary antibody can take longer, but also allows for amplification of the overall staining intensity because several enzyme-linked secondary antibodies can typically bind to a single antigen-bound primary antibody. Most importantly, the antibody of choice must be suitable for immunohistochemistry, particularly on acetone/ethanol fixed tissues. While it is true that antibodies to fresh or ethanol/acetone fixed tissues are typically more readily available than are those to formalin-fixed antigens, it is important to obtain such information if available from the manufacturer. Finally, the antibody concentration is often recommended by the manufacturer, but will have to be titrated to maximize the signal-to-noise ratio with proper negative and positive controls.

8. Hematoxylin is used as a counterstain to stain the nuclei of the endothelial cells and lymphocytes. This is particularly important in order to ensure that the integrity of the monolayer has been maintained throughout the assay period. Different batches of hematoxylin can result in different staining intensities, so try to use a single batch for staining consecutive experiments. The collagen membranes themselves do not take up color even after 40 s of staining, however different types of membranes with different coatings would have to be pre-tested.

9. Proper preservation of the monolayers is important for effective enumeration, analysis, and photography. It is strongly recommended to quantitate and photographically archive the leukocyte adhesion to HBMEC monolayers as rapidly as possible. While a non-aqueous mounting medium as applied here can be very effective following dehydration, the proper mounting medium is typically dictated by the immunostain employed. For binary substrates more typical of immunohistochemistry, aqueous based or non-aqueous mounting media can be used but should be tested experimentally with each different immunostain. Be certain that no air bubbles are introduced when adding the mounting medium. Carefully angle the cover slip, contacting the media at one edge of the slide, and lower it such that surface tension pulls the mounting media across the monolayer in an even front, until it finally rests, without the introduction of air bubbles. If bubbles are present, one can try immediately to tap them out, however this is often ineffective and it is usually best to discard the original coverslip and start with a new one.

10. The initial steps of the adhesion and migration assays are identical, and differ for the most part only in the length of the incubation time. Typically, lymphocyte adhesion to EC monolayers occurs between 30 min and 1 h, whereas the process of migration requires 3–4 h (*5*). However, the type of leukocyte, its activation state, the permeability/activation status of the monolayer, or the presence of chemoattractants can all serve to expedite or decrease the rates of leukocyte adhesion and migration.

11. 3 mL of peripheral vein blood yields 6×10^6 PMN on average. However, this will vary with the individual and the laboratory and is best determined with a few trial runs before the actual experiment. It is critical for both the blood and the Lympholyte-poly to be at room temperature, and to mix the contents of the Lympholyte-poly bottle thoroughly before use. Otherwise, the viscosity of the Lympholyte-poly will be altered and the cells will not successfully separate. This is a good time to start blocking the HBMEC with adhesion/migration blocking reagents if they are being used (*see* **Subheading 3.6.** for more detail). DAB is a carcinogen. Use in a fume hood with care. For the disposal of DAB, refer to chapter 21 (*see* **Subheading 5.8.**).

12. Fixatives, especially OsO_4, are toxic. Preferably, place cells with fixative in a well-sealed dessicator before removing from fume hood to the fridge.

Acknowledgments

This work was supported by grants from the CIHR (MOP 42534) and the Multiple Sclerosis Society of Canada.

References

1. Jutila, M. A., Berg, E. L., Kishimoto, T. K., et al. (1989) Inflammation-induced endothelial cell adhesion to lymphocytes, neutrophils, and monocytes. Role of homing receptors and other adhesion molecules. *Transplantation* **48,** 727–731.
2. Madri, J. A., and Graesser, D. (2000) Cell migration in the immune system: the evolving inter-related roles of adhesion molecules and proteinases. *Dev. Immunol.* **7,** 103–116.
3. Huynh, H., and Dorovini-Zis, K. (1993) Effects of interferon-gamma on primary cultures of human brain microvessel endothelial cells. *Am. J. Pathol.* **142,** 1265–1278.
4. Shukaliak, J., and Dorovini-Zis, K. (2000) Expression of the beta chemokines RANTES and MIP-1β by human brain microvessel endothelial cells in primary culture. *J. Neuropath. Exp. Neurol.* **59,** 339–352.
5. Wong, D., Prameya, R., and Dorovini-Zis, K. (1999) In vitro adhesion and migration of T lymphocytes across monolayers of human brain microvessel endothelial cells: regulation by ICAM-1, VCAM-1, E-selectin and PECAM-1. *J. Neuropath. Exp. Neurol.* **58,** 138–152.
6. Hickey, W. F. (1999) Leukocyte traffic in the central nervous system: the participants and their roles. *Semin. Immunol.* **11,** 125–137.
7. Raine, C. S. (1994) The Dale E. McFarlin Memorial Lecture: the immunology of the multiple sclerosis lesion. *Ann. Neurol.* **36,** S61–S72.

23

Development and Characterization of Immortalized Cerebral Endothelial Cell Lines

Pierre-Olivier Couraud, John Greenwood, Françoise Roux, and Pete Adamson

1. Introduction

Several groups have reported the isolation and in vitro pharmacological characterization of brain microvessel endothelial cells (BMECs) of various origins: bovine, porcine, murine, or human *(1–3)*. Brain capillaries are almost completely ensheathed by astrocyte processes, which are believed to provide the cerebral endothelium with specific stimuli responsible for the development and maintenance of the blood–brain barrier (BBB) phenotype *(4)*. Considering that the absence of such environmental stimuli might prevent BMECs from retaining in culture the fully differentiated phenotype of the BBB found in vivo, some investigators have proposed a co-culture system where BMECs are grown in the presence of primary astrocytes or astrocyte-conditioned medium. In these experimental conditions, in vitro models of the BBB, essentially based on bovine or porcine BMECs, have been both proposed and validated *(5,6)*. However, a number of drawbacks still limit the extensive use of these models in basic research and in drug-screening processes: 1) their use is time consuming and needs a considerable know-how which may hamper their routine use in nonexpert laboratories, 2) most available models are based on bovine or porcine BMECs, which may constitute a serious limitation for immunological studies or for studies with species-specific bacterial pathogens like *Neisseria meningitidis* or viruses like HIV and HTLV-I, and 3) BMECs rapidly de-differentiate in vitro, losing the characteristics of BBB endothelial cells after a few passages in culture, which limits their use for biochemical or pharmacological studies. In order to address these drawbacks, numerous efforts have been made over many years to establish continuous, immortalized cell lines with the capacity to retain in culture a stable phenotype reminiscent of BBB endothelium in vivo. The production and characterization of the rat RBE4 *(7)* and GP8/3.9 *(8)* cell lines will be used to illustrate the methods utilized for the immortalization of brain endothelial cell lines as a model for cerebral endothelium and as a tool for studying BBB biology.

From: *Methods in Molecular Medicine, vol. 89:*
The Blood–Brain Barrier: Biology and Research Protocols
Edited by: S. Nag © Humana Press Inc., Totowa, NJ

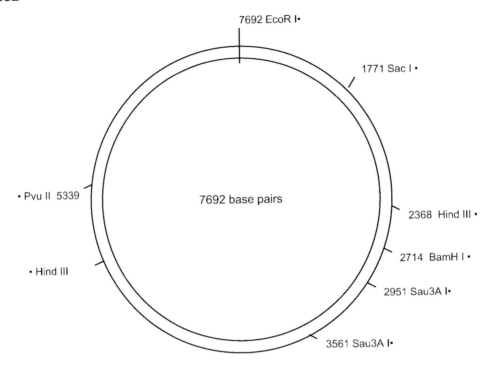

Sequences of special interest	DNA elements
6 - 1777	Adenovirus immediate early E1A region
1778 - 2367	Adenovirus DNA
3566 - 5056	Neomycin resistance (aminoglycoside-3'-O- phosphotransferase) gene

Fig. 1. The immortalizing plasmid pE1A/*neo*. The plasmid pE1A/neo carries the E1A region of Adenovirus 2 and the neomycin-resistance gene for selection by resistance to the aminoglycoside antibiotic G418.

cytes and fibroblasts were removed under the microscope with a modified glass pasteur pipet.

6. Thus primary cultures of rat brain endothelial cells were estimated to be close to 100% pure by morphological criteria (*see* **Note 3**).

7. Cells were seeded at 10^4 cells/cm² onto Type I collagen-coated dishes, in α-Medium/Ham's F10 (1:1), (*see* **step 9b** in **Subheading 2.2.**) and incubated for 3 d in a 37°C, 5% CO_2 humidified incubator, until confluence (*see* **Note 4**).

3.3. Transduction of the Immortalizing Gene

A flow diagram of the immortalization process is summarized in **Fig. 2**.

3.3.1. Transfection of Rat Brain Endothelial Cells with pE1A/Neo

Cultures of rat brain endothelial cells were transfected after two passages by the calcium phosphate precipitation procedure with the plasmid pE1A/*neo*. After transfection, cells were grown in the selective medium containing 300 µg/mL G418. Only a few

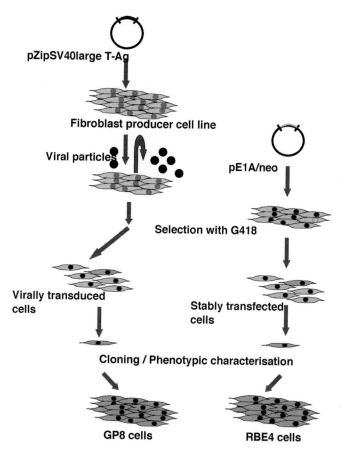

Fig. 2. Flow diagram of the immortalization procedure. The diagram presents the successive steps used for the production of the immortalized rat brain endothelial cells RBE4 and GP8/3.9 cells, respectively by transfection with the plasmid pE1A/*neo* and by infection with the replication deficient retroviral vector pZipSV40-large-T-Ag.

clones developed (1–3 per 60-mm dish), corresponding to a frequency of appearance of $1–3 \times 10^{-5}$.

3.3.2. Infection of Rat Brain Endothelial Cells with the SV40 Large T-Vector

1. Supernatant-containing retroviral particles were harvested from the producer cell line SVU19.5 following conditioning of media for 3 d.
2. The supernatant was passed through a 0.45-μm filter to remove contaminating fibroblasts.
3. The retroviral supernatant was diluted 1:10 with fresh growth media containing 8 μg/mL of polybrene and added to primary cultures of cerebral endothelial cells and incubated for 4 h at 37°C with gentle agitation.
4. Infection media was removed and 5 mL of fresh growth media added. Virally infected cells were cultured for 48 h prior to selection in 200 μg/mL G418. The GP8 parent cell line was isolated through selection in G418 (*see* **Note 1**).

3.4. Selection, Cloning, and Phenotypic Characterization of the Immortalized Cell Lines

Growing clones of RBE4 cells or GP8 cells, termed parent cell lines were individually picked using cloning cylinders. Southern blot hybridization using a SmaI-SacI 764

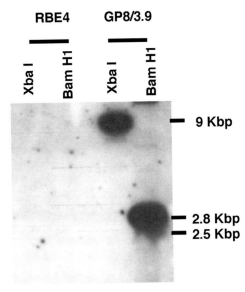

Fig. 3. Integration of the immortalizing SV40 large T-gene into the GP8 cell genome. Southern blot hybridization of GP8/3.9 and RBE4 cell DNA (10 µg), following Xba I or Bam H1 restriction. A BamH1 2.5 kbp fragment of pZipSV40-large-T-Ag was used as a probe. RBE4 cells are negative controls. Molecular weight markers are indicated on the right side.

bp E1A fragment or a BamH1 2.5 Kb fragment of pZipSV40-large-T-Ag indicated that the immortalizing SV40 large T-gene was stably integrated into the GP8cell genome (**Fig. 3**). Conversely, integration of the E1A sequence in the genome of RBE4 cells was demonstrated by Southern blot hybridization (**7**).

The parental GP8 cell line was seeded at a cell density of 0.33 cells per well to generate clonal isolates, in particular the clonal GP8/3.9 cell line that was expanded and further studied. More recently, the GPNT cell line was derived from the parental GP8 cell line by lipofectin-mediated transfection with a vector containing the puromycin resistance gene, followed by puromycin selection and cloning (**10**).

3.4.1. Expression of Endothelial Markers

1. The immortalized RBE4, GP8/3.9, and GPNT cell lines exhibit a nontransformed phenotype, on the basis of the following observations:
   a. They form regular contact-inhibited monolayers of cells (**Fig. 4**).
   b. Their proliferation is highly dependent on the presence of appropriate cell substrate, serum components, and growth factors.
   c. They do not form foci of multilayered cells, nor do they proliferate in soft agar.
   d. They do not form tumors in athymic Nude mice following subcutaneous or intrathecal injections (**7,11**).
2. Confluent monolayers of immortalized BMECs are expected to retain a whole set of endothelial markers that can be detected by immunostaining (when species cross-reactive antibodies are available) as follows:
   a. Cells were fixed in 4% paraformaldehyde made up in PBS containing 0.5 m*M* CaCl$_2$ and 0.5 m*M* MgSO$_4$.
   b. After 15 min incubation at room temperature, the cells were washed and then permeabilized by incubation for 10 min with 0.5% Triton X 100 in PBS.

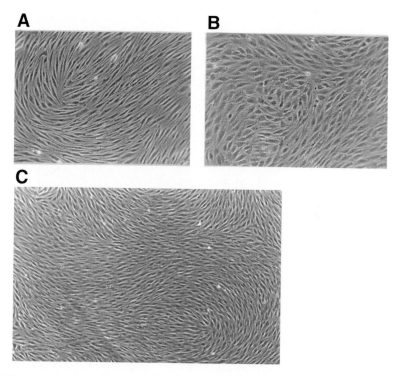

Fig. 4. Phase-contrast microscopy of RBE4 and GP8 cells compared with primary rat brain endothelial cells. Primary rat brain endothelial cells (**A**), RBE4 cells (**B**), and GP8 cells (**C**), with an endothelial characteristic morphology, form regular contact-inhibited monolayers.

c. After washing, the cells were blocked by incubation 30 min in PBS containing 3% bovine serum albumin (BSA) plus 0.1 M lysine, pH 7.4 and incubated with primary antibody in PBS containing 0.3% BSA for either 2 h at room temperature or overnight at 4°C.

d. After washing, the cells were then incubated for 30–60 min at room temperature with appropriate FITC- or TRITC-conjugated secondary antibody in PBS containing 0.3% BSA.

e. After subsequent washing steps, the slides were mounted and analyzed with a fluorescence or confocal microscope.

3. RBE4 and GP8 cells constitutively express von Willebrandt factor, the cellular junction proteins PECAM-1 (CD31), VE-cadherin (cadherin-5), α- and β-catenins, they show uptake of acetylated low density lipoprotein, and they express membrane binding sites specific for the lectins *Griffonia simplicifolia* and *Ulex Europaeus Agglutinin* (**Fig. 5**).

4. Upon treatment by inflammatory cytokines (TNF-α + IFN-γ) RBE4 and GP8/3.9 cells express the adhesion molecule VCAM-1.

5. In addition, these cells retain the capacity to release endothelin-1 and other vasoactive substances and to express a wide variety of hormone receptors known to be present on endothelial cells (*12–15*).

3.4.2. Expression of BBB Markers

Expression of multiple BBB markers has been observed in RBE4 and GP8/3.9 cells:

1. Expression of transferrin receptors can be detected by immunostaining.

2. The expression of the drug-transporting p-glycoprotein (Pgp) was detected exclusively at the luminal membrane of rat brain endothelial cells (*16*), whereas it has also been local-

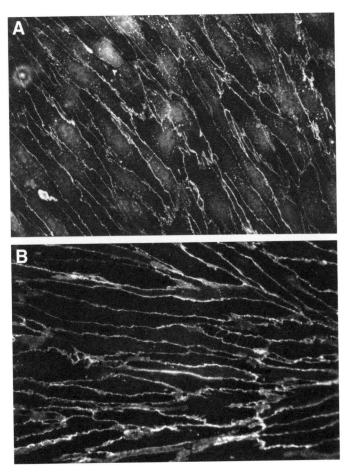

Fig. 5. Phenotypic characterization of GP8/3.9 and GPNT cells. Immunostaining of inter-cellular junction proteins expressed by GP8/3.9 cells. **(A)** ZO-1, **(B)** p100/120.

ized to astrocyte foot processes in human brain capillaries *(17)*. Indeed, Pgp expression can be documented by functional assays of [^{3}H]vinblastine uptake in (or efflux out of) rat brain endothelial cells *(10)* as follows:

a. Cellular uptake of the radiolabeled Pgp substrate [^{3}H]vinblastine was measured with endothelial cells grown to confluence in 24-multiwell plates for 5 d.

b. On the day of experiment, cells were washed three times with ice-cold PBS and preincubated for 30 min at 37°C in a shaking water bath with culture medium with or without the following Pgp inhibitors: verapamil (10 µ*M*), CsA (10 µ*M*), chlorpro-mazine (10 µ*M*).

c. [^{3}H]vinblastine (10 n*M*) was then added for 60 min to the medium.

d. The 24-multiwell plates were shaken during both preincubation and incubation periods to reduce the effect of the aqueous boundary layer on drug accumulation.

e. The reaction was stopped by rapidly removing the medium.

f. Thereafter, cells were washed three times with ice-cold PBS to eliminate the extracel-lular drug and lysed with 500 µL 0.1 *M* NaOH.

g. The amount of radio-labeled drug retained in the cells was counted in Pico Fluor by β-scintillation spectrometry.

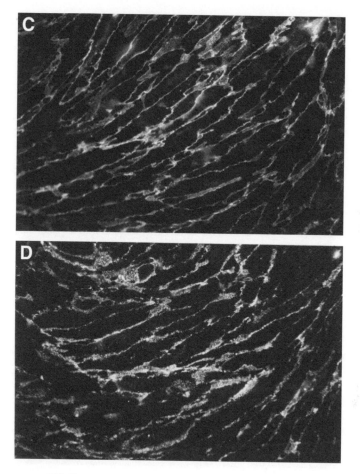

Fig. 5. *(continued)* **(C)** β-catenin, **(D)** PECAM-1. Characterization of the endothelial phenotype of GPNT cells.

h. An aliquot of cell lysate was used in parallel to determine cellular protein concentration.

i. Intracellular vinblastine concentration was expressed in pmol/mg protein.

3. Semi-permeable tissue culture filter inserts (Transwell inserts) can be used for transcellular electrical resistance (TEER) measurements, performed using an Endohm apparatus (World Precision Instruments) *(6,18)*. Resistance values are expressed in ohm-cm^2 ($\Omega.cm^2$), following subtraction of the resistance of a cell-free filter (*see* **Note 5**).

4. Cells grown on Transwell inserts can be used to measure the paracellular flux of diffusible markers through an endothelial monolayer. This is performed with radioactive tracers such as [^{3}H]inuline or [^{14}C]sucrose or fluorescent dextran polymers of increasing sizes (4–250 kDa).

a. The permeability of cell monolayers on Transwell-Clear™ was measured using FITC-labelled dextran 70 kDa (2 mg/mL) in DMEM (without phenol red) containing 0.1% BSA and 10 m*M* HEPES that was added to the upper chamber of inserts with confluent monolayers of BMECs.

b. The inserts were transferred sequentially at 5- or 10-min intervals from well to well of a tissue culture plate containing the same volume of medium.

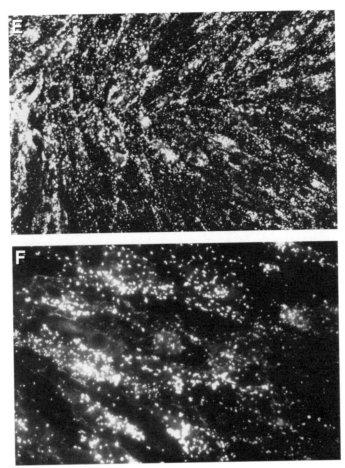

Fig. 5. *(continued)* (**E**) uptake of acetylated-LDL, (**F**) expression of von-Willebrandt factor. (A–D: ×63; E: ×20; F: ×100).

    c. Fluorescence that passed through the inserts at each time point was determined using a fluorescence multi-well plate reader and the cleared volume plotted vs time *(19)*.

    d. As illustrated in **Fig. 6**, GPNT cells displayed limited permeability coefficients for 4–150 kDa FITC-labeled dextrans, which were further decreased by dexamethasone treatment.

    Although RBE4, GP8/3.9 or GPNT cell lines display a restricted permeability to proteins, permeability of smaller molecules (lower than 10 kDa) remains much higher and TEER values much lower than expected from BBB restricted permeability. Altogether, our data indicate that these cell lines can be considered as valuable models for cerebral endothelium; however, the development of BBB permeability models based on these cell lines might require more sophisticated setups, like those proposed for primary BMECs *(5,6,16)*.

3.4.3. Characterization of Lymphocyte Migration Through Brain Endothelial Cell Lines

    1. RBE4, GP8/3.9, or GPNT cells, seeded at a density of 10^4 cells/cm^2, grow to confluence after 3–4 d in culture.

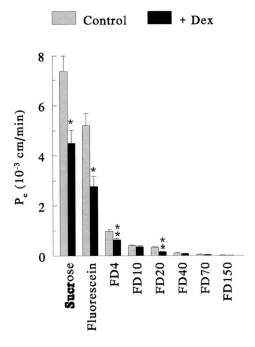

Fig. 6. Regulation of GPNT monolayer permeability. Permeability coefficients (Pe) for [^{14}C] sucrose, fluorescein and 4–150 kDa FITC-labelled dextrans (FD4–FD150) through GPNT cell monolayers. The Pe of the endothelial monolayer was calculated from the slopes of the curves for cleared volume versus time. All values are expressed as means ± SEM. GPNT cells treated with 1 *M* dexamethasone from seeding led to a decrease in transmonolayer permeability to sucrose and fluorescein (*$p < 0.001$) and FD4 and FD20 dextrans (**$p < 0.01$). Data are from four to five experiments with duplicate or triplicate filters.

2. Cell monolayers are then incubated with serum- and bFGF-free culture medium in the absence or presence of 100 U/mL IFNγ, for 48 h, to induce adhesion molecule expression.

3. The ability of immortalized endothelial cells grown to confluence in 96-well plates to support the transendothelial migration of antigen specific rat T-lymphocytes (2×10^4 cells/well) can be assessed over a 4 h migration period.

4. To evaluate the level of migration, co-cultures are placed on the stage of a phase-contrast inverted microscope housed in a temperature controlled (37°C), 5% CO_2 gassed chamber. A 200×200-µm field is randomly chosen and recorded for 10 min spanning the 4-h time point using a camera linked to a time-lapse video recorder.

5. Recordings replayed at ×160 normal speed enable lymphocytes which have either adhered to the surface of the monolayer or have migrated through the monolayer to be identified and counted.

6. Lymphocytes on the surface of the monolayer can be readily identified by their highly refractive morphology (phase-bright) and rounded or partially spread appearance.

7. In contrast, cells that have migrated through the monolayer are phase-dark, highly attenuated and are seen to probe under the endothelial cells in a distinctive manner.

8. Data is expressed as the percentage of total lymphocytes within a field that have migrated through the monolayer. All other data is expressed as a percentage of the control migrations (*20*).

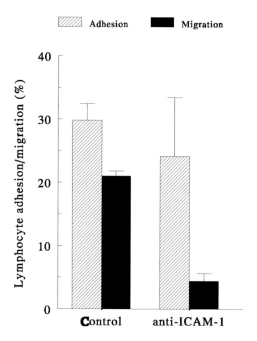

Fig. 7. Adhesion of human peripheral blood lymphocytes to and migration of antigen specific CD4+ T-lymphocytes through GPNT endothelial cell monolayers. Peripheral blood lymphocyte adhesion (hatched bars) and transendothelial migration of T-cells (solid bars) to and through GPNT monolayers in the absence and presence of anti-ICAM-1 monoclonal antibodies. Adhesion was measured after 90 min of co-culture and migration was assessed using time lapse video microscopy after 4 h of co-culture. Data is expressed as mean ± SEM from at minimum of six observations (migration) and nine independent observations (adhesion).

3.5. Application

1. Using this approach, we were able to confirm that T-lymphocyte migration is significantly prevented in the presence of antibodies to the endothelial surface adhesion molecule ICAM-1, whereas adhesion was not affected (**Fig. 7**).
2. Moreover, pretreatment of endothelial cells with actin cytoskeleton inhibitors, under the control of the GTP-binding protein Rho *(21)*, as well as by intracellular calcium chelators *(22)* also prevented T-lymphocyte migration through RBE4 and GP8 cell monolayers.
3. These data contribute to a better knowledge of the molecular mechanisms controlling the infiltration of activated T-lymphocytes through brain endothelium, as observed in inflammatory diseases of the central nervous system, such as Multiple Sclerosis.

4. Notes

1. With the rapid development of viral and chemical vectors for gene delivery in vitro and in vivo, multiple methods can now be utilized in the transduction of genes of interest, including immortalizing genes, into endothelial cells. Chemical vectors such as FuGene (Roche) or a number of other lipid or liposome mediated transfection methods have been successfully used for stable expression of transgenes in rat brain endothelial cells, including the parent cell line RBE4 *(31,32)*. HIV-based lentiviral vectors that can transduce nondividing cells are now available and have been used for the immortalization of human

endothelial cells *(28,30)*. It is important, for safety reasons, that cells that were immortalized using a retroviral vector are found to be incapable of producing retroviral particles. This possibility could arise because of the presence of pre-existing helper viruses within the cerebral endothelial cells which may complement retroviral mutations in the vector in an identical way to that used in producer cell lines such as SVU19.5. Therefore, supernatants from immortalized cerebral endothelial cell clones are used to infect an indicator cell line, such as Swiss 3T6 fibroblasts. The lack of G418-resistant 3T6 cells indicates the absence of retroviral particles in the supernatant.

2. Mouse brain endothelial cell lines have been reported, which maintain an endothelial phenotype (i.e., the MBEC4 cell line) *(23)* or exhibit a fully transformed phenotype, associated with tumorigenic potential, like the bEnd3 and bEnd4 cell lines, transformed by polyoma middle T-oncogene *(24)*. Also, human non-brain endothelial cell lines have been efficiently established, most of them on expression of the SV40-large-T-antigen *(25,26)*.

   Senescence naturally limits the proliferation of mammalian cells in culture, by shortening the telomere regions of chromosomes during cell division. Expression of the catalytic subunit of human telomerase (hTERT) has been shown to prevent telomere shortening and has been recently proposed as an alternative strategy for the extension of lifespan and in some cases immortalization of various cell types *(27)*. Immortalization of endothelial cells of different organs and species, including human endothelial cells, was observed either following hTERT transduction or concomitant transduction of cells with hTERT and SV40-large-T-antigen *(28–30)*. To our knowledge, none of the human brain endothelial cell lines produced thus far retain a stable and fully differentiated phenotype.

3. Alternatively, rat cortical grey matter free of meninges can be incubated in 0.1% collagenase/dispase solution for 2 h followed by density-dependent centrifugation on a 25% BSA solution. The pellet is resuspended in oxygen-saturated physiologic buffer (PB) (147 mM NaCl, 4 mM KCl, 3 mM CaCl$_2$, 1.2 mM MgCl$_2$, 5 mM glucose, 15 mM HEPES, pH 7.4 and 1% (wt/vol) BSA) and then passed through a 1.2 × 1.5-cm column containing 0.25-mm washed glass beads. The isolated capillaries remain attached to the beads and are collected by repeated gentle agitation in PB and decantation. The final pellet is resuspended and washed several times in PB without BSA. An aliquot is taken for protein measurement.

4. Several groups have reported the detailed experimental methods for in vitro culture of brain microvascular endothelial cells of various origins such as bovine, porcine, murine, or human *(1–3)*. The cells are grown in the presence of serum and exogenous endothelial cell growth supplements or fibroblast growth factor-2 or bFGF. Cultured cells usually reach confluence at one week and form a tight monolayer on approx d 10, before undergoing senescence and de-differentiation after approx 2 wk in culture or following limited passaging. The purity of the endothelial culture can be assessed by identifying contaminating pericytes or astrocytes, using antibodies against smooth muscle actin or glial fibrillary acidic protein (GFAP), respectively.

5. In an attempt to reconstitute in vitro the functional relationship between brain endothelial cells and astrocytes observed in vivo at the BBB, brain microvascular endothelial cells can be co-cultured with primary cultures of astrocytes, usually of rat or porcine origin. In one setup, both cell types are grown on the two opposite sides of a semi-permeable filter of culture inserts (Transwell). Alternatively, astrocytes are grown in the bottom chamber of the Transwell with the brain endothelial cells grown on the filter. In the former setup, direct contact between endothelial cells and astrocyte processes are made through the pores of the filter while, in the latter setup, only astrocyte-secreted factors can freely diffuse in the culture medium and through the pores to reach the endothelial cells. It has been shown that astrocyte co-culture or conditioned medium largely improve the in vitro BBB model, in regard to structural, metabolic, and permeability characteristics. In these experimental

conditions, in vitro models of BBB, based on bovine or porcine endothelial cells *(5,6,18)*, have been proposed and validated: (i) they display high transendothelial electrical resistance (500–1000 Ω.cm^2), whereas the electrical resistance of endothelial cells alone is generally lower than 200 Ω.cm^2); (ii) endothelial cells are highly polarised regarding the expression of receptors and transporters, and (iii) the pattern of permeability for a large number of standard molecules, over a wide range of hydrophilicity, is well correlated with the brain bio-availability of the same molecules measured in vivo.

Acknowledgments

We thank the Centre National de la Recherche Scientifique (CNRS), the Institut National de la Santé et de la Recherche Médicale (INSERM), the Association pour la Recheche sur la Sclérose en Plaques (ARSEP) and the Wellcome Trust for funding.

References

1. Bowman, P. D., Ennis, S. R., Rarey, K. E., Betz, A. L., and Goldstein, G. W. (1983) Brain microvessel endothelial cells in tissue culture: a model for study of blood-brain barrier permeability. *Ann. Neurol.* **14,** 396–402.
2. Abbott, N. J., Hughes, C. C., Revest, P. A., and Greenwood, J. (1992) Development and characterization of a rat brain capillary endothelial culture: towards an in vitro blood-brain barrier. *J. Cell Sci.* **103,** 23–37.
3. Dorovini-Zis, K., Prameya, R., and Bowman, P. D. (1991) Culture and characterization of microvascular endothelial cells derived from human brain. *Lab. Invest.* **64,** 425–436.
4. Goldstein, G. W., and Betz, A. L. (1986) The blood-brain barrier. *Sci. Am.* **255,** 74–83.
5. Rubin, L. L., Hall, D. E., Porter, S., et al. (1991) A cell culture model of the blood-brain barrier. *J. Cell Biol.* **115,** 1725–1735.
6. Cecchelli, R., Dehouck, B., Descamps, L., et al. (1999) In vitro model for evaluating drug transport across the blood-brain barrier. *Adv. Drug Deliv. Rev.* **36,** 165–178.
7. Roux, F., Durieu-Trautmann, O., Chaverot, N., et al. (1994) Regulation of gamma-glutamyl transpeptidase and alkaline phosphatase activities in immortalized rat brain microvessel endothelial cells. *J. Cell Physiol.* **159,** 101–113.
8. Greenwood, J., Pryce, G., Devine, L., et al. (1996) SV40 large T-immortalized cell lines of the rat blood-brain and blood-retinal barriers retain their phenotypic and immunological characteristics. *J. Neuroimmunol.* **71,** 51–63.
9. Jat, P. S., Cepko, C. L., Mulligan, R. C., and Sharp, P. A. (1986) Recombinant retroviruses encoding simian virus 40 large T-antigen and polyomavirus large and middle T-antigens. *Mol. Cell Biol.* **6,** 1204–1217.
10. Regina, A., Romero, I. A., Greenwood, J., et al. (1999) Dexamethasone regulation of P-glycoprotein activity in an immortalized rat brain endothelial cell line, GPNT. *J. Neurochem.* **73,** 1954–1963.
11. Lal, B., Indurti, R. R., Couraud, P. O., Goldstein, G. W., and Laterra, J. (1994) Endothelial cell implantation and survival within experimental gliomas. *Proc. Natl. Acad. Sci. USA* **91,** 9695–9699.
12. Durieu-Trautmann, O., Federici, C., Creminon, C., et al. (1993) Nitric oxide and endothelin secretion by brain microvessel endothelial cells: regulation by cyclic nucleotides. *J. Cell Physiol.* **155,** 104–111.
13. Nobles, M., Revest, P. A., Couraud, P. O., and Abbott, N. J. (1995) Characteristics of nucleotide receptors that cause elevation of cytoplasmic calcium in immortalized rat brain endothelial cells (RBE4) and in primary cultures. *Br. J. Pharmacol.* **115,** 1245–1252.

14. Karlstedt, K., Sallmen, T., Eriksson, K. S., et al. (1999) Lack of histamine synthesis and down-regulation of H1 and H2 receptor mRNA levels by dexamethasone in cerebral endothelial cells. *J. Cereb. Blood Flow Metab.* **19,** 321–330.

15. Kis, B., Szabo, C. A., Pataricza, J., et al. (1999) Vasoactive substances produced by cultured rat brain endothelial cells. *Eur. J. Pharmacol.* **368,** 35–42.

16. Beaulieu, E., Demeule, M., Ghitescu, L., and Beliveau, R. (1997) P-glycoprotein is strongly expressed in the luminal membranes of the endothelium of blood vessels in the brain. *Biochem. J.* **326,** 539–544.

17. Golden, P. L., and Pardridge, W. M. (2000) Brain microvascular P-glycoprotein and a revised model of multidrug resistance in brain. *Cell Mol. Neurobiol.* **20,** 165–181.

18. Engelbertz, C., Korte, D., Nitz, T., et al. (2000) In *The Blood-Brain Barrier and Drug Delivery to the CNS.* Begley, D., Bradbury, M. W., and Kreuter, J., eds. Marcel Dekker, New York, pp. 33–63.

19. Romero, I. A., Prevost, M. C., Perret, E., et al. (2000) Interactions between brain endothelial cells and human T-cell leukemia virus type 1-infected lymphocytes: mechanisms of viral entry into the central nervous system. *J. Virol.* **74,** 6021–6030.

20. Greenwood, J., Wang, Y., and Calder, V. L. (1995) Lymphocyte adhesion and transendothelial migration in the central nervous system: the role of LFA-1, ICAM-1, VLA-4 and VCAM-1. *Immunology* **86,** 408–415.

21. Adamson, P., Etienne, S., Couraud, P. O., Calder, V., and Greenwood, J. (1999) Lymphocyte migration through brain endothelial cell monolayers involves signaling through endothelial ICAM-1 via a rho-dependent pathway. *J. Immunol.* **162,** 2964–2973.

22. Etienne-Manneville, S., Manneville, J. B., Adamson, P., Wilbourn, B., Greenwood, J., and Couraud, P. O. (2000) ICAM-1-coupled cytoskeletal rearrangements and transendothelial lymphocyte migration involve intracellular calcium signaling in brain endothelial cell lines. *J. Immunol.* **165,** 3375–3383.

23. Shirai, A., Naito, M., Tatsuta, T., et al. (1994) Transport of cyclosporin A across the brain capillary endothelial cell monolayer by P-glycoprotein. *Biochim. Biophys. Acta* **1222,** 400–404.

24. Bussolino, F., De Rossi, M., Sica, A., et al. (1991) Murine endothelioma cell lines transformed by polyoma middle T-oncogene as target for and producers of cytokines. *J. Immunol.* **147,** 2122–2129.

25. Vicart, P., Testut, P., Schwartz, B., Llorens-Cortes, C., Perdomo, J. J., and Paulin, D. (1993) Cell adhesion markers are expressed by a stable human endothelial cell line transformed by the SV40 large T-antigen under vimentin promoter control. *J. Cell Physiol.* **157,** 41–51.

26. Schweitzer, K. M., Vicart, P., Delouis, C., et al. (1997) Characterization of a newly established human bone marrow endothelial cell line: distinct adhesive properties for hematopoietic progenitors compared with human umbilical vein endothelial cells. *Lab. Invest.* **76,** 25–36.

27. Bodnar, A. G., Ouellette, M., Frolkis, M., et al. (1998) Extension of life-span by introduction of telomerase into normal human cells. *Science* **279,** 349–352.

28. Yang, J., Chang, E., Cherry, A. M., et al. (1999) Human endothelial cell life extension by telomerase expression. *J. Biol. Chem.* **274,** 26,141–26,148.

29. Salmon, P., Oberholzer, J., Occhiodoro, T., Morel, P., Lou, J., and Trono, D. (2000) Reversible immortalization of human primary cells by lentivector-mediated transfer of specific genes. *Mol. Ther.* **2,** 404–414.

30. O'Hare, M. J., Bond, J., Clarke, C., et al. (2001) Conditional immortalization of freshly isolated human mammary fibroblasts and endothelial cells. *Proc. Natl. Acad. Sci. USA* **98,** 646–651.

31. Johnston, P., Nam, M., Hossain, M. A., et al. (1996) Delivery of human fibroblast growth factor-1 gene to brain by modified rat brain endothelial cells. *J. Neurochem.* **67,** 1643–1652.
32. Quinonero, J., Tchelingerian, J. L., Vignais, L., et al. (1997) Gene transfer to the central nervous system by transplantation of cerebral endothelial cells. *Gene Ther.* **4,** 111–119.

24

Isolation and Characterization of Retinal Endothelial Cells

David A. Antonetti and Ellen B. Wolpert

1. Introduction

Primary vascular endothelial cell cultures provide powerful systems to investigate the molecular architecture and regulation of the blood–brain and blood–retinal barriers. Most investigators agree that in vitro models of endothelial cells alone do not completely recapitulate the strong resistance barrier achieved in vivo by the blood vessels in these neural tissues. However, in vitro models provide a number of advantages that make this a highly useful system to study the transport of molecules across an endothelial monolayer. First, the system is highly defined; the investigator has control over the cell types that are present as well as the timing and degree of the perturbation applied to the system. Thus, the direct effect of a hormone or physical stress on endothelial cell transport properties can be determined and highly precise measures for time course and dose response can be made. Second, precise rate measures can be made and compared between different molecules. The effect of size and charge on solute transport rate may be determined and the rate of water, ion, and solute flux can be directly compared. Also, with the appropriate system, real-time measures of changes in transport rate after a specific perturbation may be characterized. Third, in vitro systems allow a means to rapidly dissect the molecular mechanisms employed to regulate endothelial cell barrier properties. Through the use of specific cell-signaling inhibitors, neutralizing antibodies, and transfection experiments an investigator can readily move to an understanding of the molecular mechanisms employed in endothelial cells to develop, maintain, and regulate the blood–brain and blood–retinal barrier. In combination with in vivo studies, cell culture models continue to provide an important research tool in the arsenal of the investigator.

The blood–brain and blood–retinal barriers are not absolute barriers; rather they form highly selective partitions between the neural tissue and the blood. This partition is necessary to allow proper glial–neuronal interaction, to maintain the specific neural environment necessary for nerve conduction, and to maintain tight regulation of the neurotransmitter concentration in the neural tissue, particularly amino acid neurotransmitters such as glutamate. Transport of molecules across the blood–brain and blood–retinal barriers may occur through two basic routes: either through the endothelial cells, i.e., transcellular transport, or between endothelial cells, i.e., paracellular trans-

From: *Methods in Molecular Medicine, vol. 89:*
The Blood–Brain Barrier: Biology and Research Protocols
Edited by: S. Nag © Humana Press Inc., Totowa, NJ

port. The transcellular route may be further subdivided into specific mechanisms that include channels, transporters, and general or specific vesicle transport. For example, specific glucose transporters allow the flux of this metabolite across the endothelium to the neural tissue, or receptor-mediated vesicular transport may ferry iron across the endothelium. Finally, fenestrated endothelium or the thinning of the endothelium to the point where the apical and basolateral membranes appear to touch may be points of specific transport. However, there are very few vesicles or fenestrae observed in the vasculature making up the blood–retinal barrier (1).

The junctional complex controls paracellular flux across the blood–brain and blood–retinal barriers. The junctional complex is composed of adherens junctions, tight junctions, and desmosomes to varying degrees depending on the tissue type. The blood–brain and blood–retinal barriers have a high degree of tight junctions as observed by electron microscopy (2–4). Tight junctions regulate paracellular flux to the neural tissue and define the apical and basolateral plasma membranes of the endothelial cells, allowing directional flow of molecules across the transcellular route. Recent reports reveal that tight junctions are not uniform between all tissue types and may allow varying rates of molecular transport depending on the composition of the tight junction. Specifically, tight junctions consist of at least three types of transmembrane proteins— occludin, JAM, and claudins—and each possess various isoforms (reviewed in (5) and (6)). The claudins have the greatest known diversity and are composed of at least 22 different genes (7). Expression of different claudins appears to confer specific barrier properties (8,9).

Finally, a number of reports reveal a distinction between ion flux and solute flux across a cell monolayer (10,11). When occludin is over-expressed in MDCK cells, ion flux decreases and the paracellular flux of dextran increases by 200% (12). Therefore, measures of specific molecular transport cannot be extrapolated to general statements regarding barrier integrity. However, by assessing the transport rate of a variety of different molecules the investigator may ascertain a broader scope of the barrier properties of an endothelial monolayer and the effect of specific perturbations on these barrier properties.

Bovine retinal endothelial cells (BREC) provide a powerful in vitro model for the study of the blood–retinal barrier properties. Obtaining a culture of homogeneous, high-yielding primary cells is essential for the success of subsequent experiments. Consistency and reproducibility of results depend on the quality of the endothelial cell monolayer formed in culture. Retinal pericyte or other cell type contamination may alter metabolic, proliferative, and transport properties of the endothelial cells.

This chapter will provide detailed procedures to isolate and grow in culture, primary BREC as a model for the blood–retinal barrier. In addition, the measures of flux for both ions and solute as a means to assess retinal endothelial barrier properties in vitro will be described.

2. Materials

2.1. BREC Cultures

1. Culture flasks, 25 cm^2 and 75 cm^2 (Corning).
2. Bottle top filter, cellulose acetate, 0.22 μm (Corning).
3. Conical tube, 50 mL (Falcon).

4. Calcium and magnesium free Hank's balanced salt solution (HBSS): 5 mM KCl, 0.3 mM KH$_2$PO$_4$, 138 mM NaCl, 4.0 mM NaHCO$_3$, 0.3 mM Na$_2$HPO$_4$, pH 7.4.
5. L-Glutamine 200 mM (X100).
6. 0.5% Trypsin containing 5.3 mM EDTA.
7. 10% providone-Iodine solution.
8. Supplemented MCDB-131 media (Sigma): 10% fetal calf serum (HyClone), 10 ng/mL epidermal growth factor (Sigma), 0.2 mg/mL EndoGro (VEC Technologies, Inc.), 0.09 mg/mL heparin (Fisher), and 0.01 mL/mL antibiotic/antimycotic (Life Technologies).
9. Modified Eagle's Medium D-Valine (MEM D-Val) (Sigma).
10. MEM D-Valine with 30 mM HEPES buffer, pH 7.4.
11. Enzyme cocktail: 10 mL of Ca^{++}, Mg^{++}-free HBSS, 500 μg/mL Type I Collagenase (Worthington, *see* **Note 1**), and 200 μg/mL DNase 1 (Worthington). This solution is made fresh at each isolation.
12. Standard Growth Medium: MEM D-Val, 20% fetal calf serum, 50 μg/mL of endothelial cell growth supplement (ECGS) (Collaborative Biomedical Products), 16 U/mL heparin, 0.01 mL/mL MEM vitamins (Mediatech), 0.01 mL/mL glutamine (Life Technologies), and 0.02 mL/mL antibiotic/antimycotic.
13. Nylon meshes: 53, 85, and 185 μ (Small Parts, Inc.).
14. Fibronectin (Sigma), 0.1% solution (*see* **Note 2**).
15. Ca^{++}, Mg^{++} free phosphate-buffered saline (PBS) (Fisher).
16. Porcelain Buchner funnel (Coors).
17. Con-Torque tissue homogenizer (Eberbach).
18. Microspatula (Fisher).
19. Speci-Mix mixer (Thermolyne).
20. Potter-Elvehjem tissue grinder (Wheaton).

2.2. Transendothelial Electrical Resistance Measurements

1. Transwell filters, 0.4-μm pore size, clear polyester membrane (Costar).
2. Endohm tissue resistance measurement chamber and EVOM resistance meter (World Precision Instruments).

2.3. Solute Flux Measurement

1. Transwell filters, 0.4-μm pore size, clear polyester membrane (Costar).
2. Microtest™ 96 well assay plate, optilux-plus™, black/clear bottom (Falcon).
3. Rhodamine B isothiocyante dextran (RITC Dex) at 70 kDa (Sigma).
4. Fluorescein isothiocyante bovine serum albumin (FITC-BSA) (Sigma).
5. Fluor Imager 595 (Molecular Dynamics, Sunnyvale, CA).

3. Methods

3.1. Isolation of Retinal Capillaries

The following protocol for BREC isolation and culture maintenance has been modified from the methods of Wong and associates *(13)*, Laterra and Goldstein *(14)*, and Gardner *(15)*.

Retinal capillaries are isolated from bovine eyes obtained from a local slaughterhouse. The capillary preparation is passed through a series of meshes and collagenase treatment to remove associated cells. Media with D-Valine is used when first plating the cells since endothelial cells have the isomerase to convert the D amino acid to its L isoform while contaminating pericytes do not and are, therefore, selected out. All steps are conducted under sterile conditions with gloves.

1. Ten to twenty bovine whole eyes from recently slaughtered animals are transported on ice from a local abattoir. The cell isolation procedure usually occurs up to 24 h post-mortem (*see* **Note 3**).
   a. The whole eyes are bathed in a 10% povidone-iodine solution for a minimum of 5 min.
   b. With a sterile scalpel, a circumferential cut 5 mm posterior to the limbus is made to open the eyeball for retina removal.
   c. After the vitreous and lens are extracted, the retina is gently separated and cut from the anterior portion of the eyeball using sterile tweezers.
2. The retinas are rinsed three times in ice-cold MEM D-Valine with HEPES buffer and pooled in the same solution.
3. In a laminar flow hood, the retinas are washed with the same solution through a 185-micron nylon mesh stretched over a sterile porcelain funnel placed on a vacuum flask to remove retinal pigment epithelial cells. The retinal tissue is removed from the mesh and brought to a volume of 30 mL with ice-cold MEM D-Val with HEPES.
4. Next, the retinal aliquot is homogenized on ice six times in a Teflon/glass Potter-Elvehjem type tissue grinder with 0.25 mm clearance at 250 rpm.
5. The homogenate is centrifuged at $400g$ for 10 min at 4°C. After resuspending the pelleted retinal tissue in 10 mL of 4°C Ca^{2+}, Mg^{2+}-free PBS, the suspension is shaken or inverted three to four times and kept on ice.
6. The isolated microvessel fragments are trapped on an 88-μ nylon mesh over a funnel as described in **step 3**. The nylon mesh is then cut from the funnel and placed in a glass petri dish. The microvessels are separated from the mesh by repeated rinses with Ca^{2+}, Mg^{2+}-free PBS and transferred to a 50-mL conical tube. The microvessels are then pelleted at $400g$ at 4°C for 10 min.
7. The pelleted microvessels are resuspended in 10 mL of enzyme cocktail and incubated at 37°C on a rocker for 45–60 min to separate the pericytes (*see* **Notes 4** and **5**). The digestion is halted when observation with a Nikon phase contrast microscope shows release of the pericytes.
8. The vessel preparation is passed over a 53-μ nylon mesh without suction; the mesh is transferred to a 50-mL conical tube, and the vessel fragments are separated from the mesh by washing with ice cold MEM D-Val.
9. This suspension is centrifuged at $400g$ for 5 min at 4°C, resuspended in 10 mL of MEM D-Val, and centrifuged again. The resulting pellet is resuspended in 5 mL of the standard growth medium consisting of MEM D-Val supplemented with 20% fetal calf serum, 50 μg/mL ECGS, 16 U/mL heparin, 0.01 mL/mL MEM vitamins, 0.01 mL/mL glutamine, and 0.02 mL/mL antibiotic/antimycotic.
10. The vessel fragments are plated on a 25-cm² tissue culture flask precoated with fibronectin at 2 μg/cm² (*see* **Note 2**) and are grown in a humidified incubator at 37°C with 95% CO_2, 5% O_2. The medium is removed and fresh medium is added 24 h following the plating.

3.2. BREC Cultures

1. Colonies of endothelial cells grow from the isolated microvessels after 5–7 d. They are removed with 0.05% trypsin and reseeded onto a 75-cm² tissue culture flask precoated with 1 μg/cm² fibronectin. Endothelial cells do not reach confluence with the primary seeding and should be split and reseeded when islands of endothelial cells arise, prior to the proliferation of pericytes. The cells are repeatedly subcultured with 0.05% trypsin when approximately 80% confluent and expanded for experimental use at a ratio of 1:3 (*see* **Note 6**).
2. At passage three, the cells in 10% DMSO are routinely frozen in liquid nitrogen for storage purposes.
3. BREC are used experimentally at 6 to 10 passages after isolation.

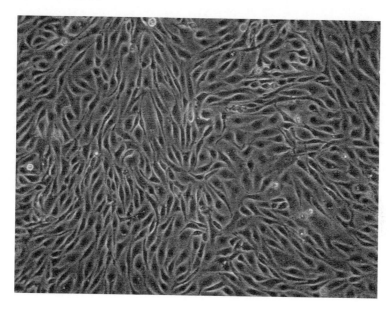

Fig. 1. Phase contrast image of bovine retinal endothelial cell culture grown to confluence. Note the uniform and cobblestone appearance of the cells.

4. Retinal cells are cultured in supplemented MCDB-131 media. A more robust and consistent cell growth occurs using this medium rather than the MEM D-Val media used in cell isolation.
5. The cell culture should appear homogeneous with a cobblestone-like appearance as seen in **Fig. 1**. Contaminating pericytes are much larger than endothelial cells and can cause areas of endothelial cell death making useful transport studies impossible.
6. To verify that the culture contains only endothelial cells the preparations are grown on glass coverslips and immunostained for the endothelial specific marker Von Willebrand factor.

3.3. Transendothelial Electrical Resistance Measurements

Transendothelial electrical resistance (TEER) is a measurement of ion flux across the endothelial monolayer. This is a rapid and simple measure of barrier integrity. Alterations to TEER most likely reflect changes in the junctional complex; however, as discussed in the notes section, the lack of change in TEER does not, by itself, prove that there are not alterations in paracellular transport (*see* **Note 7**).

1. The endothelial cells are grown to confluence on a porous transwell filter coated with 1 μg/cm^2 fibronectin.
2. To measure resistance, the transwell filters are placed in an Endohm™ chamber, which contains two concentric voltage-sensing electrodes, one at the top and one at the bottom. The Endohm is connected to an EVOM resistance meter. Ion flux is determined by applying a pulse of known amplitude across the endothelial monolayer and measuring the corresponding transendothelial voltage deflection. Ohm's law is then used to calculate resistance across the cross sectional area which is expressed in ohms (Ω) $\times$ cm^2. The investigator should be sure to subtract the resistance of a blank, fibronectin-coated, transwell filter from each sample tested.

3.4. Solute Flux Measurement

Solute flux across endothelial monolayers is determined by placing labeled sugars or proteins on the apical side of the monolayer and determining accumulation in the basolateral chamber over time. The choice of solute may reflect very different features of the endothelial barrier; dextran, for example, is often used since there are no known cell receptors for dextran that may contribute a specific transport mechanism. Also, one should be aware of the size, shape (globular or linear), and hydrophobicity of the chosen solutes (*see* **Note 8**).

Endothelial cells are grown to confluence on transwell filters. After the cells have grown to confluence, an additional 2–4 d is necessary to allow formation of a tight barrier. This may vary depending on the cell preparation.

1. Solute flux is measured by applying RITC-Dex or FITC-BSA ($10 \mu M$ to $20 \mu M$) to the apical chamber of inserts with a confluent endothelial cell monolayer. The specific experiment may dictate when to perturb the system with, for example, hormone addition, relative to the time of measuring flux. Remember to consider the time for the perturbation to effect transport rates, e.g., time for synthesis and assembly or disassembly of the junctional complex.
2. One hour after addition of fluorescent solute, 100-μL samples are taken from the basolateral chamber. This is continued on the hour for up to 4 h. The samples are placed into the 96-well black/clear bottom plate.
3. A sample is taken from the apical chamber at the last time point and also placed in the 96-well plate. In pilot studies, it is critical to demonstrate that the amount of fluorescence in the apical chamber remains essentially unchanged over the course of the experiment.
4. Fluorescence of each aliquot is quantified on a fluorescence plate reader. We use a FluorImager 595. A blank sample with media only but no fluorescent marker should be used for background subtraction.
5. The rate of diffusive flux (P_o) was calculated by the following formula at each time point (*16*):

$$1. \quad P_o = [(F_A / \Delta t)V_A] / (F_L A)$$

where, P_o = diffusive flux (cm/s); F_A = basolateral fluorescence; F_L = apical fluorescence; Δt = change in time; A = surface area of the filter (cm²); V_A = volume of the basolateral chamber (cm³). Remember that unless you replace the volume removed from the basolateral chamber then the volume in the basolateral chamber changes for each time point calculated. The data may be reduced to a rate over the total time course of the experiment if the change remains linear. This is done by plotting the change in fluorescence accumulated in the basolateral chamber corrected for fluorescence in the apical chamber, volume and area versus time and then determining the slope of the straight line or rate of diffusive flux, P_o.
6. An example of the calculation of diffusive flux is given in **Table 1**. The fluorescence obtained from each time point is normalized to the fluorescence in the apical chamber. Next this ratio is corrected for the volume in the bottom chamber and this corrected ratio is plotted verses time to determine the rate of fluorescent molecule accumulation by obtaining the slope. Finally, this slope is converted from minutes to seconds and divided by the area of the filter (0.9 cm in this case) yielding the diffusive flux in cm/s.

3.4.1. Real-Time Flux Measurement

Finally, we have recently developed a novel system to allow measures of real-time flux (*17*). This system uses a plexiglass chamber into which the transwell filter is placed. A fiber optic conducts excitation laser light to the basolateral chamber and a second fiber

Table 1
Example of the Calculation for Diffusive Flux

Condition	Name	Volume	Bottom/Top (BT)	Volume bottom	(B/T) × Vol bottom	Time (min)	Slope	Po = (slope/60)/0.9
10% serum	30m	5198.07	0.003437478	1.5	0.005156217	30	0.00020989	3.88682E-06
	60m	29213.01	0.019318533	1.4	0.027045946	60		
	90m	30186.95	0.019962598	1.3	0.025951378	90		
	120m	33406.33	0.022091571	1.2	0.026509885	120		
	top 120m	1512175.39						
10% serum	30m	8709.54	0.005378437	1.5	0.008067655	30	0.00014735	2.72872E-06
	60m	20752.62	0.012815447	1.4	0.017941626	60		
	90m	29165.5	0.018010686	1.3	0.023413892	90		
	120m	28309.7	0.017482201	1.2	0.020978641	120		
	top 120m	1619344.2						
10% serum	30m	12202.68	0.00693813	1.5	0.010407196	30	0.00013671	2.53163E-06
	60m	25155.57	0.014302811	1.4	0.020023935	60		
	90m	32432.8	0.018440457	1.3	0.023972594	90		
	120m	33360.93	0.018968168	1.2	0.022761802	120		
	top 120m	1758785.04						

optic conducts fluorescence emission light to a detector. In this manner real-time alteration in solute flux can be determined. Furthermore, the basolateral chamber is hydraulically coupled to an external reservoir, which can be lowered in order to apply a pressure gradient across the endothelial monolayer, recapitulating in vivo conditions. Thus, the effective solute flux or P_e, including both diffusive and convective components, can be determined. Application of the hydrostatic pressure gradient allows investigation of endothelial cell function under conditions that mimic normal physiology.

3.5. Concluding Remarks

Careful endothelial cell preparation from bovine retina will allow the formation and maintenance of a powerful and flexible cell culture system to study transport properties in vitro. We have successfully used this system to study retinal endothelial cell responses to both permeabilizing agents, such as vascular endothelial growth factor and agents that increase barrier properties, such as glucocorticoids. Coupled with specific inhibitors and appropriate transfection technology, the system can provide new insights into the molecular mechanisms regulating the blood–brain and blood–retinal barriers. Continued refinements of the culture environment will increase the similarity between in vitro and in vivo conditions.

4. Notes

1. Using collagenase from Worthington is highly recommended.
2. Using a 0.1% solution of fibronectin from bovine plasma to coat the tissue culture surface is recommended for successful adherence of the endothelial cells.
3. Optimal cell yields are obtained from retinas extracted from eyeballs of cows slaughtered within the previous 24 h. Keeping the eyeballs on ice and bathed in an antiseptic solution is imperative to achieve a viable, high-yielding, uncontaminated primary cell preparation.
4. The enzyme cocktail can be stored frozen in aliquots (–20°C) but only for a maximum time of 6 mo. However, optimum activity is achieved with freshly prepared cocktail.
5. Continual rocking at 37°C during the enzyme digestion step is necessary for complete separation of the pericytes.
6. Subculturing the cells at 80% confluence at a ratio of 1:3 will ensure continued proliferation and homology of the cell population. Be careful not to over-trypsinize the cells when subculturing; usually 2 min of treatment with trypsin is sufficient time to release the majority of cells.
7. The measure of electrical resistance is simple and can be quickly accomplished. In addition, the same filters can be measured repeatedly over time and after the measures are completed, the cells can also be extracted for protein analysis or fixed and stained for immunocytochemistry. However, measures of total ion flux clearly do not tell the whole story regarding barrier properties. Changes in ion channels and release of ions from the cells may affect the resistance measure. In addition, investigators have already developed evidence that the transport of different ions across the junctional complex may be regulated independently. Finally, a number of cases have revealed alterations in solute flux without changes in resistance. It is possible that a specific ion barrier exists, but more likely this apparent paradox reflects the differences in methods used for measuring large solute flux and ion flux. Ion flux determinations are instantaneous measures of resistance while solute flux measures are achieved by measuring accumulation of solute over time. Since tight junctions are a series of strands, there may not be a complete apical to basolateral pathway available at any given moment and thus a high resistance is maintained. However, if the junctions are transiently opening and closing, then transport of even relatively large

molecules may occur over time without an observed change in resistance *(18)*. A simple analogy may be the use of multiple revolving doors that allow the movement of people into and out of an air-conditioned building with minimal change in air temperature. An excellent review of TEER measures in epithelia is given in the literature *(19)*. However, it should be noted that the TEER of endothelial cells in vitro does not recapitulate the TEER of the vasculature at the blood–brain or blood–retinal barriers as well as some epithelial models are able to reflect in vivo resistance barriers.

8. A number of solutes have been used for flux rate determinations including smaller dextrans, as well as mannitol and inulin. The use of albumin is of physiologic relevance as it crosses the blood–retinal barrier in various retinal pathologies and it is useful to compare its rate of transport to that of dextran of a similar molecular weight since albumin may have a transcellular and paracellular transport component. In addition, since the relationship between fluorescence intensity and solute concentration is linear, a standard curve to determine the concentration of solute that crossed the barrier can be generated.

Acknowledgments

This research was supported by National Institutes of Health RO1 Grant EY12021, Juvenile Diabetes Foundation Career Development Award 298212, PA Lions Sight Conservation and Eye Research Foundation, and a Research to Prevent Blindness fellowship, as well as by the financial assistance of Mr. and Mrs. Jack Turner of Athens, GA.

References

1. Raviola, G. (1977) The structural basis of the blood-ocular barriers. *Exp. Eye Res.* **25 (Suppl.)**, 27–63.
2. Farquhar, M. G., and Palade, G. (1963) Junctional complexes in various epithelia. *J. Cell Biol.* **17**, 375–412.
3. Cunha-Vaz, J. G., Shakib, M., and Ashton, N. (1966) Studies on the permeability of the blood-retinal barrier. I. On the existence, development, and site of a blood-retinal barrier. *Br. J. Ophthalmol.* **50**, 441–453.
4. Shakib, M., and Cunha-Vaz, J. G. (1966) Studies on the permeability of the blood–retinal barrier. IV. Junctional complexes of the retinal vessels and their role in the permeability of the blood–retinal barrier. *Exp. Eye Res.* **5**, 229–234.
5. Fanning, A. S., Mitic, L. L., and Anderson, J. M. (1999) Transmembrane proteins in the tight junction barrier. *J. Am. Soc. Nephrol.* **10**, 1337–1345.
6. Kniesel, U., and Wolburg, H. (2000) Tight junctions of the blood–brain barrier. *Cell. Molecular Neurobiol.* **20**, 57–76.
7. Morita, K., Furuse, M., Fujimoto, K., and Tsukita, S. (1999) Claudin multigene family encoding four-transmembrane domain protein components of tight junction strands. *Proc. Natl. Acad. Sci. USA* **96**, 511–516.
8. Van Itallie, C., Rahner, C., and Anderson, J. M. (2001) Regulated expression of claudin-4 decreases paracellular conductance through a selective decrease in sodium permeability. *J. Clin. Invest.* **107**, 1319–1327.
9. Simon, D. B., Lu, Y., Choate, K. A., et al. (1999) Paracellin-1, a renal tight junction protein required for paracellular Mg2+ resorption. *Science* **285**, 103–106.
10. Calderon, V., Lazaro, A., Contreras, R. G., et al. (1998) Tight junctions and the experimental modification of lipid content. *J. Membrane Biol.* **164**, 59–69.
11. Hasegawa, H., Fujita, H., Katoh, H., et al. (1999) Opposite regulation of transepithelial electrical resistance and paracellular permeability by Rho in Madin-Darby canine kidney cells. *J. Biol. Chem.* **274**, 20,982–20,988.

12. Balda, M. S., Whitney, J. A., Flores, S., Gonzalez, M., Cereijido, M., and Matter, K. (1996) Functional dissociation of paracellular permeability and transepithelial electrical resistance and disruption of the apical-basolateral intramembrane diffusion barrier by expression of a mutant tight junction membrane protein. *J. Cell Biol.* **134,** 1031–1049.

13. Wong, H. C., Boulton, M., Marshall, J., and Clark, P. (1987) Growth of retinal capillary endothelia using pericyte conditioned medium. *Invest. Ophthalmol. Vis. Sci.* **28,** 1767–1775.

14. Laterra, J., and Goldstein, G. W. (1991) Astroglial-induced in vitro angiogenesis: requirements for RNA and protein synthesis. *J. Neurochem.* **57,** 1231–1239.

15. Gardner, T. W. (1995) Histamine, ZO-1 and blood-retinal barrier permeability in diabetic retinopathy. *Trans. Am. Ophthalmol. Soc.* **93,** 583–621.

16. Chang, Y. S., Munn, L. L., Hillsley, M. V., et al. (2000) Effect of vascular endothelial growth factor on cultured endothelial cell monolayer transport properties. *Microvasc. Res.* **59,** 265–277.

17. Antonetti, D. A., Wolpert, E. B., DeMaio, L., Harhaj, N. S., and Scaduto, R. C. (2002) Hydrocortisone decreases retinal endothelial cell water and solute flux coincident with increased content and decreased phosphorylation of occludin. *J. Neurochem.* **80,** 667–677.

18. Claude, P. (1978) Morphological factors influencing transepithelial permeability: a model for the resistance of the Zonula occludens. *J. Membrane Biol.* **39,** 219–232.

19. Madara, J. L. (1998) Regulation of the movement of solutes across tight junctions. *Annu. Rev. Physiol.* **60,** 143–159.

25

Isolation and Characterization
of Cerebral Microvascular Pericytes

Paula Dore-Duffy

1. Introduction

Although a plethora of information exists on the role of the endothelial cell (EC) in vascular hemostasis and tissue homeostasis, little is known of the role played by the microvascular pericyte (PC) **(Fig. 1)**. This lack of substantial information is most evident in the understanding of the role played by the PC in blood–brain barrier (BBB) function, and in the pathophysiology of central nervous system (CNS) disease. Development of techniques for the isolation of defined populations of CNS microvessels *(1–3)*, for the preparation of retinal PC *(3,4)*, and for the preparation of cerebrovascular PC *(5–7)* have enabled scientists to examine the function of this unique cell in the brain *(8)*. The subculture of PC from purified preparations of cerebral microvessels will be discussed below. All populations, isolated microvessels, enriched PC and EC cultures, pure primary cultures, and experimentally derived co-cultures, are suitable in experimental protocols modeling BBB function.

2. Materials

All media, instruments and glassware should be sterilized prior to use.

1. Dulbecco's Modified Eagles Medium (DMEM) (Gibco-BRL, Grand Island, NY).
2. Fetal bovine serum (FBS)/DMEM (Sigma, St. Louis, MO).
3. 17% Dextran/DMEM (85 g Dextran 500 mL final volume) (Sigma, St. Louis, MO).
4. 1% penicillin-streptomycin/1% nystatin/2.5 mM of L-Glutamine.
5. 0.1% collagenase (1 mg/mL) collagenase type II, no. LS004174 (Worthington Biochemical, Lakewood, NJ).
6. 1% bovine serum albumin (BSA) (Sigma, St. Louis, MO).
7. 70% ethanol.
8. Guillotine (Harvard Apparatus, South Natick, MA).
9. Instruments: Scalpel holder and blades, forceps, scissors, and clamp/hemostats.
10. Homogenizer and teflon pestle (Wheaton, Millville, NJ) (shaved to allow a 0.25-µm clearance between the inside radius of the glass surface and the outside radius of the teflon pestle).
11. Nitex mesh holders (Braerclif Manor, NJ) and 40-, 80-, and 118-µm Nitex meshes (Tetco, Braerclif Manor, NJ).
12. Tissue culture dishes (Falcon no. 2098, Lincoln Park, NJ), 35- or 60-mm petri dishes.

From: *Methods in Molecular Medicine, vol. 89:*
The Blood–Brain Barrier: Biology and Research Protocols
Edited by: S. Nag © Humana Press Inc., Totowa, NJ

2. Remove nonadherent cells and fragments and wash well (if incubated overnight) with tissue culture medium.
3. Replace with 2 mL defined medium and reincubate. PC adhere to noncoated plastic while EC and microvessel fragments do not adhere or adhere loosely.
4. Following an overnight incubation most cells will have adhered to the petri dish although they do not spread out to the extent of established cultures (*see* **Note 3**). EC will appear more elongated than PC. PC will appear to be comparable to those shown in **Fig. 1**.

3.2.2. Enriched EC Cultures

Non-adherent cell suspensions from **Subheading 3.2.1., step 2** can be plated on collagen-coated dishes (35-mm) and/or coverslips and incubated overnight. Non-adhered cells are removed and the plate washed. Cells can be cultured in 10% FCS following the adhesion step or in 2% FCS for maintenance. Adherent cells are greater than 90% EC which have the typical slightly elongated appearance of CNS EC.

3.3. Preparation of Pure Primary PC

The techniques utilized above yield enriched populations of either PC or EC. EC are difficult to culture without PC contamination. We have tried a number of methods to remove PC. These include antibody dependent complement mediated lysis, laser ablation, panning and the use of magnetic beads. Our most successful preparations have resulted from FACS. Details of this technique are available in the literature *(7)*. Cell yields are low, so a greater number of animals (30–40) will be required for the initial isolation of microvessels.

1. Cells from **Subheading 3.2., step 5** should be resuspended in DMEM with no serum and incubated for 60 min at 37°C with DMEM + FITC-conjugated GSA lectin in the presence of sodium azide. Cells are then washed. GSA should be used at a concentration to provide saturation density as recommended by the manufacturer. We use a dilution of 1:100.
2. Cells are washed three times then resuspended in 2–4 mL DMEM without lectin. Technicians who operate FACS machines have extensive training and are a valuable source of information and should be consulted prior to any attempt at cell preparation. The technician should set the machine to negatively sort cells, collecting GSA-populations. The density of the starting cell suspension is not important although the FACS technician will likely have a favored volume. However, total cell number is crucial, as the cell yield will be low following a sort. We have used 5×10^6 total cells derived from 30 to 40 animals. Discuss this with the FACS technician. It is important to get their recommendations as each machine differs in the amount of cell loss. Sorting must be timed to immediately follow the stain to ensure cell viability.
3. Collect GSA-cells and examine for EC or PC markers by immunocytochemistry. Take a drop of the sorted cell suspension and smear it on a clean slide. Let it dry and fix in 3% paraformaldehyde for 10 min. Wash slides and permeabilize with 0.01% Triton x-100 for 10 min at room temperature. Apply a 1:100 dilution of FITC conjugated anti-human factor VIII for 60 min at 37°C. Ideally, you should run a concentration curve but the manufacturer's recommended saturation density for most antibodies can be used. Anti Factor VIII recognizes EC. Alternatively, you can use FITC-conjugated anti-GSA lectin. GSA-negative cell suspensions should stain negatively for both markers.
4. Count cells using a microscope and a hemocytometer.
5. Resuspend GSA-viable PC in DMEM + 20% FCS and antibiotic supplements at a minimum density of 5×10^4 cells/mL and plate in tissue culture dishes of desired size. (Use

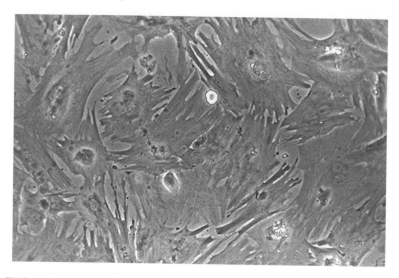

Fig. 1. CNS pericytes were isolated from cerebrovascular microvessels. One-week-old cultures display typical spreading with irregular projections as shown on light bright-field microscopy. ×100

minimums of 1-mL/35-mm petri dish.) Incubate at 37°C for 24 h before changing the medium (*see* **Note 4**).

6. We have also sorted cells after they have grown in culture. Add 10^{-4} *M* EDTA (1–2-mL/ 35-mm plate) to each dish. Resuspend cells at 1×10^6 cell/mL. Minimums of 3 mL are required. Stain and sort as detailed in **Subheading 3.3.1**.

3.4. Characterization of CNS Pericytes

The majority of markers used are not PC-specific. Not all markers described in non-CNS PC have been tested on CNS PC.

1. We have observed that very few CNS rat PC and no mouse PC are alpha muscle actin (αSMA) positive in vivo. In vitro, 100% of PC eventually become αSMA positive *(10)*.
2. Nayak and colleagues have used an antibody (3G5) directed against a ganglioside *(11)*. Expression of the ganglioside recognized by this antibody has also been found on dermal PC *(12,13)*.
3. Others have reported PC-specific aminopeptidase *(14,15)*.
4. We have reported that microvessels and freshly isolated rat PC express CD11b (αM) *(7)*. Using the OX-42 clone (Serotec, Oxford, England), we found that this marker was down regulated in cultured PC by 72 h (*see* **Note 5**).
5. PC also express the receptor for the Fc portion of the immunoglobulin molecule *(7)*.
6. Thus, the choice of marker must be determined according to the system being studied. For example:
   a. In purified microvessels, anti-CD11b and 3G5 can be used.
   b. In tissue sections, CD11b cannot be used to label PC as other cells in the CNS express this integrin. CD11b can be used to identify PC by electron microscopy where the morphological location within the basal lamina can be observed.
   c. In sections, dual and triple stains can be done to rule out astrocytes, glial cells or neuronal cells.

d. In cultured monolayers αSMA can be measured with dual label for EC markers.

e. Negative expression of EC markers such as Factor VIII, neuron specific markers, galactocerebroside (oligodendrocytes), and GFAP (astrocytes) is also helpful.

3.4.1. Differential Stain to Rule Out the Astroglial and Neuronal Contaminants in Microvessel and/or Primary Cell Cultures

1. Microvessels are dried to a standard glass or plastic coverslip then fixed with 3% paraformaldehyde for 10 min at room temperature.
2. Remove fixative and permeabilize with 0.01% Triton x100 for 10 min.
3. Add anti-GFAP antibody (1 : 1000 dilution) for 30 min at 37°C or as indicated by manufacturers recommendations. Anti-GFAP is available from numerous companies.
4. For neuronal contamination: fix and permeabilize as in **step 2**, then use the Neu (F-11) (IgG2α) clone at a 1 : 100 dilution.
5. Secondary antibody is a rabbit or goat anti-mouse IgG, which can be purchased from numerous companies.
6. Count stained cells using a fluorescent microscope.

3.4.2. Characterization of the EC Content

There is a large number of EC-specific markers. Antibodies to these markers are made by a number of companies. We routinely use anti-factor VIII antibody at the manufacturer's recommended concentration. As discussed in **Subheading 3.3.1.**, FITC GSA lectin can be used to identify EC. Divide the number of stained cells by the total cells then multiply by 100 to obtain the percentage.

3.5. Concluding Remarks

The development of techniques for the isolation of microvascular PC from a number of tissues has enabled scientists to study these intriguing cells at the cellular level. Our knowledge of PC function has largely been derived from studies of noncerebral PC *(16–27)*. As we begin to decipher the complexities of the PC's role in the brain, this information has been invaluable. We learned that the PC has immune potential *(8,23–33)*. The PC can function as an antigen-presenting cell (APC) in the CNS *(32)*. The PC does not constitutively produce major histocompatibility complex class II molecules. However, these molecules can be induced with cytokine. The PC can stimulate T-effector cells *(32)* and is phagocytic *(28)*. PC become activated during experimental autoimmune encephalomyelitis (EAE) and functionally alter T-cell cytokine-secreting phenotype *(29,30)*. Just as the PC recognizes the inflammatory milieu, it recognizes stress stimuli such as hypoxia and excitotoxic stimuli in traumatic brain injury *(34)*. The pericyte also has an important role in angiogenesis both in CNS and non-CNS tissue *(17,24,35)*.

Taken together, data from our lab and from other laboratories indicate that the PC may be a regulatory cell capable of sensing changes in the cellular environment. The production of numerous signaling molecules enables the PC to engage in complex, cross-talk mechanisms geared at maintenance of homeostasis and hemostasis.

4. Notes

1. Preparation of pure microvessel suspensions is essential. Alternative methods for their preparation have been published and may also be used. Time should be taken to master this technique first as the PC purity depends on pure microvessel starting material.

2. Microvessel preparations should be checked for cellular contaminates before enzymatic digestion. This is an additional step in determination of microvessel purity.
3. PC are not contact inhibited while in culture. Examine for purity before cultures become confluent.
4. Passaging pericytes: In our hands, using the culture medium described, the PC does not passage well. Passage of enriched cells usually results in EC survival. PC morphology changes dramatically after passage. Addition of growth factors and/or PC-conditioned medium and passaging at low density may aid in survival. It is possible that passage selects for a subset of PC. We use only primary cultures.
5. Serotec has changed the antibody it still calls OX-42. This antibody now stains PC in culture less brightly. Serotec has not been able to justify clonal changes. A number of other companies make anti-CD11b antibody. We have not yet tested all of them.

References

1. Joó, F., and Karnushina, I. (1973) A procedure for the isolation of capillaries from rat brain. *Cytobios* **8,** 41–48.
2. Bowman, P. D., Betz, A. L., Jerry, D. D. A., et al. (1981) Primary culture of capillary endothelium from rat brain. *In Vitro* **17,** 353–362.
3. Buzney, S. M., Massicotte, S. J., Hetu, N., and Zetter, B. R. (1983) Retinal vascular endothelial cells and pericytes. Differential growth characteristics. *In Vitro* **4,** 470–480.
4. Gitlin, J. D., and D'Amore, P. A. (1983) Culture of retinal capillary cells using selective growth media. *Microvas. Res.* **1,** 74–80.
5. Herman, I. M., and Jacobson, S. (1988) In situ analysis of microvascular pericytes in hypertensive rat brains. *Tissue Cell* **1,** 1–12.
6. Sussman, I., Carson, M. P., Schultz, V., et al. (1988) Chronic exposure to high glucose decreases myo-inositol in cultured cerebral microvascular pericytes but not in endothelium. *Diabetologia* **10,** 771–775.
7. Balabanov, R., Washington, R., Wagnerova, J., and Dore-Duffy, P. (1996) CNS microvascular pericytes express macrophage-like function, cell surface integrin αM, and macrophage marker ED-2. *Microvas. Res.* **52,** 127–142.
8. Balabanov, R., and Dore-Duffy, P. (1988) Role of the CNS microvascular pericyte in the blood brain barrier. *J. Neurosci. Res.* **6,** 637–644.
9. Vinters, H. V., Reavve, S., Costello, P., Girvin, J. P., and Moore, S. A. (1987) Isolation and culture of cells derived from cerebral microvessels. *Cell Tissue Res.* **3,** 657–667.
10. Nayak, R. C., Berman, A. B., George, K. L., Eisenbrth, G. S., and King, G. L. (1988) A monoclonal antibody (3G5)-defined ganglioside antigen is expressed on the cell surface of microvascular pericytes. *J. Exp. Med.* **3,** 1003–1015.
11. Helmbold, P., Wohlrab, J., Marsch, W. C., and Nayak, R. C. (2001) Human dermal pericytes express 3G5 ganglioside—a new approach for microvessel histology in the skin. *J. Cutan. Pathol.* **4,** 206–210.
12. Helmbod, P., Nayak, R. C., Marsch, W. C., and Herman, I. M. (2001) Isolation and in vitro characterization of human dermal microvascular pericytes. *Microvasc. Res.* **2,** 160–165.
13. Schlingemann, R. O., Oosterwijk, E., Wesseling, P., Rieveled, E. J., and Ruiter, D. J. (1996) Aminopeptidase a is a constituent of activated pericytes in angiogenesis. *J. Pathol.* **4,** 436–442.
14. Ramsauer, M., Kunz, J., Krause, D., and Dermietizel, R. (1998) Regulation of a blood-brain barrier- specific enzyme expressed by cerebral pericytes (pericytic aminopeptidase N/pAPN) under cell culture conditions. *J. Cereb. Blood Flow Metab.* **11,** 1270–1281.
15. Alliot, F., Rutin, J., Leeman, P. J., and Pessac, B. (1999) Pericytes and periendothelial cells of brain parenchyma vessels co-express aminopeptidase N, aminopeptidase A, and nestin. *J. Neurosci. Res.* **3,** 367–378.

16. Cameron, N. E., Eaton, S. E., Cotter, M. A., and Tesfaye, S. (2001) Vascular factors and metabolic interactions in the pathogenesis of diabetic neuropathy. *Diabetologia* **44,** 1973–1988.

17. Sieczkiewicz, G. J., Hussain, M., and Kohn, E. C. (2002) Angiogenesis and metastasis. *Cancer Treat. Res.* **107,** 353–381.

18. Provis, J. M. (2001) Development of the primate retinal vasculature. *Prog. Retin. Eye Res.* **20,** 799–821.

19. Pallone, T. L., and Silldorff, E. P. (2001) Pericyte regulation of renal medullary blood flow. *Exp. Nephrol.* **9(3),** 165–170.

20. Allt, G., and Lawrenson, J. G. (2001) Pericytes: cell biology and pathology. *Cells Tissues Organs* **169,** 1–11.

21. Sims, D. E. (2000) Diversity within pericytes. *Clin. Exp. Pharmacol. Physiol.* **27,** 842–846.

22. McLennan, S. V., Death, A. K., Fisher, E. J., Williams, P. F., Yue, D. K., and Turtle, J. R. (1999) The role of the mesangial cell and its matrix in the pathogenesis of diabetic nephropathy. *Cell Mol. Biol.* **45,** 123–135.

23. Kawada, N. (1997) The hepatic perisinusoidal stellate cell. *Histol. Histopathol.* **12,** 1069–1080.

24. Hirschi, K. K., and D'Amore, P. A. (1997) Control of angiogenesis by the pericyte: molecular mechanisms and significance. *E.X.S.* **79,** 419–428.

25. Hirschi, K. K., and D'Amore, P. A. (1996) Pericytes in the microvasculature. *Cardiovasc. Res.* **32,** 687–698.

26. Pinzani, M. (1995) Hepatic stellate (ITO) cells: expanding roles for a liver-specific pericyte. *J. Hepatol.* **22,** 700–706.

27. Shepro, D., and Morel, N. M. (1993) Pericyte physiology. *FASEB J.* **7,** 1031–1038.

28. Balabanov, R., Washington, R., Wagnerova, J., and Dore-Duffy, P. (1996) CNS microvascular pericytes express macrophage-like function, cell surface integrin αM, and macrophage marker ED-2. *Microvasc. Res.* **52,** 127–142.

29. Dore-Duffy, P., and Balabanov, R. (1998) The role of the CNS microvascular pericyte in leukocyte polarization of cytokine-secreting phenotype. *J. Neurochem.* **70,** 72.

30. Dore-Duffy, P., Balabanov, R., Rafols, J., and Swanborg, R. (1996) The recovery period of acute experimental autoimmune encephalomyelitis in rats corresponds to development of endothelial cell unresponsiveness to interferon gamma activation. *J. Neurosci. Res.* **44,** 223–234.

31. Dore-Duffy, P., Balabanov, R., Washington, R., and Swanborg, R. (1994) Transforming growth factor-β 1 inhibits cytokine-induced CNS endothelial cell activation. *Mol. Chem. Neuropathol.* **22,** 161–175.

32. Balabanov, R., Beaumon, T., and Dore-Duffy, P. (1999) Role of central nervous system microvascular pericytes in activation of antigen-primed splenic T-lymphocytes. *J. Neurosci. Res.* **55,** 578–587.

33. Dore-Duffy, P., Washington, R., and Balabanov, R. (1995) Cytokine-mediated activation of CNS microvessels: a system for examining antigenic modulation of CNS endothelial cells, and evidence for long-term expression of the adhesion protein E-selectin. *J. Cereb. Blood Flow Metab.* **14,** 43–45.

34. Dore-Duffy, P., Owen, C., Balabanov, R., Murphy, S., Beaumont, T., and Rafols, J. (2000) Pericyte migration from the vascular wall in response to traumatic brain injury. *Microvasc. Res.* **60,** 55–69.

35. Diaz-Flores, L., Gutierrez, R., and Varela, H. (1994) Angiogenesis: an update. *Histol. Histopathol.* **4,** 807–843.

V

MOLECULAR TECHNIQUES

26

Molecular Biology of the Blood–Brain Barrier

William M. Pardridge

1. Introduction

The blood–brain barrier (BBB) is formed by the microvasculature of the brain *(1)*. The permeability properties, *per se*, of the BBB are regulated by the capillary endothelial cell *(2)*. However, there are at least four different cells that comprise the brain microvasculature (**Fig. 1**), and all contribute to the regulation of the cerebral microvasculature and, indirectly, to the regulation of BBB permeability *(3)*. The endothethial cell and the pericyte share a common capillary basement membrane. There is approximately one pericyte for every two to four endothelial cells. More than 99% of the brain surface of the capillaries is invested by astrocytic foot processes *(4)*. There is innervation of the capillary by nerve endings of either intra- or extra-cerebral origin *(5,6)*. The distance between the astrocyte foot process and the capillary endothelial cell and the pericyte is only 20 nm *(7)*. Therefore, the interrelationships between the endothelium, the pericyte, and the astrocyte foot process are as intimate as any cell–cell interactions in biology. The space filled by the basement membrane and situated between the endothelium/pericyte and the astrocyte foot process forms the interface between blood and brain.

The actual transport of a ligand or substrate in either the direction of brain to blood, or blood to brain, requires movement across the capillary endothelial plasma membranes. The luminal and abluminal membranes of the capillary endothelium are separated by 100–300 nm of endothelial cytoplasm. Therefore, solute transfer across the capillary endothelial barrier is a process of transport through two membranes in series. However, in order for a molecule to move from blood to the brain interstitial space beyond the astrocyte foot process, the molecule must also escape the immediate perivascular space bordered by the plasma membranes of the capillary endothelial cell, pericyte, and astrocytic foot processes. Many "enzymatic BBB" mechanisms may operate within this space. The actual transport of nutrients or drugs across the BBB may be the result of a complex interplay between active efflux systems located on the endothelial plasma membrane, active transporters within the astrocytic foot process, and ecto-enzymes present on the pericyte plasma membrane. Although there is little direct evidence to date, it is likely that gene expression within the capillary endothelial cell, the capillary pericyte, or the capillary astrocytic foot process is highly interdependent. Expression within the capillary endothelium may be influenced by changes in either

From: *Methods in Molecular Medicine, vol. 89:*
The Blood–Brain Barrier: Biology and Research Protocols
Edited by: S. Nag © Humana Press Inc., Totowa, NJ

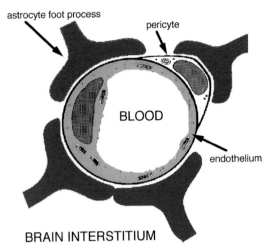

astrocyte foot process

pericyte

BLOOD

endothelium

BRAIN INTERSTITIUM

Fig. 1. The cells comprising the brain microvasculature are the capillary endothelium, the capillary pericyte, and the astrocyte foot process. In addition, nerve endings directly contact the capillary endothelial surface on the brain side of the microvasculature. From **ref. 3**.

astrocyte or pericyte signal transduction phenomena, and vice-versa. Therefore, the study of the molecular biology of the BBB is actually an investigation of the molecular biology of the cerebral microvasculature, which in turn is the study of the molecular biology of the capillary endothelial cell, the pericyte, and the astrocyte foot process all working in concert. Frequently, brain microvascular functions are ascribed to the capillary endothelial cell when, in fact, the pericyte or astrocyte foot process is actually the cellular site of the enzymatic or transport function under investigation. For example, the ecto-enzyme, aminopeptidase M, was originally localized to the brain microvasculature, and assumed to be endothelial in origin *(8)*. However, subsequent studies showed this gene product was selectively produced by the pericyte *(9)*. Other ecto-enzymes or amino-peptidases may be expressed at the endothelial surface. The active efflux system, p-glycoprotein, is localized to microvessels when examined by immunocytochemistry of brain sections *(10,11)*. However, it is often difficult with light microscopy to differentiate the astrocytic foot process from the capillary endothelium. Although this differentiation of cells comprising the microvasculature is difficult to make in tissue sections of brain, the cell differentiation can be made with isolated brain capillaries that are cyto-centrifuged to slides and immunostained with antigen-specific antibodies *(3)*. Moreover, the application of confocal microscopy and double immunolabeling of cytocentrifuged brain capillaries can lead to an unambiguous differentiation of capillary endothelial cells, pericytes, and astrocyte foot processes.

2. Gene Technologies and the BBB

There are at least two primary reasons for studying the molecular biology of the BBB. First, an understanding of the tissue-specific gene expression within the cells comprising the cerebral microvasculature can lead to a greater understanding of the pathogenesis of neurologic diseases, many of which have pathogenetic starting points at the brain microvasculature. For example, the earliest pathologic change in multiple sclerosis is

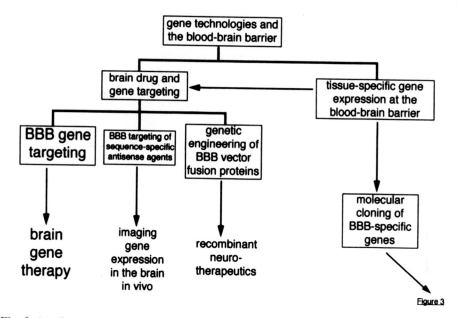

Figure 3

Fig. 2. Application of gene technologies to blood–brain barrier (BBB) research for both drug and gene targeting to the brain and for analysis of tissue-specific gene expression at the BBB.

a perivascular cuffing of lymphocytes (12). An early change in Alzheimer's disease (AD) is the elaboration of extracellular β-amyloid, owing to abnormal processing in microvascular smooth muscle cells and capillary pericytes (13). Second, progress in the molecular biology of the BBB enables the development of new strategies for drug and gene targeting through this barrier for the treatment of neurologic disease. These two platforms on which gene technologies and molecular biology are adapted to BBB biology are illustrated in **Fig. 2**.

The detection of tissue-specific gene expression at the BBB is possible with genomics and proteomics platforms as discussed later in this chapter. This work can enable the molecular cloning of BBB-specific genes as discussed below. An understanding of the tissue-specific gene expression at the BBB in neurologic disease can create new understanding of the pathogenesis of neurologic disorders. However, the understanding of tissue-specific gene expression at the BBB can also lead to the development of new approaches to targeting drugs and genes through the BBB, using endogenous receptor-mediated transport systems expressed at the BBB (see **Fig. 2**). The development of new brain drug and gene targeting technologies invariably requires the application of molecular biology (14). For example, the development of nonviral, noninvasive methods for targeting gene medicines through the BBB leads to new forms of brain gene therapy that does not require the use either of viral vectors or craniotomy for gene delivery to the brain (15). The development of gene-targeting technology requires the genetic engineering of expression plasmids, and production of recombinant proteins as targeting ligands. Gene expression in the brain can be imaged in vivo with antisense radiopharmaceuticals that are targeted through the BBB via endogenous transport systems within the capillary endothelium (16). The development of targeted antisense agents requires information on the nucleotide sequence of the target mRNA, and the genetic engineering of recombinant transport vectors (17). The genetic engineering of recombinant fusion proteins enables the development of neurotherapeutics, which have the dual function of transport through the BBB via an endogenous transport system and binding to specific receptors on brain cells to evoke intended therapeutic effects. In summary, brain gene therapy, imaging gene expression of the brain in vivo with targeted antisense radiopharmaceuticals, and the use of genetic engineering to produce recombinant neurotherapeutics are all various applications of molecular biology applied to BBB transport problems (*see* **Fig. 2**).

3. Molecular Cloning of BBB-Enriched Genes

3.1. Overview

BBB-enriched genes are genes that are expressed in brain selectively at the microvasculature, either capillary endothelial cells, pericytes, or at astrocytic foot processes. BBB-enriched genes may be expressed in peripheral tissues, but in brain, the cellular location of the gene expression is primarily at the microvasculature (*14*). Some BBB-specific genes are expressed only at the brain microvasculature and are not expressed in either brain cells or in cells in peripheral organs. There are dual benefits to the discovery of tissue-specific gene expression at the BBB. First, the identification of tissue-specific gene expression at the brain microvasculature provides a platform for understanding the pathogenesis of neurologic diseases that originate from the brain microvasculature. Second, the molecular analysis of tissue-specific gene expression at the BBB can lead to the identification of new targets for the delivery of genes, recombinant proteins, antisense agents, or drugs through the BBB in vivo.

3.1. Brain Capillary RNA Isolation

The cloning of BBB-enriched genes must start with the initial isolation of microvessels from either fresh animal or human brain (**Fig. 3**). An intact capillary preparation can be obtained from human autopsy brain and this preparation can be used for analysis of receptors at the BBB using methodology directed at the protein (*18*). However, it is unlikely that investigations of the ribonucleic acid (RNA) products at the microvasculature can be routinely performed on capillaries obtained from autopsy human brain that has undergone many hours of postmortem autolysis. Conversely, brain removed at neurosurgery can be used for the isolation of human brain capillary-derived mRNA (*19*).

Microvessels can be isolated from fresh brain with either a mechanical homogenization or an enzymatic homogenization technique. Either methodology results in a metabolic impairment of the capillaries. Freshly isolated brain capillaries obtained with either a mechanical or an enzymatic homogenization technique do not exclude trypan blue. It is often claimed that capillaries obtained with the enzymatic homogenization technique do exclude trypan blue but, to date, no laboratory has demonstrated that brain capillaries obtained with an enzymatic homogenization technique have normal cellular adenosine triphosphate (ATP) levels. The ATP content in freshly isolated brain capillaries obtained with either a mechanical or enzymatic homogenization technique is less than 10% of the expected cellular ATP content, indicating the capillaries are metabolically impaired (*20*). Nevertheless, if the capillaries are isolated with an efficient procedure using ribonuclease (RNase)-free conditions, it is possible to obtain brain capillary-derived polyA+ RNA in high yield without degradation (*21*). Unfortunately, few laboratories have established protocols for the isolation of polyA+ mRNA from iso-

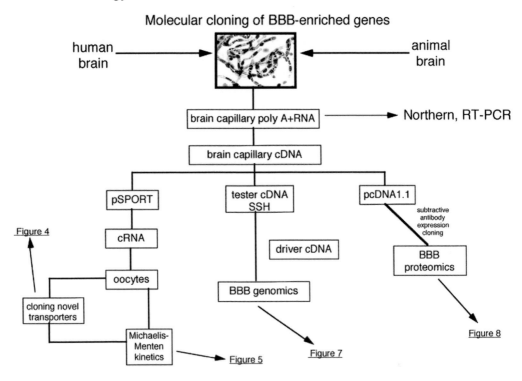

Fig. 3. Outline of pathways for molecular cloning of blood–brain barrier (BBB)-enriched genes. All methodologies start with the purification of polyA+ RNA from capillaries isolated from fresh human or animal brain. The cloning of novel transporters is outlined further in **Fig. 4**. The use of cloned RNA (cRNA) and the frog oocyte expression system to investigate the Michaelis-Menten kinetics of BBB transporters is outlined further in **Fig. 5**. The results of BBB genomics programs are outlined in **Fig. 7**. The methodology for subtractive antibody expression cloning and a BBB proteomics program is shown in **Fig. 8**.

lated brain capillaries, although these methods were described more than 10 yr ago *(21)*. It is straightforward to isolate polyA+ RNA from whole brain or other organs. However, a more demanding task is to first isolate the capillaries from brain and then isolate the capillary-derived polyA+ RNA. The reason this task is more difficult is that the capillary component of the brain is such a small fraction of the entire organ. The intraendothelial water volume in the brain is approx 0.8 µL/g brain, which is 1:1000 parts of the total water volume in brain. Therefore, more than 99% of the brain tissue is discarded during the process of isolating brain capillary-derived polyA+ RNA. Approximately 15 to 20 µg of polyA+ RNA can be isolated from 1 g of whole brain *(21)*. However, in order to obtain a comparable amount of brain capillary derived polyA+ RNA, it is necessary to first isolate the microvessels from 200 g brain. Nevertheless, it is possible to obtain amounts of brain capillary-derived polyA+ RNA that are adequate for many molecular biology studies from the pooled brains of 20 rats. Once the brain capillary polyA+ RNA is obtained, then capillary-derived cDNA can be generated with reverse transcriptase (RT) using priming with oligodeoxythymidine (ODT). This complementary deoxyribonucleic acid (cDNA) can then be inserted into

7. Paulson, O. B. and Newman, E. A. (1987) Does the release of potassium from astrocyte endfeet regulate cerebral blood flow? *Science* **237,** 896–898.

8. Solhonne, B., Gros, C., Pollard, H., and Schwartz, J. C. (1987) Major localization of aminopeptidase *M* in rat brain microvessels. *Neuroscience* **22,** 225–232.

9. Kunz, J., Krause, D., Kremer, M., and Dermietzel, R. (1994) The 140-kDa protein of blood-brain barrier-associated pericytes is identical to aminopeptidase *M. J. Neurochem.* **62,** 2375–2386.

10. Cordon-Cardo, C., O'Brien, J. P., Casals, D., et al. (1989) Multi-drug-resistance gene (P-glycoprotein) is expressed by endothelial cells at blood-brain barrier sites. *Proc. Natl. Acad. Sci. USA* **86,** 695–698.

11. Thiebaut, F., Tsuruo, T., Hamada, H., Gottesman, M. M., Pastan, I., and Willingham, M. C. (1989) Immunohistochemical localization in normal tissues of different epitopes in the multidrug transport protein P170: evidence for localization in brain capillaries and crossreactivity of one antibody with a muscle protein. *J. Histochem. Cytochem.* **37,** 159–164.

12. Raine, C. S., Lee, S. C., Scheinberg L. C., Duijvestin, A. M., and Cross, A. H. (1990) Adhesion molecules on endothelial cells in the central nervous system: an emerging area in the neuroimmunology of multiple sclerosis. *Clin. Immunol. Immunopathol.* **57,** 173–187.

13. Wisniewski, H. M., Wegiel J., Vorbrodt, A. W., Mazur-Kolecka, B., and Frackowiak, J. (2000) Role of perivascular cells and myocytes in vascular amyloidosis. *Ann. N. Y. Acad. Sci.* **903,** 6–18.

14. Pardridge, W. M. (2001) *Brain Drug Targeting; The Future of Brain Drug Development.* Cambridge University Press, Cambridge, United Kingdom.

15. Shi, N., Zhang, Y., Boado, R. J., Zhu, C., and Pardridge, W. M. (2001) Brain-specific expression of an exogenous gene following intravenous administration. *Proc. Natl. Acad. Sci. USA* **98,** 12,754–12,759.

16. Shi, N., Boado, R. J., and Pardridge, W. M. (2000) Antisense imaging of gene expression in the brain in vivo. *Proc. Natl. Acad. Sci. USA* **97,** 14,709–14,714.

17. Li, J. Y., Sugimura, K., Boado, R. J., et al. (1999) Genetically engineered brain drug delivery vectors—cloning, expression, and in vivo application of an anti-transferrin receptor single chain antibody-streptavidin fusion gene and protein. *Protein Engineering* **12,** 787–796.

18. Pardridge, W. M., Eisenberg, J., and Yang, J. (1985) Human blood-brain barrier insulin receptor. *J. Neurochem.* **44,** 1771–1778.

19. Shusta, E. V., Boado, R. J., Mathern, G. W., and Pardridge, W. M. (2002) Vascular genomics of the human brain. *J. Cereb. Blood Flow Metabol.* **22,** 245–252.

20. Lasbennes, R., and Gayet, J. (1983) Capacity for energy metabolism in microvessels isolated from rat brain. *Neurochem. Res.* **9,** 1–9.

21. Boado, R. J., and Pardridge, W. M. (1991) A one-step procedure for isolation of poly A+ mRNA from isolated brain capillaries and endothelial cells in culture. *J. Neurochem.* **57,** 2136–2139.

22. Boado, R. J., Li, J. Y., Nagaya, M., Zhang, C., and Pardridge, W. M. (1999) Selective expression of the large neutral amino acid transporter (LAT) at the blood-brain barrier. *Proc. Natl. Acad. Sci USA* **96,** 12,079–12,084.

23. Pardridge, W. M. (1983) Brain metabolism: A perspective from the blood-brain barrier. *Physiol. Rev.* **63,** 1481–1535.

24. Shusta, E. V., Boado, R. J., and Pardridge, W. M. (2002) Vascular proteomics and subtractive antibody expression cloning. *Molec. Cellular Proteomics*, in press.

25. Yoshida, K., Seto-Ohshima, A., and Sinohara, H. (1997) Sequencing of cDNA encoding serum albumin and its extrahepatic synthesis in the Mongolian gerbil, Meriones unguiculatus. *DNA Res.* **4,** 351–354.

26. Boado, R. J., and Pardridge, W. M. (1994) Measurement of blood-brain barrier GLUT1 glucose transporter and actin mRNA by a quantitative polymerase chain reaction assay. *J. Neurochem.* **62,** 2085–2090.

27. Boado, R. J., and Pardridge, W. M. (1990) Molecular cloning of the bovine blood-brain barrier glucose transporter cDNA and demonstration of phylogenetic conservation of the 5′-untranslated region. *Mol. Cell. Neurosci.* **1,** 224–232.

28. Li, J. Y., Boado, R. J., and Pardridge, W. M. (2001) Cloned blood-brain barrier adenosine transporter is identical to the rat concentrative Na+ nucleoside cotransporter CNT2. *J. Cereb. Blood Flow Metabol.* **21,** 929–936.

29. Gao, B., Stieger, B., Noe, B., Fritschy, J. M., and Meier, P. J. (1999) Localization of the organic anion transporting polypeptide 2 (Oatp2) in capillary endothelium and choroid plexus epithelium of rat brain. *J. Histochem. Cytochem.* **47,** 1255–1264.

30. Li, J. Y., Boado, R. J., and Pardridge, W. M. (2001) Blood-brain barrier genomics. *J. Cereb. Blood Flow Metabol.* **21,** 61–68.

31. Boado, R. J., Golden, P. L., Levin, N., and Pardridge, W. M. (1998) Upregulation of blood-brain barrier short form leptin receptor gene products in rats fed a high fat diet. *J. Neurochem.* **71,** 1761–1764.

32. Choi, T., and Pardridge, W. M. (1986) Phenylalanine transport at the human blood-brain barrier. Studies in isolated human brain capillaries. *J. Biol. Chem.* **261,** 6536–6541.

33. Santoni, V., Molloy, M., and Rabilloud, T. (2000) Membrane proteins and proteomics: Un amour possible? *Electrophoresis* **21,** 1054–1070.

34. Pardridge, W. M., Yang, J., Eisenberg, J., and Mietus, L. J. (1986) Antibodies to blood-brain barrier bind selectively to brain capillary endothelial lateral membranes and to a 46K protein. *J. Cereb. Blood Flow Metab.* **6,** 203–211.

35. Shusta, E. V., Zhu, C., Boado, R. J., and Pardridge, W. M. (2002) Subtractive expression cloning reveals high expression of CD46 at the blood-brain barrier. *J. Neuropathol. Exp. Neurol.* **61,** 597–604.

36. Pardridge, W. M., Triguero, D., Yang, J., and Cancilla, P. A. (1990) Comparison of in vitro and in vivo models of drug transcytosis through the blood-brain barrier. *J. Pharmacol. Exp. Ther.* **253,** 884–891.

27

Blood–Brain Barrier Genomics

Ruben J. Boado

1. Introduction

The application of the suppressive subtraction hybridization (SSH) technique to blood–brain barrier (BBB) genomics has accelerated the discovery and identification of BBB-specific genes in humans and in experimental animals (*1–3*). This procedure allows for the development of gene arrays based on gene products derived from isolated brain capillaries, which represents the BBB in vivo (*4*). The brain capillary volume is 10^{-3} or less than 0.1% of the total brain (*4*). Because the sensitivity of a typical gene microarray of the brain approximates 10^{-4} (*5*), the identification of BBB-specific genes from gene arrays derived from whole brain is statistically unlikely, even for BBB highly expressed transcript with relative abundance ranging 0.01–0.02% (i.e., actin, LAT1) (*6,7*). In addition, the levels of low abundant BBB-specific genes, like the p80 BBB-GLUT1 mRNA-binding protein, are less than 0.0005% of total cell proteins (*8*), which is several log orders below the limit of detection of deoxyribonucleic acid (DNA) microarrays (*5*). However, BBB-specific genes can be identified with a gene array derived from gene products initially obtained from isolated brain capillaries (**Fig. 1**).

Serial gene expression analysis (SAGE) was used for identification of genes expressed in tumor endothelium of human malignant colorectal tissue (*9*), and it may also be applicable to BBB genomics. However, SAGE requires a comprehensive sequence analysis to compare gene expression profiles in two different tissue samples (*9*), i.e., normal and tumor endothelium. On the contrary, the SSH protocol only analyzes differentially expressed genes (*10*); for example, BBB-specific genes. The SSH procedure was successfully used in this laboratory to identify and isolate BBB-specific genes in humans and rats (*1–3*). The identified clones encode proteins of known and unknown functions that are involved in angiogenesis, neurogenesis, molecular transport, and maintenance of BBB tight junctions or cytoskeleton. Because only a small percentage of BBB-SSH cDNA libraries has been screened (i.e., 0.1–5%) (*1–3*), it is anticipated that BBB genomics programs based on SSH cloning may provide a complete BBB-specific gene array to investigate mechanisms of brain pathology at the microvascular level.

The complete BBB-SSH protocol is described in this chapter as summarized in **Fig. 1**. The BBB-SSH procedure begins with the isolation of brain capillaries (**Fig. 1, II**). Poly-A$^+$/mRNA is purified from isolated brain capillaries, kidney, and liver

From: *Methods in Molecular Medicine, vol. 89:*
The Blood–Brain Barrier: Biology and Research Protocols
Edited by: S. Nag © Humana Press Inc., Totowa, NJ

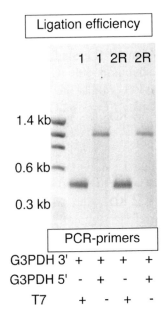

Fig. 4. Ligation efficiency of adaptors 1 and 2R into blood–brain barrier (BBB)-tester cDNAs. The efficiency of the ligation was determined by PCR amplification using G3PDH and T7 primers. Aliquots of 5 μL were resolved by electrophoresis in a 2% agarose gel, and the ethidium bromide staining of the gel is shown in the figure. The combination of primers used in the PCR protocol are indicated in the bottom of the figure, and the migration of DNA standards on the left side. The PCR amplification of BBB-cDNA 1 or 2R with G3PDH primers produced the expected approx 1.2-kb band corresponding to an internal cDNA fragment of rat G3PDH. Amplification with T7 and G3PDH-3′ primers yielded an approx 0.5-kb PCR product in both samples, which corresponds to the G3PDH-adaptor cDNA fragment. The intensity of the bands are comparable, suggesting that ligation of adaptors was successfully performed.

3.2.3. Efficiency of Ligation

The efficiency of the ligation of adaptors is determined by PCR amplification using G3PDH and T7 primers. cDNA samples are amplified using forward and reverse G3PDH primers (G3PDH-5′, 5′-ACCACAGTCCATGCCATCAC-3′; and -3′, 5′-TCCACCACCCTGTTGCTGTA-3′, respectively), which amplify a cDNA fragment of approx 1.2 kb rat G3PDH. In addition, samples are also amplified with G3PDH reverse primer and the T7 primer (also known as PCR primer 1) located at the ligated adaptor. Using the latter combination of primers, PCR products would be seen only if the ligation of adaptors was successful.

1. PCR amplification is performed using the Advantage 2 PCR enzyme system in a total volume of 25 μL Advantage 2 PCR buffer containing 2 μL tester cDNA ligated to adaptor 1 or 2R, 0.2 mM dNTPs, 0.4 μM primers of interest (i.e., G3PDH-3′ and G3PDH-5′ or T7) and 0.5 μL Advantage 2 polymerase mix.
2. Samples are extended for 5 min at 75°C in a temperature cycler, and the PCR amplification is carried out for 30 cycles of denaturing at 95°C for 30 s, annealing at 65°C and 30 s, and extension at 68°C for 2.5 min. In a representative experiment, the PCR amplification of rat BBB-cDNA produced two expected major bands corresponding to the approx 1.2 kb rat G3PDH and the approx 0.5 kb G3PDH-T7 (linker/adaptor) fragment (**Fig. 4**).

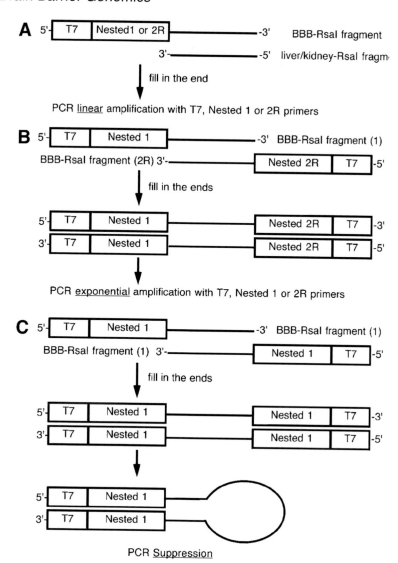

Fig. 5. Diagram of different blood–brain barrier (BBB)-tester and liver/kidney-driver cDNA hybrids. **(A)** The BBB-tester is subjected to linear amplification in a tester-driver cDNA hybrid following PCR amplification with either T7 or Nester 1 or 2R primers. **(B)** Double stranded tester cDNA hybrids possessing both adaptors (i.e., 1 and 2R) are exponentially amplified in two successive PCR amplification steps performed with T7, and nested 1 and 2R primers, respectively. **(C)** BBB-tester-tester hybrids carrying the same adaptor (i.e., 1 or 2R) are suppressed during the PCR amplification because the kinetics of annealing favors the formation intramolecular panlike structures over annealing a shorter primer. Pool of hybrids comprised of driver-driver double stranded cDNA lacking adaptors, as well as single stranded tester cDNA, are not amplified (not shown).

The intensity of the bands are comparable, suggesting that ligation of adaptors was successfully performed. If the intensity of the 0.5 kb G3PDH-T7 fragment represents less than 25% of the 1.2 kb G3PDH, ligation of adapter should be repeated to ensure a successful SSH procedure.

3.2.4. Hybridization of Tester (BBB) and Driver (Liver–Kidney) cDNAs

To enrich for differentially expressed sequences, the SSH protocol proceeds with the hybridization of the two populations of tester BBB-cDNA prepared as described above with adaptor 1 or 2R with the driver liver and kidney cDNA.

1. Aliquots of 1.5 µL BBB-tester-adaptor 1 and -adaptor 2R cDNAs are independently hybridized with 1.5 µL of driver cDNA in a total volume of 4 µL hybridization buffer in a temperature cycler at 98°C for 1.5 min followed by 8 h at 68°C.
2. The hybridization step is completed in a second incubation wherein a new aliquot of 1 µL RsaI-digested cDNA is heat-denatured for 1.5 min at 98°C, immediately mixed with the already hybridized samples tester-adaptor 1-driver and tester-adaptor 2R-driver, and incubated overnight at 68°C.
3. Following incubation, the sample is diluted with 200 µL 20 m*M* HEPES-HCl, pH 8.3, 50 m*M* NaCl, 0.2 m*M* EDTA, and stored at –20°C.

3.2.5. Suppressive Subtraction PCR

The hybridization procedure creates tester-driver cDNA hybrids with transcripts expressed in both tissues (i.e., BBB, liver, and kidney) (**Fig. 5A**). Because the suppressive PCR step begins by fill-in of the ends followed by amplification with the T7 primer, tester-driver hybrids are just subjected to linear amplification. On the contrary, double stranded tester cDNA hybrids possessing both adaptor 1 and 2R sequences are exponentially amplified in two successive PCR amplification steps performed with T7, and nested 1 and 2R primers, respectively (**Fig. 5B**). This procedure results in the selective amplification of BBB-specific transcripts and suppression of mRNAs commonly expressed in BBB, liver, and kidney. Pools of hybrids comprised of driver-driver double stranded cDNA lacking adaptors, as well as single stranded tester cDNA, are not amplified. Tester-tester hybrids carrying the same adaptor (i.e., 1 or 2R) are suppressed during the PCR amplification because the kinetics of annealing favors the formation intramolecular panlike structures over annealing a shorter primer (**Fig. 5C**) (*10*). The suppressive subtraction procedure is performed in two successive PCR steps.

1. A first PCR run is performed to amplify differentially expressed BBB-tester sequences with the T7 primer. The PCR reaction is carried out in 0.5-mL nuclease-free microtubes in a total volume of 25 µL PCR buffer (i.e., Advantage 2 PCR kit, Clontech) containing 1 µL subtracted (hybridized) or unsubtracted (obtained above immediately after ligation of adaptors) tester cDNA, 0.2 m*M* dNTPs, 0.4 µ*M* T7 primer and 0.5 µL Advantage 2 polymerase mix.
2. The ends of double stranded cDNAs are filled by incubation at 75°C for 5 min, and PCR amplification is performed for 30 cycles (denaturation, 94°C for 30 s; annealing, 68°C for 30 s; extension, 72°C for 1.5 min).
3. An aliquot of the samples is taken (i.e., 8 µL) and analyzed by agarose gel electrophoresis (*see* below).
4. A second PCR run is performed to further enrich differentially expressed BBB-cDNA sequences and to suppress background of linearly amplified cDNAs (**Fig. 5A**). The second PCR is performed for 15 cycles with 1 µL 1/10 v/v diluted samples obtained in the first PCR run in a total volume of 25 µL as described above for the first PCR run, but with 0.4 µ*M* Nested PCR 1 and 2R primers in lieu of the T7 primer. **Figure 6** shows the expected increase in the PCR products following first and second PCR amplification of both BBB and a control subtracted cDNA, respectively.

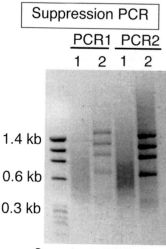

Sample 1: BBB-SSH
Sample 2: control

Fig. 6. Blood–brain barrier (BBB)-PCR products following SSH. BBB-tester cDNA 1 and R2 were hybridized with liver/kidney-driver cDNA as described in text, and subjected to two successive PCR amplification steps with T7, and nested 1 and 2R primers, respectively. Aliquots of 5 μL were resolved by electrophoresis in a 2% agarose gel, and the ethidium bromide staining of the gel is shown in the figure. The migration of DNA standards is indicated on the left side of the figure. Samples were (1) BBB-tester cDNA following SSH, and (2) control subtracted cDNA (Clontech). The expected increase in the PCR products following first and second PCR amplification of both BBB and control subtracted cDNA is seen.

5. The efficiency of the subtraction procedure may be determined by PCR analysis of glyceraldehyde-3-phosphate dehydrogenase (G3PDH) as previously described in this laboratory *(2)*. In a typical BBB-SSH experiment, G3PDH PCR products are seen at 18 and 33 cycles of PCR with unsubtracted and subtracted BBB-tester cDNA, respectively *(2)*.

3.2.6. Construction of Subtracted cDNA Libraries

Subtracted cDNA libraries are constructed with aliquots of 1 μL of products of the second PCR products in the pCR 2.1 or pCR II vectors (Invitrogen) using standard cloning techniques in the *E. coli* INVαF competent cells *(1–3)*. A typical yield approximates 500 to several thousand recombinants, with a nonrecombinant background activity of less than 10% *(1–3)*.

3.3. Differential Hybridization

Positive clones are identified by a differential hybridization procedure in which randomly selected colonies are hybridized with subtracted and unsubtracted ^{32}P-labeled BBB-cDNA probes *(1–3)*. Clones that show a strong signal with the subtracted cDNA probe compared to the unsubtracted one (i.e., ratio >2–10, **Fig. 7**, LK3) are selected for DNA sequencing and confirmation of its BBB-specificity by Northern blot analysis.

1. Bacterial colonies are randomly selected and cultured in 96-well plates in LB/amp medium.
2. Colonies (i.e., 100-μL culture) are individually blotted on the GenScreen Plus membrane using a 96-well dot-blot system. Duplicate membranes may allow for simultaneously probing with subtracted and unsubtracted BBB-cDNA.

3. DNA is denatured by incubation of the membranes (face up) on Whatman 3-mM paper soaked in 0.5 M NaOH for 2 min at 22°C, followed by neutralization with 1 M Tris-HCl, pH 7.4.

4. Membranes are prehybridized with 20 mL hybridization solution [1.5 × SSPE, 1% SDS, 0.5% nonfat dry milk and 10 µL/mL PCR select differential screening blocking solution (Clontech) for 2 h at 65°C in a hybridization incubator.

5. Following prehybridization, the solution is discarded and the membranes are hybridized in 5 ml fresh hybridization solution containing 1 × 10⁶ cpm/mL ³²P-labeled subtracted or unsubtracted BBB-tester cDNA probe for 16 h at 65°C.

6. ³²P-labeled probes and blocking solutions are heat-denatured immediately before use (*see* **Note 5**). cDNAs are labeled with α³²P-dCTP to a specific activity = 1 × 10⁹ cpm/µg using the random primer labeling technique (*see* **Note 6**).

7. Membranes are successively washed with increasing stringency washing solutions (*see* **Note 7**).

8. Positive clones are identified by film autoradiography (*see* **Note 8**). In a representative experiment, visual analysis of the subtracted and unsubtracted autoradiograms allows for the unequivocal identification of clones with hybridization signals that are markedly augmented with the subtracted probe compared with the unsubtracted one (*see* **Fig. 7**, LK-1 to LK-6). Autoradiograms may also be quantified using the NIH Image program 1.54 in a Macintosh G4 computer, and the subtracted/unsubtracted ratio may be calculated for each clone using the integrated density values. In order to validate the differential hybridization protocol, the specific activity of the subtracted and unsubtracted ³²P-cDNAs probes ought to be similar (*see* **Note 9**). Probing of membranes with ³²P-labeled subtracted and unsubtracted cDNAs of non-comparable specific activities may result in isolation of false positive clones.

3.4. Isolation and Characterization of Clones

Clones of interest (i.e., LK3, **Fig. 7**) are amplified in LB/amp medium and plasmid DNA isolated using the Qiagen mini-prep kit (*see* **Note 11**). Clones are sequenced in either one or both directions with M13 forward and reverse primers (see below). The size of the cDNA inserts are determined by digestion with EcoRI, sites located at both 5′- and 3′-flanking region of the insertion site in pCR vectors, followed by 1.5% agarose gel electrophoresis and staining with ethidium bromide (*15*). cDNA inserts are then isolated from the gel using the QIAquick PCR purification kit (*see* **Note 11**), and used to probe Northern blot filters containing a panel of tissue RNA, including BBB, brain, and peripheral tissues (**Fig. 2**) (*see* **Subheading 3.6.**).

3.5. DNA Sequencing and BLAST Analysis

DNA sequencing of isolated clones is performed in either one or both directions using M13 forward and reverse primers (*1–3*). Sequences are performed at university or private DNA sequencing Core Facilities using DNA sequencing protocols provided by those laboratories. Similarities with other genes deposited in GenBank are investigated using the BLAST program (NCBI, http://www.ncbi.nlm.nih.gov/BLAST/). Searching of the nonredundant (NR) database is used to identify known genes, whereas screening of the expressed sequence tags database (EST) is used to identify genes of unknown function. If screening of both NR and EST databases, as well as genomic survey sequence (GSS) and high throughout genomic sequences (HTGS), is negative, the sequence is considered a novel BBB-specific transcript (**Table 1**). BBB-SSH novel clones may be used for screening of a rat BBB cDNA library for isolation of full length

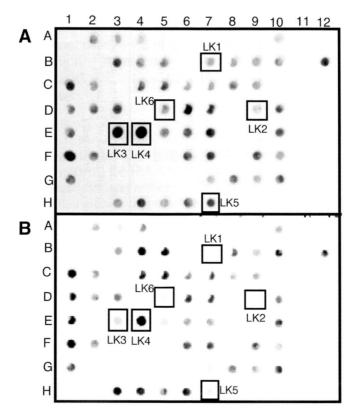

Fig. 7. Autoradiogram of brain–blood barrier (BBB)-specific clones following Southern blot hybridization with **(A)** ^{32}P-labeled subtracted and **(B)** unsubtracted BBB-tester cDNA. Bacterial colonies were randomly selected and cultured in 96-well plates in LB/amp medium. Aliquots of 100 µL bacterial culture were individually blotted on the GenScreen Plus membrane using a 96-well dot-blot system. Membranes were hybridized with ^{32}P-labeled subtracted or unsubtracted blood–brain barrier (BBB)-tester cDNA, and the autoradiogram was exposed for 16 h at 22°C. Comparison of the subtracted and unsubtracted autoradiograms allows for the identification of BBB-specific clones showing augmented signals with the subtracted cDNA probe (i.e., LK-1 to LK-6).

cDNA clones *(1)*. A partial list of clones isolated in a representative rat BBB-SSH is shown in **Table 1**. Clone K2 was novel and not found in any databases. This clone was used for isolation of a full-length cDNA from a rat BBB cDNA library, and later named BBB-specific anion transporter type 1 (BSAT1) *(1)*. The BSAT1 gene expression is specific to the BBB and is not detected in brain or peripheral tissues **(Fig. 2)**. Other isolated clones were expressed in brain only at the BBB and in peripheral tissues, i.e., organic anion transporter type 2 (Oatp2, clone K1), vascular endothelial growth factor receptor (Flt-1, clone LK-7) and a rat EST (clone LK-9) **(Table 1, Fig. 7)**.

3.6. Northern Blotting

Northern blot analysis is performed with 2 µg Poly-A⁺ RNA isolated from rat brain capillaries, rat brain, and peripheral tissues including rat heart, lung, liver, and kidney.

values determined by the plate reader were less than 1000, were not included in this initial analysis as per the recommendation of the manufacturer.

3. Criterion of consistency across different observations: All experiments were repeated in triplicate on comparable cell populations isolated from different patients or sources. Only changes greater than three times the SD and consistently occurring in these cell populations were considered significant (**Fig. 3C–E**). EC were obtained from aneurysm and epileptic surgeries ($n = 3$ each) and exposed to flow (3 or 20 dyne/cm^2) or grown under static conditions. The families of genes that showed the most consistent changes according to our criteria were related to cell cycle, glucose metabolism, and cytoskeletal functions (**Fig. 3A,B**). The data obtained were then confirmed by Western blotting (inset, **Fig. 3A**), demonstrating that the observed changes correlate well with protein expression.

4. Notes

1. In a co-culture, both cell populations are fed the same media by design. If a cell line has been grown in a different media it must be weaned over to the new formula before it is added to the cartridge. Generally, the feed media of choice is the one in which the more fastidious cell population grows. In this example it is the human endothelial cells lining the lumen that require their own specific medium, therefore, astrocytes, which are a less demanding population, are weaned over prior to loading in the ECS. As a rule, feeding the astrocytic flasks first with one-third endothelial medium and two-thirds glial medium followed by two-thirds endothelial medium and one-third glial medium and finally 100% endothelial medium will prepare astrocytes to grow in the endothelial medium present in the cartridge.

2. It is essential to be meticulous about maintaining the sterility of the system. Wipe all connections with alcohol swabs before and after disconnection and use sterile syringes for each sample. Do all sampling in a laminar flow hood if at all possible. Once contaminated, the cartridge is useless as antibiotics alone are usually not sufficient to kill bacteria or fungi.

3. Care must be taken to avoid bubble formation in the tubing, as bubbles travel to the lumen of the capillaries and strip off cells, leaving bare areas. Bubble formation leads to cell loss, capillary drying, and also decreases the flow rate which deprives cells of adequate nourishment, and thus interferes with normal cell growth and differentiation. All connections must be tightened to prevent air leaks. A several-day priming of the system with complete media allows all spaces to fill and bubbles can then be chased into the bottle before any cells are added. Pre-warming media to 37°C is helpful both for the comfort of the cells and to lessen bubble formation.

4. Cells need to be monitored frequently for glucose consumption when first set up in order to establish a satisfactory feeding schedule, as the cell demands in a cartridge may be very different from those in a flask. Generally the medium is replaced two to three times per week. Extra glucose may be added to help survival when media replacement is not possible. Starved cells may begin to die, fall off the capillary walls, and disrupt the barrier formation. A sign of underfeeding is the presence of an acid pH shift (solutions turn yellow), but hypoglycemia may occur even in the absence of any significant pH change.

5. When removing bubbles from between a membrane and the hybridization tube, it is best to grasp the corner of the membrane and remove the bubbles by lifting or pulling the membrane around the tube. Care should be taken not to scratch the membrane's surface.

6. Utilizing ^{32}P labeled probes could cause blossoming or bleeding to occur when hybridized with GENEFILTERS microarrays. Because of this phenomenon, ^{32}P hybridized GENE-FILTERS membranes should not be analyzed using Pathways™ software.

7. Avoid drying of the membranes between hybridizations and washes since this increases the stripping efficiency. Wrap the moist membranes with plastic sheets (e.g., Cling Wrap) but not with mylar, due to the low energy emittance of ^{33}P.

8. It is recommend that you determine the exposure time empirically based on the specific activity of your probe. To maximize the effectiveness of Pathways™ analysis software we recommend that you compare only those images that have a similar range of intensity.

Acknowledgments

This work was supported by the following Grants: NIH-2RO1 HL51614, NINDS RO1 43284, and NINDS RO1 NS38195.

References

1. Cancilla, P. A., Bready, J., and Berliner, J. (1993) Brain endothelial-astrocyte interactions. In *The Blood-Brain Barrier Cellular and Molecular Biology*. Pardridge, W. M., ed. Raven, New York, pp. 25–47.

2. Grant, G. A., Abbott, N. J., and Janigro, D. (1998) Understanding the Physiology of the Blood-Brain Barrier: In Vitro Models. *News Physiol. Sci.* **13,** 287–293.

3. Desai, S., Marroni, M., Cucullo, L., et al. (2002) Mechanisms of endothelial survival under shear stress. *Endothelium* **9,** 89–102.

4. Janigro, D., Leaman, S. M., and Stanness, K. A. (1999) Dynamic modeling of the blood-brain barrier: a novel tool for studies of drug delivery to the brain. *Pharm. Sci. Technol. Today* **2,** 7–12.

5. Janigro, D., Stanness, K. A., Soderland, C., and Grant, A. G. (2000) Development of an In Vitro Blood-Brain Barrier. In *Biochemical and Molecular Neurotoxicology*. Maires, M., Costa, L., Reed D., et al., eds., John Wiley & Sons, Inc., New York, 12.2.1–12.2.10.

6. McAllister, M. S., Krizanac-Bengez, L., Macchia, F., et al. (2001) Mechanisms of glucose transport at the blood-brain barrier: an in vitro study. *Brain Res.* **904,** 20–30.

7. Salvetti, F., Ceci, F., Janigro, D., Lucacchini, A., Benzi, L., and Martini, C. (2002) Insulin permeability across an in vitro model of the endothelium. *Pharm. Res.* **19,** 445–450.

8. Sinclair, C. J., Krizanac-Bengez, L., Stanness, K. A., Janigro, D., and Parkinson, F. E. (2001) Adenosine permeation of a dynamic in vitro blood-brain barrier inhibited by dipyridamole. *Brain Res.* **898,** 122–125.

9. Stanness, K. A., Guatteo, E., and Janigro, D. (1996) A dynamic model of the blood-brain barrier "in vitro." *Neurotoxicology* **17,** 481–496.

10. Stanness, K. A., Neumaier, J. F., Sexton, T. J., et al. (1999) A new model of the blood-brain barrier: co-culture of neuronal, endothelial and glial cells under dynamic conditions. *Neuroreport* **10,** 3725–3731.

11. Stanness, K. A., Westrum, L. E., Fornaciari, E., et al. (1997) Morphological and functional characterization of an in vitro blood-brain barrier model. *Brain Res.* **771,** 329–342.

12. Janigro, D., Stanness, K. A., Nguyen, T.-S., Tinklepaugh, D. L., and Winn, H. R. (1995) Possible role of glia in the induction of CNS-like properties in aortic endothelial cells: ATP-activated channels. In *Adenosine and Adenine Nucleotides*. Bellardinelli, L., and Pelleg, A., eds. Nijhoff, Boston, pp. 85–96.

13. Janigro, D., Strelow, L., Grant, G. A., and Nelson, J. A. (1998) Development of an in vitro BBB model for neuroAIDS. *NeuroAIDS* **1**.

14. Parkinson, F. E., Stanness, K. A., Anderson, C. M, and Janigro, D. (1998) *Nucleoside transporter subtypes in rat brain endothelial cells and astrocytes*. Society for Neuroscience Abstracts, abstract.

15. Pekny, M., Stanness, K. A., Eliasson, C., Betsholtz, C., and Janigro, D. (1998) Impaired induction of blood-brain barrier properties in aortic endothelial cells by astrocytes from GFAP-deficient mice. *Glia* **22,** 390–400.
16. Stanness, K. A., Janigro, D., and Neumayer, J. (1998) *A new blood-brain barrier model: culturing of serotoninergic neurons with endothelium and glia.* Society for Neuroscience Abstracts, abstract.
17. Strelow, L., Janigro, D., and Nelson, J. A. (2001) The blood-brain barrier and AIDS. *Adv. Virus Res.* **56,** 355–388.
18. Ballermann, B. J., Dardik, A., Eng, E., and Liu, A. (1998) Shear stress and the endothelium. *Kidney Int. Suppl.* **67,** S100–S108.
19. Meyer, J., Tauh, J., and Galla, H. G. (1991) The susceptibility of cerebral endothelial cells to astroglial induction of blood-brain barrier enzymes depends on their proliferative state. *J. Neurochem.* **57,** 1971–1977.
20. Ott, M. J., Olson, J. L., and Ballermann, B. J. (1995) Chronic in vitro flow promotes ultrastructural differentiation of endothelial cells. *Endothelium* **3,** 21–30.

Phage Display Technology for Identifying Specific Antigens on Brain Endothelial Cells

Jamshid Tanha, Arumugam Muruganandam, and Danica Stanimirovic

1. Introduction

The development of efficient ways to deliver large molecules such as peptides, proteins, and nucleic acids across the blood–brain barrier (BBB) is crucial to future therapeutic strategies for treatment of central nervous system (CNS) disorders. The principal approach to deliver macromolecules across the BBB is the development of chimeric peptides (1). Ligands to various receptors that undergo transcytosis across brain capillary endothelium and are essential for physiological transport of proteins, including transferrin, insulin growth factor, and low-density lipoprotein, into the brain are used as vectors to deliver drugs or therapeutic peptides chemically linked to the ligand (1,2). This process is known as receptor-mediated endocytosis/transcytosis. An anti-transferrin receptor antibody (OX-26), for example, has been used to deliver endorphin, vasoactive intestinal peptide, and brain-derived neurotrophic factor (1), as well as oligonucleotides and plasmid deoxyribonucleic acid (DNA) (2) into the brain parenchyma. Further development of this approach requires rapid discovery of other suitable receptors/antigens expressed on human BBB endothelium that undergo transcytosis upon ligand binding.

In this chapter, we will describe an application of antibody-phage display technology in discovery of new antigens that undergo transcytosis across the BBB endothelium. The same approach can be used to develop novel vectors for brain drug delivery.

1.1. General Principles of the Phage Display Approach

Phage display libraries of peptides, proteins, and antibodies have previously been used to identify binders to biological targets in vitro and in vivo (3), including purified antigens, whole cells, and tissues. The technology is based on the observation that a polypeptide (capable of performing a function, typically the specific binding to a target of interest) can be displayed on the filamentous bacteriophage surface by inserting the gene coding for the polypeptide in the phage genome (4). Various formats and sizes of polypeptides have been displayed in phage libraries including peptides, small proteins, or antibody fragments (4). If one is able to select and purify a phage particle that displays binding specificity to a target from a large repertoire of phage particles (e.g., phage

From: *Methods in Molecular Medicine, vol. 89:*
The Blood–Brain Barrier: Biology and Research Protocols
Edited by: S. Nag © Humana Press Inc., Totowa, NJ

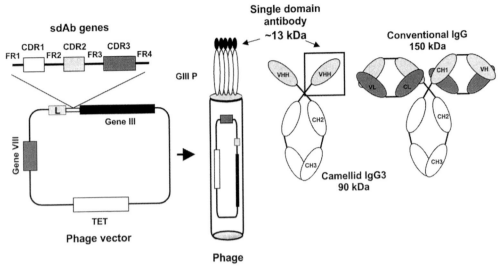

Fig. 1. Llama single-domain antibody (sdAb) phage display library. A nonimmunized llama sdAb phage displayed library is derived from the V$_H$H of the heavy chain IgGs that occur naturally in the absence of light chain. The sdAb library genes are constructed as described (*6*) and fused to Gene III and leader sequence gene (L) as shown. Only two genes, III and VIII, out of 11 genes in phage genome are shown. TET-tetracycline resistance gene cassette. The sdAbs are expressed as *N*-terminal fusion to GIIIp and exposed for binding at the tip of the phage. The library is constructed using a phage vector allowing for a multivalent display of sdAbs. The average molecular weight of sdAbs is about 13 kDa, as compared to 150 kDa for an IgG. The library contains 5.6×10^8 different sdAb species.

library), the genetic information (i.e., sequence) coding for the binding protein can be extracted and the corresponding phage can be amplified by bacterial infection.

The process of selection of specific binders from a library is called panning. The selection can be done against purified immobilized baits, cells, tissues, or by in vivo injection of the library. In a typical round of panning, a phage display library is applied onto a support coated with the antigen of interest; only some binding specificities are retained on the surface after washing and these can be selectively eluted (*3*). The eluted phage is then amplified by bacterial infection and used for a further round of panning. Typically, several rounds of panning are necessary to select and amplify rare binding specificities present in large repertoires.

In protocols described here, we used a non-immunized llama single-domain antibody (sdAb) phage display library derived from the V$_H$H of the heavy chain IgGs that occur naturally in the absence of light chain (*5*). The library was constructed and characterized at the Institute for Biological Sciences (NRC, Ottawa) as described in detail in Tanha et al. (*6*). The average molecular weight of sdAbs displayed in the library is about 13 kDa, approximately half the size of a scFv, and 10 times smaller than conventional IgG (150 kDa). The library exists in an fd-based filamentous phage vector and contains 5.6×10^8 antibody species. Each antibody is expressed in three to five copies on the surface of the phage. The main properties of the phage display library used in the described protocols are shown (**Fig. 1**). Other antibody- or peptide-phage display libraries (some of which are commercially available) can be used instead.

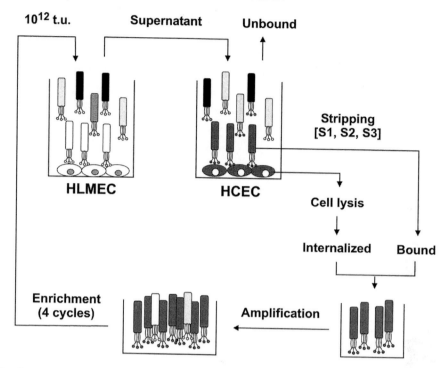

Fig. 2. A subtractive panning protocol to enrich llama sdAb phage display library for antibodies that bind selectively to human cerebromicrovascular endothelial cells (HCEC). The library (10^{12} pfu) was first applied to human lung microvascular endothelial cells (HLMEC) to subtract common endothelial binders; non-bound phage was then applied to HCEC. Phage bound to HCEC was dislodged by three stripping washes, and internalized phage was recovered by cell lysis. Each stripping and internalized fraction was analyzed by plaque PCR; the fraction containing the highest percent of full-length clones was amplified and used as input phage in the next round of panning. Enrichment was done in four rounds of panning.

The selection in this protocol was done against two endothelial cell types: human lung microvascular endothelial cells (HLMEC) and human cerebromicrovascular endothelial cells (HCEC). Protocols for isolation, culturing, and maintenance of human brain microvascular endothelial cells are described elsewhere (7). Other cell types can be used instead. The phage library was first preabsorbed onto HLMEC to remove common endothelial binders, and unbound phage was then applied to HCEC to identify BBB-specific binders (**Fig. 2**). HCEC express BBB-specific antigens such as γ-glutamyl transpeptidase that are not expressed in peripheral endothelial cells (7). After removing unbound phage, phage bound to HCEC was dislodged by three rounds of highly stringent stripping washes (S1, S2, and S3) to favor the selection of sdAbs with higher affinity. HCEC were then lysed to capture the internalized (i.e., endocytosed) phage (Int). This approach is called subtractive selection, and is repeated for four rounds (*see* **Fig. 2**).

After enriching the library for HCEC-specific binders, this enriched library was passed through an in vitro BBB model to identify phage that transmigrate HCEC monolayer. Phage clones that transmigrated HCEC were sequenced and sdAbs coded by these sequences were expressed, purified, and tagged with markers that allow for their purification and detection in in vitro and in vivo experimental systems.

For the purpose of this chapter, the llama sdAb phage display library and human cell cultures used in selection procedures are considered starting materials.

2. Materials
2.1. Cells

1. HLMEC (Mandel Scientific Company Ltd., Guelph, ON, Canada).
2. HCEC (passage 2-3) *(7)*.
3. Fetal human astrocytes (FHA) *(8)*.
4. *E. coli* strain TG1 (Stratagene, La Jolla, CA).

2.2. Phage Display Library

Llama single-domain antibody phage display library (or other antibody or peptide phage display library).

2.3. Media and Buffers

1. Phosphate-buffered saline (PBS).
2. Luria-Bertani (LB), $2 \times$ YT and Induction media (Terrific Broth with no salts).
3. Agarose top media: 10 g bacto-tryptone, 5 g yeast extract, 10 g NaCl, 1 g $MgCl_2 \cdot 6H_2O$, and 7 g agarose (per liter).
4. Protein purification starting buffer: 10 mM HEPES (N-[2-hydroxyethyl]piperazine-N'-[2-ethanesulfonic acid]) buffer, 10 mM imidazole, 500 mM NaCl, pH 7.0.
5. Protein purification elution buffer: 10 mM HEPES buffer, 500 mM imidazole, 500 mM NaCl, pH 7.0 in Hank's balanced salt solution (HBSS).
6. Stripping buffer: 50 mM glycine pH 2.8, 0.5 M NaCl, 2 M urea, 2% polyvinyl pyrolydine.
7. Neutralizing buffer: 1 M Tris-HCl buffer pH 7.4.
8. Triethanolamine.
9. Autoclaved PEG/NaCl solution: 20% (w/v) polyethylene glycol 6000 or 8000, 2.5 M NaCl).
10. Transport buffer: 10 mM HEPES pH 7.4, 5 mM $MgCl_2$ and 0.05% bovine serum albumin (BSA) in HBSS.
11. Autoclaved deionized water.

2.4. Antibodies

1. ALP- or HRP-conjugated monoclonal anti-c-myc antibody (Sigma Chemical Co., Oakville, ON, Canada).
2. HRP-conjugated anti-M13 monoclonal antibody (Amersham Pharmacia Biotech, Baie d'Urfé, QC, Canada).

2.5. Other Chemicals, Columns, and Kits

1. Ampicillin.
2. Isopropyl-β-D-thio-galactopyranoside (IPTG).
3. TMB peroxidase substrate and H_2O_2 (KPL, Gaithersburg, MD).
4. Oligonucleotide primers.
5. Restriction enzymes and T4 DNA ligase (New England Biolabs, Inc., Mississauga, ON, Canada) and AmpliTaq™ DNA polymerase (Perkin Elmer, Mississauga, ON, Canada).
6. pSJF2 expression vector **(Fig. 3)** *(19)*.
7. Durex™ 13×100-mm borosilicate glass tubes (VWR Scientific, Mississauga, ON, Canada).

A

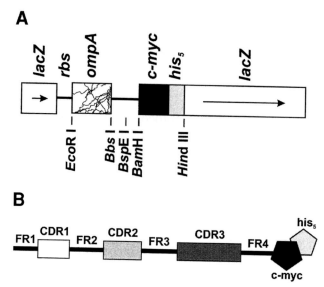

B

Fig. 3. Periplasmic expression of V_HHs in *E. coli*. (**A**) Schematic drawing of the expression vector pSJF2 cloning site. The vector is constructed by replacing the *Eco*R I-*Bam*H I polylinker fragment, located within *lac* Z gene, in pUCE8 *(18)* with the *Eco*R I-*Bam*H I expression cassette shown in **A**. Cloning between the Bbs I and BamH I sites places the sdAb genes precisely between the ompA signal peptide gene *(ompA)* and c-myc-His$_5$ genes (*c-myc* and *His$_5$*). The arrow in the diagram shows the direction of *lac* Z expression in pUCE8. (**B**) Schematic drawing of the mature V_HH. Following translation, the signal peptide directs the transport of sdAb-c-myc-His$_5$ to periplasm where the signal peptide is cleaved and the tagged protein folds into its native state.

8. QIAquick Gel Extraction™ kit (QIAGEN Inc., Mississauga, ON, Canada).
9. Hi-Trap™ chelating column (Amersham Pharmacia Biotech).
10. His Microspin™ purification module (Amersham Pharmacia Biotech).
11. Ni-NTA HisSorb strips (QIAGEN Inc., Mississauga, ON, Canada).

2.6. Equipment

1. Agarose gel electrophoresis equipment.
2. Sodium dodecyl sulfate polyacrylamide gel electrophoresis (SDS-PAGE) equipment.
3. DNA sequencing equipment.
4. Chromatography equipment.
5. Thermal cycler.
6. Multiwell enzyme-linked immunosorbent assay (ELISA) plate reader.
7. Sorval high-speed and swinging bucket bench-top (RT6000B Refrigerated) centrifuges or their equivalents.
8. Spectrophotometer.
9. Gene Pulser™ electroporator (Bio-Rad Laboratories, Mississauga, ON, Canada).

3. Methods

The methods described in this chapter will outline: 1) procedures for enrichment of llama sdAbs phage display library for HCEC-specific binders by subtractive panning, 2) methods for monitoring the enrichment of HCEC-specific phage clones, 3) proce-

from the previous round (**Fig. 2**) (*see* **Note 6**). Typically, three to four rounds of panning are performed to enrich for the binders, although occasionally more rounds may be necessary to obtain enrichment.

3.2. Monitoring the Enrichment

Subheadings 3.2.1.–3.2.3. describe three approaches employed during panning experiments to monitor the enrichment of HCEC-specific sdAbs: 1) determining the titer of the recovered phage, 2) performing plaque polymerase chain reaction (PCR), and 3) performing phage ELISA on the plaques from the recovered phage (*see* **Subheading 3.1.1.**). Enrichment monitoring is done at the end of each round of panning. The input phage fraction for the subsequent round of panning was selected based on the evaluation of enrichment (i.e., panning progress).

3.2.1. Phage Titer

1. Prepare exponentially growing TG1 cells as described in **Subheading 3.1.2.**
2. Make 10^2, 10^4, and 10^6 dilutions of infected cells from **step 3** (**Subheading 3.1.2.**) in 500 µL exponentially growing cells and use 200 µL to determine titres of output phage fractions as described above (**steps 4** and **5, Subheading 3.1.4.**).
3. Keep the plates at 4°C as the plaques from these plates, are also used for plaque PCR and phage ELISA (*see* **Subheadings 3.2.2.–3.2.3.**).

In a typical panning experiment the titer drops after the first round, gradually increases after the second round, and reaches a plateau in later rounds. The higher titer in the first round is attributed to the higher background resulting from a very high non-binder phage to binder phage ratio. In cases where the input phage varies from round to round, the ratio of output phage titer to input phage titer is a more accurate measure of enrichment.

3.2.2. Plaque PCR

During each round of panning plaques from titer plates, prepared in **Subheading 3.2.1.**, were randomly chosen for PCR analysis.

1. Cut the agarose-top layer around the individual plaques (by gently pressing, with a disposable loop, against the agarose top) and transfer plaques to tubes containing 50 µL of autoclaved H_2O (*see* **Note 7**).
2. After a brief vortexing, use 1 µL of the plaque solution as template for PCR amplification using vector-specific primers fdTGIII, 5′(GTGAAAAAATTATTATTCGCAATTCCT)3′ and -96GIII, 5′(CCCTCATAGTTAGCGTAACG)3′ in a total volume of 10 µL. The PCR mixture contains 10 pmol/µL each of the two primers, 1 X buffer (Perkin Elmer, Mississauga, ON, Canada), 200 µM each of the four dNTPs and 0.05 unit/µL AmpliTaq DNA polymerase (Perkin Elmer). PCR protocol includes an initial denaturation step at 95°C for 5 min followed by 30 cycles at 94°C for 30 s, at 53°C for 30 s, and at 72°C for 1 min, and a final extension step at 72°C for 10 min.
3. Resolve the amplified products on 1% agarose gel against the appropriate DNA molecular weight marker. Enrichments are assessed by the percent of plaques that yield an approx 550 bp full-length gene product.

3.2.3. Phage ELISA

ELISA against phage coat protein P8 (phage ELISA) was used to determine the binding of selected phage clones to various human endothelial cells in culture.

1. Transfer the individual plaques from titer plates, prepared in **Subheading 3.2.1.**, to separate wells of a sterile 96-well microtiter plate containing 200 μL of LB in each well (*see* **Note 7**). Incubate overnight at 37°C in a rotary shaker at 100 rpm.

2. Grow monolayers of HCEC and HLMEC in 96 well plates to confluence, and wash two times in 300 μL PBS. Block by adding 300 μL 4% BSA-PBS and incubate for 1 h at room temperature. Aspirate the contents of the wells and add 75 μL 4% BSA-PBS using a multichannel pipet.

3. Spin the infected TG1 cells from **step 1** at 600*g* for 10 min in a swinging bucket benchtop centrifuge. Transfer 75 μL aliquots from each supernatant to duplicate wells of HLMEC and HCEC monolayers in 4% BSA-PBS (prepared in **step 2**). Mix gently on a belly dancer and incubate at 37°C for 1 h.

4. Remove the supernatants and wash the cells six times with PBS-0.05% (v/v) Tween-20. Add 100 μL of a 1:1000 dilution of horseradish peroxidase (HRP)-conjugated anti-M13 monoclonal antibody in 2% BSA-PBS to each well and incubate at 37°C for 1 h.

5. Wash the wells as described in **step 4** and blot the plates on a paper to remove any remaining liquid. Detect the binding of phage to the cells colorimetrically by adding 100 μL of the TMB peroxidase substrate and H_2O_2 mixture at room temperature for 5–10 min. A blue color should appear.

6. Terminate the reaction by adding 100 μL of 1 M H_3PO_4 (the blue will change to yellow) and read the optical density at 450 nm using an ELISA plate reader.

Each subsequent round of panning should produce higher percentage of clones that bind selectively to HCEC and do not bind to HLMEC.

3.3. Selection by Functional Criteria

After four rounds of panning, a library of sdAbs that selectively bind and internalize HCEC was obtained. Because the goal of the study was to identify species that transmigrate across the BBB, the next round of selection from the enriched library was done based on the ability of phage clones to transmigrate HCEC monolayers. For this purpose, a compartmentalized in vitro BBB model consisting of HCEC monolayer grown on a porous membrane positioned to separate two media compartments as shown in **Fig. 4** was used. HCEC were seeded at 3×10^5 cells/cm^2 on a 0.5% gelatin coated Falcon tissue culture inserts (pore size 1 μm; surface area 0.83 cm^2) in 1 mL of growth medium. The bottom chamber of the insert assembly contained 2 mL of growth medium supplemented with the fetal human FHAs-conditioned media *(10)*. The FHAs-conditioned medium was obtained by incubating confluent FHAs in a serum free DME for 72 h. The model has been characterized in detail previously *(10)*.

3.3.1. Phage Transmigration

1. Replace growth media in both chambers of the in vitro BBB model with the transport buffer and equilibrate the cells for 30 min at 37°C (*see* **Note 8**).

2. Add 10^{11} pfu of phage library enriched for HCEC binding (fourth round S3 and Int fractions) clones to the upper compartment (50 μL). Phage expressing unrelated sdAb and empty phage (same pfu) are used as negative controls **(Fig. 4)**. At specified times **(Fig. 4)**, collect 10-μL aliquots from the bottom compartment and infect 200 μL of TG1 cells as described in **Subheading 3.1.4.** Grow infected TG1 cells as plaques overnight as described in **Subheading 3.1.4.**

3. In the morning count the plaques and determine the titer. Keep the plates at 4°C as the plaques from these are also used for performing plaque PCR (as described in **Subheading 3.2.2.**).

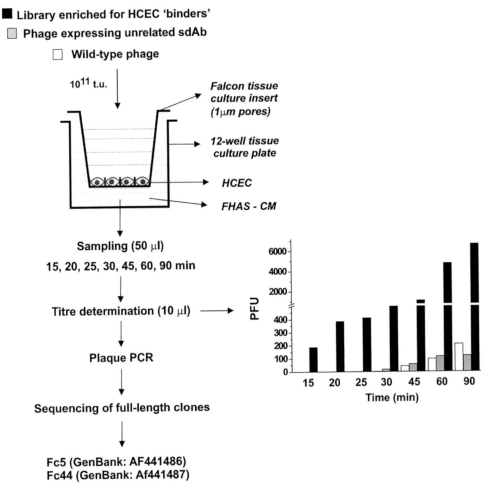

■ Library enriched for HCEC 'binders'
▨ Phage expressing unrelated sdAb
□ Wild-type phage

Fig. 4. Selection of sdAbs that transmigrate HCEC monolayer from the phage-display library enriched for HCEC-binding sdAbs. An in vitro BBB model consisting of HCEC grown as a monolayer on the membrane of the tissue culture insert was used. Media conditioned by FHAs (FHAs-CM) are applied to the bottom compartment to induce BBB phenotype of HCEC. 10^{11} pfu of the wild type phage (open bars), the unrelated sdAb, NC11 displaying phage (grey bars) and phage library enriched for HCEC binding and HCEC internalizing phage (black bars) were applied to the top chamber of triplicate BBB assemblies and the phage titer was determined from 10-μL aliquots of the bottom chamber at the indicated time points. Plaque PCR was done on more than 100 colonies from these titer plates to identify clones with full-length inserts. These clones were then sequenced yielding two distinct sdAb species, named FC5 and FC44.

4. Plaques containing full-length clones (determined by plaque PCR from 5–15 min titer-plates when no wild-type phage transmigration was observed) were then sequenced (*see* **Subheading 3.4.**).

3.4. Sequencing

1. Prepare sequencing templates for selected phage clones by PCR using the same set of primers and condition used for plaque PCR.
2. Purify the amplified products with QIAquick PCR Purification™ kit (*see* **Note 9**), and sequence the amplified template by the dideoxy method *(11)* using the AmpliTaq DNA

Polymerase FS kit, fdTGIII and -96GIII primers and 373A DNA Sequencer Stretch (PE Applied Biosystems) or its equivalent (*see* **Note 10**).

Sequencing revealed that all full-length clones recovered from the bottom chamber of the in vitro BBB model belonged to two different sequences, FC5 and FC44 **(Fig. 4)**, indicating that these two antibodies are capable of transmigrating HCEC monolayer.

3.5. Expression, Tagging, and Purification

3.5.1. Expression and Tagging

All the cloning steps were performed essentially as described previously (*12*).

1. Amplify the sdAb genes out of the phage vector in a total volume of 50 µL by plaque PCR using the primers, VHBbs, 5′(TAT<u>GAAGAC</u>ACCAGGCCGATGTGCAGCT GCAGGCG)3′ and VHBam, 5′(TAT<u>GGATCC</u>TGAGGAGACGGTGACCTG)3′ which introduce *Bbs* I and *Bam*H I sites at the ends of the amplified fragments (underlined).
2. Purify the sdAb genes with QIAquick PCR Purification kit in a final volume of 50 µL water.
3. Cut the purified DNA sequentially with *Bbs*I and *Bam*HI restriction endonucleases and purify again with QIAquick Gel Extraction™ kit in 50 µL water.
4. Ligate the cut fragment to the *Bbs* I/*Bam*H I-treated pSJF2 expression vector that, at the protein level, adds C-terminal c-myc and His$_5$ tags **(Fig. 3)**.
5. Prepare electrocompetent *E. coli* strain TG1 cells (*13*) and use small aliquot of the ligated product to transform the cells using the BIO-RAD Gene Pulser™ or its equivalent. Alternatively, cells can be transformed employing chemical transformation method (*12*).
6. Following transformation, spread 100 µL of cells on LB plates containing 100 µg/mL ampicillin and leave the plates with lids half open for 5–10 min on the clean bench. Cover, invert and incubate overnight at 32°C.
7. In the morning, identify the positive clones (i.e., clones with sdAb gene) by colony PCR and agarose gel electrophoresis. Transfer the cells to the PCR mixture by gently touching the surface of the colonies with a sterile pipet tip and swirling the tip directly in PCR mixture. Perform colony PCR with primers, RP, 5′(CAGGAAACAGCTATGAC)3′ and FP, 5′(GTAAAACGACGGCCAGT)3′ using the same conditions as for plaque PCR. Determine the size of the amplified product on a 1% agarose gel.
8. Perform a second colony PCR on the positive clones in a total volume of 25 µL. Purify and sequence the PCR template as described in **Subheading 3.4.**, using RP and FP as primers.
9. For expression, use a single positive clone to inoculate 25 mL of LB containing 100 µg/mL ampicillin. Incubate in a rotary bacterial shaker at 240 rpm overnight at 37°C.
10. Transfer the entire overnight culture to 1 L of M9 medium supplemented with 5 µg/mL vitamin B1, 0.4% casamino acid and 100 µg/mL ampicillin. Incubate while shaking the culture at 180 rpm for 30 h at room temperature, and then supplement the culture with 100 mL of 10 × induction medium and 100 µL of 1 *M* IPTG and incubate for another 60 h.
11. Extract the periplasmic fraction by osmotic shock method (*14*) and verify expression by detecting the presence of sdAb in the extract by Western blotting against the c-myc tag (*15*). Dialyze the periplasmic fraction extensively against 10 m*M* HEPES buffer, pH 7.0, 0.5 *M* NaCl. Following dialysis, add imidazole to a final concentration of 10 m*M*.

3.5.2. Purification

The presence of the C-terminal His$_5$ tag in sdAbs allows for one-step protein purification by immobilized metal affinity chromatography using 5-ml HiTrap chelating column **(Fig. 5)**.

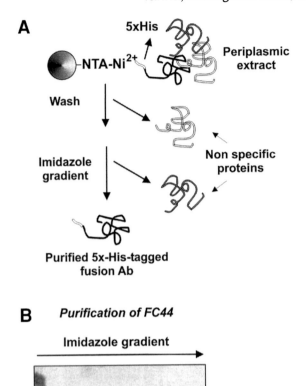

Fig. 5. Purification of sdAbs using immobilized metal affinity chromatography using HiTrap™ chelating column. **A,** The 5-mL column was charged with Ni^{2+} by applying 30 mL of a 5 mg/mL $NiCl_2 \times 6H_2O$ solution and subsequently washed with 15 mL deionized water. The periplasmic extract of bacterial cultures expressing sdAb-c-myc-His$_5$ fusion protein was applied onto the column. His$_5$ tag allows the sdAb to bind to the Ni^{2+} charged column. Unbound fraction (non-specific proteins) was washed out from the column, and bound protein (sdAb) was eluted from the column by applying 10–500 mM imidazole gradient. **B,** Each eluate fraction was resolved on 12% SDS-PAGE, transferred to nitrocellulose membrane, and detected in a Western blot analysis with primary mouse anti-c-myc IgG and secondary anti-mouse IgG antibody conjugated to ALP. Molecular weight of sdAb-c-myc-His$_5$ fusion protein is approx 15 kDa.

1. Charge the column with Ni^{2+} by applying 30 mL of a 5 mg/mL $NiCl_2 \cdot 6H_2O$ solution and subsequently wash with 15 mL deionized water.
2. Carry out purification as described *(15)* using the starting buffer (*see* **Subheading 2.**), and elute the bound protein with a 10–500 mM imidazole gradient (**Fig. 3**).
3. Examine the fractions corresponding to the eluted peaks on the chromatogram for the presence/purity of the sdAb proteins by SDS-PAGE *(16)*. Pool the sdAb fractions and dialyze extensively against PBS. Add sodium azide at a final concentration of 0.02% and keep the proteins stored at 4°C.

3.6. Antibody Characterization

Purified sdAbs can be used and/or further characterized in various in vitro and in vivo systems. For example, binding and/or internalization of antibodies (in)to various cells or tissues in vivo can be tested *(17)*. *C-myc* tag allows for sdAb detection using either indirect immunocytochemistry or ELISA with anti-*c-myc* antibodies. SdAbs can also be labeled with fluorescent or radiolabeled tag(s) to facilitate quantitative pharmacokinetics studies.

sdAbs derived from naïve phage-display library are considered lead targeting moieties. Further improvement of their affinity, avidity, and pharmacodynamic properties can be done by genetic engineering and structural studies.

3.7. Concluding Remarks

We have used the described protocols (see flow chart in *17*) to identify two sdAbs, FC5 and FC44, that selectively bind to HCEC, transmigrate across the in vitro BBB model, and target the brain in vivo *(17)*. The principles of the technique described here are also applicable to in vivo biopanning of antibody or peptide phage-display libraries to select species that home to specific tissues, including the brain.

Described protocols can be coupled with down-stream chemical and genetic engineering of selected antibodies to design appropriate linkers for nonpermeable small molecules, peptides, and oligonucleotides targeted to the brain. In addition, selected sdAbs could be used to vector other transport vehicles such as liposomes and polymers encapsulating large particles (e.g., plasmids, vaccines). Drug and gene delivery could be targeted specifically to the cerebrovascular compartment to, for example, prevent stroke by stimulating anticoagulant properties of endothelium, treat migraine, transiently disrupt tight junctions, reduce the expression of adhesion molecules, stimulate re-vascularization, and so on. sdAbs coupled to appropriate radioactive ligand(s) may find application in cerebrovascular imaging or can be used in laboratory research protocols for a large-scale affinity purification of brain microvascular endothelial cells or immunochemical staining of brain endothelium in vitro and in vivo.

Finally, identification of binding partners for selected sdAbs and their functional roles in endocytosis, transcytosis, and BBB permeability will be of seminal importance in future development of strategies to modulate BBB functions.

4. Notes

1. Always keep the phage on ice to prevent a possible enzymatic cleavage of the gIIIp, due to protease contamination. gIIIp cleavage results in the loss of infectivity as well as binding activity of the phage particles.
2. To grow colonies on the stock plate, streak out frozen stock of TG1 onto a minimal plate *(12)* supplemented with thiamine. Incubate at 37°C for at least 24 h. Seal the plate with parafilm and store at 4°C for up to 1 mo. It is recommended to grow the TG1 cells on minimal media to ensure that F pilus, which mediates phage infection, is maintained within the cells. Thiamine is added to the media since TG1 cells are auxotrophic for thiamine.
3. Agarose top media are stored at room temperature, in 50–100 mL aliquots, and, when needed, can be melted in a microwave oven. The media can be re-solidified and reused many times. Before pouring onto the plates, however, the molten agarose top needs to be equilibrated to 50°C. This is done by transferring the media, in 3-mL aliquots, to sterile borosilicate glass tubes, closing the tubes with metal caps and placing them in a 50°C-temp block for 30–60 min.
4. Following the overnight growth, discrete plaques are formed in the background of the uninfected cells. Very frequently, however, lawn of plaques is formed especially during the later rounds of panning. This has a rough appearance in contrast to the lawn of the uninfected cells (negative control plate), which has a smooth appearance.

5. Unlike cell pellets, phage pellets are loose and streak along the side of the tube. Therefore, during removal of the supernatant care must be taken to minimize phage losses.

6. Second and third round of panning in this study was done using S3 and Int fractions from previous rounds. The fourth round of panning was performed using S1 fraction. The input phage fraction for each round of panning was chosen from the output fraction that contained the highest percent of phage with the full-length V_HH. However, a more general rule can be applied for choosing an input fraction for the subsequent round of panning: increasing the stringency of stripping will reduce non-specific binders and favor selection of higher affinity binders. However, overly stringent stripping protocols may result in a loss of many specific binders with low affinity.

7. Alternatively, a toothpick, which has touched the surface of the plaque, can be used to transfer phage by dipping the toothpick directly in PCR mixture. In the case of phage ELISA, however, the toothpick method occasionally fails to initiate the overnight growth in microtiter culture plates possibly because it does not transfer enough phage-infected cells to the wells.

8. The barrier integrity was assessed by measuring the passage of sodium fluorescein (MW=376 Da, 25 μg/mL), fluorescently labeled dextran (MW=10 kDa), and ^{14}C dextran-carboxyl (MW=70 kDa) across triplicate HCEC monolayers, and across 0.5% gelatin-coated inserts without cells. Samples were collected from the bottom chambers over a 5- to 90-min period. The fluorescence was measured using Cytofluor 2350 (Millipore) and the radioactive counts were measured in Wallac microbeta liquid scintillation counter. Clearance values were determined using previously described protocols *(18)*.

9. In instances where the PCR preparation contains a significant amount of non-specific products, the sdAb gene can be purified away from nonspecific products by QIAquick Gel Extraction™ kit or equivalent kits. Alternatively, single-stranded phage DNA, purified by standard procedures *(12)*, can be used as a template for manual or automated sequencing.

10. The AmpliTaq DNA Polymerase FS kit contains instructions for performing cycle sequencing. Cycle sequencing is the preferred method for sequencing PCR templates as traditional methods of sequencing gives poor results with respect to PCR template denaturation. The method involves repeated cycles of template denaturation, annealing and extension using *Taq* DNA polymerase. This provides ample opportunities for template denaturation and extension resulting in adequate detection signals. For manual sequencing, cycle sequencing method can be used in conjunction with radiolabeled dNTPs.

Acknowledgments

The authors wish to express appreciation to Ginette Dubuc, Marguerite Ball, Tammy Herring, and Kevan Mcrae for expert technical assistance, Joseph Michniewicz for DNA sequencing, and Simon Foote for constructing pSJF2 expression vector.

References

1. Bickel, U., Yoshikawa, T., and Pardridge, W. M. (2001) Delivery of peptides and proteins through the blood-brain barrier. *Adv. Drug Deliv. Rev.* **46**, 247–279.
2. Shi, N., and Pardridge, W. M. (2000) Noninvasive gene targeting to the brain. *Proc. Natl. Acad. Sci. USA* **97**, 7567–7572.
3. Hoogenboom, H. R., de Bruine, A. P., Hufton, S. E., Hoet, R. M., Arends, J. W., and Roovers, R. C. (1998) Antibody phage display technology and its applications. *Immunotechnology* **4**, 1–20.
4. Nilsson, F., Tarli, L., Viti, F., and Neri, D. (2000) The use of phage display for the development of tumor targeting agents. *Adv. Drug Del. Rev.* **43**, 165–196.

5. Muyldermans, S., and Lauwereys, M. (1999) Unique single-domain antigen binding fragments derived from naturally occurring camel heavy-chain antibodies. *J. Mol. Recog.* **12,** 131–140.

6. Tanha, J., Dubuc, G., Hirama, T., Narang, S. A., and MacKenzie, C. R. (2002) Selection by phage display of llama conventional V_H fragments with heavy chain antibody V_HH properties. *J. Immunol. Meth.* **263,** 97–109.

7. Stanimirovic, D., Morley, P., Ball, R., Hamel, E., Mealing, G., and Durkin, J. P. (1996) Angiotensin II-induced fluid phase endocytosis in human cerebromicrovascular endothelial cells is regulated by the inositol-hosphate signaling pathway. *J. Cell. Physiol.* **169,** 455–467.

8. Yong, V. W., Tejada-Berges, T., Goodyear, C. G., Antel, J. P., and Yong, F. P. (1992) Differential proliferative responses of human and mouse astrocytes to gamma interferon. *Glia* **6,** 269–280.

9. Narang, S. A., Yao, F.-L., Michniewicz, J. J., Dubuc, G., Phipps, J., and Somorjai, R. L. (1987) Hierarchical strategy for protein folding and design: synthesis and expression of T4 lysozyme gene and two putative folding mutants. *Protein Eng.* **1,** 481–485.

10. Muruganandam, A., Herx, L. M., Monette, R., Durkin, J. P., and Stanimirovic, D. B. (1997) Development of immortalized human cerebromicrovascular endothelial cell line as an in vitro model of the human blood-brain barrier. *FASEB J.* **13,** 1187–1197.

11. Sanger, F., Nicklen, S., and Coulson, A. R. (1977) DNA sequencing with chain terminating inhibitors. *Proc. Natl. Acad. Sci. USA* **74,** 5463–5467.

12. Sambrook, J., Fritsch, E. F., and Maniatis, T. (1989) *Molecular Cloning: A Laboratory Manual*, 2nd ed. Cold Spring Harbor Laboratory, Cold Spring Harbor, NY.

13. Tung, W. L., and Chow, K. C. (1995) A modified medium for efficient electrotransformation of *E. coli. Trends Genet.* **11,** 128, 129.

14. Anand, N. N., Dubuc, G., Phipps, J., et al. (1991) Synthesis and expression in *Escherichia coli* of cistronic DNA encoding an antibody fragment specific for a *Salmonella* serotype B O-antigen. *Gene* **100,** 39–44.

15. MacKenzie, C. R., Sharma, V., Brummell, D., et al. (1994) Effect of C lambda-C kappa domain switching on Fab activity and yield in *Escherichia coli*: synthesis and expression of genes encoding two anti-carbohydrate Fabs. *Biotechnology NY* **12,** 390–395.

16. Laemmli, U. K. (1970) Cleavage of structural proteins during the assembly of the head of bacteriophage T4. *Nature* **227,** 680–685.

17. Muruganandam, A., Tanha, J., Narang, S., and Stanimirovic, D. (2001) Selection of phage-displayed llama single-domain antibodies that transmigrate human blood-brain barrier endothelium. *FASEB J.* **16,** 240–242.

18. Pardridge, W. M., Triguero, D., Yang, J., and Cancilla, P.A. (1990) Comparison of in vitro and in vivo models of drug transcytosis through the blood-brain barrier. *J. Pharmacol. Exp. Therap.* **253,** 884–891.

labeled probes have gained popularity, have become widely accepted, and solve many of the problems associated with radioactive probes *(4-9,13–17,19,22–24)*.

Thus, nonradioactive labeling of cDNA and oligonucleotide probes has been used for rapid visualization of mRNA with ISH histochemistry, because it lacks many of the aversive effects of the radioactive method and gives a stable signal. We routinely use digoxigenin (DIG)-labeled oligonucleotide probes combined with immunological amplification of the probe signal to significantly enhance the sensitivity of mRNA detection *(22–24)*.

2. Materials
2.1. Tissue Samples

The tissues used were obtained from patient biopsies removed for diagnosis following appropriate consent. Tissues examined were

1. Twenty gliomas diagnosed and graded according to the WHO grading system *(26)*:
   a. Seven glioblastomas multiforme (GBMs; astrocytoma, WHO Grade IV/IV).
   b. Four anaplastic astrocytomas (AAs; astrocytoma WHO Grade III/IV).
   c. Four low-grade diffuse astrocytomas (LGAs; astrocytoma, WHO Grade II/IV).
   d. Five juvenile pilocytic astrocytomas (JPAs; astrocytoma WHO Grade I/IV).
2. Three cases of cerebral infarction. In these cases, tissue was obtained 2, 5, and 7 d after the onset of the infarct.
3. Three samples of histologically normal brain removed in the course of surgical exposure were used as controls. When present, normal tissue adjacent to the tumors in the same sections served as an internal control.

2.2. ISH

Most of the chemicals were purchased from Sigma Chemical Co. (St Louis, MO) unless otherwise noted.

2.2.1. Tissue Preparation

1. M-1 Embedding Matrix (Shandon Lipshaw, Pittsburgh, PA).
2. Solvents: xylene, ethanol.
3. Paraffin.
4. γ-methacryloxypropyltrimethoxysilane coated slides.
5. Cover slips.
6. Nuclear fast red, 1% (Arleco, Gibbstown, NJ).
7. Glycerin jelly.
8. Rubber cement.
9. Diethylpyrocarbonate (DEPC)-treated sterile water: use as instructed by Sigma. Added primarily to distilled water, phosphate-buffered saline (PBS) and saline sodium citrate (SSC) as needed to control ribonuclease (RNase) contamination. Use in other solutions as needed (*see* **Note 1**).
10. PBS: 0.1 M phosphate buffer (pH 7.2) + 0.9% NaCl.
11. Paraformaldehyde (4%): 4.0 g of paraformaldehyde in 100 mL of 0.1 M PBS.
12. 20× Standard saline citrate (SSC) solution: 175 g of NaCl, 88 g of Na citrate dihydrate. Make up to 100 mL in distilled water and adjust pH to 7.0.
13. Sodium decasulfate: 0.5% in 1× SSC.
14. Sodium bisulfite: 20% in 2× SSC.
15. Triton X-100 (0.3%): 300 µL of Triton X-100 and bring to 100 mL with 1 × SSC.

16. Proteinase K (Boehringer Mannheim, Indianapolis, IN), Stock Solution 5 mg/mL: Dilute the stock solution using 2 × SSC to a final concentration of 250 μg/mL.
17. Blocking solution: denatured/sheared salmon sperm DNA (1 mg/mL) and bovine serum albumin (BSA, prepared from Fraction V Albumin, nitrogen content 15.5%, fatty acid-free) 50 mg/mL in 2 × SSC.
18. 15% sucrose in DEPC-treated PBS.

2.2.2. Probe Preparation

1. XbaI/EcoRI of pKS+ (Stratagene Cloning Systems, La Jolla, CA).
2. EcoRI/HindIII of pKS+ (Stratagene Cloning Systems).
3. Not1 (New England Biolabs, Beverly MA).
4. pBluescript II SK+/– (Stratagene Cloning Systems).
5. Sequenase (US Biochemicals, Cleveland, OH).
6. Plasmids pKS+/hTL1 and pKS+/TL2.
7. Phenol, Chloroform.
8. T3 and T7 polymerase.

2.2.3. Labeling and Testing of Probe and Hybridization

1. Digoxigenin RNA labeling kit (Boehringer Mannheim, Indianapolis, IN).
2. Total brain RNA (Stratagene Cloning Systems, La Jolla, CA).
3. Hybridization buffer: 4 × SSC, 50% formamide, 1 × Denhardt's solution, 0.3% Triton X-100, 1% N-laurosyl sarcosine, 1 mg/mL sheared salmon sperm DNA, 1 μg/mL tRNA, 10% Dextran sulfate, 1 : 1000 β-mercaptoethanol (*see* **Note 2**).
4. Probe: Add 5–10 ng of probe to 20 μL of hybridization buffer to make a working probe solution.

2.2.4. Posthybridization and Digoxigenin Detection

1. 0.2 × SSC: 1 : 100 dilution of 20 × SSC in sterile water.
2. Mung bean nuclease (New England Biolabs, Beverly, MA): 3 U per slide in buffer supplied by the manufacturer.
3. 1 Maleate buffer, pH 7.2 (Boehringer Mannheim): add 0.3% Triton X-100, use as washing buffer.
4. Posthybridization blocking solution (Boehringer Mannheim): 4% containing 0.3% Triton X-100.
5. Sheep anti-DIG-alkaline phosphatase antibody (anti-DIG-AP, Boehringer Mannheim): 1 : 500 dilution with 2% blocking buffer (Boehringer Mannheim) and 0.3% Triton X-100.
6. Alkaline phosphatase substrate buffer: 100 mM Tris-HCl (pH 9.5), 150 mM NaCl, 50 mM MgCl$_2$.
7. Substrate: 45 μL nitroblue tetrazolium salt (NBT, 75 mg/mL) in 70% dimethylformamide and 35 μL bromo-4-chloro-3-indolylphosphate toluidinium salt (BCIP, 50 mg/mL) in 100% dimethylformamide per 10 mL alkaline phosphatase substrate buffer.

3. Methods

ISH by its name is hybridization adapted to tissue sections. The theory of hybridization, whether it is northern or southern, is performed basically unchanged, with an adjustment only to help gain access to the target nucleotides within the section. Two important parameters that determine the success of the procedure are the quality of the sample (largely determined by the time elapsed between tissue excision from the patient and subsequent fixation) and the efficiency of labeling of the probe.

An optimal ISH protocol should 1) have a high sensitivity, 2) have cell-to-subcellular structural resolution, 3) be compatibile with immunocytochemical detection of proteins, and 4) allow routinely prepared paraffin-embedded tissue to be used.

The methods described below outline 1) tissue preparation, 2) probe preparation, 3) digoxigenin labeling of probes, 4) hybridization, 5) posthybridization washing, and 6) the application of the ISH procedure.

3.1. Tissue Preparation

Tissue processing is different for paraffin embedding and for frozen sections (*see* **Note 3**). However, the first few steps are identical.

1. Fresh surgical specimens are immediately retrieved from the operating room and are rapidly rinsed in DEPC-treated PBS for 5–10 min.
2. The tissue is then fixed overnight in 4% paraformaldehyde in PBS at 4°C (*see* **Note 4**).

3.1.1. Paraffin-Embedded Tissue

1. After fixation, the tissue is dehydrated in ethanols and xylene and embedded in paraffin using standard methods.
2. Cut 5 μm-thick sections and place on γ-methacryloxypropyltrimethoxysilane-coated slides.
3. Dry slides overnight in a 60°C oven.
4. Deparaffinize slides in 100% xylene followed by absolute ethanol, rinse in DEPC-treated sterile water, and then in 2 × SSC.

3.1.1.1. PRETREATMENT OF SECTIONS

The main purpose of the pretreatment is to make the target nucleotides in the section more accessible to the probes. Detergents, sodium bisulfide, and proteinase K are used to permeate and loosen some of the crosslinks created during tissue fixation (*see* **Note 5**).

1. Place slides in 1 × SSC containing 0.3% Triton X-100 for 10 min at room temperature.
2. Transfer slides to 0.5% sodium decasulfate in 1 × SSC for 5 min at 37°C, rinse in 2 × SSC, and incubate in 20% sodium bisulfite in 2 × SSC for 20 min at 45°C.
3. Rinse slides three times in 2 × SSC, treat with proteinase K (250 μg/mL) in 2 × SSC for 20 min at 37°C and rinse three times in 2 × SSC.
4. Block sections with denatured/sheared salmon sperm DNA (1 mg/mL) and BSA (50 mg/mL) in 2 × SSC at 37°C for 20 min, rinse briefly in 2 × SSC, and in DEPC-treated sterile water.
5. Dehydrate in a graded series of ethanols, and air-dry.

3.1.2. Frozen Tissue

1. After fixation, cryoprotect the tissue by incubation in 15% sucrose in DEPC-treated PBS for at least 24 h.
2. Embed in M-1 Embedding Matrix, freeze on dry ice, and store at –80°C.
3. Cut 12-μm sections in a cryomicrotome at –20°C, mount on γ-methacryloxypropyl-trimethoxysilane-coated slides, and store at –80°C until use.
4. Immediately after removing the sections from –80°C, process for ISH starting with incubation in 20% sodium bisulfite/2 × SSC for 15 min at 45°C.
5. Wash the slides once in DEPC-treated water and three times in 2 × SSC for 3 min.
6. Treat sections with 250 μg/mL proteinase K in 2 × SSC for 15 min at 37°C and wash once in DEPC-treated water.

7. Wash three times in $2 \times$ SSC.
8. Block sections with 1 mg/mL denatured/sheared salmon sperm DNA in 4% blocking buffer at 37°C for 2 min and then wash as given in **step 7**.
9. Dehydrate the slides through a graded series of ethanols, and air-dry.

3.2. Probe Preparation

The development of biotinylated probes that can be detected using standard immuno-histochemical methods has enabled ISH techniques to be adopted by the Histopathology Laboratory *(22–24,27)*. For example, the use of these probes has now become well-established for the routine detection of repetitive genomic or viral DNA sequences in paraffin sections. Three examples of probe preparation are provided for the detection of mRNA sequences of molecules related to angiogenesis in neoplastic and reactive conditions of the CNS: angiopoietins *(22)*, tenascin-C (TN-C) *(23)*, and vascular endothelial growth factor (VEGF) *(24)*.

3.2.1. Angiopoietins

Angiopoietin-1 and its naturally occurring antagonist angiopoietin-2 are novel ligands that regulate tyrosine phosphorylation of Tie2/Tek receptor on endothelial cells. Proper regulation of Tie2/Tek is absolutely required for normal vascular development, seemingly by regulating vascular remodeling and endothelial cell interactions with supporting pericytes/smooth muscle cells.

1. The plasmid pKS+/hTL1 used contains a 570-base pair (bp) SpeI-EcoRI fragment of Ang-1 subcloned into XbaI/EcoRI of pKS+. In addition, it contains a 70-bp 5′UT and a 500-bp coding region ending at amino acid 166.
2. The plasmid pKS+/TL2 contains a 640-bp EcoRI-HindIII fragment from human Ang-2 subcloned into EcoRI/HindIII of pKS+ *(22)*. The plasmid contains a 360-bp 5′UT and a 280-bp coding region ending at amino acid 99.
3. The plasmid DNA was linearized with EcoRI for antisense probe production and HindIII for sense RNA probe production.
4. To synthesize DIG-labeled RNA probes, 1 μg of the plasmid DNA was linearized with NotI for antisense probe production and EcoRI for sense RNA probe production. The digested plasmid DNA was purified by phenol/chloroform extraction and ethanol precipitation *(21)*.

3.2.2. Tenascin-C (TN-C)

TN-C is an extracellular matrix glycoprotein, expressed in the developing CNS, cartilage and mesenchyme that is upregulated in tumors, wound healing, and inflammation. TN-C is important for cellular adhesion, migration, and proliferation.

1. Oligonucleotide primers complementary to a 5′ region of exon 1 of the human TN gene (primer 1: 5′- CTA GAA TTC CAG CAG CAC CCA GC-3′ and primer 2: 5′- CTC AAG CTT CAC CGA ACA CTG G-3′) were designed based on the published sequence *(23,28)*.
2. With human genomic DNA as a template for these primers, polymerase chain reaction (PCR) was used to amplify a 231 bp product.
3. The PCR product was cloned into pBluescript II SK+/–. Two resulting clones each containing a 0.23-kb insert were sequenced according to the Sanger method *(29)* using Sequenase and were found to match exactly a 231-bp fragment of the published sequence of exon 1 of the human TN gene *(23,28)*.

3.2.3. VEGF

VEGF is an endothelial cell mitogen that increases microvascular permeability and is angiogenic in vivo. A probe of the published whole sequence of VEGF/VPF (980 bp) was used. The sequence product was introduced into pBluescript II SK, as described for TN-C *(24)*.

3.2.4. Control Probes

For each probe, a control sense probe was always prepared and labeled as previously described *(22–24)*.

3.3. Digoxigenin Labeling of Probes

An important parameter is the efficiency of labeling with the probe. Nick translation will label both strands of a DNA probe. Although mRNA probes or riboprobes are more specific and tend to bind more tightly, they are also less stable, especially during storage and handling. However, an mRNA duplex after hybridization is stable. Each probe was labeled with DIG.

1. TN-C, antisense and sense riboprobes were prepared using the DIG RNA Labeling Kit. The specificity of the probes was verified by Northern hybridization of human fetal brain total RNA. Using the antisense probe, we detected a 6–7-kb band which corresponds to the size of human TN mRNA. No signals were generated using the sense probe.
2. The same DIG RNA Labeling Kit was used for labeling of the angiopoietin probes. Here antisense and sense DIG-labeled RNA probes were transcribed by T3 and T7 RNA polymerase, respectively.
3. The DIG-labeled RNA was ethanol precipitated in order to remove unincorporated DIG-dNTP.
4. Probes that are labeled are tested in appropriate gel or blot assays to determine the sensitivity to picogram amounts of the target. (*See* **Note 6**.)
5. Probes were stored at –80°C.

3.4. Hybridization

1. Heat 10 ng of the digoxigenin-labeled sense and antisense probes separately in 20 μL of hybridization buffer (*see* **Subheading 2.2.4.**) at 80°C for 5 min.
2. Add 17 μL of probe to the treated slides, which are then covered with a glass coverslip, sealed with rubber cement, and incubated at 42°C overnight.

3.5. Posthybridization Washing and Blockage

1. Remove the cover slip, and wash briefly in 2 × SSC at 60°C.
2. Wash twice in 0.2 × SSC at 65°C for 15 min total followed by a final wash in 2 × SSC at room temperature.
3. Treat sections with Mung Bean Nuclease, using three U per slide for 2 min at 37°C.
4. Rinse slides in Maleate buffer with 0.3% Triton X-100.
5. Place sections in 4% blocking buffer containing 0.3% Triton X-100 for 30 min at 37°C.

3.6. Detection of the DIG-Labeled Probes

1. Detection is performed using anti-DIG-alk phos antibody at 1:500 dilution in 2% blocking buffer with 0.3% Triton X-100 for 60 min at 37°C, followed by overnight incubation at 4°C.

2. Wash slides with Maleate buffer containing 0.3% Triton X-100, three times for 3 min each, and rinse in 1 × SSC briefly.
3. The slides are then incubated in alkaline phosphatase substrate buffer for 5 min at room temperature (*see* **Subheading 2.2.5.**).
4. Incubate the slides in substrate for 3 h at 37°C (*see* **Subheading 2.2.5.**) (*see* **Notes 7** and **8**).
5. Rinse slides vigorously in water, counterstain with nuclear fast red, and rinse in water again.
6. Mount with glycerin jelly and examine slides using a light microscope.

Modifications of the above-described protocol are necessary to tailor the procedure to each specific probe. For example, ISH for VEGF was performed using a similar protocol as for TN-C, with a few modifications. The concentration of the VEGF probe was 6 ng/μL. Baker's yeast was added to the hybridization buffer. Hybridization was achieved by applying 125 μL of the probe with incubation at 56°C. Washes following hybridization were done using 2 × SSC at 56°C and 0.2 × SSC at room temperature. The alkaline phosphatase was used at 1:5000 dilution. Incubation with NBT/BCIP was done at room temperature in the dark *(22)*. Before mounting, the slides were washed with Tris-ethylenediaminetetraacetic acid (EDTA) and counterstained with methylene green.

Although combinations of ISH with radioactive probes and immunohistochemistry have been proposed *(30–34)*, combinations that are completely nonradioactive have also been developed *(35–43)*. All of these methods include either immunoperoxidase- or immunofluorescence-based detection systems. Adequate suppression of endogenous peroxidase is a concern with immunoperoxidase techniques.

3.7. Applications

To illustrate the utility of nonradioactive ISH, the detection of RNAs for angiogenic proteins in pathological conditions in the CNS including human gliomas and hypoxic/ischemic lesions are shown in **Figs. 1–4**.

Angiopoietin-1 mRNA was localized in tumor cells **Fig. 1** and angiopoietin-2 mRNA was detected in endothelial cells of hyperplastic and nonhyperplastic tumor vessels (**Fig. 2**). Angiopoietin-2 was also expressed in partially sclerotic vessels and in vascular channels surrounded by tumor cells in brain adjacent to the tumor. Neither angiopoietin-1 nor angiopoietin-2 were detected in normal brain. These results suggest that angiopoietins are involved in the early stages of vascular activation and in advanced angiogenesis, and identify angiopoietin-2 as an early marker of glioma-induced neovascularization. No staining was observed when sense probes were used.

ISH of astrocytomas using a digoxigenin-labeled antisense riboprobe detected strong staining for TN mRNA in vascular cells, especially in hyperplastic vessels (**Fig. 3**) and JPAs (data not shown), including those at the invasive edge of the tumors, but not in vessels of normal brains. It was seen in endothelial cells adjacent to the vascular lumina and within the walls of the vascular complexes. TN-C mRNA was detected in vessels beyond the tumor margin in the brain tissue adjacent to the tumor in GBMs but in none of the JPAs. TN-C mRNA was also detected in tumor cells in the GBMs, including in the pseudopalisading cells around foci of necrosis (data not shown). No message was detected in the tumor cells in the JPAs. No staining was observed when the sense probe was used.

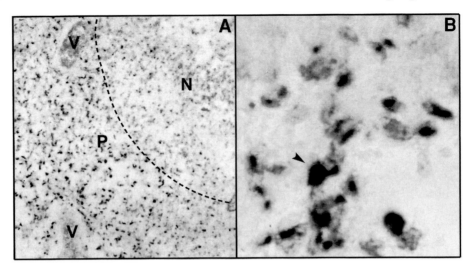

Fig. 1. ISH for angiopoietin-1 mRNA in GBM. **(A)** There is a diffuse signal in tumor cells, but no remarkable upregulation in pseudopalisading (P) areas adjacent to necrotic zones (N) demarcated by the dashed line. Blood vessels (V) are not stained. **(B)** Higher magnification of panel A showing strong tumor cell (arrowhead) staining for Ang-1. A × 50; B × 400.

VEGF was localized in pseudopalisading tumor cells around areas of necrosis and in hyperplastic vessels of GBMs, in areas adjacent to tumor infarction, microcysts and vascular hyperplasia in JPAs and in the walls of macrocysts in both GBMs and JPAs as previously described *(18,44)*. In infarcts, VEGF mRNA was detected in macrophages and vascular and astrocytic cells. In addition, neurons adjacent to the infarct showed strong staining for VEGF mRNA **(Fig. 4)**. VEGF staining was strong in 5- and 7-d-old infarcts. Several labeled cells had irregular and elongated nuclei consistent with microglial cells (data not shown). No signal was seen with the sense probe.

In summary, ISH detection of mRNAs using nonradioactive methods is an invaluable tool for research and diagnostics, dramatically advancing the study of cell- and tissue-specific expression of many genes including those implicated in pathophysiological processes in the CNS. This approach can identify the cells involved with the production of various proteins. It is reliable, safe, and reproducible.

4. Notes

1. RNase is especially relevant to mRNA hybridization using riboprobes. If there is no RNase contamination, the use of standard sterile technique will suffice. By contrast, if RNase contamination does exist, an RNase-free technique is recommended. We recommend DEPC treatment of all solutions where appropriate followed by autoclaving whenever possible. After the hybrid is formed, contamination by RNase is less relevant.

2. Dextran sulfate acts as a space taker. It is, in a sense, increasing the actual concentration of the probe by absorbing the water in the probe solution. This is also why it is important to seal the coverslip of the sections with rubber cement after the probe is added, otherwise, the probe concentration may become diluted during the overnight hybridization. The BSA, transfer RNA (tRNA), and Salmon sperm DNA are all components of the blocking reagents in the hybridization buffer. The combination of all these, and formamide, in the hybridization buffer reduces background and increases the specificity of the hybridization process.

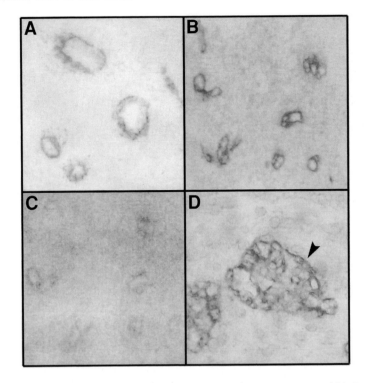

Fig. 2. ISH for angiopoietin-2 mRNA in vessels of astrocytomas. (**A**) Angiopoietin-2 signal is seen in vessels with early hyperplastic changes. (**B**) Stronger signal for angiopoietin-2 is seen in hyperplastic vascular complexes. (**C**) No signal is seen with the sense probe. (**D**) Angiopoietin-2 mRNA is detected in hyperplastic vessels (arrowhead). A × 100; B × 50; D × 200.

Probe buffer containing detergents increases the accessibility and also reduces background. It is also helpful to spread the probe over the section.

3. The shortcomings of using frozen sections are the difficulty in cutting the sections that tend to be thicker. Thicker sections are difficult to read and are more likely to detach from the slide during the ISH. By contrast, paraffin embedding allows sections to be in the range of 4–5 µm. On the other hand, although nonradioactive ISH can be performed on paraffin-embedded tissue, frozen sections, if preserved properly, require less pretreatment to access the target nucleotides.

4. Fixation stabilizes the target nucleotides but optimal hybridization in fixed tissue requires pretreatment with proteinase K because of the crosslinks that are formed during fixation. Our results confirm that in routinely fixed biopsy tissue, saturated signals can be obtained. In unfixed sections, mRNA is less stable and prone to degradation more readily than DNA targets. It is recommended that a tissue block be stored at a low temperature and cut no more than a few days before the actual *in situ* procedure. In addition, it is suggested that tissues that are to be used for mRNA projects be kept as cold as possible, in an RNase-free environment, and fixed as soon as possible, but with no more than the standard overnight fixation since any excess fixation will render the target nucleotides inaccessible. Several studies have shown that formalin-based fixatives retain cellular RNAs with good tissue morphology (*13,17,27,29*).

5. If the section can survive the pretreatment steps, one should be able to get abundant information from the hybridization. If it does not or if the morphology is substandard, some of

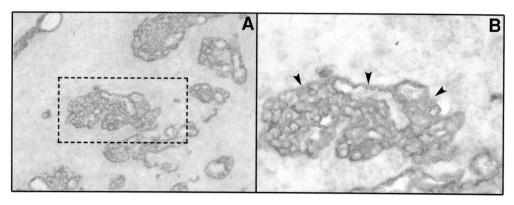

Fig. 3. ISH for TN mRNA in a GBM using a digoxigenin labeled riboprobe transcribed from a 231 bp PCR product corresponding to a segment of exon 1 of human TN. It is detected with anti-digoxigenin antibodies conjugated to an alkaline phosphatase enzyme after 3 h exposure to NBT/BCIP substrate. (A) A strong staining is seen in hyperplastic vascular channels with the antisense riboprobe. (B) Higher magnification of panel A: the staining is seen in cells lining the lumen as well as in other cells of a glomeruloid vessel *(arrowheads)*. No staining was detected with the sense probe (not shown). A, × 100; B, × 200.

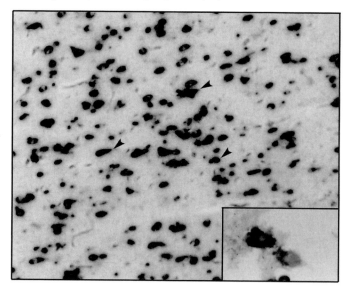

Fig. 4. ISH for VEGF in a cerebral infarct shows signal within cortical neurons (arrowheads) adjacent to an infarcted area (NBT/BCIP; × 50). Inset: Higher magnification of a pyramidal neuron labeled for VEGF mRNA. (NBT/BCIP). × 50, inset × 200.

the time the pretreatment steps may have to be reduced, or concentrations of pre-treatment solution may have to be lowered or steps may have to be eliminated altogether, or a combination of these changes may have to be applied. Several of the reagents used for the pretreatment of the sections have been studied in depth. For example, with high concentrations of proteinase K, there is deterioration of morphology due to excessive digestion.

6. When preparing the probe and evaluating it with the dot blot technique, it is important to use filter paper hybridization and develop the signal using the same substrates and detection that will be used in the actual in situ procedure. Development time should take no longer

then 20 min, otherwise the probe may not be suitable for *in situ* runs. As an alternative, there are vendors able to perform labeling.

7. For tissue sections, colorimetric detection (NBT/BCIP, Diaminobenzidine) is the method of choice, but it is more time consuming because of the detection procedure. However, increasing the detection time allows for a greater sensitivity, but this must be balanced against the increase in background. Colorimetric procedures can take overnight to one week and should be performed in the dark.

8. The detection procedure using NBT/BCIP allows for a relatively long development time, thereby enabling the detection of mRNA present at relatively low expression levels. For highly expressed targets, a quicker procedure may use fluorescent-labeled antibodies, or even better, direct fluorescent probes. Of course, one must have a dedicated fluorescent scope system with image capturing and manipulating capabilities before considering this route. This may be cost prohibitive for some laboratories.

Acknowledgments

The authors are indebted to Drs Marty Grumet, Jocelyn Holash, Stan Wiegand, George Yancopoulos, Virginia Capo, and Kevin Claffey. We thank Dr. D.C. Miller for his comments. This work was supported by a grant from the American Cancer Society (RPG-00-060-01-CCE) (to DZ).

References

1. Baldino, F. Jr., Ruth, J. L., and Davis, L. G. (1989) Nonradioactive detection of vasopressin mRNA with in situ hybridization histochemistry. *Exp. Neurol.* **104,** 200–207.
2. Hoefler, H., Childers, H., Montminy, M. R., Lechan, R. M., Goodman, R. H., and Wolfe, H. J. (1986) In situ hybridization methods for the detection of somatostatin mRNA in tissue sections using antisense RNA probes. *Histochem. J.* **18,** 597–604.
3. Landry, M., and Hokfelt, T. (1998) Subcellular localization of preprogalanin messenger RNA in perikarya and axons of hypothalamo-posthypophyseal magnocellular neurons: an in situ hybridization study. *Neuroscience* **84,** 897–912.
4. Kiyama, H., and Emson, P. C. (1990) Distribution of somatostatin mRNA in the rat nervous system as visualized by a novel non-radioactive in situ hybridization histochemistry procedure. *Neuroscience* **38,** 223–244.
5. Meltzer, J. C., Sanders, V., Grimm, P. C., et al. (1998) Production of digoxigenin-labelled RNA probes and the detection of cytokine mRNA in rat spleen and brain by in situ hybridization. *Brain Res. Brain Res. Protoc.* **2,** 339–351.
6. Le Moine, C., Normand, E., and Bloch, B. (1995) Use of non-radioactive probes for mRNA detection by in situ hybridization: interests and applications in the central nervous system. *Cell Mol. Biol.* **41,** 917–923.
7. Springer, J. E., Robbins, E., Gwag, B. J., Lewis, M. E., and Baldino, F. Jr. (1991) Non-radioactive detection of nerve growth factor receptor (NGFR) mRNA in rat brain using in situ hybridization histochemistry. *J. Histochem. Cytochem.* **39,** 231–234.
8. Tsukamoto, T., Kusakabe, M., and Saga, Y. (1991) In situ hybridization with non-radioactive digoxigenin-11-UTP-labeled cRNA probes: localization of developmentally regulated mouse tenascin mRNAs. *Int. J. Dev. Biol.* **35,** 25–32.
9. Wang, D., and Cutz, E. (1994) Simultaneous detection of messenger ribonucleic acids for bombesin/gastrin-releasing peptide and its receptor in rat brain by nonradiolabeled double in situ hybridization. *Lab. Invest.* **70,** 775–780.
10. Key, M., Wirick, B., Cool, D., and Morris, M. (2001) Quantitative in situ hybridization for peptide mRNAs in mouse brain. *Brain Res. Protoc.* **8,** 8–15.

11. Bloch, B., Guitteny, A. F., Normand, E., and Chouham, S. (1990) Presence of neuropeptide messenger RNAs in neuronal processes. *Neurosci. Lett.* **109,** 259–264.

12. Bloch, B. (1993) Biotinylated probes for in situ hybridization histochemistry: use for mRNA detection. *J. Histochem. Cytochem.* **41,** 1751–1754.

13. Relf, B. L., Machaalani, R., and Waters, K. A. (2002) Retrieval of mRNA from paraffin-embedded human infant brain tissue for non-radioactive in situ hybridization using oligonucleotides. *J. Neurosci. Methods* **115,** 129–136.

14. Boissin-Agasse, L., de Bouard, V., Roch, G., and Boissin, J. (1992) In situ hybridization of GnRH mRNA in the rat and the mink hypothalamus using biotinylated synthetic oligonucleotide probes. *Brain Res. Mol. Brain Res.* **14,** 57–63.

15. Breitschopf, H., Suchanek, G., Gould, R. M., Colman, D. R., and Lassmann, H. (1992) In situ hybridization with digoxigenin-labeled probes: sensitive and reliable detection method applied to myelinating rat brain. *Acta Neuropathol. (Berl.)* **84,** 581–587.

16. McQuaid, S., and Allan, G. M. (1992) Detection protocols for biotinylated probes: optimization using multistep techniques. *J. Histochem. Cytochem.* **40,** 569–574.

17. Fleming, K. A., Evans, M., Ryley, K. C., Franklin, D., Lovell-Badge, R. H., and Morey, A. L. (1992) Optimization of non-isotopic in situ hybridization on formalin-fixed, paraffin-embedded material using digoxigenin-labelled probes and transgenic tissues. *J. Pathol.* **167,** 9–17.

18. Shweiki, D., Itin, A., Soffer, D., and Keshet, E. (1992) Vascular endothelial growth factor induced by hypoxia may mediate hypoxia-initiated angiogenesis. *Nature* **359,** 843–845.

19. Beck, H., Acker, T., Wiessner. C., Allegrini, P. R., and Plate, K. H. (2000) Expression of angiopoietin-1, angiopoietin-2, and tie receptors after middle cerebral artery occlusion in the rat. *Am. J. Pathol.* **157,** 1473–1483.

20. Plate, K. H., Beck, H., Danner, S., Allegrini, P. R., and Wiessner, C. (1999) Cell type specific upregulation of vascular endothelial growth factor in an MCA-occlusion model of cerebral infarct. *J. Neuropathol. Exp. Neurol.* **58,** 654–666.

21. Holash, J., Maisonpierre, P. C., Compton, D., et al. (1999) Vessel cooption, regression, and growth in tumors mediated by angiopoietins and VEGF. *Science* **284,** 1994–1998.

22. Zagzag, D., Hooper, A., Friedlander, D. R., et al. (1999) In situ expression of angiopoietins in astrocytomas identifies angiopoietin-2 as an early marker of tumor angiogenesis. *Exp. Neurol.* **159,** 391–400.

23. Zagzag, D., Friedlander, D. R., Dosik, J., et al. (1996) Tenascin-C expression in angiogenic vessels in human astrocytomas and by human brain endothelial cells in vitro. *Cancer Res.* **56,** 182–189.

24. Zagzag, D., and Capo, V. (2002) Angiogenesis in the central nervous system: a role for vascular endothelial growth factor/vascular permeability factor and tenascin-C. Common molecular effectors in cerebral neoplastic and non-neoplastic angiogenic diseases. *Histol. Histopathol.* **17,** 301–321.

25. Baldino, F. Jr., Chesselet, M. F., and Lewis, M. E. (1989) High-resolution in situ hybridization histochemistry. *Methods Enzymol.* **168,** 761–777.

26. Kleihues, P., Louis, D. N., Scheithauer, B. W., et al. (2002) The WHO classification of tumors of the nervous system. *J. Neuropathol. Exp. Neurol.* **61,** 215–225.

27. Pringle, J. H., Primrose, L., Kind, C. N., Talbot, I. C., and Lauder, I. (1989) In situ hybridization demonstration of poly-adenylated RNA sequences in formalin-fixed paraffin sections using a biotinylated oligonucleotide poly d(T) probe. *J. Pathol.* **158,** 279–286.

28. Nies, D. E., Hemesath, T. J., Kim, J.-H., Gulcher, J. R., and Stefanson, K. (1991) The complete cDNA sequence of human hexabrachion (tenascin). A multidomain protein containing unique epidermal growth factor repeats. *J. Biol. Chem.* **266,** 2818–2823.

29. Sanger, F., Nicken, S., and Coulson, A. R. (1977) DNA sequencing with chain-terminating inhibitors. *PNAS* **74,** 5463–5467.

30. Bugnon, C., Bahjaoui, M., and, Fellmann D. (1991) A simple method for coupling in situ hybridizion and immunocytochemistry: application to the study of peptidergic neurons. *J. Histochem. Cytochem.* **39,** 859–862.

31. Sollberg, S., Peltonen, J., and Uitto, J. (1991) Combined use of in situ hybridization and unlabeled antibody peroxidase anti-peroxidase methods: simultaneous detection of type I procollagen mRNAs and factor VIII-related antigen epitopes in keloid tissue. *Lab. Invest.* **64,** 125–129.

32. Trimmer, P. A., Phillips, L. L., and Steward, O. (1991) Combination of in situ hybridization and immunocytochemistry to detect messenger RNAs in identified CNS neurons and glia in tissue culture. *J. Histochem. Cytochem.* **38,** 891–898.

33. Mitchell, V., Beauvillain, J. C., and Mazzuca, M. (1992) Combination of immunocytochemistry and in situ hybridization in the same semi-thin sections: detection of Met-enkephalin and pro-enkephalin mRNA in the hypothalamic magnocellular dorsal nucleus of the guinea pig. *J. Histochem. Cytochem.* **40,** 581–585.

34. Kiyama, H., McGowan, E. M., and Emson, P. C. (1991) Coexpression of cholecystokinin mRNA and tyrosine hydroxylase mRNA in populations of rat substantia nigra cells; a study using a combined radioactive and non-radioactive in situ hybridization procedure. *Brain Res. Mol. Brain Res.* **9,** 87–93.

35. Biffo, S., Verdun di Cantogno, L., and Fasolo, A. (1992) Double labeling with non-isotopic in situ hybridization and BrdU immunohistochemistry: calmodulin (CaM) mRNA expression in post-mitotic neurons of the olfactory system. *J. Histochem. Cytochem.* **40,** 535–540.

36. Smith, M. D., Parker, A., Wikaningrum, R., and Coleman, M. (2000) Combined immunohistochemical labeling and in situ hybridization to colocalize mRNA and protein in tissue sections. *Methods Mol. Biol.* **123,** 165–175.

37. Bursztajn, S., Berman, S. A., and Gilbert, W. (1990) Simultaneous visualization of neuronal protein and receptor mRNA. *Biotechniques* **9,** 440–449.

38. Liang, J. D., Liu, J., McClelland, P., and Bergeron, M. (2001) Cellular localization of BM88 mRNA in paraffin-embedded rat brain sections by combined immunohistochemistry and non-radioactive in situ hybridization. *Brain Res. Protoc.* **7,** 121–130.

39. Heppelmann, B., Senaris, R., and Emson, P. C. (1994) Combination of alkaline phosphatase in situ hybridization with immunohistochemistry: colocalization of calretinin-mRNA with calbindin and tyrosine hydroxylase immunoreactivity in rat substantia nigra neurons. *Brain Res.* **635,** 293–299.

40. Kriegsmann, J., Keyszer, G., Geiler, T., Gay, R. E., and Gay, S. (1994) A new double labeling technique for combined in situ hybridization and immunohistochemical analysis. *Lab. Invest.* **71,** 911–917.

41. Larsen, P. J., and Mikkelsen, J. D. (1994) Simultaneous detection of neuropeptides and messenger RNA in the magnocellular hypothalamo-neurohypophysial system by a combination of non-radioactive in situ hybridization histochemistry and immunohistochemistry. *Histochemistry* **102,** 415–423.

42. Oh, Y., and Waxman, S. G. (1995) Differential Na+ channel beta 1 subunit mRNA expression in stellate and flat astrocytes cultured from rat cortex and cerebellum: a combined in situ hybridization and immunocytochemistry study. *Glia* **13,** 166–173.

43. Trembleau, A., Roche, D., and Calas, A. (1993) Combination of non-radioactive and radioactive in situ hybridization with immunohistochemistry: a new method allowing the simultaneous detection of two mRNAs and one antigen in the same brain tissue section. *J. Histochem. Cytochem.* **41,** 489–498.

44. Leung, S. Y., Chan, A. S., Wong, M. P., Yuen, S. T., Cheung, N., and Chung, L. P. (1997) Expression of vascular endothelial growth factor and its receptors in pilocytic astrocytoma. *Am. J. Surg. Pathol.* **21,** 941–950.

31

Proteomics of Brain Endothelium

Separation of Proteins by Two-Dimensional Gel Electrophoresis and Identification by Mass Spectrometry

Reiner F. Haseloff, Eberhard Krause, and Ingolf E. Blasig

1. Introduction

The first complete deoxyribonucleic acid (DNA) sequence of an organism was published in 1977 *(1)*, marking a starting point for what is today called genomics. In the meantime, there was a rapid development in scientific equipment and data processing capabilities, which led to a strong acceleration in the aquisition of knowledge on DNA sequences. There is no doubt that the sequencing of the human genome was one of the most important driving forces in that development. Based on the DNA sequence information, the next step is to determine the cellular expression levels of all genes and the expression of the complete set of proteins within a cell. The proteomics approach represents a further challenge for the completion of our knowledge on the molecular composition of cells.

The method of two-dimensional (2-D) gel electrophoresis, introduced in the mid-1970s *(2,3)* is a key technique for studying protein expression on a larger scale. In principle, the application of any two methods providing separation of proteins based on different molecular properties can result in a 2-D resolution of the information on protein expression. The most common way of analyzing protein mixtures is by separation in the first dimension according to their charge (isoelectric focusing, IEF) followed by a separation in the second dimension using sodium dodecyl sulfate polyacrylamide gel electrophoresis (SDS-PAGE). The resulting gel patterns may consist of more than 10,000 protein spots *(4)*.

For further analysis, mass spectrometric techniques have obtained wide importance and acceptance because of their speed, sensitivity, and reliability. In particular, matrix-assisted laser desorption/ionization mass spectrometry (MALDI-MS) has become a useful tool for identifying proteins in large-scale proteome investigations. Digestion of a protein by an endoproteinase in-gel followed by MALDI-MS of the resulting peptide mixture provides a peptide mass fingerprint, which is characteristic for the protein *(5,6)*. The method has been optimized *(7)* and used for the identification of large numbers of proteins from 2-D gel electrophoresis *(8,9)*.

From: *Methods in Molecular Medicine, vol. 89:*
The Blood–Brain Barrier: Biology and Research Protocols
Edited by: S. Nag © Humana Press Inc., Totowa, NJ

18. The gels were carefully removed from the tubes by pumping water with a syringe (without needle, but with an Eppendorf pipet tip attached) through the tube.
19. Ampholytes and urea were removed by incubation for 10 min in equilibration solution.
20. Gels were either immediately used for the SDS-PAGE or stored in Petri dishes at –80°C.

3.3. Separation of Proteins by 2-D Electrophoresis: Second Dimension

SDS-PAGE is a more common technique as compared to IEF. For the experiments described here, a casting and electrophoresis unit DESAVOR VA was used, which allows casting and running of two high-resolution gels simultaneously under identical conditions.

1. The SDS separation gel solution was poured into the casting unit and covered for approx 30 min with 2-propanol to obtain a plain surface.
2. After removing the 2-propanol, the gels were covered with overlay buffer and allowed to stand overnight.
3. Before loading the IEF gel onto the 2-D gel, the overlay buffer was removed and the surface of the SDS gel was dried carefully using filter paper.
4. Using a spatula, the first-dimensional gel was positioned on the SDS gel. This procedure has to be done very carefully to prevent any damage or stretching of the IEF gel.
5. Avoid trapping air bubbles between both gels.
6. Finally, cover the gel with completely melted agarose gel solution using a glass pasteur pipet.
7. Gels of 1.5-mm thickness were run for 15 min at 120 mA and for approx 6.5 h at 150 mA. The corresponding current values for gels of 0.9 mm thickness are 75 mA and 110 mA, respectively.
8. The temperature of the buffer was kept at 15°C using a cooler thermostat.
9. After the run, the gels were removed from the electrophoresis unit and the glass plates were carefully separated from the gels, which were transferred into dishes containing the fixation solution.
10. Gels were shaken for at least 2 h or overnight on a horizontal shaker.

3.4. Procedures for Protein Detection

There are various techniques to visualize proteins in a 2-D gel, including autoradiography (which requires labeling of the proteins with radioactive isotopes) and different staining techniques. Silver staining techniques are very sensitive and detect 1–10 ng protein per spot. However, the latter technique has limited application for MS analysis of the stained proteins (*see* **Note 4**). In this method, separated proteins were stained using Coomassie Brilliant Blue G-250, which does not affect the subsequent MS identification of protein. The protocol described below is based on the method of Neuhoff et al. *(23)*.

1. The gels were kept overnight in CBBG fixation solution on a horizontal shaker.
2. After three washes in distilled water, the gels were shaken in CBBG incubation solution for 2 h followed by the transfer into the staining solution.
3. The gels were kept in the staining solution (covered with aluminum foil) for 5 d.
4. After a very short washing step, the gels were sealed in transparent foil and stored at 4°C.
5. Application:
   A 2-D gel obtained by the procedures described here is shown in **Fig. 2**. Differences between gels from normoxic and hypoxic cells were assessed by visual examination and

├───▶pl

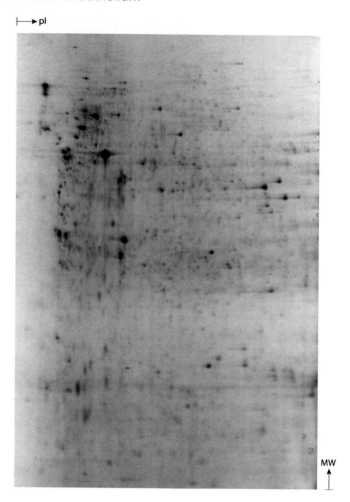

MW
↑

Fig. 2. Gel obtained by 2-D electrophoresis of soluble proteins of brain capillary endothelial cells (staining method, Coomassie Brillaint Blue G250; original gel size, 23–30 cm).

by using computer-based evaluation of the data obtained after scanning of the gels. Results were analyzed using the program Phoretix®. It is beyond the scope of this chapter to describe this software. There are different products available commercially for computer-assisted analysis of 2-D gels such as Melanie II (BIO-RAD, Munich, Germany). The criterion for considering that proteins were differentially expressed (**Fig. 3**) was a difference of at least 100% between the volume of the protein spots (determined using the Phoretix® software) in gels obtained from at least three different preparations.

3.5. Identification of Proteins by Mass Spectrometry

3.5.1. In-Gel Digestion

1. The protein spots were excised from the stained gel, washed twice with 50% (v/v) acetonitrile in 25 mM ammonium bicarbonate, shrunk by dehydration in acetonitrile, and dried in a vacuum centrifuge.
2. Disulfide bonds were reduced by incubation in 30 μL of 10 mM DTT in 100 mM ammonium bicarbonate for 45 min at 55°C.

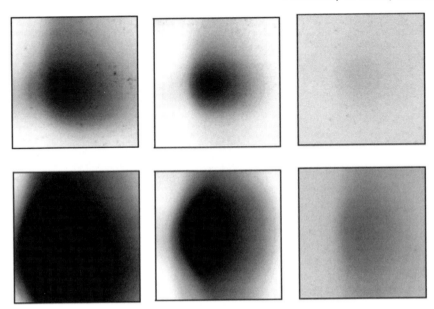

Fig. 3. Details of gels demonstrating differential expression of proteins: spots originate from control (**upper panel**) and hypoxic cells (**lower panel**); mass spectrometric analysis revealed the following identities: glyceraldehyde 3-phosphate dehydrogenase (GAPDH, left pair), α-enolase (middle pair), and phosphoglycerate kinase (right pair).

3. Alkylation was performed by replacing the DTT solution with 55 m*M* iodoacetamide in 100 m*M* ammonium bicarbonate.
4. After 20 min incubation at 25°C in the dark, the gel pieces were washed with 50–100 μL of 50% (v/v) acetonitrile in 25 m*M* ammonium bicarbonate, shrunk by dehydration in acetonitrile, and dried in a vacuum centrifuge.
5. The gel pieces were reswollen in 10 μL of 5 m*M* ammonium bicarbonate, containing 300 ng trypsin.
6. After 15 min, 5 μL of 5 m*M* ammonium bicarbonate was added to keep the gel pieces wet during enzymatic cleavage (37°C, overnight).
7. To extract the peptides, 15 μL of 0.5% (v/v) trifluoroacetic acid (TFA) in acetonitrile was added, and the samples were sonicated for 5 min.
8. The separated supernatant was dried under vacuum and redissolved in 8 μL of 0.1% (v/v) TFA in water.
9. The peptides were purified with C18 reversed-phase minicolumn filled in a micropipet tip (ZipTip C18), before MS analysis.
10. The purification was done according to the manufacturer's instructions, except that peptides were eluted with 3 μL of 60% (v/v) acetonitrile, 0.3% (v/v) TFA for MALDI-MS.

3.5.2. MALDI-MS

MALDI-MS measurements were performed on a Voyager-DE STR BioSpectrometry Workstation MALDI-TOF MS.

1. 1 μL of the analyte solution was mixed with 1 μL of alpha-cyano-4-hydroxycinnamic acid matrix solution consisting of 10 mg of matrix dissolved in 1 mL of 0.3% TFA in acetonitrile-water (1:1, v/v) (*see* **Note 6**).

2. From the resulting mixture 1 µL was applied to the sample plate.
3. Samples were air-dried at ambient temperature (24°C).
4. Measurements were performed in the reflection mode at an acceleration voltage of 20 kV, 70% grid voltage and a delay of 200 ms.
5. Each spectrum obtained was the mean of 256 laser shots. Mass spectra were calibrated using known trypsin fragments as internal standards **(Fig. 4A)**.
6. Interpretation:

   Although much work has been done to optimize experimental conditions, peak intensities of peptides differ significantly. Thus, a sequence coverage of 40–60% is typical. The results obtained depend largely on the amount of protein available and the staining procedure *(24)* (*see* **Note 4**). In general, the sensitivity of MALDI-MS for peptide detection is in the lower fmol range, which should be sufficient for an analysis of even weakly Coomassie blue stained protein spots by peptide mass fingerprinting. External factors such as sample preparation methods *(25,26)* have been shown to affect the peak intensity of peptides. Detection problems might also be caused by intrinsic properties of the peptide, which influence its ionization behavior. Charged side chains *(27,28)* and particularly the guanidino group of arginine *(29)* may influence signal intensity. Despite trypsin cleaves with similar propensity at the C-terminus of lysine and arginine we have recently demonstrated that arginine-containing peptides generally provide stronger signals. The fact that the most intense MALDI-MS peaks of tryptic mass fingerprints denote peptides bearing arginine at the C-terminal end provides an additional criterion for the reliable assignment of a peptide mass fingerprint to a protein in database searches *(29)*.

3.5.3. Database Searches for Protein Identification

Proteins were identified using the search program MS-FIT. The peptide masses determined were matched with the theoretical peptide masses of proteins from the SWISS-PROT or NCBI database. The parameters used for the search were as follows: modifications were considered (oxidation of methionine, cysteine as carbamidomethyl derivative, modification of cysteine by reaction with acrylamide), partial enzymatic cleavages leaving one cleavage site, a protein mass range estimated from the gel ± 20%, a mass accuracy of 0.05 Da **(Fig. 4B)**. If no protein matched the mass window was extended. The criteria, which have to be met for the protein to be unambiguously identified, depend on mass measurement accuracy and the molecular weight of the protein *(30)*. Within a mass accuracy of 0.05 Da a protein was accepted as identified if at least 25% of the complete protein sequence matched (sequence coverage >25%).

4. Notes

1. The quality of the chemicals used for two-dimensional electrophoresis determines the quality of the results. Especially for the IEF, the best reagents available should be used (electrophoresis purity). The use of double distilled water (or Millipore Milli-Q water) is recommended. Batch-to-batch variations (e.g., of the ampholytes) can be minimized by purchasing large amounts and preparing aliquots for long-term use.
2. The ideal sample consists of proteins only. Any contamination of the protein solution by other cell constituents, such as nucleic acids and lipids, or by high salt concentrations may interfere with the performance of the two-dimensional electrophoresis. Therefore, it is recommended to apply appropriate methods of purification (e.g., treatment with nucleic acid-degrading enzymes, dialysis) when the sample is suspected to be contaminated. Of course, this applies also to studies of the effect of chemicals on protein expression.

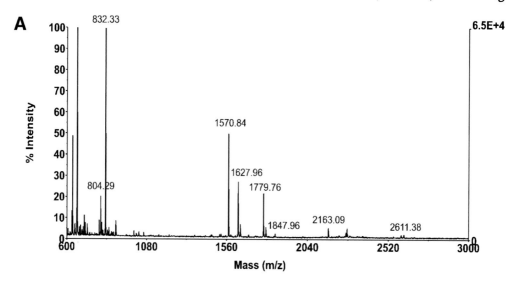

MS-Fit Search Results

B

Sample ID (comment): **Spot 26**
Database searched: **SwissProt.9.30.2001**
Molecular weight search (**1000 - 100000 Da**) selects **94637** entries.
Species search (**MAMMALS**) selects **18496** entries

Parameters Used in Search

Min # Peptides to Match: 13; Peptide Mass Tolerance (+/-): 0.05 Da;
Digest Used: Trypsin; Max. # Missed Cleavages: 1

Results

1. 17/73 matches (18%). 35836.4 Da, pI = 8.43. Acc. # P04797. RAT.
GLYCERALDEHYDE 3-PHOSPHATE DEHYDROGENASE (GAPDH) (38 KDA BFA-
DEPENDENT ADP-RIBOSYLATION SUBSTRATE) (BARS-38).

m/z submitted	MH⁺ matched	Delta Da	start	end	Peptide Sequence	Modifications
795.4296	795.4187	0.010	226	232	(K)LTGMAFR(V)	
805.4368	805.4321	0.004	4	11	(K)VGVNGFGR(I)	
811.4174	811.4136	0.004	226	232	(K)LTGMAFR(V)	1Met-ox
1319.6771	1319.7211	-0.044	247	257	(R)LEKPAKYDDIK(K)	
1356.6588	1356.6543	0.004	322	333	(R)VVDLMAYMASKE(-)	
1369.7733	1369.7440	0.029	199	213	(R)GAAQNIIPASTGAAK(A)	
1570.8408	1570.8263	0.014	233	246	(R)VPTPNVSVVDLTCR(L)	
1627.9584	1627.9536	0.004	65	78	(K)LVINGKPITIFQER(D)	
1779.7647	1779.7978	-0.033	308	321	(K)LISWYDNEYGYSNR(V)	
1847.9554	1847.9359	0.019	144	160	(K)IVSNASCTTNCLAPLAK(V)	
2213.1478	2213.1099	0.038	117	137	(R)VIISAPSADAPMFVMGVNHEK(Y)	
2363.2336	2363.2215	0.012	226	246	(K)LTGMAFRVPTPNVSVVDLTCR(L)	1Met-ox
2369.2217	2369.2110	0.011	116	137	(K)RVIISAPSADAPMFVMGVNHEK(Y)	
2595.3843	2595.3605	0.024	161	184	(K)VIHDNFGIVEGLMTTVHAITATQK(T)	
2611.3833	2611.3554	0.028	161	184	(K)VIHDNFGIVEGLMTTVHAITATQK(T)	1Met-ox
2933.4781	2933.4541	0.024	117	143	(R)VIISAPSADAPMFVMGVNHEKYDNSLK(I)	
2949.4863	2949.4490	0.037	117	143	(R)VIISAPSADAPMFVMGVNHEKYDNSLK(I)	1Met-ox

The matched peptides cover **49%** (164/333AA's) of the protein.

Fig. 4. Identification of proteins (separated by 2-D gel electrophoresis) by peptide mass fingerprint analysis and database search. **A**, MALDI-MS of the peptide mixture resulting from tryptic in-gel digestion. **B**, MS-FIT search result, showing the search criteria, sequence of the
matching peptides, deviation of the observed peptide mass determined from theoretical mass
values (delta), and the sequence coverage.

3. Clean glass and plasticware is a prerequisite for reproducible results of the experiments. IEF tubes should be rinsed with water immediately after removing the gels. Then, solutions of decontaminating agents, such as Deconex (Borer Chemie AG, Switzerland) or RBS-35 (Pierce Chemical Co, USA) should be used according to the manufacturer's instructions. Dirty glass plates will strongly influence the polymerization of the gels for SDS-PAGE. It is recommended to clean them after and immediately before casting the gels (any dust particles must be removed).

4. The commonly used stains for visualizing proteins on two-dimensional gels have benefits and drawbacks. Silver staining procedures are most sensitive while Coomassie stains exhibit lower sensitivity. However, silver stains give rise to reduced recovery of peptides from in-gel digests for mass spectrometry, and this decreases the overall sensitivity of the MALDI mass fingerprint analysis. Although omission of glutaraldehyde increased the sequence coverage in MALDI-MS the result is still below the sequence coverage of Coomassie Brilliant Blue stained spots *(24)*. The limited dynamic range frequently prevents a reliable determination of subtle differences in protein quantities. The novel fluorescent dye (SYPRO Ruby) has properties like a broad linear dynamic range and enhanced recovery of peptides, which makes it well suited to proteomics projects *(23)*.

5. The use of standards for 2-D electrophoresis is recommended for assessment of pI and molecular weight values of the protein spots, which can be helpful for further analysis. More precise information on the apparent molecular weight can be obtained by placing a small piece of agarose containing appropriate molecular weight standards adjacent to the IEF gel before running the SDS-PAGE.

6. Although the dried-droplet technique, in which sample and matrix are premixed before spotting, was found to give somewhat lower signal intensities than the thin layer technique it provided better reproduction of peak intensities from different spots *(31)*. For this reason the dried-droplet technique was used throughout.

Acknowledgments

The authors thank B. Eilemann for assistance in the cultivation of endothelial cells. They also acknowledge the help of Ch. Köberle and H. Lerch in two-dimensional electrophoresis and mass spectrometry, respectively.

References

1. Sanger, F., Air, G. M., Barrell, B. G., et al. (1977) Nucleotide sequence of bacteriophage phi X174 DNA. *Nature* **265,** 687–695.
2. O'Farrell, P. H. (1975) High resolution two-dimensional electrophoresis of proteins. *J. Biol. Chem.* **250,** 4007–4021.
3. Klose, J. (1975) Protein mapping by combined isoelectric focusing and electrophoresis of mouse tissues. A novel approach to testing for induced point mutations in mammals. *Humangenetik* **26,** 231–243.
4. Klose, J., and Kobalz, U. (1995) Two-dimensional electrophoresis of proteins: an updated protocol and implications for a functional analysis of the genome. *Electrophoresis* **16,** 1034–1059
5. Henzel, W. J., Billeci, T. M., Stults, J. T., Wong, S. C., Grimley, C., and Watanabe, C. (1993) Identifying proteins from two-dimensional gels by molecular mass searching of peptide fragments in protein sequence databases. *Proc. Natl. Acad. Sci. USA* **90,** 5011–5015.
6. Pappin, D. J. C., Hojrup, P., and Bleasby, A. J. (1993) Rapid indentification of proteins by peptide-mass fingerprinting. *Curr. Biol.* **3,** 327–332.
7. Gevaert, K., Demol, H., Puype, M., et al. (1997) Peptides adsorbed on reverse-phase chromatographic beads as targets for femtomole sequencing by post-source decay matrix assisted

laser desorption ionization-reflectron time of flight mass spectrometry. *Electrophoresis* **18,** 2950–2960.

8. Shevchenko, A., Jensen, O. N., Podteljnikov, A. V., et al. (1996) Linking genome and proteome by mass spectrometry: Large-scale identification of yeast proteins from two dimensional gels. *Proc. Natl. Acad. Sci. USA* **93,** 14,440–14,445.

9. Müller, E.-C., Thiede, B., Zimny-Arndt, U., et al. (1996) High-performance human myocardial two-dimensional electrophoresis database: Edition 1996. *Electrophoresis* **17,** 1700–1712.

10. Faller, D. V. (1999) Endothelial cell responses to hypoxic stress. *Clin. Exp. Pharmacol. Physiol.* **26,** 74–84.

11. Abbruscato, T. J., and Davis, T. P. (1999) Protein expression of brain endothelial cell E-cadherin after hypoxia/aglycemia: influence of astrocyte contact. *Brain Res.* **842,** 277–286.

12. Oehler, R., Schmierer, B., Zellner, M., Prohaska, R., and Roth, E. (2000) Endothelial cells downregulate expression of the 70 kDa heat shock protein during hypoxia. *Biochem. Biophys. Res. Commun.* **274,** 542–547.

13. Stins, M. F., Gilles, F., and Kim, K. S. (1997) Selective expression of adhesion molecules on human brain microvascular endothelial cells. *J. Neuroimmunol.* **76,** 81–90.

14. Park, J. H., Okayama, N., Gute, D., Krsmanovic, A., Battarbee, H., and Alexander, J. S. (1999) Hypoxia/aglycemia increases endothelial permeability: role of second messengers and cytoskeleton. *Am. J. Physiol., Cell Physiol.* **277,** C1066–C1074.

15. Dreher, D., Vargas, J. R., Hochstrasser, D. F., and Junod, A. F. (1995) Effects of oxidative stress and Ca2+ agonists on molecular chaperones in human umbilical vein endothelial cells. *Electrophoresis* **16,** 1205–1214.

16. Portig, I., Pankuweit, S., Lottspeich, F., and Maisch, B. (1996) Identification of stress proteins in endothelial cells. *Electrophoresis* **17,** 803–808.

17. Herbert, B. (1999) Advances in protein solubilisation for two-dimensional electrophoresis. *Electrophoresis* **20,** 660–663.

18. Molloy, M. P. (2000) Two-dimensional electrophoresis of membrane proteins using immobilized pH gradients. *Anal. Biochem.* **280,** 1–10.

19. Pasquali, C., Fialka, I., and Huber, L. A. (1997) Preparative two-dimensional gel electrophoresis of membrane proteins. *Electrophoresis* **18,** 2573–2581.

20. Rabilloud, T. (1998) Use of thiourea to increase the solubility of membrane proteins in two-dimensional electrophoresis. *Electrophoresis* **19,** 758–760.

21. Blasig, I. E., Giese, H., Schroeter, M. L., et al. (2001) ˙NO and oxyradical metabolism in new cell lines of rat brain capillary endothelial cells forming the blood-brain barrier. *Microvasc. Research* **62,** 114–127.

22. Rickwood, D., Chambers, J. A. A., and Spragg, S. P. (1990) Two-dimensional gel electrophoresis. In *Gel Electrophoresis of Proteins: A Practical Approach.* Hames, B. D., and Rickwood, D., eds. IRL Press, Oxford, New York, Tokyo, pp. 217–272.

23. Neuhoff, V., Arold, N., Taube, D., and Ehrhardt, W. (1988) Improved staining of proteins in polyacrylamide gels including isoelectric focusing gels with clear background at nanogram sensitivity using Coomassie Brilliant Blue G-250 and R-250. *Electrophoresis* **9,** 255–262.

24. Scheler, C., Lamer, S., Pan, Z., Li, X.-P., Salnikow, J., and Jungblut, P. (1998) Peptide mass fingerprint sequence coverage from differently stained proteins on two-dimensional electrophoresis patterns by matrix assisted laser desorption/ionization mass spectrometry (MALDI-MS). *Electrophoresis* **19,** 918–928.

25. Kussmann, M., Nordhoff, E., Rahbek-Nielsen, H., et al. (1997) Matrix-assisted laser desorption/ionization mass spectrometry sample preparation techniques designed for various peptide and protein analytes. *J. Mass Spectrom.* **32,** 593–601.

26. Cohen, S. L., and Chait, B. T. (1996) Influence of matrix solution conditions on the MALDI-MS analysis of peptides and proteins. *Anal. Chem.* **68,** 31–37.

27. Liao, P. C., and Allison, J. (1995) Enhanced detection of peptides in matrix-assisted laser desorption ionization mass spectrometry through the use of charge-localized derivatives. *J. Mass Spectrom.* **30,** 511–512.

28. Juhasz, P., and Biemann, K. (1994) Mass spectrometric molecular-weight determination of highly acidic compounds of biological significance via their complexes with basic polypeptides. *Proc. Natl. Acad. Sci. USA* **91,** 4333–4337.

29. Krause, E., Wenschuh, H., and Jungblut, P. R. (1999) The dominance of arginine-containing peptides in MALDI-derived tryptic mass fingerprints of proteins. *Anal. Chem.* **71,** 4160–4165.

30. Clauser, K. R., Baker, P., and Burlingame, A. L. (1999) Role of accurate mass measurement in protein identification strategies employing MS or MS/MS and database searching. *Anal. Chem.* **71,** 2871–2882.

31. Wenschuh, H., Halada, P., Lamer, P., Jungblut, P., and Krause, E. (1998) The ease of peptide detection by matrix-assisted laser desorption/ionization mass spectrometry: the effect of secondary structure on signal intensity. *Rapid Commun. Mass Spectrom.* **12,** 115–119.

32. Lopez, M. F., Berggren, K., Chernokalskaya, E., Lazarev, A., Robinson, M., and Patton, W. F. (2000) A comparison of silver stain and SYPRO Ruby Protein Gel Stain with respect to protein detection in two-dimensional gels and identification by peptide mass profiling. *Electrophoresis* **21,** 3673–3683.

32

Immunoblot Detection of Brain Vascular Proteins

Christian T. Matson and Lester R. Drewes

1. Introduction

There continue to be many new technical developments that propel advances in our understanding of biological events taking place at the cellular and molecular level. One method that developed early in the expansion of cell and molecular biology and still remains a valuable tool and major core laboratory protocol is immunoblotting. Immunoblot detection of proteins, originally and often called Western blotting, is a method that takes advantage of antibodies that are raised in animals following immunization with the target protein (immunogen). The ability of the generated antibodies to recognize the target protein with high specificity and bind with high affinity makes the method very powerful. When a cell- or tissue-derived mixture of proteins containing the target is first separated based on size or charge and then immobilized on an inert support, the method becomes very informative. This chapter presents aspects of the immunoblotting technique relevant to preparations of brain vascular proteins, standard blotting protocols for optimal detection and performance, and generation of chicken-raised polyclonal antibodies.

The most common method to obtain brain vascular proteins is to isolate brain microvessels, a preparation enriched in cells and structures that form the brain vascular wall. There are many published reports on brain microvessel isolation that are modifications or improvements of the first published method (1). It is generally accepted that brain microvessels are highly enriched in endothelial cells, but also contain significant numbers of pericytes, some smooth muscle cells and remnants of astrocytic endfeet. Also included are the proteins of the basal lamina that form a structurally strong framework among the endothelial cells, pericytes, and astrocytes. The method for brain microvessel isolation that our laboratory has used for several years (2) has proven very useful for other investigators and is available at the internet site www.d.umn.edu/medweb/biochem/dreweslab/Brain_Microvessel_Isolation.pdf.

When brain vascular proteins are immobilized by adsorption onto a membrane support, individual proteins may be detected using protein-specific antibodies. Virtually any type of antibody, polyclonal raised in any of several species or monoclonal, is suitable provided that an appropriate system for detecting the presence of the antibody-antigen complex is available. Hundreds of different polyclonal antibodies that specifically recognize a different protein are commercially available or may be obtained from

From: *Methods in Molecular Medicine, vol. 89:*
The Blood–Brain Barrier: Biology and Research Protocols
Edited by: S. Nag © Humana Press Inc., Totowa, NJ

other investigators. Similarly, monoclonal antibodies are available through vendors, a nonprofit hybridoma bank (www.uiowa.edu/~dshbwww/), or other scientists.

Frequently, however, antibodies against uncommon or novel proteins are not commercially available and must be generated by the investigator. Several approaches are available in the literature *(3)*. Our laboratory has found polyclonal antibodies raised in chickens to be useful reagents for detecting brain vascular proteins *(4)*. The primary advantages to using chickens for antibody production are 1) chickens have a rapid and robust immune response to mammalian proteins, 2) the chicken immunoglobulin (IgY) is easily collected from egg yolks, and 3) the proportion of IgY from yolk proteins represents a very high proportion of the total egg yolk proteins.

2. Materials

2.1. Sodium Dodecyl Polyacrylamide Gel Electrophoresis (SDS-PAGE)

1. Sample buffer: 60 mM Tris-HCl, pH 6.8, 5% SDS, 10% glycerol.
2. Mercaptoethanol solution (5 ×): 20% mercaptoethanol, 80% sample buffer, and 0.1% bromophenol blue.
3. Electrophoresis buffer (5 ×): 15.1 g Tris base, 72.0 g glycine, 5.0 g SDS, distilled H$_2$O to 1000 mL.
4. Pre-cast polyacrylamide gel (fixed concentration or gradient), Ready Gel Tris-HCL 4–10% (BioRad, Hercules, CA).
5. Electrophoresis unit (chamber and gel holder), Mini-Protean 3 Electrophoresis Cell (BioRad).
6. Constant voltage power supply, Power Pac 300 (BioRad).
7. Molecular weight standards, Kaleidoscope Pre-stained Standards (BioRad).
8. Gel loading pipet tips.
9. BCA protein assay reagents (Pierce Chemical, Rockford, IL).

2.2. Protein Transfer to Nitrocellulose Membrane

1. Nondenaturing transfer buffer: 25 mM Tris base, 190 mM glycine, 20% methanol (store at 4°C).
2. TBS-T buffer (10 ×): 1% Tween 20, 0.5 M Tris-HCl, pH 7.5, 1.5 M NaCl.
3. Nitrocellulose membrane.
4. Whatman No. 2 filter paper.
5. Transfer electrophoresis unit with coolant (chamber, sandwich holder, Scotchbrite™-like pads), Mini Trans-Blot Electrophoretic Transfer Cell (BioRad).

2.3. Immunoblotting

1. Small container for membrane incubation and washing.
2. Benchtop shaker/rotating platform.
3. Primary antibody, polyclonal.
4. Secondary antibody, horseradish peroxidase conjugated, species specific.
5. Blocking solution, Sea Block™ (Pierce Chemical) or BlokHen™ (AVES Labs, Tigard, OR).
6. Chemiluminescent detection kit, Super Signal (Pierce Chemical).
7. Clear plastic wrap.
8. Scientific imaging film (Eastman Kodak, Rochester, NY).
9. Film cassette.
10. Manual or automated film processing.
11. Alternative to film: CCD digital chemiluminescence imaging system, Fluorchem™ 8000 Advanced Fluorescence (Alpha Innotech Corp., San Leandro, CA).

2.4. Antibody Preparation

2.4.1. Immunization

1. Immunogen, target peptide conjugated to Keyhole limpet hemocyanin (Sigma, St. Louis, MO).
2. Sterile phosphate-buffered saline (PBS).
3. Microcentrifuge tubes, 1.5-mL.
4. Freund's complete adjuvant (Sigma).
5. Motorized Teflon pestle made for use with microcentrifuge tubes.
6. Tuberculin syringe with needle, 1-mL.

2.4.2. Chicken IgY Recovery and Affinity Purification

1. Phosphate-buffered saline (PBS).
2. Polyethylene glycol (PEG 8,000) (Sigma).
3. 7% PEG 8,000 in PBS.
4. 24% PEG 8,000 in PBS.
5. Equilibration Buffer: 50 mM Tris-HCl, pH 7.4.
6. Salt Buffer: 4.5 M MgCl$_2$, 0.1% bovine serum albumin (BSA) in buffer T. Prepare by adding 91.5 g of MgCl$_2$•6H$_2$O to 40 mL Equilibration Buffer containing 2.5 mg/mL BSA.
7. CNBr-activated Sepharose 4B coupled to antigen (Amersham/Bioscience, Piscataway, NJ).
8. 6 M Guanidine-HCl.

3. Methods

3.1. SDS-PAGE

The first step in performing an immunoblot is separating solubilized proteins by electrophoresis through a uniformly crosslinked matrix (polyacrylamide gel). Sample proteins are dissolved in detergent (SDS) under reducing conditions, thus forming polyanionic randomly flexible polymers that migrate toward the positive pole in the electrophoretic field under constant voltage (5). The migration distance of proteins is inversely proportional to their log molecular weight. Essentially, all brain vascular cell proteins are solubilized by standard sample buffers. However, basement membrane that separates vascular cells from neural cells may contain significant quantities of crosslinked proteins such as collagen and other extracellular matrix proteins that resist solubilization.

1. Dissolve isolated microvessels from one rat brain (approx 250 µg protein) in 50–100 µL sample buffer. Brief sonication may be required for complete solubilization. Centrifuge and retain clear supernatant.
2. Assay sample for total protein. Sample buffer is compatible with the BCA protein assay. Dilute the sample to 2.5 mg/mL in sample buffer.
3. Assemble pre-cast polyacrylamide gel and buffer chamber. (Wear gloves to avoid exposure to toxic acrylamide and to avoid contaminating the gel.) Place assembly into the electrophoresis box and fill the lower chamber with SDS electrophoresis buffer to a level that covers the electrodes. Fill the upper buffer chamber to cover the tops of the sample. Check for leaks (*see* **Note 1**).
4. Flush the sample wells of the gel with electrophoresis buffer using a 1-mL Eppendorf pipetor and ejecting buffer into each well.
5. Add 5 × mercaptoethanol solution to each sample in a 1:4 (v/v) proportion and heat for 5 min in a boiling bath.

Anode (+)

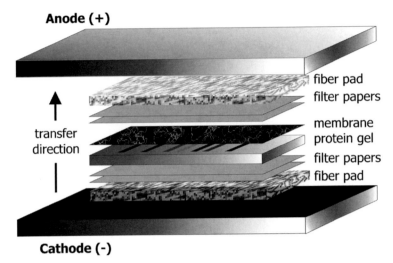

transfer
direction

fiber pad
filter papers

membrane
protein gel
filter papers
fiber pad

Cathode (-)

Fig. 1. Arrangement of transfer sandwich holder and components. Components should be stacked within the sandwich holder in the order shown. Note the nitrocellulose membrane is between the gel and the anode.

6. Load each lane of the gel with sample solution or molecular weight standard using a pipettor equipped with a gel-loading tip.
7. Connect the power source to the electrophoresis chamber and run at 200 V for 1.5 h for a 4–20% gradient gel. The voltage and time may vary depending on the size and composition of the gel (*see* **Note 2**).
8. Turn off and disconnect the power supply when the bromophenol blue reaches the bottom of the gel. Disassemble the gel apparatus according to the manufacturer's instructions and transfer the gel into a container with transfer buffer. Mark the gel for orientation by diagonally cutting a small portion of one corner.

3.2. Protein Transfer to Nitrocellulose Membrane

Once the solubilized proteins have been separated, they must be immobilized on a stable surface before antibodies and other reagent solutions can be applied. This section describes the electrophoretic transfer of proteins from the gel to a nitrocellulose membrane. The method takes advantage of the fact that the proteins remain negatively charged from the SDS. For further reading, see Bittner et al. (**6**).

1. Immerse all components of the transfer sandwich in transfer buffer to eliminate air bubbles (*see* **Note 3**).
2. Assemble the transfer sandwich (**Fig. 1**) by sequentially laying down each component on the open sandwich holder while keeping all components submerged. The order of assembly is Scotchbrite™-like pad, two sheets of Whatman filter paper, acrylamide, gel, nitrocellulose membrane, two sheets Whatman filter paper, and a second Scotchbrite™-like pad.
3. Close the transfer sandwich holder and place it in the transfer chamber with the nitrocellulose membrane between the positive electrode and the gel. Fill the chamber with chilled transfer buffer.
4. Connect the chamber to the power source and initiate the run (200 mA; 1.5 h, 4°C) (*see* **Note 5**).

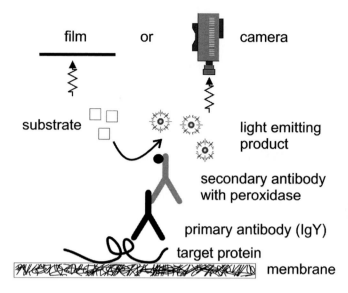

Fig. 2. Chemiluminescent immunodetection scheme. The immobilized protein is exposed sequentially to the primary and secondary antibodies. Incubation with a peroxidase substrate that yields a light-emitting product allows detection with a film of light-sensitive CCD camera. The light detected is proportional to the amount of target protein present on the membrane.

3.3. Immunoblotting and Detection

After the vascular proteins are immobilized on the nitrocellulose membrane, the protein of interest (target protein) is visualized using an antibody against the target protein *(7)*. Nonspecific protein binding to the membrane is blocked with a mixture of non-mammalian proteins, followed by successive exposures to primary antibody, horseradish peroxidase (HRP)-linked anti-IgY secondary antibody, and chemiluminescence substrate as described in the following steps.

1. Disassemble the sandwich and transfer the nitrocellulose membrane to the container with TBS-T solution. Cut the corner of the membrane corresponding to the previously cut portion of the gel. With a pencil, mark the position of the molecular weight markers.
2. Wash the membrane with agitation (rotating platform) once with TBS-T for 20 min, then three additional times for 5 min each.
3. After the last wash, add blocking agent (Sea Block(tm) or BlokHen(tm)) or alternative and incubate for 1 h with agitation. Pour off blocking solution and rinse once for 5 min with TBS-T.
4. Add primary antibody diluted in TBS-T, cover, and incubate for at least 1 h. Alternatively, incubation may be overnight (*see* **Note 6**).
5. Wash the membrane once with TBS-T for 20 min, then four additional times for 5 min each.
6. Dilute the secondary antibody (HRP-linked anti-IgY) in TBS-T, and incubate with agitation for 1 h.
7. Repeat **step 5**.
8. Apply luminescence substrates according to vendor instructions.
9. Place membrane on Whatman filter paper to absorb excess solution. Transfer the membrane and wrap in plastic film, and expose membrane to X-ray film or other light detection system (**Fig. 2**).

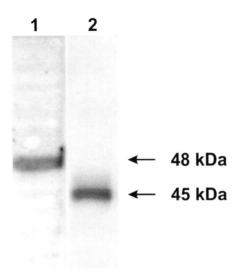

Fig. 3. Immunoblot detection of rat brain proteins. MCT1 (lane 1) and LAT1 (lane 2) were detected using chicken antibodies as described in **Subheading 3.4.** The apparent molecular masses as determined with molecular weight standards are shown.

3.3.1. Application

Several examples illustrate application of the protocols described in this chapter. They include immunoblot detection of transporter proteins such as a monocarboxylic acid transporter (MCT1) and a large neutral amino acid transporter (LAT1) expressed by cells of the rat cerebral vasculature. Representative immunoblots of MCT1 and LAT1 from rat brain membranes indicate major immunoreactive bands at apparent molecular masses of 48 kDa and 45 kDa, respectively **(Fig. 3)**. The observed molecular masses are consistent with the expected size and electrophoretic mobility of membrane proteins. The cellular location of MCT1 and LAT1 in the brain vasculature has been confirmed and reported previously *(4,8)*.

The light emitted by the immunoreactive band during chemiluminescence detection may be used to obtain relative quantities of the target protein. For example, by applying equal quantities of microvessel protein from control and experimental treatments to adjacent lanes, the signal from each lane may be compared and relative amount of immunoreactive protein calculated. Also, the presence of proteins that associate strongly with the target protein may be detected by electrophoresis under nonreducing conditions and observing a shift in apparent molecular weight. This mobility shift may be indicative of association with a second protein or may indicate association with other identical transporter molecules forming dimers or multimeric oligomers.

3.4. Antibody Preparation

3.4.1. Immunization

To generate primary antibodies adult chickens are immunized with the target immunogen suspended in appropriate adjuvant.

1. Dissolve Keyhole limpet hemocyanin-linked peptide (immunogen) in sterile PBS at a concentration of 0.1 to 2 mg/mL and transfer 0.4 mL to a 1.5 mL microcentrifuge tube (*see* **Note 7**).
2. Add 0.4 mL Freund's complete adjuvant and emulsify with a motorized Teflon pestle made for use with microcentrifuge tubes. Continue the emulsification for 3 min. The effectiveness of the emulsification can be determined by placing a drop onto the surface of a beaker of water. If the drop floats and does not disperse, the emulsion is complete.
3. Draw the emulsion into a 1-mL tuberculin syringe and attach a 22-gage needle.
4. Expulse the air and inject 0.1–0.2 mL into each breast muscle of an adult (14–52 wk) female chicken at two different sites.
5. Booster injections of similarly prepared immunogen are given at biweekly intervals for up to 2 mo beginning 2 wk after the initial injection. Maximum antibody production is usually achieved within 6 wk.

3.4.2. Chicken IgY Preparation

To generate primary antibodies adult chickens are immunized with the target immunogen suspended in appropriate adjuvant. Antigen-specific IgY is isolated from yolks and affinity purified.

1. Collect control eggs from the chicken before immunization and antibody producing eggs at least 4 wk after the initial immunization (*see* **Note 8**).
2. Separate the yolk from the white and rinse the yolk three times with deionized H_2O.
3. Pierce the yolk membrane with a sharp object and drain the contents into a graduated cylinder. Add 2 vol of PBS with mixing.
4. Add dropwise an equal volume of 7% PEG in PBS with stirring. A yellow precipitate should form.
5. Centrifuge at greater than 2,000g for 10 min.
6. Filter the supernatant using a Whatman filter with vacuum or a coffee filter. Measure the filtrate volume and calculate the amount of PEG that is present. Add solid PEG with stirring to achieve a final concentration of 12% PEG and continue to stir until all the PEG is dissolved. A white precipitate should form.
7. Collect the precipitate by centrifugation (>2,000g for 20 min). Decant the supernatant and dissolve the white pellet in approx 20 mL PBS. This will take about 30 min on a rocking platform and is facilitated by mechanical disruption of the pellet.
8. Re-precipitate the IgY fraction by adding an equal volume of 24% PEG in PBS dropwise with stirring. Continue to stir for 30 min. Centrifuge again at 2,000g for 20 min. A large white pellet should result.
9. Collect the precipitate by centrifugation (>2000g for 20 min) and dissolve the pellet in 40 mL of PBS. Aliquot the solution and store it frozen at –20°C.
10. Prepare a 1-mL column of antigen-coupled Sepharose.
11. Wash the column successively with the following solutions at a flow rate of approx 20 mL/h:
    a. 15 mL 6 *M* guanidine-HCl.
    b. 25 mL Equilibration Buffer.
    c. 20 mL Salt Buffer.
    d. 50 mL Equilibration Buffer.
12. Load the column with 5–30 mL crude chicken IgY extract (from **step 8**). Maximize antibody binding by using a flow rate of less than 2.5 mL/h. The eluate may be saved for verification of antibody depletion.
13. Wash the column with 20 mL Equilibration Buffer, 40 mL of 1.0 *M* guanidine-HCl, and 20 mL Equilibration Buffer.

MCT1:   KLH-C-LQNSSGDPAEEESPV-COOH

LAT1:   NH₂-MAVAGAKRRAVAAPA-C-KLH

Fig. 4. Immunogens used for raising chicken antibodies are shown. Each polypeptide corresponds to the C-terminal or N-terminal end of the respective transporter protein. A cysteine is added to the end to facilitate conjugation to KLH. The amino acid single letter code is used.

14. Elute with Salt Buffer and collect 1-mL fractions. Immediately dialyze fractions against PBS and assay fractions by spotting a nitrocellulose membrane with the immunogen and detecting the spots with diluted fractions as in **Subheading 3.3.** Use fractions from pre-immune egg yolks for controls.

3.4.3. Application

The antibodies used for immunoblot detection of MCT1 and LAT1 (*see* **Fig. 3**) were affinity purified from egg yolks laid by chickens immunized with 15 amino acid peptides linked to Keyhole limpet hemocyanin (KLH). The amino acid sequences of the two peptides (**Fig. 4**) correspond to the C-terminal end of MCT1 and the N-terminal end of LAT1. It was found that antibodies raised against the C-terminal end of LAT1 were not immunoreactive to the transporter on immunoblots or in tissue sections (immunocytochemistry). Thus, alternative peptides corresponding to different hydrophilic regions of the protein may be necessary for successful antibody preparation.

4. Notes

1. If the electrophoresis apparatus contains a leak, the level in the upper buffer chamber will drop below the level of the wells and current will be lost. This will arrest the migration of proteins. If this happens, adding buffer to the upper chamber will restore the current. Continue to monitor the buffer level and add buffer as needed.

2. In addition to the gel thickness and content, time management can determine the voltage at which to run the gel. It is perfectly acceptable to run a gel at a lower voltage so that the researcher is available to carry out the next step. Running at a lower voltage will continue the separation, prevent diffusion of bands, and prevent bands from running off the bottom of the gel into the buffer solution.

3. The membrane should be handled using gloves or clean forceps to prevent contamination and to prevent the oils from hands from transferring onto the membrane.

4. Minimizing the time that the components are out of the buffer reduces the presence of air pockets that disrupt transfer of protein and lead to areas that are devoid of proteins on the membrane.

5. Electrophoresis buffer should be stirred with a magnetic bar and cooled by inserting a cooling block or setting the chamber in slushed ice.

6. The dilution of antibody is empirically determined. Normally, initial primary antibody dilutions of 1:1000, 1:5000, 1:10,000, and 1:20,000 are tested to determine which concentration yields the best signal-to-noise ratio. Then the best concentration is used in subsequent incubations. Covering the container tightly with a lid or plastic wrap prevents evaporation of the incubation solution and drying out of the gel.

7. The immunogen consists of a short polypeptide (10–20 amino acids) linked to a large carrier protein, KLH. The sequence of the polypeptide is identical to the amino acid sequence of a hydrophilic segment of the target protein because these sequences are found to elicit

the most robust immune responses and produce antibodies with highest affinities. This frequently is the carboxyl or amino terminal end of the protein, but may also be an internal sequence. Assuming the target protein gene has been cloned, the protein sequence may be downloaded from GenBank (www.ncbi.nlm.nih.gov/Genbank/GenbankSearch.html) and examined by Internet available freeware (us.expasy.org/tools/#primary) to generate a hydrophobic profile of the protein. Once a hydrophilic 10–20 amino acid sequence is identified it is chemically synthesized and an additional cysteine residue on either the carboxyl or amino terminal end is included to facilitate conjugation to the KLH. A recommended method for conjugation involves maleimide chemistry *(9)*. If the peptide represents an internal target protein segment, then the carboxyl or amino terminal end is blocked with amidation or acetylation, respectively.

8. IgY from pre-immunization eggs serves as a control. Although target-specific IgY may be detectable within two weeks following immunization, it is recommended to wait at least four weeks to achieve a robust immune response.

Acknowledgments

This work is supported by grants from the National Institutes of Health (NS37762), the Duluth Clinic education and Research Foundation, and the Minnesota Medical Foundation.

References

1. Goldstein, G. W., Wolinsky, J. S., Csejtey, J., and Diamond, I. (1975) Isolation of metabolically active capillaries from rat brain. *J. Neurochem.* **25,** 715–717.
2. Gerhart, D. Z., Broderius, M. A., and Drewes, L. R. (1988) Cultured human and canine endothelial cells from brain microvessels. *Brain Res. Bull.* **21,** 785–793.
3. Harlow, E., and Lane, D. (1988) *Antibodies, A Laboratory Manual.* Cold Spring Harbor Laboratory, Cold Spring Harbor, NY, pp. 53–239.
4. Gerhart, D. Z., Enerson, B. E., Leino, R. L., Zhdankina, O., and Drewes, L. R. (1997) Expression of monocarboxylate transporter MCT1 by brain endothelium and glia in adult and suckling rats. *Am. J. Phys.* **273,** E207–E213.
5. Peterson, S. (1998) One-dimensional SDS gel electrophoresis of proteins. In *Current Protocols in Molecular Biology*, Vol. 2. Ausubel, F. M., Brent, R., Kingston, R. E., et al., eds. John Wiley and Sons, New York, unit 10.2.
6. Bittner, M., Kupferer, P., and Morris, C. F. (1980) Electrophoretic transfer of proteins and nucleic acids from slab gels to diazobenzyloxymethyl cellulose or nitrocellulose sheets. *Anal. Biochem.* **102,** 459–471.
7. Harper, D. R., and Murphy, G. (1991) Nonuniform variation in band pattern with luminal/horseradish peroxidase western blotting. *Anal. Biochem.* **192,** 59–63.
8. Duelli, R., Enerson, B. E., Gerhart, D. Z., and Drewes, L. R. (2001) Expression of large amino acid transporter LAT1 in rat brain endothelium. *J. Cereb. Blood Flow Metab.* **20,** 1557–1562.
9. Harlow, E., and Lane, D. (1988) *Antibodies: A Laboratory Manual.* Cold Spring Harbor Laboratory, Cold Spring Harbor, New York, pp. 82–83.

33

Immunohistochemical Detection of Endothelial Proteins

Sukriti Nag

1. Introduction

Immunohistochemistry is a widely used research technique in blood–brain barrier (BBB) research being used for the cellular localization of proteins of interest in normal vessels and documentation of altered expression following disease states, for the identification of cultured cells and for the spatial localization of the products of novel genes. Antigens of interest are detected by the binding of specific antibodies to small unique regions on these molecules called epitopes. This antigen-antibody complex is directly or indirectly labeled with a marker, which may be an enzyme or fluorochrome that can be visualized by light and/or electron microscopy or fluorescence microscopy, respectively. Multiple labeling allows identification of more than one antigen in the same cell or tissue section.

Major technological advances have occurred in the field of immunohistochemistry since the introduction of the direct method in which antigen was detected using primary antibody conjugated to a fluorescent marker (1). Introduction of the indirect immuno-labeling methods with enzyme-tagged antibodies allowed amplification of the immuno-logical signal by catalyzing an increasing deposition of detectable products at antigen-antibody sites in cells and tissues. The enzyme most frequently used is horse-radish peroxidase (HRP) mainly because of its low molecular weight (40 kDa), its high catalytic activity, and its fairly easy availability. The HRP is detected by the classic method of 3,3′-diaminobenzidine (DAB) tetrahydrochloride which is oxidized by the enzyme giving rise to a brown-colored insoluble reaction product which can be localized by light and electron microscopy (2). Indirect immunoenzymatic methods include the peroxidase-antiperoxidase technique (3), the multiple bridge technique (4), the avidin-biotin-peroxidase complex method (5), and the labeled avidin binding method (6), to name just a few. Avidin has a high isoelectric point, close to 10, and contains carbohydrate chains and both features are thought to contribute to its nonspecific binding to many cellular constituents (7). These characteristics are overcome with the use of streptavidin, which is derived from Streptomyces avidinii and has the same strong binding avidity of avidin for biotin. The indirect streptavidin-biotin peroxidase method is one of the most popularly used methods for the detection of single or multiple proteins including those in cerebral vessels (8–10). A standard protocol used to detect a single antigen will be described in this chapter. The main steps in this protocol are sum-

From: *Methods in Molecular Medicine, vol. 89:*
The Blood–Brain Barrier: Biology and Research Protocols
Edited by: S.Nag © Humana Press Inc., Totowa, NJ

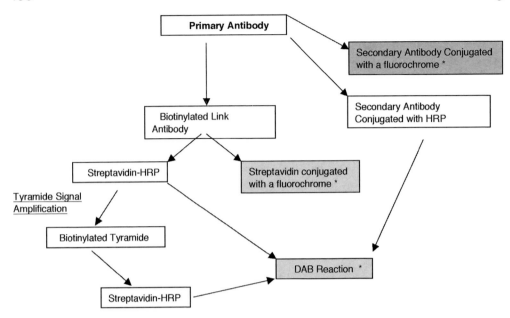

Fig. 1. The main steps in the different single-labeling immunohistochemical protocols described in the text are shown. '*' denotes the end of the reaction.

marized in **Fig. 1**. The main advantage of this method is that slides can be stored indefinitely and no specialized equipment is required to evaluate results other than a light microscope.

In addition, protocols for dual labeling using a peroxidase label detected by two different substrates, and detection of two antigens by different fluorochromes, are also given. Readers are also referred to other helpful reviews of immunohistochemical techniques *(7,11,12)*.

2. Materials

2.1. Standard Protocol for Detection of a Single Antigen

2.1.1. Tissue Preparation Techniques

1. 3% paraformaldehyde in 0.1 M phosphate buffer, pH 7.4.
2. Solvents: Ethanol, Xylene, Acetone.
3. Paraplast Plus Tissue Embedding medium (Oxford Labware, St. Louis, MO).
4. Fisher Colorfrost slides, other glass slides.
5. Type A Gelatin, Chromium potassium sulfate (Chrome alum), poly-L-lysine hydrobromide, MW 300,000 (Sigma Chem. Co., St Louis, MO).
6. Tissue-Tek® O.C.T. Compound 4583 (Fisher Scientific, Nepean, ON, Canada), 30% sucrose solution.
7. Sta-On Tissue Section Adhesive (Surgipath Medical Industries Inc., Richmond, IL).
8. 3-aminopropyltriethoxysilane, 0.15 M sodium borate, pH 8.5.
9. 0.01 M Na Citrate buffer, pH 6.0, 0.01 M Tris-HCl, pH 8.0.
10. 0.01 M HCl solution, 2 N HCl solution.
11. Pepsin.
12. Microwave, hotplate, and vortexer.

2.1.2. Antigen-Antibody Reaction

1. A humid chamber.
2. Glass staining dishes that hold 10–50 slides or plastic dishes with plastic racks that hold up to 20 slides.
3. Pipetors: 1–10 μL, 10–100 μL with tips.
4. Shaker for slide washes.
5. Methanol, 30% H_2O_2 Hydrogen Peroxide.
6. Phosphate-buffered saline (PBS): Dissolve 8.0 g of NaCl, 0.2 g of KCl, 1.44 g of Na_2HPO_4 and 0.24 g of KH_2PO_4 in 800 mL of distilled water. Adjust pH to 7.4 and the volume to 1 L. Sterilize by autoclaving. This solution can be made as a 10 × stock.
7. 0.05 M Tris-HCl, pH 7.6.
8. Antibody-diluting buffer (Dimension Laboratories, Mississauga, ON, Canada), or PBS containing 1.0% bovine serum albumin or PBS containing 0.025% Tween-20.
9. Normal serum from the same species as the secondary biotinylated link antibody, streptavidin conjugated peroxidase antibody.
10. Chromogen: 3,3′ DAB, (Sigma Chem. Co., St. Louis, MO) 0.05% in 0.05 M Tris-HCl, pH 7.6.
11. 1% hydrogen peroxide solution.
12. Commercial Bleach: Javex, Kimwipes® EX-L (Fisher Scientific, Nepean, ON, Canada).

2.1.3. Counterstaining and Mounting

1. Surgipath Hematoxylin, Harris Formula (Surgipath Medical Industries Inc., Richmond, IL).
2. Acid alcohol solution: add 2.4 mL of 10% HCl to 93.6 mL of 95% ethanol.
3. Permount mounting medium (Fisher Scientific, Nepean, ON, Canada).
4. Coverslips.

2.1.4. Amplification Technique

1. Tyramide Signal Amplification Kit (NEN™ Life Science Products, Boston, MA) contains Streptavidin-HRP, blocking reagent, amplification diluent, and biotinyl tyramide. At 4°C, the components of this kit are stable for 6–8 mo.
2. Tris-HCl Saline buffer: 0.1 M Tris-HCl, 0.15 M sodium chloride buffer, pH 7.5.
3. TNT buffer: Tris-HCl Saline buffer with 0.05% Tween-20.
4. TNB Buffer: Tris-HCl Saline buffer with 0.5% DuPont blocking reagent. To dissolve blocking reagent, heat TNB buffer to 60°C for 1 h while stirring.
5. Biotinyl Tyramide Stock Solution is supplied with the kit. Add 1.25 mL of dimethylsulfoxide to the biotinyl tyramide and store at 4°C. This solution is stable for at least 6–8 mo.

2.2. Dual Labeling Using Peroxidase and Different Substrates

1. AEC Chromogen Kit (Sigma Chemical, St. Louis, MO) contains 2.5 M acetate buffer, pH 5.0, 3-amino-9 ethylcarbazole (AEC) in N,N-dimethyl-formamide and 3% H_2O_2 in deionized water.
2. Crystal Mount® (Biomeda, Foster City, CA).

2.3. Dual Labeling Using Different Fluorochromes

1. Rubber Cement.
2. 5-mL plastic syringe with a 22-gage needle.
3. Fluorochrome conjugated secondary antibody or fluorochrome conjugated streptavidin.
4. Mowiol mounting medium: add 2.4 g of Mowiol (Calbiochem, Hoechst) to 6 g of glycerol. Stir to mix. Add 6 mL of water and leave for several hours at room temperature. Add

3.1.6.2. Enzyme-Induced Epitope Retrieval

The mode of action of proteases may include cleavage of crosslinks between proteins and embedding media with removal of macromolecules that hinder access, thus exposing higher numbers of antigenic epitopes *(14)*.

1. In our laboratory 0.5% pepsin alone for 30 min at 37°C is used for 6 μ sections for detection of all antibodies, and if this treatment does not give an optimal result other options are explored (*see* **Note 1**).
2. Heat 0.01 *M* HCl in a microwave at full power for 10 s, add the required amount of pepsin, and stir.
3. When the temperature of the pepsin is 39°C, add the slides to the container and place in a 37°C water bath for 30 min.
4. Wash with three changes of distilled water for 3 min each after the enzyme treatment.

3.1.7. Suppression of Endogenous Peroxidase Activity

1. Place sections in 0.3% methanolic peroxide for 20–30 min depending on the number of red blood cells present in the tissue.
2. Wash twice with PBS for 3 min each.

3.1.8. Antigen-Antibody Reaction

1. Shake off excess buffer and circle the tissue section with a PAP pen and apply normal serum diluted 1:20 in PBS for 15 min at room temperature in a humid chamber (*see* **Notes 2–4**).
2. Shake off the normal serum and cover sections with primary antibody diluted in diluting buffer for 2–4 h at room temperature or overnight at 4°C (approx 16 h) (*see* **Notes 5–9**).
3. Rinse three times with PBS for 3 min each (*see* **Note 10**).
4. Shake off excess buffer and apply a biotinylated link antibody diluted in diluting buffer to sections for 30 min at room temperature (*see* **Notes 11–12**).
5. Repeat **step 3**.
6. Shake off excess buffer and apply peroxidase-conjugated streptavidin, 1:300 (Dako Diagnostics) for 30 min at room temperature.
7. Wash with two changes of 0.05 *M* Tris-HCl, pH 7.6 for 3 min each.
8. Add 0.75 ml of 1% H_2O_2 to 50 mL of DAB solution and stir. Add slides for 5 min. Precautions have to be used when handling DAB (*see* **Note 13**).
9. If a brownish-black reaction product is required instead of the brown reaction product produced by DAB alone, then add nickel salts to the substrate given in **step 8** (*see* **Note 14**).
10. Rinse with flowing tap water for 3 min with intermittent shaking.

3.1.9. Counterstaining and Mounting

1. Stain sections with Surgipath Hematoxylin (Harris Formula) for 10–20 s depending on the type of tissue and wash with tepid tap water.
2. If nuclei are too dark then dip slides in the Acid alcohol solution and wash with tap water.
3. Wash in three changes of distilled water for 5 min each.
4. Dehydrate sections by placing in two changes of ethanol for 2 min each, and in two changes of xylene for 2 min each.
5. Place a large drop of permount on the section and apply a coverslip.

3.1.10. Tyramide Signal Amplification Technique

This amplification technique is reported to increase the sensitivity and efficiency of antigenic detection varying from 8–10,000 fold *(16)* and is useful when the antigen to be detected is in very small quantity. In this technique the catalytic action of HRP, present at the site of the antigen-antibody reaction, results in the deposition of biotinylated tyramine at or near the site of the enzyme *(16)*. The biotin sites on the bound tyramine act as further binding sites for streptavidin-biotin complexes (**Fig. 1**). The protocol recommended by the manufacturer follows.

1. Follow the steps given in **Subheading 3.1.1.** to **step 3** of **Subheading 3.1.8.** given in the standard protocol.
2. Apply biotinylated link antibody at 1:700 dilution (Dako Diagnostics) for 30 min.
3. Wash slides in TNT buffer three times for 5 min each.
4. Block slides with TNB buffer for 30 min.
5. Drain off the TNB buffer and apply streptavidin-HRP 1:500 (Dako Diagnostics) in TNB buffer for 30 min.
6. Wash slides three times with TNT buffer for 5 min each.
7. Add Biotinyl tyramide diluted 1:50 in 1 × amplification buffer and incubate for 8 min.
8. Wash slides three times with TNT buffer for 5 min each.
9. Repeat **steps 5** and **6**.
10. Follow **steps 7–10** in **Subheading 3.1.8.** of the standard protocol to demonstrate HRP reaction product, and the steps in **Subheading 3.1.9.** for staining, dehyration and mounting of sections.

3.1.11. Interpretation

In both the standard technique and the tyramide amplification technique, a positive reaction at the antigen site is indicated by a chocolate brown color in the cells of tissue sections when DAB is used as a substrate (**Fig. 2A,B**). The nuclei are light blue.

Differences in immunoreactivity between test and control groups can be expressed semiquantitatively by assigning a score of 1+ to 5+ as follows: 1+, sparse immunostaining; 2+, mild immunostaining; 3+, moderate immunostaining; 4+, marked immunostaining; 5+, maximal immunostaining present *(17)*. Sections should be assessed in a blinded manner and the assessor should be unaware to which group the sections belong.

Computerized densitometry has also been used to obtain a quantitative estimate of differences in immunoreactivity between test and control tissue *(18,19)*.

3.2. Dual Labeling Using Peroxidase and Different Substrates

Dual labeling using the indirect immunoperoxidase technique is given in which two different chromogenic substrate reactions such as DAB and AEC are used. The latter precipitates as an insoluble bright-red substance when exposed to peroxidase and hydrogen peroxide. Its drawback is that it is soluble in ethanol, therefore, an aqueous mounting medium has to be used and may decrease resolution at high magnifications. This is overcome by the use of Crystal Mount®, which is baked to form an insoluble plastic layer which can then be covered by a layer of permount mounting medium before placing a cover slip.

Two proteins can be localized using antibodies synthesized in different species or the same species. The antigen which is expected to be sparse is localized first by an overnight incubation in primary antibody at 4°C followed by detection of the second antigen by incubating in primary antibody for a few hours at room temperature or overnight at 4°C.

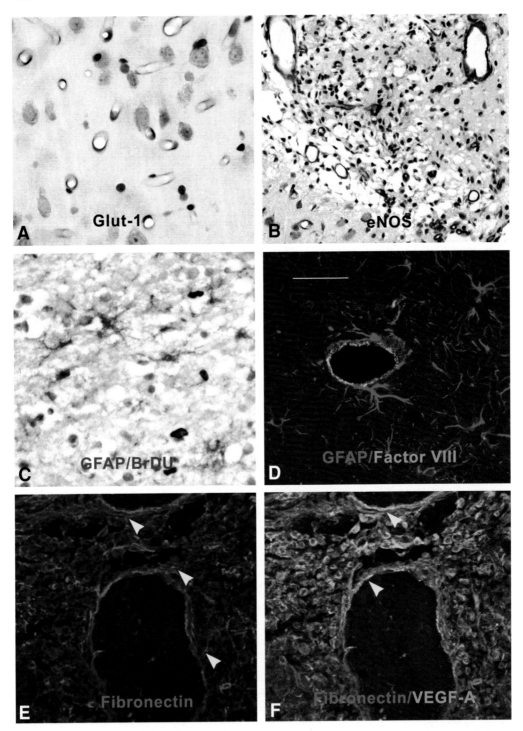

Fig. 2. **(A)** Endothelial Glut-1 in cerebral cortical vessels is demonstrated using the standard indirect streptavidin-biotin peroxidase technique. **(B)** Tyramide amplification was used to demonstrate endothelial nitric oxide synthase in vessels at the site of a cortical cold lesion and in vessels in the adjacent brain. Without the tyramide amplification, weak immuno-reactivity was present in only few vessels. **(C)** Dual labeling of a frozen section using the indirect

3.2.1. Detection of the First Antigen

Follow the standard protocol for the detection of the first antigen as outlined in **Subheadings 3.1.1.** to **3.1.8.** and use DAB as a substrate.

3.2.2. Detection of the Second Antigen

1. Wash twice with two changes of PBS for 5 min each.
2. Follow **steps 1–6** given in **Subheading 3.1.8.**
3. Wash twice with PBS for 5 min each.
4. Carefully dry the moisture around the section with a Kimwipe and apply two drops of AEC Substrate Reagent for 8–10 min. The AEC Substrate Reagent is prepared according to the manufacturer's instructions. Two drops of Acetate Buffer, one drop of AEC Chromogen and one drop of 3% H_2O_2 are added to 4 mL of deionized water.
5. Wash with a distilled water spray and then place in water for 5 min.
6. Stain with hematoxylin and wash in tepid tap water.
7. Wipe slide around section with a paper towel, place a thin film of Crystal Mount® over the section, and dry in an oven for 20 min at 60°C.
8. Mount in permount mounting medium.

3.2.3. Interpretation

The first antigen detected will appear chocolate brown while the second antigen will appear a bright red. Result using DAB as a substrate to demonstrate bromodeoxyuridine in proliferating endothelial cells and AEC as a substrate to demonstrate glial fibrillary acidic protein (GFAP) in astrocytes is shown (**Fig. 2C**).

3.3. Dual Labeling Using Different Fluorochromes

Fluorescence microscopy is a high-resolution method for detection of antigens in tissues. Fluorescein isothiocyanate has been used since the introduction of immunohistochemistry by Coons and coworkers in 1941 (*1*). A number of newer fluorochromes are now available that are considered to be more photostable, and produce less quenching of fluorescence and have higher number of fluorochromes per antibody. For dual labeling select fluorochromes with minimal overlap of their excitation and emission spectra. This technique is most useful to detect co-localization of two antigens in a single cell.

Fig. 2. *(continued)* streptavidin-biotin peroxidase technique is shown. DAB was used as a substrate to demonstrate bromodeoxyuridine in proliferating cells *(brown)* and AEC was used to demonstrate GFAP in astrocytes. (**D**) Merged confocal image showing double labeling for Factor VIII (1 : 100 dilution, Dako Diagnostics) in cerebral endothelium and GFAP (1 : 1000) in astrocytes using goat-antirabbit conjugated with Alexa Fluor 448 (1 : 150, Molecular Probes Inc., Eugene, OR) and goat antirabbit conjugated with Indocarbocyanine (1 : 200, Cy3, Jackson Immunoresearch Laboratories Inc, Westgrove, PA) respectively. (**E**) Confocal image showing blood-brain barrier breakdown to fibronectin *(arrowheads)* from two cerebral veins at the site of a cortical cold injury at d 4. (**F**) Merged confocal image showing both fibronectin (red) and vascular endothelial growth factor A (VEGF-A, green). Proliferating endothelial cells at the lesion site show cytoplasmic VEGF-A. Endothelium of veins permeable to fibronectin show co-localization of VEGF-A and fibronectin (yellow, arrowheads). **A**, ×200; **B**, ×240; **C**, ×250; **D–F**, Scale Bar = 50 µm.

Few disadvantages of fluorescence microscopy are that the background is difficult to appreciate, autofluorescence may sometimes make interpretation a problem, and preparations are not permanent. In addition, one must have access to a fluorescence microscope with the necessary filters.

3.3.1. Detection of the First Antigen

1. Follow steps in **Subheading 3.1.1.** to **step 3** in **Subheading 3.1.8.** of the standard method **(Subheading 3.1)**.
2. Instead of a PAP pen, rubber cement dispensed in a syringe using a 22-gage needle is used to circle the tissue section.
3. Shake off excess buffer and apply biotinylated secondary antibody (1:200) to sections for 30 min at room temperature. If the antigen is abundant one can use a secondary link antibody conjugated with a fluorochrome at this stage in the dark (*see* **Note 15**).
4. Wash slides three times with PBS for 5 min.
5. Shake off excess buffer and apply Streptavidin linked to a fluorochrome and incubate at room temperature for 15–60 min.
6. Wash slides three times with PBS for 5 min.

3.3.2. Detection of the Second Antigen

1. Shake off excess buffer and apply 1:20 dilution of normal serum for 15 min at room temperature.
2. Shake off normal serum and apply primary antibody for 2–4 h at room temperature.
3. Wash slides three times with PBS for 5 min each.
4. Follow **steps 3–6 in Subheading 3.3.1.**
5. Sections are mounted using mowiol mounting medium.
6. Slides are saved in folders with covers at 4°C. When the mounting medium dries, slides can be stored in a plastic box at 4°C.

3.3.3. Interpretation

Sections are examined in a fluorescence microscope or a laser scanning confocal microscope using the appropriate optical filter sets. Digital images obtained using the different filters can be merged using Adobe Photoshop 7.0 (**Figs. 2D,F**).

4. Notes

1. The concentration of pepsin can be varied from 0.05 to 0.5% and the period of exposure to pepsin can be varied from 15–30 min. Both parameters have to be established for demonstration of a specific protein. Other enzymes used for antigen retrieval are 0.1–0.025% pronase and 0.025% Proteinase K. For demonstration of bromodeoxyuridine uptake in cells, sections are placed in 2 N HCl for 30 min and then in 0.15 M Na Borate, pH 8.5 for 10 min prior to pepsinization.
2. A humid chamber can be purchased or made by gluing glass rods to the bottom of a plastic box. Use a leveler to ensure that the rods are level. If slides are not level immunoglobulins may flow with gravity away from the specimen location. Strips of paper towel moistened with an excess of water are placed at the bottom of the chamber. Once the primary antibody has been applied to the test slide, the section should never be allowed to dry.
3. Blocking with normal serum allows protein to bind to charged sites on the specimen. An alternative to blocking with normal serum is a universal blocking solution which is available commercially (Dako Diagnostics) and is used undiluted. The latter can be used only if the primary antibodies are derived from mouse or rabbit.

4. The volume of normal serum and antibodies added to the section should be sufficient to form a puddle with a slightly convex contour over the section including a 2-mm margin around the section. If reagents are not applied evenly staining artifacts may occur, such as an edge effect or a rim of false positivity around the edge of the tissue section.

5. Purchased antibodies have data sheets stating the recommended working dilution or concentrations, which are usually in the range of 10 to 20 µg/mL. It is usual to set up serial dilutions of the antibody to determine the optimum dilution for the test tissue. Aim for the highest dilution giving the least amount of non-specific background staining.

6. Incubation times in primary antibody varies from few hours at room temperature to overnight incubations at 4°C. Overnight incubations allow the use of higher dilutions of antibody thus decreasing nonspecific background staining.

7. Negative control sections are set up in which (a) nonimmune serum from the same species that was used for preparation of the primary antibody and at the same dilution, (b) antiserum that has been preincubated with the appropriate blocking peptide for 1 h, and (c) PBS in place of the primary antibody are applied.

8. It is helpful to set up positive controls as well. Select another tissue in which the test antigen is known to be present and/or select an antibody derived from the same species as the test antibody. Positive reaction in the latter will indicate that the secondary antibody, the enzyme label, and histochemical reaction for the demonstration of HRP reaction product are working.

9. Antibodies are stable indefinitely when stored concentrated in aliquots in ultralow freezers at –70° to –80°C. Once they are thawed, they should not be refrozen. Prior to use, make sure that the antibody is fully dissolved or reconstituted. Slight vortexing before application helps to ensure proper solution.

10. Thorough rinsing is essential for reducing background. Shake off the antibody and use a wash bottle to direct a jet of PBS directly above the section so it flows down over the section, washing it for 1 min. Then place the sections in a staining dish containing PBS and place on a shaker for 3 min. Decant the PBS and replace with fresh PBS for two more washes.

11. The optimum dilution of the secondary link antibodies used in our laboratory are biotinylated goat anti-rabbit antibody 1:400 (Sigma Chem Co., St. Louis, MO), biotinylated goat anti-mouse antibody 1:200 (Dako Diagnostics) or biotinylated rabbit anti-goat 1:160 (Vector Laboratories Inc., Burlingame, CA).

12. If the primary antibody is made in mouse or rabbit one can use a synthetic polymer conjugate with HRP for detection of the signal and eliminate the steps requiring a biotinylated link antibody and streptavidin HRP antibody. Commercially available polymers include the Zymed Polymer Detection System (Zymed Laboratories, South San Francisco, CA) and the Dako EnVision™ Labeled Polymer, Peroxidase (Dako Diagnostics). These polymers are used undiluted. The amplification capacity of polymers derives from the fact that the polymeric conjugates can contain up to 100 enzyme molecules and up to 20 antibody molecules per dextran backbone *(20)*. These polymers are very sensitive and they eliminate nonspecific staining due to the binding of avidin with endogenous biotin in the brain, and they also reduce the staining time by over 1 h.

13. The chromogen DAB is a potential carcinogen and should be handled with gloves. It is made fresh daily and dissolved in a fume hood. Hydrogen peroxide is added just before use as oxidation begins directly after its addition. After the reaction, bleach is added to the DAB to allow oxidation to a white precipitate.

14. DAB reaction product can be converted from a brown to a black color by performing the substrate reaction in the presence of nickel salts *(21)*. This is useful particularly when dual labeling is done to clearly differentiate brown from red. Add 0.16% nickel chloride solu-

tion to the substrate given in **Subheading 3.1.8., step 8** and incubate sections in this substrate for 8 min.

15. Once antibodies conjugated with fluorochromes are used, antibody incubations and washes on shakers are done in the dark by covering the humid chamber or container with a cardboard box. In addition, the overhead fluorescent lights should be switched off. A table lamp may be used if necessary.

Acknowledgments

This work is supported by the Heart and Stroke Foundation of Ontario and the Ontario Neurotrauma Foundation. Thanks are expressed to Dan Kilty and Andrew Morrison for technical assistance.

References

1. Coons, A. H., Creech, H. H., and Jones, R. N. (1941) Immunological properties of an antibody containing a fluorescent group. *Proc. Soc. Exp. Biol. Med.* **47,** 200–202.
2. Graham, R. C. Jr., and Karnovsky, M. J. (1966) The early stages of absorption of injected horseradish peroxidase in the proximal tubules of mouse kidney: Ultrastructural cytochemistry by a new technique. *J. Histochem. Cytochem.* **14,** 291–302.
3. Sternberger, L. A., Hardy, P. H. Jr., Cuculis, J. J., and Meyer, H. G. (1970) The unlabeled antibody-enzyme method of immunohistochemistry. Preparation and properties of soluble antigen-antibody complex (horseradish peroxidase-antihorseradish peroxidase) and its use in the identification of spirochetes. *J. Histochem. Cytochem.* **18,** 315–333.
4. Hsu, S. M., and Ree, H. J. (1980) Self-sandwich method. An improved immunoperoxidase technic for the detection of small amount of antigens. *Am. J. Clin. Pathol.* **74,** 32–40.
5. Hsu, S. M., Raine, L., and Fanger, H. (1981) A comparative study of the peroxidase antiperoxidase method and an avidin-biotin complex method for studying polypeptide hormones with radioimmunoassay antibodies. *Am. J. Clin. Pathol.* **75,** 734–738.
6. Elias, J., Margiotta, M., and Gabore, D. (1989) Sensitivity and detection efficiency of the peroxidase antiperoxidase (PAP), avidin-biotin peroxidase complex (ABC), and the peroxidase-labeled avidin-biotin (LAB) methods. *Am J. Clin. Pathol.* **92,** 62–67.
7. Mayer, G., and Bendayan, M. (2001) Amplification methods for the immunolocalization of rare molecules in cells and tissues. *Progr. Histochem. Cytochem.* **36,** 3–85.
8. Nag, S. (1996) Cold-injury of the cerebral cortex: immunolocalization of cellular proteins and blood-brain barrier permeability studies. *J. Neuropathol. Exp. Neurol.* **55,** 880–888.
9. Nag, S., Takahashi, J. T., and Kilty, D. (1997) Role of vascular endothelial growth factor in blood-brain barrier breakdown and angiogenesis in brain trauma. *J. Neuropathol. Exp. Neurol.* **56,** 912–921.
10. Nag, S., Picard, P., and Stewart, D. J. (2001) Expression of nitric oxide synthases and nitrotyrosine during blood-brain barrier breakdown and repair after cold injury. *Lab. Invest.* **81,** 41–49.
11. Harlow, E., and Lane, D., eds. (1999) *Using antibodies. A Laboratory Manual.* Cold Spring Harbor Laboratory Press, Cold Spring Harbor, NY.
12. Javois, L. C., ed. (1999) *Immunocytochemical Methods and Protocols.* Humana, Totowa, NJ.
13. Simmons, D. A., Arriza, J. L., and Swanson, L. W. (1989) A complete protocol for *in situ* hybridization of messenger RNAs in brain and other tissues with radio-labeled single-stranded RNA probes. *J. Histotechnol.* **12,** 169–180.
14. Pileri, S.A., Roncador, G., Ceccarelli, C., et al. (1997) Antigen retrieval techniques in immunohistochemistry: comparison of different methods. *J. Pathol.* **183,** 116–123.

15. Shi, S. R., Cote, R. J., and Taylor, C. R. (1997) Antigen retrieval immunohistochemistry: past, present, and future. *J. Histochem. Cytochem.* **45,** 327–343.
16. Sabattini, E., Bisgaard, K., Ascani, S., et al. (1998) The EnVision™+ system: a new immunohistochemical method for diagnostics and research. Critical comparison with the APAAP, ChemMate™, CSA, LABC, and SABC techniques. *J. Clin. Pathol.* **51,** 506–511.
17. Adams, J. C. (1981) Heavy metal intensification of DAB-based HRP reaction products. *J. Histochem. Cytochem.* **40,** 1457–1463.
18. Adams, J. C. (1982) Biotin amplification of biotin and horseradish peroxidase signals in histochemical stains. *J. Histochem. Cytochem.* **40,** 1457–1463.
19. Nag, S. (1996) Immunohistochemical localization of extracellular matrix proteins in cerebral vessels in chronic hypertension. *J. Neuropathol. Exp. Neurol.* **55,** 381–388.
20. Knerlich, F., Schilling, L., Görlach, C., Wahl, M., Ehrenreich, H., Sirén, A.-L. (1999) Temporal profile of expression and cellular localization of inducible nitric oxide synthase, interleukin-1β and interleukin converting enzyme after cryogenic lesion of the rat parietal cortex. *Mol. Brain Res.* **68,** 73–87.
21. Vilaplana, J., and Lavialle, M. (1999) A method to quantify glial fibrillary acidic protein immunoreactivity on the suprachiasmatic nucleus. *J. Neurosci. Methods* **88,** 181–187.

VI

GENETICALLY ALTERED MICE

34

Cerebral Vascular Function in Genetically Altered Mice

Frank M. Faraci

1. Introduction

Manipulation of genes in the mouse genome to produce transgenic or gene-targeted animals represents a powerful experimental tool to study the role of specific gene products in complex physiological systems. Because of the power of studying genetically altered mice, many laboratories have begun to incorporate the use of these models in studies of blood vessels including the cerebral circulation *(1)*. Perhaps the biggest effort in this regard relates to studies of cerebral ischemia and mechanisms of brain injury and cell death. The present review will summarize methods and progress that has been made in studies of one aspect of vascular biology—vascular function in the carotid artery and cerebral circulation in mice.

2. Advantages and Limitations of Genetically Altered Mice

Mice that overexpress, or are lacking in expression of, selected genes are excellent models to establish the functional importance of a particular gene product. In addition to studies of loss of gene function, these animals often represent models of human inherited or acquired diseases *(1)*. One of the great strengths of the gene-targeting technique is that it can eliminate many problems present in other, more commonly used models. This includes the limited specificity of pharmacological agents (e.g., enzyme inhibitors or receptor antagonists). When such inhibitors are used, there are often uncertainties regarding tissue or cellular access and the extent of enzyme or receptor inhibition. A major strength of the gene-targeting approach is that it allows the use of a precise genetic alteration to study complex responses in blood vessels.

With the data that are available to date, cerebral vascular responses in the mouse appear generally very similar to those seen in other species, including humans. For example, stimuli that produce vasoconstriction and vasodilatation are generally the same in murine and non-murine species *(1,2)*. Mechanisms that produce endothelium-dependent relaxation in cererbal blood vessels are similar in human and mouse *(2–6)* and other species *(7)*.

There are limitations in the use of genetically altered mice for experimental studies. One limitation related to the small size of blood vessels, particularly cerebral blood vessels, in the mouse (*see* **Subheading 3.1.** to **3.3.**). A second potential limitation exists in that compensation may occur in the animal either during development or later, in

From: *Methods in Molecular Medicine, vol. 89:*
The Blood–Brain Barrier: Biology and Research Protocols
Edited by: S.Nag © Humana Press Inc., Totowa, NJ

response to deletion or overexpression of a selected gene. For example, another redundant gene product may replace the function of the gene that was disrupted. As a result, deletion of one gene product may not result in any detectable change in phenotype. Expression of compensatory mechanisms is not unique to genetically altered animals, however, as it may occur with other experimental approaches including traditional pharmacological agents, antisense oligonucleotides, or RNA interference.

3. How Does One Study Vascular Function in Mice?

Three basic approaches have been developed, and all are being modifications of similar methods used to study blood vessels from larger species. These methods are studies of carotid artery, or intracranial arteries in vitro, and measurements of vascular diameter or local cerebral blood flow in vivo.

3.1. Carotid Artery

The number of studies of vascular function using the carotid artery from genetically altered mice are increasing rapidly. The following approach is used most commonly to study this artery. Following anesthesia, both common carotid arteries are quickly removed and placed in Krebs buffer. Loose connective tissue is removed and vessels are cut into small rings (generally 3–4 mm in length). Vascular rings are suspended in organ baths containing Krebs maintained at 37°C. The rings are connected to force transducers to measure isometric tension. Resting tension is increased stepwise, generally to a final tension of 0.25 g. To study responses to vasoconstrictor stimuli, agonists are applied to the vessels at this level of resting tension. To study relaxation responses, vessels are contracted submaximally (generally to about 50% of maximum). One of the best agonists for this purpose is the thromboxane A_2 analogue U46619, which produces concentration-dependent and stable contraction of mouse carotid and cerebral arteries. After reaching a stable contraction plateau, vessels are exposed to increasing concentrations of agonists or antagonists.

In relation to endothelial function, studies with mice deficient in expression of endothelial NO synthase (eNOS) have shown that relaxation of the carotid artery in response to acetylcholine, the classic endothelium-dependent agonist, is mediated exclusively by eNOS *(8,9)*. It was also found that vasoconstrictor responses to serotonin are normally inhibited by eNOS, particularly in females, as responses to serotonin are augmented greatly in eNOS-deficient mice *(9)*. These studies have also been insightful as they were the first to demonstrate that vasomotor effects of eNOS exhibit gene dosing effects *(8,9)*. For example, vasoconstrictor responses to serotonin were enhanced moderately in heterozygous eNOS-deficient mice (eNOS +/–) but were markedly enhanced in homozygous eNOS-deficient mice (eNOS –/–) *(9)*.

In addition to such studies of normal vascular physiology, studies of carotid artery have been particularly valuable in providing insight into mechanisms of vascular dysfunction in disease states. This includes vascular dysfunction during inflammation *(10–12)*, ceramide-induced oxidative stress *(13)*, diabetes *(14)*, hypercholesterolemia *(15)*, and chronic hypertension *(16)*. In another major area of vascular biology, genetically altered mice are being used increasingly for studies of intimal proliferation following injury of the carotid artery.

Because the common carotid artery is not a resistance blood vessel, data obtained using this artery may not always be representative of mechanisms present in smaller

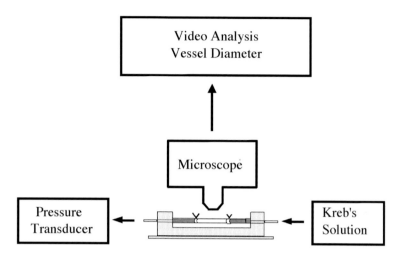

Fig. 1. Schematic illustration of the system used to measure responses of isolated cerebral arteries from mice. Arteries are cannulated with glass micropipets (30–80 μm in diameter) and exposed to intraluminal pressure of 60 mm Hg. Drugs can then be applied either extraluminally or intraluminally as vascular responses are measured with a video system.

intracranial blood vessels. However, studies of the carotid artery in non-murine species have generally been very informative over the years and have often been reliable predictors of function in other blood vessels, including cerebral arteries. Moreover, studies of the carotid artery are very relevant in relation to atherosclerosis, a disease that is predominantly localized in large arteries, including the carotid artery of humans. Carotid artery disease is a major risk factor for stroke and recent findings suggest that carotid artery disease is associated with local inflammation and impairment of endothelium-dependent relaxation. Such findings reaffirm the importance of studying mechanisms of vascular dysfunction in carotid arteries under pathophysiological conditions.

3.2. Cerebral Arteries

After anesthesia, the brain is rapidly removed and placed in ice-cold Krebs buffer. Cerebral arteries (typically the basilar or middle cerebral arteries) are isolated using a dissecting microscope. Arteries are then cannulated onto glass micropipets filled with Krebs in an organ chamber and secured with nylon suture (**Fig. 1**). Oxygenated and warmed (37°C) Krebs buffer is continuously circulated through the organ chamber. Vessels are typically pressurized to 60 mmHg pressure to mimic in vivo conditions. In most studies, arteries are pressurized but are observed under conditions of no-flow. Video analysis is used to measure lumen diameter. Arteries are either studied after development of spontaneous tone or after submaximal precontraction with agents such as U46619.

This approach has been used by us and others to examine mechanisms related to receptor and endothelial cell signaling *(5,17,18)* as well as the role of potassium channels, calcium, and calcium sparks in regulation of cerebral vascular tone *(19–24)*. An example of responses of the basilar artery to acetylcholine in wild-type and M5 muscarinic acetylcholine receptor deficient mice is shown in **Fig. 2**. Acetylcholine produces marked dilatation of the basilar artery in controls. This response was almost completely absent in the M5 muscarinic acetylcholine receptor deficient mice.

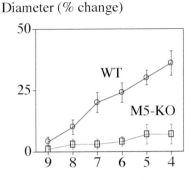

Fig. 2. Responses of the basilar artery to acetylcholine in control, wild-type (WT) and M5-receptor knockout mice (M5-KO). Basilar arteries were precontracted submaximally with U46619 before application of acetylcholine. Data redrawn from Yamada et al. *(5)*.

3.3. Cerebral Arterioles In Vivo

This method has been used most extensively in mice to study cerebral blood vessels. A detailed description can be found in Sobey et al. and Dalkara et al. *(3,25)*.

Following anesthesia, the most common current approach is to ventilate the mice mechanically. An artery (typically the femoral) and vein are cannulated for measurement of systemic pressure, to sample arterial blood, and to provide intravenous access. End-tidal CO_2 can be continuously monitored using a microcapnometer and maintained in the normal range by adjusting minute ventilation. Samples of arterial blood can be drawn into capillary tubes for measurement of blood gases. Rectal temperature is continually monitored and maintained at 37–38°C. A cranial window is placed over the parietal cortex to expose the pial microcirculation of the cerebrum with methods used extensively in many laboratories on larger species. The craniotomy is suffused with artificial cerebrospinal fluid (CSF) (temperature maintained at 37°C) that is bubbled with gases to maintain normal gases and pH. Diameter of cerebral arterioles is then measured using a microscope equipped with a TV camera coupled to a video monitor and an image-shearing device.

Prior to the age of producing genetically altered mice, this approach to study cerebral arterioles in mice was performed in essentially only one laboratory. Rosenblum used mice and this method exclusively over the past three decades and published many papers dealing with cerebral vascular function *(2)*. In relation to study with genetically altered mice, this approach has been used by us and others to examine mechanisms related to endothelium-dependent relaxation and NO signaling *(3–5,26–29)*. In addition, studies of vascular dysfunction in genetic models of Alzheimer's disease *(30)* and hyperhomocysteinemia (Faraci and Lentz, unpublished observations) have also been performed.

A variation of this in vivo approach is to measure changes in local cerebral blood flow (instead of microvascular diameter) in the exposed parietal cortex using laser-Doppler flowmetry *(17,31,32)*. An additional technique to study cerebral arteries in the mouse is to measure the caliber of the vessels following perfusion fixation in vivo. This approach has been used in transgenic mice to examine the hypothesis that overexpression of the CuZn isoform of superoxide dismutase (SOD) protects against vasospasm following subarachnoid hemorrhage *(33)*. Finally, in vivo studies are now being performed using genetically altered mice to examine other aspects of cerebral vascular biology including mechanisms of vascular hypertrophy *(34)* as well as disruption of the blood-brain barrier during meningitis *(35,36)*.

4. Can Molecular Biology Be Done with Intracranial Vessels from Mice?

Data on levels of expression of modified ribonucleic acid (mRNA) and protein in cerebral blood vessels can often be very valuable, particularly when combined with complementary functional data. Immunocytochemistry and in situ hybridization have both been used to localize protein and mRNA (respectively) on cerebral vessels from mice *(27,37–39)*. Because of their relatively small size, quantitative measurements of mRNA or protein in any blood vessel from mice are challenging. This is particularly true if one wishes to measure these variables in any blood vessel smaller than the aorta. To date, only a few studies have attempted such measurements. Reverse transcriptase polymerase chain reaction (RT-PCR) has been performed to document mRNA expression in cerebral *(5,37,39)* and carotid arteries *(10,11)* and Western blotting has been used in mouse brain vascular preparations *(18)*. In addition, immunoflourescence has been used to quantify relative levels of CREB (cAMP-responsive element binding protein) in cerebral arteries *(37)*.

A new methodology that should be very useful for quantification of mRNA levels in murine blood vessels is real time RT-PCR. We recently developed a real-time polymerase chain reaction (PCR) method to quantify eNOS mRNA and found that the method could be used to reproducibly measure mRNA levels for eNOS in carotid and basilar arteries from C57BL/6 mice *(40)* This method should be a powerful tool for studies of vascular biology in mice.

5. Future Directions

Transgenic and gene-targeted mice have now been used by several laboratories as a new approach for studies related to cerebral vascular biology. This application of current first generation genetically altered mice will certainly continue. As this field of research evolves, however, it will be important for studies to incorporate newer and more sophisticated genetically altered mouse models. Rather than simply producing lifelong deletion or lifelong systemic overexpression of a particular gene, newer models will increasingly allow spatial and temporal control over the genetic alteration. For example, by producing genetically altered mice in which the transgene is driven by the eNOS promoter *(41,42)*, expression of the gene product can be directed selectively to vascular endothelium including cerebral endothelium. These and other novel approaches should provide powerful tools that can be utilized to gain greater insight into cerebral vascular biology.

Acknowledgments

Work that is summarized in this review was supported by National Institutes of Health grants NS-24621, HL-38901, and HL-62984.

References

1. Faraci, F. M., and Sigmund, C. D. (1999) Vascular biology in genetically-altered mice: smaller vessels, bigger insight. *Circulation Res.* **85,** 1214–1225.

2. Rosenblum, W. I. (1998) A review of vasomotor responses of arterioles on the surface of the mouse brain: the necessary prelude to studies using genetically manipulated mice. *Microcirculation* **5,** 129–138.

3. Sobey, C. G., and Faraci, F. M. (1997) Effects of a novel inhibitor of guanylyl cyclase on dilator responses of mouse cerebral arterioles. *Stroke* **28,** 837–843.

4. Faraci, F. M., Heistad, D. D., and Lamping, K. G. (2002) ADP-induced dilatation of cerebral arterioles is mediated by EDHF: evidence from wild-type and eNOS deficient mice (abstract). *FASEB J.* **16,** A1125.

5. Yamada, M., Lamping, K. G., Duttaroy, A., et al. (2001) Cholinergic dilation of cerebral blood vessels is abolished in M_5 muscarinic acetylcholine receptor knockout mice. Proc. Natl. Acad. Sci. USA **98,** 14,096–14,101.

6. Elhusseiny, A., and Hamel, E. (2000) Muscarinic—but not nicotinic—acetylcholine receptors mediate a nitric oxide-dependent dilation in brain cortical arterioles: A possible role for the M5 receptor subtype. *J. Cerebral Blood Flow Metabol.* **20,** 298–305.

7. Faraci, F. M., and Heistad, D. D. (1998) Regulation of the cerebral circulation: Role of endothelium and potassium channels. *Physiological Reviews* **78,** 53–97.

8. Faraci, F. M., Sigmund, C. D., Shesely, E. G., Maeda, N., and Heistad, D. D. (1998) Responses of carotid artery in mice deficient in expression of the gene for endothelial NO synthase. *Am. J. Physiol.* **274,** H564–H570.

9. Lamping, K. G., and Faraci, F. M. (2001) Role of sex differences and effects of endothelial NO synthase deficiency in responses of carotid arteries to serotonin. *Arterioscler. Thromb. Vasc. Biol.* **21,** 523–528.

10. Gunnett, C. A., Berg, D. J., and Faraci, F. M. (1999) Vascular effects of lipopolysaccharide are enhanced in interleukin-10-deficient mice. *Stroke* **30,** 2191–2196.

11. Gunnett, C. A., Chu, Y., Heistad, D. D., Loihl, A., and Faraci, F. M. (1998) Vascular effects of LPS in mice deficient in expression of the gene for inducible nitric oxide synthase. *Am. J. Physiol.* **275,** H416–H421.

12. Gunnett, C. A., Heistad, D. D., Berg, D. J., and Faraci, F. M. (2000) IL-10 deficiency increases superoxide and endothelial dysfunction during inflammation. *Am. J. Physiol.* **279,** H1555–H1562.

13. Didion, S. P., and Faraci, F. M. (2002) Overexpression of CuZn-SOD protects against ceramide-induced endothelial dysfunction. *FASEB J.* **16,** A1136.

14. Gunnett, C. A., Heistad, D. D., Berg, D. J., and Faraci, F. M. (2002) Interleukin-10 protects endothelium-dependent relaxation during diabetes: role of superoxide. *Diabetes* **51,** 1931–1937.

15. d'Uscio, L. V., Smith, L. A., and Katusic, Z. S. (2001) Hypercholesterolemia impairs endothelium-dependent relaxations in common carotid arteries of apolipoprotein E-deficient mice. *Stroke* **32,** 2658–2664.

16. Didion, S. P., Sigmund, C. D., and Faraci, F. M. (2000) Impaired endothelial function in transgenic mice expressing both human renin and human angiotensinogen. *Stroke* **31,** 760–765.

17. Niwa, K., Porter, V. A., Kazama, K. Cornfield, D., Carlson, G. A., and Iadecola, C. (2001) Aβ-peptides enhance vasoconstriction in cerebral circulation. *Am. J. Physiol.* **281,** H2417–H2424.

18. Geary, G. G., McNeill, A. M., Ospina, J. A., Krause, D. N., Korach, K. S., and Duckles, S. P. (2001) Genome and hormones: Gender differences in physiology. Selected contribution: cerebrovascular nos and cyclooxygenase are unaffected by estrogen in mice lacking estrogen receptor-alpha. *J. Appl. Physiol.* **91,** 2391–2399.

19. Brenner, R., Perez, G. J., Bonev, A. D., et al. (2000) Vasoregulation by the β1 subunit of the calcium-activated potassium channel. *Nature* **407,** 870–876.

20. Lohn, M., Jessner, W., Furstenau, M., et al. (2001) Regulation of calcium sparks and spontaneous transient outward currents by RyR3 in arterial vascular smooth muscle cells. *Circ. Res.* **89,** 1051–1057.

21. Lohn, M., Lauterbach, W. B., Haller, H., Pongs, O., Luft, F. C., and Gollasch, M. (2001) β_1-subunit of BK channels regulates arterial wall $[Ca^{2+}]$ and diameter in mouse cerebral arteries. *J. Appl. Physiol.* **91,** 1350–1354.

22. Pluger, S., Faulhaber, J., Furstenau, M., et al. (2000) Mice with disrupted BK channel β1 subunit gene feature abnormal Ca2+ spark/STOC coupling and elevated blood pressure. *Circ. Res.* **87,** e53–e60.

23. Wellman, G. C., Santana, L. F., Bonev, A. D., and Nelson, M. T. (2001) Role of phospholamban in the modulation of arterial Ca^{2+} sparks and Ca^{2+}-activated K^+ channels by cAMP. *Am. J. Physiol.* **281,** C1029–C1037.

24. Zaritsky, J. J., Eckman, D. M., Wellman, G. C., Nelson, M. T., and Schwarz, T. L. (2000) Targeted disruption of Kir2.1 and Kir2.2 genes reveals the essential role of the inwardly rectifying K+ current in K+-mediated vasodilation. *Circ. Res.* **87,** 160–166.

25. Dalkara, T., Irikura, K., Huang, Z., Panahian, N., and Moskowitz, M. A. (1995) Cerebrovascular responses under controlled and monitored physiological conditions in the anesthetized mouse. *J. Cerebral Blood Flow Metabol.* **15,** 631–638.

26. Fujii, M., Hara, H., Meng, W., Vonsattel, J. P., Huang, Z., and Moskowitz, M. A. (1997) Strain-related differences in susceptibility to transient forebrain ischemia in SV-129 and C57Black/6 mice. *Stroke* **28,** 1805–1811.

27. Irikura, K., Huang, P. L., Ma, J., et al. (1995) Cerebrovascular alterations in mice lacking neuronal nitric oxide synthase gene expression. *Proc. Natl. Acad. Sci. USA* **92,** 6823–6827.

28. Meng, W., Ayata, C., Waeber, C., Haung, P. L., and Moskowitz, M. A. (1998) Neuronal NOS-cGMP-dependent Ach-induced relaxation in pial arterioles of endothelial NOS knockout mice. *Am. J. Physiol.* **274,** H411–H415.

29. Meng, W., Ma, J., Ayata, C., et al. (1996) ACh dilates pial arterioles in endothelial and neuronal NOS knockout mice by NO-dependent mechanisms. *Am. J. Physiol.* **271,** H1145–H1150.

30. Christie, R., Yamada, M., Moskowitz, M., and Hyman, B. (2001) Structural and functional disruption of vascular smooth muscle cells in a transgenic mouse model of amyloid angiopathy. *Am. J. Pathol.* **158,** 1065–1071.

31. Iadecola, C., Zhang, F., Niwa, K., et al. (1999) SOD1 rescues cerebral endothelial dysfunction in mice overexpressing amyloid precursor protein. *Nature Neuroscience* **2,** 157–161.

32. Niwa, K., Haensel, C., Ross, M. E., and Iadecola, C. (2001) Cyclooxygenase-1 participates in selected vasodilator responses of the cerebral circulation. *Circ. Res.* **88,** 600–608.

33. Kamii, H., Kato, I., Kinouchi, H., et al. (1999) Amelioration of vasospasm after subarachnoid hemorrhage in transgenic mice overexpressing CuZn-superoxide dismutase. *Stroke* **30,** 867–872.

34. Baumbach, G. L., Sigmund, C. D., and Faraci, F. M. (2002) Deficiency of endothelial NO synthase promotes cerebral vascular hypertrophy (abstract). *FASEB J.* **16,** A1113.

35. Koedel, U., Paul, R., Winkler, F., Kastenbauer, S., Haung, P. L., and Pfister, H. W. (2001) Lack of endothelial nitric oxide synthase aggravates murine pneumococcal meningitis. *J. Neuropath. Exp. Neurol.* **60,** 1041–1050.

36. Winkler, F., Koedel, U., Kastenbauer, S., and Pfister, H. W. (2001) Differential expression of nitric oxide synthases in bacterial meningitis: role of the inducible isoform for blood-brain barrier breakdown. *J. Infectious Disease* **183,** 1749–1959.

37. Cartin, L., Lounsbury, K. M., and Nelson, M. T. (2000) Coupling of Ca2+ to CREB activation and gene expression in intact cerebral arteries from mouse. Roles of ryanodine receptors and voltage-dependent Ca2+ channels. *Circ. Res.* **86,** 760–767.

38. Demas, G. E., Kriegsfeld, L. J., Blackshaw, S., et al. (1999) Elimination of aggressive behavior in male mice lacking endothelial nitric oxide synthase. *J. Neurosci.* **19,** RC30:1–5.

39. Saito, A., Kamii, H., Kato, I., (2001) Transgenic CuZn-superoxide dismutase inhibits NO synthase induction in experimental subarachnoid hemorrhage. *Stroke* **32,** 1652–1657.

40. Chu, Y., Heistad, D. D., Lamping, K. G., and Faraci, F. M. (2002) Quantification by real mRNA for endothelial nitric oxide synthase (eNOS) in mouse blood vessels by real time polymerase chain reaction. *Arterioscler. Thromb. Vasc. Biol.* **22,** 611–616.

41. Guillot, P. V., Guan, J., Liu, L., et al. (1999) A vascular bed-specific pathway regulates cardiac expression of endothelial nitric oxide synthase. *J. Clin. Invest.* **103,** 799–805.

42. Teichert, A.-M., Miller, T. L., Tai, S. C., et al. (2000) In vivo expression profile of an endothelial nitric oxide synthase promoter-reporter transgene. *Am. J. Physiol.* **278,** H1352–H1361.

35

Generation of Transgenic Mice by Pronuclear Injection

Annette Damert and Heike Kusserow

1. Introduction

1.1. General Considerations on Transgenic Mice

Over the past decades, transgenic mice have become a valuable tool in investigating gene expression, regulation, and function. Mouse models have been established for various human diseases to determine gene expression patterns and to elucidate gene function in tissues and organ systems. However, the outcome of a particular transgenic experiment is affected by a variety of factors that are only partly understood. Expression of transgenes may occur at ectopic sites, mouse lines generated might not express the transgene at all, or expression levels may decrease with increasing number of generations (*see* **Note 1**). Some of these difficulties can be solved by carefully planning the experiment. Thus the following chapter will not only deal with practical aspects of the microinjection but intends to provide guidelines for setting up a transgenic experiment.

We use the term "transgenic" animals for mice generated by microinjection of a designed deoxyribonucleic acid (DNA) fragment (transgene) into one of the pronuclei of a fertilized oocyte. This method results in multi copy, tandem-array (in rare cases a single copy) and random integration of the transgene into the genome. Generation of transgenic mice can also be done by homologous recombination in embryonic stem cells (knockout- or knockin-mice). The latter will be described in Chapter 36.

From a theoretical point of view transgenic mice can be grouped according to the aim of the experiment: the first group encompasses transgenics used for the investigation of gene expression patterns and the function of regulatory sequences while the second group comprises transgenics designed to gain more insight into gene function. In the former group gene regulatory sequences are used in combination with a reporter gene (e.g., *LacZ*) to determine expression patterns. In the latter case gene coding sequences are expressed under the control of either tissue specific regulatory sequences or ubiquitously active promoters. New techniques now allow transgene expression to be inducible. Details of the latter methods are beyond the scope of this chapter, but Ryding and colleagues (*1*) offer a review on this subject.

1.2. Design of a Transgene-Containing Vector

Elements necessary for transgene expression comprise a promoter and (optional) additional regulatory elements, a coding sequence (occasionally as a fusion protein or

From: *Methods in Molecular Medicine, vol. 89:*
The Blood–Brain Barrier: Biology and Research Protocols
Edited by: S. Nag © Humana Press Inc., Totowa, NJ

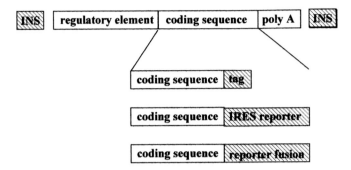

Fig. 1. Schematic representation of the elements necessary for expression of transgenes in mice is shown. Optional elements are shaded. INS, insulator; IRES, internal ribosomal entry site; tag, epitope tag.

with an attached epitope tag for easy detection of the gene product) or reporter gene and a polyadenylation signal (**Fig. 1**).

1.2.1. Selection of Promoters and Additional Regulatory Sequences for Tissue-Specific Expression

Depending on the objective of the investigation either ubiquitous or tissue-specific expression of a given transgene may be desired. Expression of a specific transgene in defined tissues or cell populations with the required strength of expression should be based on previous knowledge about the regulatory sequences (*see* **Note 2**). If such data are not available, experiments using reporter gene constructs should be performed in tissue culture and/or in mice. Transfection of promoter/enhancer deletion mutants coupled to a reporter gene into differentiated cells of the desired tissue provide useful hints on regulatory elements. Nevertheless, the suitability of the sequences identified has to be proven in vivo. The simplest approach uses combinations of promoter/enhancer deletions with the *LacZ* or fluorescent protein reporter genes. Beta-galactosidase activity driven by the tissue-specific regulatory sequences can easily be detected by standard histochemistry, whereas green fluorescent protein (GFP) fluorescence is detectable under ultraviolet light. Once tissue specificity has been established, the regulatory sequences identified can be used to express the desired transgene.

The large genomic fragments contained in artificial chromosomes from yeast, phages, or bacteria (YAC, PAC, or BAC) provide the opportunity to introduce a large amount of gene-specific regulatory sequences into transgenic mice and thus increase the chances of reproducible tissue-specific expression. However, molecular cloning, purification and injection of fragments derived from these vectors require special techniques that are beyond the scope of this chapter (*see* **Note 3**).

1.2.2. Selection of Promoters and Additional Regulatory Sequences for Ubiquitous Expression

To achieve ubiquitous and in most cases high-level expression of the transgene a couple of promoters are available. These are mostly derived from housekeeping genes like β-actin or phosphoglyceratekinase (PGK) (*2*). Chicken β-actin regulatory sequences comprising not only the core promoter but also sequences from the first intron and exon in combination with the CMV_{IE} enhancer, have also been used for ubiquitous expression of exogenous sequences in mice *(3)*. More recently Kisseberth and colleagues *(4)* demonstrated evenly distributed reporter gene expression using an 800-bp promoter fragment derived from the ROSA 26 locus.

1.2.3. Insulators

Careful evaluation of regulatory sequences in vitro and in vivo sometimes does not produce reproducible expression patterns of a given transgene in mice. Dependent upon the integration site, certain genomic sequences surrounding the transgene construct can exert effects on transgene expression. To reduce variability in transgene expression so-called "insulators" (e.g., the chicken β-globin 5′ hypersensitive site 4 *[5]*) have recently been applied to transgenic technology. Another approach is to use flanking sequences isolated from a high-level expressing transgenic line to overcome position effects *(6)*. However, in the process of these investigations these authors noticed that the function of a given insulator sequence is, at least partly, specific for a given transgene.

1.2.4. Selection and Design of the Coding Sequence

In most cases a cDNA of the protein of interest is the most convenient genetic information available for expression in transgenic animals. While designing the transgenic vector, care has to be taken that no additional ATG sequence is present between the transcription initiation site and the start codon of the complementary deoxyribonucleic acid (cDNA). This can result in expression of an inappropriate protein either terminating in the coding sequence or forming a fusion protein with the product of the transgene. In many cases transgenes display high homology to endogenous gene products and consequently detection of transgene-specific expression may be difficult. At the ribonucleic acid (RNA) level, primers or probes specific for the (often heterologous) polyadenylation cassette can be designed. For assessment of protein expression, however, transgene specific antibodies are required. As many antibodies crossreact between species this might present certain difficulties for the investigator. There are several strategies available to label transgene products and thus to overcome these problems. First, the transgene can be designed to yield a dicistronic RNA from which both the target protein and an additional reporter are translated. Insertion of the EMCV internal ribosomal entry site (IRES) *(7,8)*, in front of the reporter gene enables expression in the same cell types as observed for the target protein *(9)*. Second, an epitope tag can be attached. Although frequently used in tissue culture experiments epitope tags are only rarely found in transgenics. Several authors report interference with protein function for transgenes marked with *myc (10)* or FLAG *(11)* epitope tags. Selection of the tag itself and its location with respect to the coding sequence (N or C terminally, internally) should be based on experiments in tissue culture to exclude interference of the tag with protein function. Finally, transgene products can be expressed as fusion proteins coupled to, e.g., β-galactosidase *(12)* or GFP *(13)* to facilitate detection.

1.2.5. Selection of the Polyadenylation Cassette

Although the endogenous polyadenylation signal of the gene of interest can be used for expression in transgenic mice, most investigators prefer polyadenylation cassettes with known activity for reliable termination of the transgene RNA. Influences of tissue-specific use of polyadenylation signals or of multiple polyadenylation sites resulting in a variety of RNAs can be avoided. Available polyadenylation cassettes include, for instance, the SV40 small t (available in many commercial expression vectors) and rabbit β-globin sequences (*see* **Note 4**).

2. Materials

2.1. Material for Construction of the Transgene-Containing Vector

1. Cloning vector, e.g., pBluescript (Stratagene) or an equivalent.
2. cDNA of the gene of interest or reporter gene.
3. Regulatory sequences and polyadenylation cassette.
4. Restriction and modifying enzymes.
5. Agarose gel equipment.
6. *E. coli* strains and media for propagation of plasmids.
7. DNA purification: QIAquick™ (Qiagen) or GeneClean™ (GeneClean, Bio 101) system.
8. Commercial plasmid isolation kit (Qiagen or equivalent).

2.2. Material for the Generation of Transgenic Mice

1. Animals (outbred strains as NMRI or CD-1/ICR, or F1-Hybrids such as C6H3; *see* **Note 5**).
   a. Female mice (virgin, 4 wk old) as oocyte donors.
   b. Female mice (>10 wk old) as recipients.
   c. Male mice (>10 wk old) for vasectomized males and mating.
2. Hormones for superovulation (local suppliers for drugs used in veterinary medicine). Follicle-stimulating hormone (pregnant mare serum gonadotropin [PMSG]), and human chorionic gonadotrophin (hCG) (Cal-Biochem, CA).
3. Anesthetic: Avertin™ (2,2,2 tribromoethanol (Aldrich), stock solution 1.6 g/mL in tert amyl alcohol, store in the dark at room temperature (RT); working solution 20 mg/mL in normal saline, kept at 4°C) (*see* **Note 6**).
4. Syringes (1 mL disposable), needles (26 G hypodermic).
5. Surgical equipment: Fine forceps, fine scissors, two pairs of watchmaker's no. 5 forceps, wound clips (Clay Adams® Autoclips® and applicator, Becton Dickinson, Sparks, MD), suture with curved needle swagged on, serrefine clamps.
6. Microscopes:
   a. Stereomicroscope with understage illumination and 20–40× magnification (Zeiss SV series).
   b. Inverted Microscope with fixed stage (Zeiss Axiovert or Leica DM IRE2), fitted with micromanipulators (Märzhäuser [distributed by Zeiss] or Eppendorf) (*see* **Note 7**).
7. Microinjector (FemtoJet®, Eppendorf).
8. Manual microinjector for control of the holding pipet (CellTram®, Eppendorf).
9. Drawn-out pasteur pipets.
10. Needle puller (David Kopf Instruments, Tujunga, CA, e.g., model 720).
11. Microforge (e.g., David Kopf Instruments, Tujunga, CA).
12. Borosilicate glass capillaries (Clark Electromedical Instruments, UK).
    a. For injection needles with internal filament, GC 100 TF-15.
    b. For holding pipets, GC 100 TF-10.
13. Hyaluronidase (from ovine testes, Calbiochem, La Jolla, stock solution 10 mg/mL in M2, store at –20°C, working solution 300 µg/mL).
14. Media: M2 and M16 (Sigma, either liquid medium or powder that has to be supplemented with lactic acid, $NaHCO_3$ and penicillin/streptomycin).
15. Mineral oil (Sigma, embryo-tested).

16. 35-mm tissue culture dishes (Falcon® easy grip).
17. Embryological watchglasses.
18. Depression slide injection chamber (Dynamic Aqua Supply Ltd., Surrey, Canada).
19. Humidified incubator set to 37°C and 5% CO_2.
20. Gas burner.
21. Injection buffer: 10 mM Tris-HCl, 0.2 mM EDTA, pH 7.4.
22. 10-cm culture dishes with plasticine for storing holding- and injection pipets.

3. Methods

The methods described outline 1) the construction of the transgene containing vector and purification of the injection fragment, 2) vasectomy of male mice and breeding of pseudopregnant females, 3) superovulation, mating, and preparation of zygotes, 4) microinjection of the DNA into the pronucleus, and 5) retransfer into pseudopregnant females.

3.1. Construction of the Transgene-Containing Vector and Purification of the Injection Fragment

3.1.1. Construction of the Transgene-Containing Vector

The elements necessary for transgene expression have been discussed in detail in **Subheading 1.2.**

1. All necessary elements are combined in a cloning vector like pBluescript (Stratagene) by standard cloning techniques *(14)*.
2. To allow easy excision of the transgene cassette for injection, recognition sites for two (or a single) restriction enzymes should be left 5′ and 3′ of the expression cassette. These should be recognition sites for enzymes cleaving only once to prevent digestion of the transgene itself (**Fig. 2A**).
3. In addition, for later detection of transgene integration into the genome by Southern analysis of genomic DNA, a recognition site for another enzyme cleaving only once within the transgene sequence should be present.
4. Digestion with this enzyme in combination with a suitable probe will then yield a distinct, integration site-dependent pattern for each established transgenic line (**Fig. 2B**).

3.1.2. Purification of the Transgene

Plasmid DNA is prepared using commercial isolation kits (e.g., Qiagen). The injection fragment is prepared as follows:

1. Approximately 30 µg of plasmid DNA is digested with the appropriate restriction enzyme(s) (as outlined in **Subheading 3.1.1., Fig. 2A**).
2. The resulting fragment is separated from the vector backbone on a 0.8% agarose gel, visualized by ethidium bromide staining and excised under UV-light.
3. Purification from the agarose matrix can be performed using either the GeneClean or the Qiagen QIAquick kit according to the manufacturer's instructions.
4. DNA is eluted in injection buffer. The final concentration after elution should be approximately 100 ng/µL (*see* **Note 8**).
5. The eluted fragment can be stored at –20°C for several months.
6. Dilutions of 1.5, 2 and 2.5 ng/µL should be prepared to determine the optimal DNA concentration for injection. Test injections are performed as described in **Subheadings 3.2.2.5.** and **3.2.3.2.**

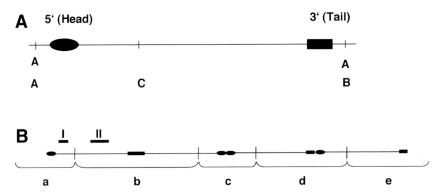

Fig. 2. Construction of a transgene-containing vector is shown. (**A**) Schematic representation of the insert and vector (multicloning site). *A* and *B* represent flanking restriction enzyme recognition sites for excision of the transgene. A recognition site for a third enzyme, *C*, cleaving only once within the fragment, should be present for Southern analysis of transgene integration as shown in **B**. (**B**) Schematic drawing of possible tandem array transgene integration. 5′ and 3′ ends of the transgene are depicted as in **A**. a–e represent restriction fragments obtained after digestion with restriction endonuclease C which are detectable with probes I and II, respectively.

3.2. Generation of Transgenic Mice

The generation of transgenic mice by pronuclear injection is described in **Subheadings 3.2.1.–3.2.4.**

3.2.1. Vasectomy of Male Mice and Breeding for Pseudopregnant Recipients

3.2.1.1. ANESTHESIA

1. Vasectomy and retransfer of injected oocytes are performed under Avertin™ anesthesia.
2. The animal to be anesthetized is weighed, then placed on the grid of a cage lid and grasped firmly by the skin of the neck as close to the ears as possible. The tail is fixed with the little finger. Avertin (0.4–0.6 mg/g body weight) is administered with a 26-gage hypodermic needle into the peritoneum.
3. After injection the animal is placed back into the cage to minimize stress.
4. Successful anesthesia can be checked by the absence of the toe pinch reflex.

3.2.1.2. VASECTOMY

1. Vasectomy is performed on mice at 10 weeks of age.
2. The mouse is weighed and anesthetized and then placed on its back (*see* **Note 9**).
3. The area around the scrotum is disinfected with 70% ethanol and a 1-cm incision is made in the skin above the scrotum.
4. Care should be taken not to injure the bladder and the penis musculature when opening the peritoneum.
5. By applying slight pressure on the scrotum the fat pad attached to the testis becomes visible.
6. The testis, epididymis and vas deferens are carefully pulled out from the body cavity by grasping the fat pad with anatomical forceps (**Fig. 3A**). Since the testes are very sensitive they should not be touched with the forceps.
7. The vas deferens can be recognized by a blood vessel running along one side. It is lifted with a pair of forceps and the ends of the loop that forms are squeezed together with for-

ceps that have been flamed on a gas burner. This results in resection of a segment of the vas deferens and sealing of the cut ends (*see* **Fig. 3B**).

8. The testis is carefully eased back into the abdominal cavity and the same procedure is repeated with the second testis.
9. After sealing the second vas deferens, the body wall is sutured and the skin is closed with two wound clips.
10. The mouse is kept on a 37°C-heated stage until recovery from anesthesia.
11. These vasectomized mice are then tested for sterility after 2–3 wk when the wound clips have been removed. They are mated with two or three females.
12. Successfully vasectomized animals should be able to generate a vaginal plug in females and the females should not become pregnant.
13. Vasectomized males are usually kept for up to 1 yr.

3.2.1.3. MATING FOR PSEUDOPREGNANT RECIPIENTS

On the day before preparation of oocytes and microinjection, mating is set up between vasectomized males and 10-wk-old female mice in a ratio of 1:2 to 1:3. Examination next morning shows a vaginal plug indicative of pseudopregnancy (*see* **Note 10**).

3.2.2. Superovulation, Mating, and Preparation of Zygotes

3.2.2.1. SUPEROVULATION AND MATING

Since female mice produce only small numbers of fertilized oocytes in a natural mating, superovulation has been established as the method of choice to increase zygote yields for pronuclear injection.

1. Superovulation is achieved by timed application of follicle stimulating hormone or PMSG and luteinizing hormone (hCG) to virgin 4-wk-old mice.
2. Intraperitoneal application of PMSG (5–10 I.E.) is carried out 72 h before preparation of the oocytes, while hCG (10 I.E.) is applied 48 h later. The time point for hCG administration is crucial. It must be given approx 6 h before the beginning of the dark cycle in the animal facility. Assuming that mating occurs in the middle of the dark cycle, fertilization is then ensured.
3. Superovulated females are mated with stud males in a 1:1 ratio.
4. Successful mating is indicated by the formation of a vaginal plug, which is evident next morning.
5. Successfully mated females are then sacrificed by cervical dislocation for preparation of zygotes.

3.2.2.2. PREPARATION OF MICRODROP CULTURES

1. Microdrop cultures for cultivation of zygotes before and after injection have to be set up prior to isolating fertilized oocytes.
2. Small drops (20–40 µL) of M16 are set onto a 35-mm tissue culture plate using a micropipet.
3. These drops are overlayed with mineral oil to prevent evaporation. This can be done by either carefully pouring the oil on top of the drops or overlaying with a pasteur pipet.

3.2.2.3. HANDLING OF OOCYTES

A prerequisite for the successful preparation of fertilized oocytes and injection is the skilled handling of a mouth-controlled drawn-out pasteur pipet.

1. Pipets are prepared by holding the thin part of a 25-cm glass pasteur pipet in the flame of a gas burner until it starts to melt.

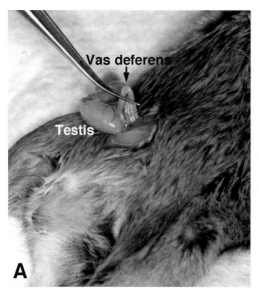
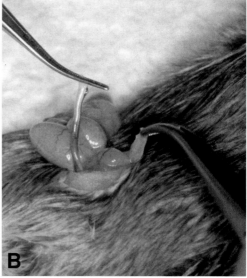

Fig. 3. Vasectomy: The vas deferens is lifted up (**A**), and after a segment has been resected, the two heat-sealed ends are separated (**B**).

2. The pipet is then quickly withdrawn from the flame and both ends are pulled.
3. Bending of the resulting thin part should then give a clean and blunt breakpoint.
4. The quality of the breakpoint should be checked by microscopy.
5. The pipet can then be attached to a rubber tube fitted with a mouthpiece.

3.2.2.4. Preparation of Zygotes

1. Fertilized female mice are sacrificed by cervical dislocation.
2. The abdomen is sprayed with 70% ethanol for disinfection.
3. A lateral incision is made in the skin and subcutaneous tissue and the margins of the incision are retracted toward the head and tail respectively.
4. The peritoneum is incised and the gut is pushed aside.
5. The bipartite uterus with attached oviducts and ovaries is now visible on both sides.
6. Each uterine horn is now carefully lifted with one pair of forceps, a second pair is inserted into the mesometrium directly beneath the oviduct in the closed position and then opened, thus separating the uterus and oviduct from the attached mesometrium.
7. The oviduct can now be excised with a pair of fine scissors at the junction with the ovary and in the upper part of the uterus, respectively (**Fig. 4A,B**).
8. The isolated oviducts are then transferred to an embryological watch glass containing M2.
9. Each oviduct is then transferred to another embryological watch glass containing hyaluronidase (**Fig. 4C**).
10. Using a stereomicroscope (20 or 40× magnification) the ampulla (the upper part of the oviduct where the oocytes are found) is located and then torn open with watchmaker's forceps and the zygotes (forming a clump with the cumulus cells) are released from the surrounding tissue.
11. The oviduct is removed from the dish.
12. To separate oocytes from the cumulus cells, the clump is kept in the hyaluronidase solution for a couple of minutes at room temperature. Digestion is complete when fertilized oocytes are clearly separated from the cumulus cells (**Fig. 4D**).

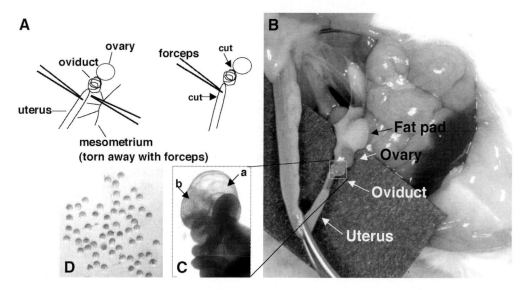

Fig. 4. Oocyte Preparation: (**A**) Schematic overview of the dissection of oviducts for the preparation of fertilized oocytes. (**B**) After opening the peritoneal cavity, the uterus with oviduct and ovary are taken out of the body cavity. The oviduct is cut between ovary and uterus (*see* A). (**C**) The ampulla (a) is clearly visible. It is filled with internal oocytes (b)—that are surrounded by cumulus cells. (**D**) Isolated oocytes after hyaluronidase-treatment.

13. The reaction is then stopped by passaging the oocytes with a mouth-driven pipet through consecutive drops of M2.
14. Oocytes are then transferred to a drop culture of M16 and kept in the 37°C incubator (5% CO_2) until needed for injection (*see* **Note 11**).
15. When isolated in the morning after mating, fertilized oocytes are approx 0.5 d post coitum (dpc) at mid-day. At this stage both pronuclei are clearly visible for a couple of hours. This is a prerequisite for successful microinjection.

3.2.2.5. CONTROL OF MEDIA AND CONDITIONS

Culture media (if prepared from powder) and the concentration and quality of the DNA used for injection have to be tested to ensure high efficiency in the generation of transgenic mice.

1. Oocytes isolated as described in **Subheading 3.2.2.4.** are cultured overnight in microdrops of M16 to test the suitability of the media prepared.
2. Approximately 80% of the oocytes should have reached the two-cell-stage on the next day.
3. To assess the quality and optimal concentration of the DNA fragment three different dilutions in injection buffer are prepared (*see* **Subheading 3.1.2.**).
4. Thirty oocytes are injected for each concentration as described in **Subheading 3.2.3.2.** and cultured overnight in microdrops of M16.
5. The suitable concentration should yield about 50% two-cell-stage embryos.
6. Higher numbers of surviving and developing embryos indicate that the DNA concentration used is too low, while a lower percentage suggests that the DNA concentration is too high or that the quality of the fragment preparation is poor.

3.2.3. Pronuclear Injection

3.2.3.1. PREPARATION OF HOLDING PIPETTES AND INJECTION NEEDLES

1. Holding pipets and injection needles can be produced with either a horizontal or vertical pipet puller.
2. Depending on the set values for the various parameters in a pulling sequence the length, taper, and tip of the pipet will vary. The values of the parameters have to be determined empirically and will vary depending on the device used.
3. Injection needles should be short and very tapered. They are made from capillary tubes containing a thin inner filament (*see* **Subheading 3.2.3.2.**).
   a. Injection needles should be prepared on the day that they will be used.
   b. Since injection needles tend to clog or break, sufficient needles should be prepared for use for the whole day.
   c. Needles can be stored in culture dishes supported by two rows of plasticine.
   d. The blunt end of the needle should not be touched since this end is immersed in the DNA solution (*see* **Subheading 3.2.3.2.**).
   e. The thin filament inside the pipet allows the DNA solution to enter the pipette by capillary force.
4. Filament-less capillaries are used for preparation of the holding pipets.
   a. They are pulled out much longer, to obtain an opening wide enough to hold the oocyte.
   b. The long-pulled holding pipet is placed on the glass ball of the micro forge (*see* **Note 12**).
   c. All the following steps are done using the oculars of the micro forge.
   d. In the area touching the surface of the glass the diameter of the pulled-out capillary should be approx 100 µm.
   e. By briefly heating the filament (until it starts glowing) the capillary will break due to stress, and a perfectly blunt end should form.
   f. Subsequently, the blunt end of the pipet and the filament are brought to the same focal plane and the filament is heated gradually.
   g. The increase in temperature induces melting of the blunt end of the capillary and an accurately shaped holding pipet is formed.
   h. The optimal diameter of the holding pipet has to be determined empirically (*see* **Note 13**).
   i. Holding pipets can be made in sufficient quantities and saved in culture dishes.
5. Alternatively, both holding pipets and injection needles can be obtained commercially (e.g., Eppendorf).

3.2.3.2. DNA INJECTION INTO OOCYTES

1. The oocytes are kept in the incubator until the pronuclei are clearly visible (usually around mid-day).
2. Before starting the preparations on the injection stage the injection pipet is placed with its blunt end in the DNA solution (which is stored on ice).
3. For injection a drop of M2 medium is pipeted into the depression of a depression slide-injection chamber on the stage of the injection microscope and covered with mineral oil.
4. The holding pipet is fastened into the fitting of the micromanipulator, adjusted to the center of the slide depression, and immersed into the medium.
5. Up to 15 oocytes are then transferred into the M2 droplet by the mouth-controlled pipet using the lowest magnification of the injection microscope.
6. Focus on the oocytes.
7. Divide the slide into two areas (e.g., above and below the holding pipet) to separate the uninjected and injected oocytes and move the oocytes to their respective areas by gently

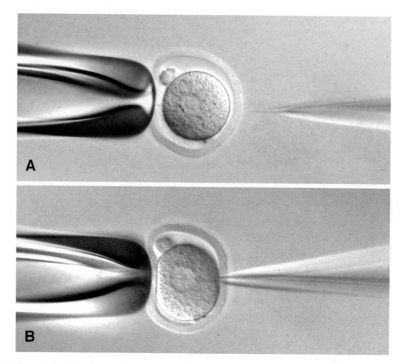

Fig. 5. Microinjection Into the Pronucleus: **(A)** The pronucleus is in the same focus as the injection needle. **(B)** The pronucleus is swollen due to the injected DNA solution.

    touching with the holding pipette, which should be adjusted to the same focal plane as the oocytes.

8. By applying gentle suction, one oocyte at a time is picked up and fixed to the holding pipet (*see* **Note 13**). The zona pellucida should be pulled slightly into the pipet.
9. Then the injection needle is fixed to the micromanipulator, carefully immersed into the medium, and adjusted to the same focal plane as the oocytes and holding pipet.
10. Change the magnification to high power and adjust the focus so that the tip of the injection needle is in the same focal plane as one of the pronuclei (which should be clearly visible with a sharp border, **Fig. 5A**).
11. With a firm move to break through the zona pellucida, the needle is pushed into the pronucleus (*see* **Note 14**).
12. The DNA solution is pressed into the pronucleus by pushing the foot-pedal of the injector.
13. Successful injection is clearly visible by a swelling produced in the injected pronucleus (**Fig. 5B**). Then the needle is quickly but cautiously removed.
14. The injected oocyte is then sorted to the designated area on the slide and the next oocyte can be injected.
15. Oocytes that survive microinjection, are transferred back to the incubator and kept until implantation into a pseudopregnant recipient.

3.2.4. Retransfer of Injected Oocytes

1. A pseudopregnant mouse is weighed and anesthetized with Avertin (*see* **Subheading 3.2.1.1.**).
2. While waiting for completion of anesthesia the microdrop culture is taken from the incubator and oocytes are carefully drawn up into a mouth-driven pulled-out pasteur pipet (*see* **Note 15**).

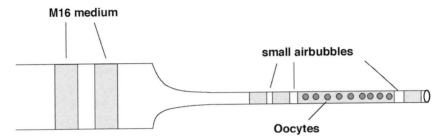

Fig. 6. Schematic drawing of a pasteur pipet for retransferring oocytes.

3. To facilitate visual control of the retransfer, the pipet is loaded as follows: medium–air bubble–medium–air bubble–medium with oocytes–and facultative another air bubble near the tip (**Fig. 6**). It is important to keep the volume as small as possible and to arrange the oocytes close to each other.

4. The maximum number of oocytes, which can be retransferred to each uterine horn, is about 10 to 15.

5. The loaded pipet should be stored in a safe corner of the working bench, away from the area where the animal is handled.

6. The anesthetized animal is placed on its stomach on a piece of tissue paper.

7. A 1-cm skin incision is made to the left of the spinal cord, at the level of the last rib.

8. A fat pad should become visible through the peritoneum, which indicates the correct position.

9. After opening the peritoneum with another small incision, the fat pad attached to the ovary is carefully grasped with a pair of serrated forceps and pulled out.

10. The ovary, oviduct, and upper part of the uterus become visible.

11. The fat pad is fixed with a serrefine clamp and brought into a position in which the ovary and oviduct are easily accessible (**Fig. 7A**).

12. The mouse is then transferred to the stage of a stereomicroscope. With a pair of fine forceps the bursa is carefully opened (**Fig. 7B**; *see* **Note 16**).

13. The funnel-shaped infundibulum is usually situated at the border between ovary and oviduct (*see* **Note 17**).

14. The tip of the oocyte-loaded pipet (supported by a spread pair of fine forceps) can now be inserted into the infundibulum. The oocytes are transferred with a steady blow into the pipet.

15. The appearance of the air bubbles within the oviduct indicates successful delivery of oocytes.

16. The mouse is now removed from the stereomicroscope stage.

17. By gently pushing the fat pad, the ovary is carefully placed back into the body cavity.

18. Since the incision in the peritoneum is only small, suturing is usually not required.

19. The skin is closed with a wound clip.

20. The same procedure is repeated with the second oviduct.

21. After surgery the mouse is kept on a 37°C-heated stage and observed until recovery from anesthesia.

3.3. Concluding Remarks

Since the procedures described require technical expertise (especially in handling oocytes at the various stages), we strongly recommend that a beginner visit a laboratory where all these techniques are properly established. Alternatively, you might attend

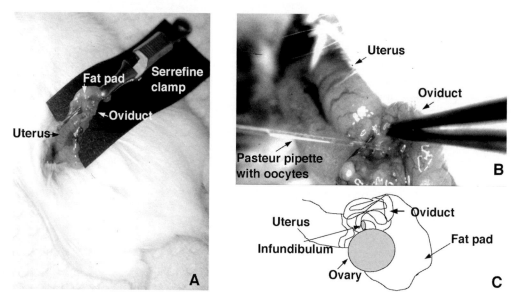

Fig. 7. Retransfer of Injected Oocytes into a Pseudopregnant Recipient. (**A**) Ovary and uterus are carefully taken out and fixed at the fat pad with a serrefine clamp. (**B**) Oocytes are retransferred into the infundibulum. (**C**) Schematic overview.

one of the "mouse courses" offered by several laboratories or companies (*see* for example: http://nucleus.cshl.org/meetings/ or http://www.jax.org/courses/index.html).

A very comprehensive guide on establishing transgenic mouse lines and analysing embryos at all stages of development is available *(15)*. A discussion group has been established at http://www.med.ic.ac.uk/db/dbbm/tgunit.html, that may provide further helpful information.

4. Notes

1. Mechanisms leading to decreased expression or complete silencing of transgenes have been the focus of intense investigations for the past few years. A good overview of these mechanisms and possible strategies to overcome these influences is given by Cranston et al. *(6)*.
2. Although there are regulatory sequences available for transgene expression in non-cerebral endothelial cells to date there are no cis-acting elements described confining expression to the specialized endothelium of the blood-brain barrier. Genes selectively expressed at the blood-brain barrier include, e.g., the large amino acid transporter LAT *(16)*, mdr1a/3 *(17)* or transcripts recently identified by substractive hybridization *(18)*. Once regulatory sequences for these genes have been identified these may serve as a source for blood-brain barrier specific transgene expression.
3. Artificial chromosomes have not only been used to provide all necessary regulatory sequences to engineered transgenes and reporter genes but also to express highly complex genes spanning several hundred kilobases of sequence. Good reviews on applications and techniques for the introduction of YACs into the genome are available *(19,20)*.
4. If a heterologous polyadenylation signal is used, the presence of a small intron, as in the case of the SV40 cassette, may enhance expression levels on one hand but it may also cause aberrant splicing to cryptic splice donors in the coding sequence. There is no way one can predict whether and to what extent this process might occur in vivo. Thus careful analysis of the mRNA resulting from the transgene should be performed.

5. Most laboratories prefer using outbred or F1-hybrid strains instead of genetically uniform inbred strains for virtue of large oocyte yields and good breeding behavior. From the point of view of uniformity in the genetic background F1-hybrids are certainly the better choice, but they are also more expensive. Apart from the oocyte yield and breeding behavior, other factors such as suitability for the analysis of the expected phenotype may be considered in the selection of the mouse strain used for genetic manipulation. The highest amount of oocytes are obtained from approximately 4-wk-old virgin mice. Hormone stimulation in older females interferes with the endogenous cycle, which may lead to asynchronous ovulation and degeneration of oocytes.

6. There are several anesthetics available for rodents. A comprehensive listing with the doses recommended can be found at http://oacu.od.nih.gov/ARAC/tablesspecies.pdf.

7. Suppliers like Leica (in collaboration with Eppendorf) or Zeiss offer complete microinjection equipment. For setting up the instrument, refer to the supplier's instructions.

8. The amount of DNA to be digested has to be calculated according to the final fragment concentration and elution volume desired. Digested DNA should be run using a preparative comb for the agarose gel to avoid overloading; alternatively several lanes can be connected using adhesive tape. Agarose for purification of injection fragments should be of the best available quality. We recommend Life Technologies Ultra Pure™.

9. During all surgery the eyes of the animals should be protected with eye balm to prevent them from drying out.

10. The number of pseudopregnants obtained can vary considerably from day to day. To compensate for these variations we recommend setting up more matings than are actually needed for the calculated amount of oocytes. In addition, we routinely record mating performance. This allows early detection of failure of individual males that can then be replaced.

11. Do not introduce any air bubbles into the drop culture. This makes it very difficult to draw the oocytes into the pasteur pipet for the retransfer. If air bubbles occur, the oocytes should be transferred into a new drop.

12. The filament of the micro forge can melt if it is heated abruptly. Therefore it is important to control all steps using the binoculars. The glass ball has to be melted onto the filament. To do this, a long-pulled capillary is put on the filament and broken. With increasing heat, the broken part of the capillary will melt onto the filament. By repeating this procedure, a glass ball is formed on the filament.

13. If the opening of the holding pipet is too small the oocyte cannot be thoroughly fixed and might rotate when it is touched with the injection needle. If openings are too wide there is the risk of the oocyte being sucked into the pipet. Therefore it is best to test holding pipets for their performance. Since the risk of damage and blocking is rather low, holding needles rarely have to be changed.

14. Usually the male pronucleus is larger and therefore easier to inject. During the injection one should try to avoid touching the nucleoli in the pronucleus with the tip of the injection needle. This often causes plugging of the needle, which then has to be replaced. Withdrawing the needle should be as quick as possible. Slow retraction favors attachment of nuclear components to the injection needle. Sometimes the whole nucleus can be withdrawn with the injection needle.

15. The narrow end of the mouth-driven pipet should not be too long, since shorter ends are easier to handle. The opening should be flame-polished to obtain perfectly blunt edges otherwise the risk of injuring the tissue of the infundibulum or oviduct is high.

16. The pseudopregnant ovary is well perfused. Bleeding from accidentally injured vessels complicates access to the infundibulum. When bleeding occurs, small drops of adrenaline can be applied to the wound, which should then stop bleeding.

17. Locating the infundibulum is sometimes difficult and requires practice. If the infundibulum cannot be located on one side and the number of pseudopregnant females available is limited, the double amount of oocytes can be transferred to one side since the oocytes can move from one uterine horn to the other.

References

1. Ryding, A. D., Sharp, M. G., and Mullins, J. J. (2001) Conditional transgenic technologies. *J. Endocrinol.* **171**, 1–14.
2. McBurney, M. W., Staines, W. A., Boekelheide, K., Parry, D., Jardine, K., and Pickavance, L. (1994) Murine PGK-1 promoter drives widespread but not uniform expression in transgenic mice. *Dev. Dyn.* **200**, 278–293.
3. Ikawa, M., Yamada, S., Nakanishi, T., and Okabe, M. (1998) 'Green mice' and their potential usage in biological research. *FEBS Lett.* **430**, 83–87.
4. Kisseberth, W. C., Brettingen, N. T., Lohse, J. K., and Sandgren, E. P. (1999) Ubiquitous expression of marker transgenes in mice and rats. *Dev. Biol.* **214**, 128–138.
5. Potts, W., Tucker, D., Wood, H., and Martin, C. (2000) Chicken beta-globin 5′HS4 insulators function to reduce variability in transgenic founder mice. *Biochem. Biophys. Res. Commun.* **273**, 1015–1018.
6. Cranston, A., Dong, C., Howcroft, J., and Clark, A. J. (2001) Chromosomal sequences flanking an efficiently expressed transgene dramatically enhance its expression. *Gene* **269**, 217–225.
7. Jang, S. K., Krausslich, H. G., Nicklin, M. J., Duke, G. M., Palmenberg, A. C., and Wimmer, E. (1988) A segment of the 5′ nontranslated region of encephalomyocarditis virus RNA directs internal entry of ribosomes during in vitro translation. *J. Virol.* **62**, 2636–2643.
8. Takeuchi, T., Yamazaki, Y., Katoh-Fukui, Y., et al. (1995) Gene trap capture of a novel mouse gene, jumonji, required for neural tube formation. *Genes Dev.* **9**, 1211–1222.
9. Li, X., Wang, W., and Lufkin, T. (1997) Dicistronic LacZ and alkaline phosphatase reporter constructs permit simultaneous histological analysis of expression from multiple transgenes. *Biotechniques* **23**, 874–878, 880, 882.
10. Previtali, S. C., Quattrini, A., Fasolini, M., et al. (2000) Epitope-tagged P(0) glycoprotein causes Charcot-Marie-Tooth-like neuropathy in transgenic mice. *J. Cell Biol.* **151**, 1035–1046.
11. Oster-Granite, M. L., McPhie, D. L., Greenan, J., and Neve, R. L. (1996) Age-dependent neuronal and synaptic degeneration in mice transgenic for the C terminus of the amyloid precursor protein. *J. Neurosci.* **16**, 6732–6741.
12. Wight, P. A., Duchala, C. S., Readhead, C., and Macklin, W. B. (1993) A myelin proteolipid protein-LacZ fusion protein is developmentally regulated and targeted to the myelin membrane in transgenic mice. *J. Cell Biol.* **123**, 443–454.
13. Young, W. S. III, Iacangelo, A., Luo, X. Z., King, C., Duncan, K., and Ginns, E. I. (1999) Transgenic expression of green fluorescent protein in mouse oxytocin neurones. *J. Neuroendocrinol.* **11**, 935–939.
14. Sambrook, J., and Russell, D. W., eds. (2001) *Molecular cloning: A laboratory manual, 3ed.* Cold Spring Harbor Laboratory Press, Cold Spring Harbor, NY.
15. Hogan, B., Beddington, R., Costantini, F., and Lacy, E., eds. (1994) *Manipulating the mouse embryo: A laboratory manual, 2ed.* Cold Spring Harbor Laboratory Press, Cold Spring Harbor, NY.
16. Boado, R. J., Li, J. Y., Nagaya, M., Zhang, C., and Pardridge, W. M. (1999) Selective expression of the large neutral amino acid transporter at the blood-brain barrier. *Proc. Natl. Acad. Sci. USA* **96**, 12,079–12,084.

17. Qin, Y., and Sato, T. N. (1995) Mouse multidrug resistance 1a/3 gene is the earliest known endothelial cell differentiation marker during blood-brain barrier development. *Dev. Dyn.* **202,** 172–180.

18. Li, J. Y., Boado, R. J., and Pardridge, W. M. (2001) Blood-brain barrier genomics. *J. Cereb. Blood Flow Metab.* **21,** 61–68.

19. Lamb, B. T., and Gearhart, J. D. (1995) YAC transgenics and the study of genetics and human disease. *Curr. Opin. Genet. Dev.* **5,** 342–348.

20. Peterson, K. R., Clegg, C. H., Li, Q., and Stamatoyannopoulos, G. (1997) Production of transgenic mice with yeast artificial chromosomes. *Trends Genet.* **13,** 61–66.

36

Mouse Genome Modification

Richard Rozmahel

1. Introduction

The ability to alter the mouse genome through homologous recombination in their embryonic stem (ES) cells, and propagate the modification through their germ-line, has revolutionized biomedical research. Such gene-targeted mice have afforded researchers unprecedented opportunities to analyze gene function in vivo, and provided models for disease studies.

1.1. History and Outline of Gene-Targeting Experiments

Over 20 years ago, two independent reports documented the isolation of pluripotent ES cells from explanted preimplantation mouse embryos *(1,2)*. In 1984, Bradley et al. *(3)* demonstrated that these ES cells, following their transfer to the inner cell mass of host blastocysts, could contribute to the germline of resultant animals. Shortly thereafter, Smithies et al. *(4)* showed that homologous recombination with an introduced deoxyribonucleic acid (DNA) fragment could specifically alter the genome of cultured mammalian cells. The gene targeting methodology was then successfully applied by Thomas and Capecchi *(5)* and Doetschman et al. *(6)* to disrupt the selectable *Hprt* gene in ES cells, resulting in the production of HPRT-deficient mice, an animal model for Lesch-Nyhan disease *(7)*. Soon after, the first mouse line containing a targeted disruption of nonselectable gene *(cAbl)* was reported *(8)*. Since then, the number of gene-targeted mouse lines has risen exponentially, such that presently more than 1000 lines exist (*see* http://www.bioscience.org/knockout/alphabet.htm for a partial list). Conventional gene targeting strategies have also evolved to include manipulations that can result in specific spatial and/or temporal effects. Thus, contemporary experiments can produce mouse lines having gene disruptions, mutations or replacements, chromosome rearrangements, and distinct gene expression patterns.

This chapter outlines considerations and practical aspects of a basic gene-targeting experiment. Before beginning such an experiment, the strategy to arrive at the desired outcome requires careful planning. A gene-targeting experiment relies on successful execution of several different technical procedures; however, the general sequence and methodologies are similar among experiments. A portrayal of a basic gene targeting experiment is given in **Fig. 1**. Each experimental step is expanded upon in **Subheading 3.**

From: *Methods in Molecular Medicine, vol. 89:*
The Blood–Brain Barrier: Biology and Research Protocols
Edited by: S. Nag © Humana Press Inc., Totowa, NJ

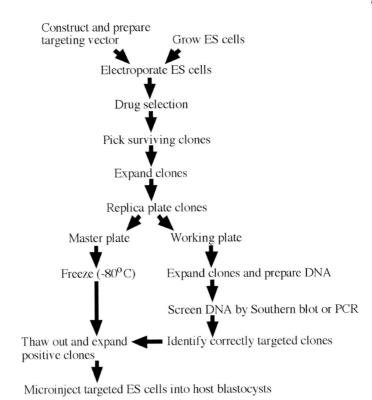

Fig. 1. Schematic outline of the progression of a conventional gene-targeting experiment.

1.2. Design of a Gene Targeting Experiment

Introduction of DNA molecules into mammalian cells can result in their recombination with homologous sequence within the genome. Molecular processes responsible for maintenance of the cell's chromosomes and sequence integrity are thought to mediate the homologous recombination process (reviewed in **ref. 9**). Specific alteration of the genome by recombination with an in vitro-constructed DNA vector is known as gene targeting; hereafter referred to as targeting.

The design of the targeting vector is of primary importance to success of the experiment, and thus requires much consideration. The basic components required of a targeting vector are sequences homologous to the endogenous sequence that is to be modified, selectable markers (positive, and optionally negative), and plasmid sequences for bacterial amplification. The positive selectable marker (typically the prokaryotic neomycin phosphotransferase *[Neo]* gene expressed under the ubiquitous phospho-glycerate kinase-1 (PGK-1) promoter) normally serves two functions: it confers viability to cells containing the vector, and it can serve to disrupt the endogenous gene.

There are two general types of targeting vectors. As depicted in **Fig. 2**, insertion-type vectors integrate into their homologous sequences through their ends, while replacement-type vectors undergo recombination internal to their homologous sequences. The site of linearization dictates the difference in recombination sites between the two vector types; whereas insertion-type vectors are linearized within the

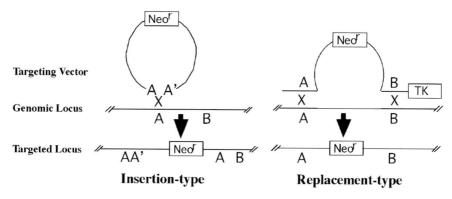

Fig. 2. Schematic representation of insertion-type and replacement-type gene-targeting experiments.

homologous sequence, linearization of replacement-type vectors is outside this sequence. Although insertion-type vectors tend to undergo homologous recombination at higher frequencies *(10)*, their usage is limited by the kind of genomic modification they can introduce. Replacement-type vectors are more versatile in their ability to modify the genome and are more widely used in targeting experiments. A brief description of the two vector types and considerations for their use is given in **Subheadings 1.2.1.** and **1.2.2.** Important considerations in their design are outlined in **Notes 1–4**.

1.2.1. Insertion-Type Vectors

As depicted in **Fig. 2**, an insertion-type vector undergoes a single reciprocal recombination mediated through a double-stranded break in the vector, resulting in its insertion into the genome. The insertion results in partial or complete disruption of the gene due to the duplication of the homologous sequences and addition of the vector's heterologous sequences to the targeted allele. However, because of multiple factors such as exon skipping, in-frame exon additions, alternative splicing, whose description is beyond this chapter, the actual outcome of an insertion-type targeting experiment cannot be determined in advance.

Required elements of an insertion-type targeting vector are 1) region of homology to the locus to be targeted with a unique restriction enzyme site for linearization; 2) a positive selectable marker used to identify clones having the vector; and 3) plasmid sequences for bacterial amplification.

1.2.2. Replacement-Type Vectors

In contrast to insertion-type vectors that duplicate endogenous sequences, replacement-type vectors replace the sequences with those contained in the vector, thus making these vectors more versatile and commonly used in targeting experiments. Required elements of replacement-type vectors are 1) two segments of homology to the locus to be targeted separated by a positive selectable marker; 2) a unique restriction site for linearization outside of the homologous sequence; and 3) plasmid sequences for bacterial amplification. Targeting by replacement-type vectors occurs by double reciprocal recombination between the vector sequences and their homologous endogenous

3. Vacuum source.
4. Variable temperature water bath (37°C–56°C).
5. –80°C and –20°C freezers.
6. 4°C refrigerator.
7. Liquid Nitrogen storage tank.
8. Inverted microscope with phase contrast objectives (4× to 50×).
9. Dissecting microscope with transmitted light source.
10. Refrigerated tabletop centrifuge.
11. Electroporator with capacitance extender.
12. Single channel and multichannel pipeters.
13. Electric Pipet-Aid.
14. Hemacytometer.

2.2. Solutions and Reagents

1. All reagents should be filter-sterilized.
2. Phosphate-buffered saline (without Ca^{2+} and Mg^{2+}).
3. Trypsin Solution (0.05%) dissolved in double distilled water and filtered (5 μ*M*).
4. 0.1% Gelatin Solution (prepared in double distilled water and autoclaved).
5. Lysis Buffer: 100 m*M* Tris-HCl pH 8.5, 5 m*M* EDTA, 200 m*M* NaCl, 0.2% SDS and 100 μg Proteinase K/mL.
6. Tissue Culture Grade dimethyl sulfoxide (DMSO).
7. Isopropanol.
8. 70% ethanol.

2.3. Disposables

1. Plates: 100-m*M*, 60-m*M*, and 96 well tissue culture plates.
2. 50- and 500-mL disposable filtration units (0.5 μm).
3. Pipets: 1, 2, 5, 10, and 25 mL.
4. Electroporation cuvettes.
5. Reagent reservoirs for multichannel pipeters.
6. Parafilm.

2.4. ES Cell Media

Each ES cell line has its own specific requirements for optimal growth and maintenance of pluripotency, the conditions we normally use for the R1 ES cell line *(48)* are provided here.

1. ***ES Cell Culture Media:*** Dulbecco's minimal Eagle's medium (DMEM) with high glucose containing 2 m*M* L-glutamine, 0.1 m*M* non-essential amino acids (NEAA), 1 m*M* sodium pyruvate, 10^{-4} *M* β-mercaptoethanol, 2000 U/mL leukemia inhibitory factor (LIF), 20% ES Cell-Qualified Fetal Calf Serum (heat inactivated for 30 min at 56°C), penicillin/streptomycin (optional).
2. ***Geneticin (G418)-Selection Media:*** ES cell culture media with 170 μg/mL active Geneticin.
3. ***G418/Gancyclovir-Double Selection Media:*** ES Cell Culture Media with 170 μg/mL active Geneticin and 2 m*M* Gancyclovir.
4. ***ES Cell Freezing Medium:*** ES Cell Culture Media (containing 20% fetal calf serum) with 1 : 10 final volume DMSO and 1 : 10 final volume fetal calf serum.

Table 1
Germline Competent Mouse Embryonic Stem (ES)
Cell Lines

ES cell line	Mouse strain	Reference
E14	129/Ola	*(53)*
D3	129/Sv	*(54)*
AB2.2	129/Sv	*(55)*
R1	(129/SvX129/J) F1	*(48)*
W4	129/SvEv	*(56)*
BL/6-III	C57BL/6	*(57)*
WB6d	C57BL/6	*(56)*
BALB/c-I	BALB/c	*(58,59)*

2.5. Embryonic Stem Cells

Successful generation of gene-targeted mouse lines relies on the quality of the ES cells, measured by their pluripotency (ability to give rise to different cell types in the mature animal). ES cells are established from the inner cell mass of mouse blastocysts *(1,2)*, and are maintained pluripotent by the differentiation-inhibiting agent LIF *(49,50)* and co-culture with mouse embryonic fibroblast cells (feeder cells). When maintained pluripotent, ES cells reintroduced into blastocysts will contribute to normal embryonic development and give rise to all three germ layers, including the germline of resultant animals *(1,51)*. Although ES cells can be established from different mouse strains, strain-inherent differences have important implications to their pluripotency, particularly with respect to germ cells *(52)*. The most common ES cell lines have been established from the 129 and its derivative mouse strains *(52)*. However, requirements for mouse lines of alternative genetic backgrounds have necessitated development of ES cells from different strains. A compendium of some germline-competent ES cell lines of different mouse strains is given in **Table 1**.

3. Methods

The conditions and protocols we normally use for the R1 ES cell line *(48)* using *Neo* for positive and *TK* for negative selection are given here. Although the protocols would be similar to those used for other ES cell lines, the reader is referred to the laboratory in which the cell lines were derived to determine optimal conditions. For alternative selectable markers the reader is likewise referred to protocols describing the different drug selection schemes. The protocols and reagents provided here have been used to maintain pluripotent ES cells independent of a feeder layer. Different conditions should be employed accordingly if the ES cells are to be grown on feeders. All manipulations must be carried out using sterile protocols.

3.1. Preparation of Gelatin-Coated Plates

Mouse ES cells are grown and maintained on gelatin-coated plates. In our hands, NUNC brand (Nalge NUNC International) plates are the preferred substrate for optimal growth of R1 ES cells. Plates are prepared as follows:

1. Dispense to each plate a sufficient volume gelatin solution (*see* **Subheading 2.2.**) to cover the bottom of the plate.
2. Aspirate off gelatin solution and let plate air-dry in a laminar flow hood (about 15 min).
3. Use plates immediately or store under sterile conditions at room temperature.

3.2. Preparation of ES Cells for Electroporation

1. 7–10 d before the planned electroporation thaw two vials of frozen ES cells (5×10^6 cells/vial) onto two gelatin-treated 60-mm plates containing ES cell culture media (*see* **Subheading 2.4.**). Change media next day.
2. 2–3 d after plating (80% confluency) trypsinize (*see* **Subheading 2.2.**) and pass each 60-mm plate onto two gelatin-treated 100-mm plates containing ES cell culture media.
3. Two to three days after passage (80% confluency) trypsinize and transfer cells from each 100-mm plate onto five gelatin-treated 100-mm plates containing ES cell culture media. Plates should be ready for electroporation (80% confluent) 2–3 d later.

3.3. Preparation of Targeting Vector for Electroporation

Gene-targeting experiments require electroporation of a linearized vector. The vector should be linearized at the desired site of targeted insertion (insertion-type experiments), or outside the homologous segments region of homology (replacement-type experiments) (*see* **Subheading 1.3.1.**). Preparation of the vector for a single experiment (electroporation of 10 cuvettes) is performed as follows:

1. Linearize 500 μg of the targeting vector (assuming a total vector size of 5–10 Kb, if vector larger than 10 Kb scale up quantity appropriately) by digestion for more than 2 h with the appropriate restriction enzyme.
2. Run digested vector on an ethidium bromide-stained agarose gel to verify complete linearization. If not completely linearized, add additional enzyme and continue digestion until complete.
3. Precipitate vector DNA by addition of 1 : 10 volume 3 *M* sodium-acetate and 2 vol ethanol at room temperature, centrifuge at 10,000 RPM for 1 min and resuspend in sterile double distilled water (or alternatively phosphate-buffered saline [PBS]) at 1 μg/μL (500 μL total vol for a 5- to 10-Kb vector).
4. Verify final concentration of vector sample by running 0.5 μL on an ethidium bromide-stained agarose gel alongside marker DNA of known concentration.

3.4. Electroporation of ES Cells

On the day planned for electroporation, eight 100-mm plates containing the ES cells at 80% confluency should be available (*see* **Subheading 3.2.**). These plates are treated as follows:

1. Remove media from plates and wash with 10 mL of PBS prewarmed to 37°C.
2. Aspirate PBS and cover bottom of plate with 2 mL trypsin solution for 5 min.
3. Pipet trypsin solution up and down at least 10× to dislodge and disaggregate cells, and transfer to a 50-mL tube containing 40 mL of ES cell media. The contents of four plates are transferred to each of two 50-mL tubes.
4. Centrifuge cells in media for 5 min at 1000 rpm and room temperature.
5. Carefully remove supernatant and resuspend each of the two cell pellets in 4 mL ice-cold PBS. Combine the two cell suspensions and place on ice.
6. Ascertain cell concentration using a hemacytometer (count clumps of cells as one cell, do not count dead cells).

7. Dilute suspension to a final concentration of 10^7 cells/mL with ice-cold PBS and transfer 800 µL into each of 10 electroporation cuvettes along with 50 µL of linearized vector (*see* **Subheading 3.3.**). Mix thoroughly and keep on ice.
8. Pulse cuvettes with 0.24 kV with capacitor at 500 µF and let stand on ice for 15 min.
9. Combine cell suspensions from all cuvettes into 100 mL of ES cell culture media, mix thoroughly and plate 10 mL of final cell suspension onto each of 10 gelatin-treated plates. Place into incubator.
10. Change media with ES cell culture media next day.

3.5. Drug Selection

Two days after electroporation, the cells are ready to undergo drug selection. For experiments not using double selection, replace media on all plates with 10 mL of G418-selection media (*see* **Subheading 2.4.**). Continue to change media with G418-selectable media daily for 7–10 d. For experiments using double selection (incorporation of TK gene in the targeting vector):

1. Separate one plate from the rest, remove media and replace with 10 mL of G418-selection media (*see* **Subheading 2.4.**). Change media daily for 5–7 d.
2. For the remaining plates, replace media with G418/Ganc-double-selection media (*see* **Subheading 2.4.**).
3. Continue to change media daily with G418/Ganc-double selection media for 5–7 d.

3.6. Selection of Clones

Five to seven days after initiation of drug selection, all non-transfected cells should have died and drug-resistant colonies should be visible to the naked eye.

1. Count the number of total surviving colonies. For double selection experiments, ascertain the enrichment factor by dividing the number of colonies on the G418-selection media plate by the average number on the G418/Ganc-double selection media plates. A typical experiment should yield an enrichment factor between 5 and 20×.
2. Under the dissecting microscope in an open hood, pick the surviving colonies (if using double selection pick from only the G418/Ganc-double selection media plates) by scraping them from the bottom and sucking them up using a 200-µL pipet tip. Transfer each colony to a well of a gelatin-treated 96-well plate containing 200 µL of ES cell culture media. Place plates into incubator.
3. The following day trypsinize the colonies as follows:
   a. Aspirate off the media from each well.
   b. Wash each well with 200 µL of 37°C PBS and aspirate off.
   c. Add 10 µL of 37°C trypsin solution to each well and let stand for 5 min.
   d. Add 200 µL of 37°C ES cell culture media and pipet up and down 8–10×. Place plate back into incubator. Change media with ES cell culture media every second day.
4. When the majority of wells are 80% confluent (usually 3–4 d after trypsinization) replicate each 96-well plate by trypsinization (as in **step 3**) and transfer half (100 µL) of each well's cell suspension to a new gelatin-treated 96-well plate (**Working plate**). The remaining cell suspension is kept on the same plate (**Master Plate**) with an additional 100 µL media added to each well.

3.7. ES Cell Freezing on 96-Well Plates

Two to three days after plate replication (the majority of wells should be at least 80% confluent) the **Master Plate** is ready to be frozen, as follows:

1. Replace media from each well with ice-cold ES Cell-Freezing Media (*see* **Subheading 2.4.**), seal edges of plate with parafilm and place plates on ice.
2. Transfer plates to a Styrofoam box and place into –80°C freezer to allow slow freezing.
3. The next day remove plates from box and place back into –80°C freezer. Cells stored in this way are viable for at least 6 mo.

3.8. Genomic DNA Preparation and Screening of Clones

Continue to incubate the **Working plate** until the majority of wells are 100% confluent (usually 3–4 d). At maximum confluency treat each well as follows:

1. Aspirate off the media and wash wells with 200 µL PBS.
2. Add 100 µL of lysis buffer (*see* **Subheading 2.2.**) to each well, seal edges of plate with parafilm and place plate into incubator for a minimum of 8 h.
3. Centrifuge plates in a swing bucket rotor with microtiter plate carriers at 500 rpm for 1 min to remove condensate from lid.
4. Add 100 µL of isopropanol to each well, place on low speed orbital shaker for 10 min to mix, and centrifuge plates at 1000 rpm for 5 min to pellet DNA at bottom of well.
5. Carefully aspirate off supernatant with a 200-µL pipet tip (pellet should be clearly visible) and wash pellet with 200 µL 70% ethanol.
6. Carefully aspirate off the 70% ethanol with a 200-µL pipet tip and let pellet air-dry for at least 15 min in the laminar flow hood.
7. Resuspend pellet in 50 µL sterile double distilled water or TE solution. Use 5 µL or 20 µL for PCR or Southern blot assay, respectively.

Two types of screening procedures can be used to identify targeted ES cell clones: PCR and Southern blot analysis. Both types of screens exploit the targeted juxtaposition of vector-introduced exogenous or modified endogenous sequences, with specific target locus sequences. The PCR is the most sensitive assay for identifying targeted clones. This assay relies on amplification of a novel fragment derived from the desired homologous recombination, where one primer is complementary to the target sequence outside of the vector and the other is complementary to unique sequences within the vector. Thus, amplification of the correct fragment will only occur when the two primer complementary sequences are juxtapositioned by homologous recombination. Because of false positive signals and potential rearrangements of the targeted locus, follow-up Southern blot analysis should always be used to confirm correct targeting. Since each experiment has its own specific characteristics for screening recombinants, an outline of these procedures is not warranted here. For specific details regarding experimental protocols, the reader is referred to the laboratory manual by Sambrook et al. *(40)*.

3.9. Thawing, Expansion, and Freezing in Vials

Following identification of recombinants, the individual clones need to be thawed, expanded, and frozen in vials, as follows:

1. Place the **Master Plate** containing the targeted clone(s) into a 37°C incubator or, alternatively, hold on the surface of a 37°C water bath, until completely thawed (3–4 min).
2. Remove parafilm from plate and rinse outside surface with 70% ethanol moistened tissue.
3. Aspirate off the freezing media from the selected wells (and adjacent wells to be safe) and replace with 200 µL prewarmed ES cell culture media. Aspirate freezing media from all remaining wells and leave empty. Place plate into incubator until next day.

4. Next day remove media from each well, wash with 200 µL PBS, add 10 µL of prewarmed trypsin solution, and let stand for 5 min.

5. Add 200 µL of prewarmed ES cell culture media, pipet up and down at least eight to 10 times, and transfer plate back into incubator.

6. When wells are 80% confluent (generally 2–3 d), pass 90% of contents onto a gelatin-treated 60-mm plate and add 200 µL of fresh media back to original well (refreeze when 80% confluent).

7. Let 60-mm plate grow for 3–4 d (changing media with ES cell culture media every 2 d) until 80% confluent, then pass onto two gelatin-treated 100-mm plates.

8. When 100 mM plates are 80% confluent (2–3 d) trypsinize both plates (2 mL trypsin each) for 5 min, add 10 mL of ES cell culture media to stop reaction, and pipet up and down at least five times to break up cell clumps.

9. Lyse 50% of the cell suspension in a 10-mL tube (as described above for 96-well plates) for DNA preparation to confirm proper targeting. Transfer remaining 50% of cell suspension into 10-mL centrifuge tubes.

10. Pellet cells by centrifugation for 5 min at 1000 rpm at room temperature, aspirate off media.

11. Resuspend cell pellet in 5 mL ice-cold Freezing Medium and aliquot the suspension into five freezing vials (about 5×10^6 cells/vial). Keep on ice.

12. Transfer freezing vials to sealed Styrofoam box and place into –80°C freezer overnight. Next day transfer vials to liquid nitrogen storage.

3.10. Preparation of ES Cells for Microinjection

Two days before scheduled ES cell microinjection into blastocysts, prepare cells as follows:

1. Place vial of frozen ES cells into a 37oC water bath (with lid just above surface) until thawed (2–3 min). Transfer thawed contents into 10 mL of ES cell culture media prewarmed to 37°C and centrifuge at 1000 rpm at room temperature for 5 min.

2. Remove media, resuspend pellet in 5 mL of prewarmed ES cell culture media, plate onto a gelatin-treated 60-mm plate, and place into incubator. Change media with fresh ES cell culture media the next day.

3. On day of microinjection, the plate should be 50–80% confluent. Aspirate off media and wash plate with 5 mL PBS. Remove PBS and add 1 mL of trypsin solution to cover bottom of plate and let stand 5 min, pipet up and down several times to detach and disaggregate the cells, add 5 mL ES cell culture media. Pellet the cells by centrifugation at 1000 rpm for 5 min at room temperature.

4. Resuspend the cell pellet in 1 mL of injection media (ES cell media with 20% FCS and 20 mM HEPES). Keep cells on ice until microinjection.

3.11. Concluding Remarks

The ability to modify the mouse genome by homologous recombination of ES cells and pass these alterations through the germline has greatly facilitated the generation and applicability of so-called "designer mice." Engineered modifications of the mouse genome using current technology include complete gene disruptions, introduction of subtle mutations, gene replacements, and chromosome inversions, deletions, and rearrangements. Moreover, with the development of tissue-specific and inducible systems, both temporal and spatial control of the above gene-targeted modifications is now also feasible. With current and emerging technological advancements, it is, and will continue to become, increasingly possible to devise and implement virtually any genetic alteration in the mouse toward the development of novel mouse models for biomedical research.

4. Notes

1. Considerations in design of insertion-type vectors:
   a. Genomic DNA incorporated into the vector should be isogenic (from the same mouse strain) to the ES cell line.
   b. The vector should contain at least 5 Kb of DNA homologous to the locus to be targeted.
   c. A unique restriction site for vector linearization is required within the homologous sequence.
   d. If polymerase chain reaction (PCR) is to be used to screen for recombinants, the linearization site should be less than 1.5 Kb from the unique priming sequence in the vector.
   e. A single-copy probe outside of the vector and recognizing a unique restriction band for the targeted allele on Southern blots should be available before the vector is completed.
2. Considerations for complete gene ablation by insertion-type vectors:
   a. If the vector contains only one exon, the sequences near its splice sites should not be altered by the positive selectable marker (or other means) that can interfere with exon splicing.
   b. If only one exon is present in the vector, the positive selectable marker should not be inserted into it, since the elongation will usually result in its exclusion from the final spliced product.
   c. The exon(s) that will be duplicated should cumulatively not be a multiple of 3 bp, thereby ensuring a frame-shift mutation when incorporated into the transcript.
   d. Introduction of a nonsense or frame-shift mutation into one of the exons of the vector will increase the likelihood of complete loss of gene function.
3. Considerations in design of replacement-type vectors:
   a. Genomic DNA incorporated into the vector should be isogenic (from the same mouse strain) to the ES cell line.
   b. The vector should contain at least 5 Kb of DNA homologous to the locus to be targeted.
   c. Addition of a negative selectable marker such as TK at end of the homologous sequence can be used to enrich for recombinant clones.
   d. A unique restriction site for vector linearization is required within the plasmid backbone (outside of the sequence homologous to the mouse genome).
   e. If PCR is to be used to screen for recombinants, one of the homologous sequence arms should be less than 1.5 Kb in length to ensure efficient amplification.
   f. A single-copy probe outside of the vector and recognizing a unique restriction band for the targeted allele on Southern blots should be available before the vector is completed.
4. Considerations for complete gene ablation by replacement-type vectors:
   a. If the gene contains multiple exons, disrupt one or more of the 5′-most exons.
   b. Replacement of an exon with the positive selectable marker can be used to disrupt the gene.
   c. Ensure that the exon or exons being disrupted are not cumulatively a multiple of 3 bp, to ensure a frame-shift mutation.
   d. Disrupt an exon known to code for an integral function of the protein product.
5. Online databases of existing Cre-expressing transgenic mouse lines
   a. Jackson Laboratories: http://tbase.jax.org
   b. The European Mutant Mouse Archive: http://www.emma.rm.cnr.it
   c. Dr. A. Nagy: http://www.mshri.on.ca/nagy/Cre-pub.html
6. Inducible Targeting Systems

The most common inducible system is based on transcriptional regulation by a mutant tetracycline repressor fused to a viral transactivator *(34)*. This hybrid molecule requires a tetracycline derivative for specific DNA binding and transcriptional activation. Administration of the tetracycline derivative results in Cre expression mediated from an upstream tetracycline operator, resulting in LoxP-mediated recombination of the target locus. Other inducible systems that have been developed utilize ecdysone *(35)*, tamoxifen *(36,37)*, heat shock *(38)*, and insulin *(39)* inducible promoters. Nevertheless, inducible systems have shortcomings, including low-level activity independent of the inducer, incomplete recombination, and toxic effects of the inducer. A comprehensive review of inducible systems and their application is provided in **ref. *40***.

References

1. Evans, M. J., and Kaufman, M. H. (1981) Establishment in culture of pluripotential cells from mouse embryos. *Nature,* **292,** 154–156.
2. Martin, G. R. (1981) Isolation of a pluripotent cell line from early mouse embryos cultured in medium conditioned by teratocarcinoma stem cells. *Proc. Natl. Acad. Sci. USA* **78,** 7634–7638.
3. Bradley, A., Evans, M., Kaufman, M. H., and Robertson, E. (1984) Formation of germ-line chimaeras from embryo-derived teratocarcinoma cell lines. *Nature* **309,** 255–256.
4. Smithies, O., Gregg, R. G., Boggs, S. S., Koralewski, M. A., and Kucherlapati, R. S. (1985) Insertion of DNA sequences into the human chromosomal beta-globin locus by homologous recombination. *Nature* **317,** 230–234.
5. Thomas, K. R., and Capecchi, M. R. (1987) Site-directed mutagenesis by gene targeting in mouse embryo-derived stem cells. *Cell* **51,** 503–512.
6. Doetschman, T., Maeda, N., and Smithies, O. (1988) Targeted mutation of the Hprt gene in mouse embryonic stem cells. *Proc. Natl. Acad. Sci. USA* **85,** 8583–8587.
7. Koller, B. H., Hagemann, L. J., Doetschman, T., et al. (1989) Germ-line transmission of a planned alteration made in a hypoxanthine phosphoribosyltransferase gene by homologous recombination in embryonic stem cells. *Proc. Natl. Acad. Sci. USA* **86,** 8927–8931.
8. Schwartzberg, P. L., Goff, S. P., and Robertson, E. J. (1989) Germ-line transmission of a c-abl mutation produced by targeted gene disruption in ES cells. *Science* **246,** 799–803.
9. Thompson, L. H., and Schild, D. (1999) The contribution of homologous recombination in preserving genome integrity in mammalian cells. *Biochimie* **81,** 87–105.
10. Hasty, P., Rivera-Perez, J., Chang, C., and Bradley, A. (1991) Target frequency and integration pattern for insertion and replacement vectors in embryonic stem cells. *Mol. Cell Biol.* **11,** 4509–4517.
11. Mansour, S. L., Thomas, K. R., and Capecchi, M. R. (1988) Disruption of the proto-oncogene int-2 in mouse embryo-derived stem cells: a general strategy for targeting mutations to non-selectable genes. *Nature* **336,** 348–352.
12. Hasty, P., Abuin, A., and Bradley, A. (2000) Gene targeting, principles, and practice in mammalian cells. In *Gene Targeting—A Practical Approach.* 2 ed. Joyner, A. L., ed. Oxford University Press, New York, NY, pp. 1–35.
13. Lurquin, P. F. (1997) Gene transfer by electroporation. *Mol. Biotechnol.* **7,** 5–35.
14. Andreason, G. L., and Evans, G. A. (1989) Optimization of electroporation for transfection of mammalian cell lines. *Anal. Biochem.* **180,** 269–275.
15. Lupton, S. D., Brunton, L. L., Kalberg, V. A., and Overell, R. W. (1991) Dominant positive and negative selection using a hygromycin phosphotransferase-thymidine kinase fusion gene. *Mol. Cell Biol.* **11,** 3374–3378.
16. Chen, Y. T., and Bradley, A. (2000) A new positive/negative selectable marker, puDeltatk, for use in embryonic stem cells. *Genesis* **28,** 31–35.

17. Karreman, C. (1998) New positive/negative selectable markers for mammalian cells on the basis of Blasticidin deaminase-thymidine kinase fusions. *Nucleic Acids Res.* **26,** 2508–2510.

18. Selfridge, J., Pow, A. M., McWhir, J., Magin, T. M., and Melton, D. W. (1992) Gene targeting using a mouse HPRT minigene/HPRT-deficient embryonic stem cell system: inactivation of the mouse ERCC-1 gene. *Somat. Cell Mol. Genet.* **18,** 325–336.

19. Vasquez, K. M., Marburger, K., Intody, Z., and Wilson, J. H. (2001) Manipulating the mammalian genome by homologous recombination. *Proc. Natl. Acad. Sci. USA* **98,** 8403–8410.

20. Yagi, T., Ikawa, Y., Yoshida, K., et al. (1990) Homologous recombination at c-fyn locus of mouse embryonic stem cells with use of diphtheria toxin A-fragment gene in negative selection. *Proc. Natl. Acad. Sci. USA* **87,** 9918–9922.

21. Kobayashi, K., Ohye, T., Pastan, I., and Nagatsu, T. (1996) A novel strategy for the negative selection in mouse embryonic stem cells operated with immunotoxin-mediated cell targeting. *Nucleic Acids Res.* **24,** 3653–3655.

22. Donehower, L. A., Harvey, M., Slagle, B. L., et al. (1992) Mice deficient for p53 are developmentally normal but susceptible to spontaneous tumours. *Nature* **356,** 215–221.

23. Lindberg, R. L., Porcher, C., Grandchamp, B., et al. (1996) Porphobilinogen deaminase deficiency in mice causes a neuropathy resembling that of human hepatic porphyria. *Nat. Genet.* **12,** 195–199.

24. te Riele, H., Maandag, E. R., and Berns, A. (1992) Highly efficient gene targeting in embryonic stem cells through homologous recombination with isogenic DNA constructs. *Proc. Natl. Acad. Sci. USA* **89,** 5128–5132.

25. Mortensen, R. M., Conner, D. A., Chao, S., Geisterfer-Lowrance, A. A., and Seidman, J. G. (1992) Production of homozygous mutant ES cells with a single targeting construct. *Mol. Cell Biol.* **12,** 2391–2395.

26. Mortensen, R. M., Zubiaur, M., Neer, E. J., and Seidman, J. G. (1991) Embryonic stem cells lacking a functional inhibitory G-protein subunit (alpha i2) produced by gene targeting of both alleles. *Proc. Natl. Acad. Sci. USA* **88,** 7036–7040.

27. te Riele, H., Maandag, E. R., Clarke, A., Hooper, M., and Berns, A. (1990) Consecutive inactivation of both alleles of the pim-1 proto-oncogene by homologous recombination in embryonic stem cells. *Nature* **348,** 649–651.

28. Utomo, A. R., Nikitin, A. Y., and Lee, W. H. (1999) Temporal, spatial, and cell type-specific control of Cre-mediated DNA recombination in transgenic mice. *Nat. Biotechnol.* **17,** 1091–1096.

29. Sauer, B. (1998) Inducible gene targeting in mice using the Cre/lox system. *Methods* **14,** 381–392.

30. Torres, R. M., and Kuhn, R., eds. (1997) *Laboratory protocols for conditional gene targeting.* Oxford University Press, Oxford, UK.

31. Rajewsky, K., Gu, H., Kuhn, R., et al. (1996) Conditional gene targeting. *J. Clin. Invest.* **98,** 600–603.

32. Wilson, T. J., and Kola, I. (2001) The LoxP/CRE system and genome modification. *Methods Mol. Biol.* **158,** 83–94.

33. Le, Y., and Sauer, B. (2000) Conditional gene knockout using cre recombinase. *Methods Mol. Biol.* **136,** 477–485.

34. Gossen, M., Freundlieb, S., Bender, G., Muller, G., Hillen, W., and Bujard, H. (1995) Transcriptional activation by tetracyclines in mammalian cells. *Science* **268,** 1766–1769.

35. No, D., Yao, T. P., and Evans, R. M. (1996) Ecdysone-inducible gene expression in mammalian cells and transgenic mice. *Proc. Natl. Acad. Sci. USA* **93,** 3346–3351.

36. Danielian, P. S., Muccino, D., Rowitch, D. H., Michael, S. K., and McMahon, A. P. (1998) Modification of gene activity in mouse embryos in utero by a tamoxifen-inducible form of Cre recombinase. *Curr. Biol.* **8,** 1323–1326.

37. Hayashi, S., and McMahon, A. P. (2002) Efficient recombination in diverse tissues by a tamoxifen-inducible form of Cre: a tool for temporally regulated gene activation/inactivation in the mouse. *Dev. Biol.* **244,** 305–318.

38. Dietrich, P., Dragatsis, I., Xuan, S., Zeitlin, S., and Efstratiadis, A. (2000) Conditional mutagenesis in mice with heat shock promoter-driven cre transgenes. *Mamm. Genome* **11,** 196–205.

39. Gannon, M., Shiota, C., Postic, C., Wright, C. V., and Magnuson, M. (2000) Analysis of the Cre-mediated recombination driven by rat insulin promoter in embryonic and adult mouse pancreas. *Genesis* **26,** 139–142.

40. Rossant, J., and McMahon, A. (1999) "Cre"-ating mouse mutants-a meeting review on conditional mouse genetics. *Genes Dev.* **13,** 142–145.

41. Fiering, S., Epner, E., Robinson, K., et al. (1995) Targeted deletion of 5′HS2 of the murine beta-globin LCR reveals that it is not essential for proper regulation of the beta-globin locus. *Genes Dev.* **9,** 2203–2213.

42. Valancius, V., and Smithies, O. (1991) Testing an "in-out" targeting procedure for making subtle genomic modifications in mouse embryonic stem cells. *Mol. Cell Biol.* **11,** 1402–1408.

43. Hasty, P., Ramirez-Solis, R., Krumlauf, R., and Bradley, A. (1991) Introduction of a subtle mutation into the Hox-2.6 locus in embryonic stem cells. *Nature* **350,** 243–246.

44. Askew, G. R., Doetschman, T., and Lingrel, J. B. (1993) Site-directed point mutations in embryonic stem cells: a gene-targeting tag-and-exchange strategy. *Mol. Cell Biol.* **13,** 4115–4124.

45. Reid, L. H., Shesely, E. G., Kim, H. S., and Smithies, O. (1991) Cotransformation and gene targeting in mouse embryonic stem cells. *Mol. Cell Biol.* **11,** 2769–2677.

46. Davis, A. C., Wims, M., and Bradley, A. (1992) Investigation of coelectroporation as a method for introducing small mutations into embryonic stem cells. *Mol. Cell Biol.* **12,** 2769–2776.

47. Sambrook, J., Fritsch, E. F., and Maniatis, T. (1989) *Molecular Cloning—A Laboratory Manual.* 2ed. Cold Spring Harbor Laboratory Press, Cold Spring Harbor, NY.

48. Nagy, A., Rossant, J., Nagy, R., Abramow-Newerly, W., and Roder, J. C. (1993) Derivation of completely cell culture-derived mice from early-passage embryonic stem cells. *Proc. Natl. Acad. Sci. USA* **90,** 8424–8428.

49. Smith, A. G., and Hooper, M. L. (1987) Buffalo rat liver cells produce a diffusible activity which inhibits the differentiation of murine embryonal carcinoma and embryonic stem cells. *Dev. Biol.* **121,** 1–9.

50. Williams, R. L., Hilton, D. J., Pease, S., et al. (1988) Myeloid leukaemia inhibitory factor maintains the developmental potential of embryonic stem cells. *Nature* **336,** 684–687.

51. Wood, S. A., Allen, N. D., Rossant, J., Auerbach, A., and Nagy, A. (1993) Non-injection methods for the production of embryonic stem cell-embryo chimaeras. *Nature* **365,** 87–89.

52. Kawase, E., Suemori, H., Takahashi, N., Okazaki, K., Hashimoto, K., and Nakatsuji, N. (1994) Strain difference in establishment of mouse embryonic stem (ES) cell lines. *Int. J. Dev. Biol.* **38,** 385–390.

53. Doetschman, T., Gregg, R. G., Maeda, N., et al. (1987) Targeted correction of a mutant HPRT gene in mouse embryonic stem cells. *Nature* **330,** 576–578.

54. Doetschman, T. C., Eistetter, H., Katz, M., Schmidt, W., and Kemler, R. (1985) The in vitro development of blastocyst-derived embryonic stem cell lines: formation of visceral yolk sac, blood islands and myocardium. *J. Embryol. Exp. Morphol.* **87,** 27–45.

55. Soriano, P., Montgomery, C., Geske, R., and Bradley, A. (1991) Targeted disruption of the c-src proto-oncogene leads to osteopetrosis in mice. *Cell* **64,** 693–702.

56. Auerbach, W., Dunmore, J. H., Fairchild-Huntress, V., et al. (2000) Establishment and chimera analysis of 129/SvEv- and C57BL/6-derived mouse embryonic stem cell lines. *Biotechniques* **29,** 1024–1028, 1030, 1032.

57. Ledermann, B., and Burki, K. (1991) Establishment of a germ-line competent C57BL/6 embryonic stem cell line. *Exp. Cell Res.* **197,** 254–258.

58. Noben-Trauth, N., Kohler, G., Burki, K., and Ledermann, B. (1996) Efficient targeting of the IL-4 gene in a BALB/c embryonic stem cell line. *Transgenic Res.* **5,** 487–491.

59. Dinkel, A., Aicher, W.K., Warnatz, K., Burki, K., Eibel, H., and Ledermann, B. (1999) Efficient generation of transgenic BALB/c mice using BALB/c embryonic stem cells. *J. Immunol. Methods* **223,** 255–260.

Index